Analytical Applications of NMR

CHEMICAL ANALYSIS

A SERIES OF MONOGRAPHS ON ANALYTICAL CHEMISTRY AND ITS APPLICATIONS

Editors

P. J. ELVING · J. D. WINEFORDNER

Editor Emeritus: I. M. KOLTHOFF

Advisory Board

Fred W. Billmeyer, Jr.
Eli Grushka
Barry L. Karger
Viliam Krivan

Victor G. Mossotti
A. Lee Smith
Bernard Tremillon
T. S. West

VOLUME 48

A WILEY-INTERSCIENCE PUBLICATION

JOHN WILEY & SONS
New York / London / Sydney / Toronto

Analytical Applications of NMR

D. E. LEYDEN

Department of Chemistry
University of Georgia

R. H. COX

Department of Chemistry
The University of Georgia

A WILEY-INTERSCIENCE PUBLICATION

JOHN WILEY & SONS
New York / London / Sydney / Toronto

Copyright © 1977 by John Wiley & Sons, Inc.

All rights reserved. Published simultaneously in Canada.

No part of this book may be reproduced by any means, nor transmitted, nor translated into a machine language without the written permission of the publisher.

Library of Congress Cataloging in Publication Data:

Leyden, Donald E   1938–
   Analytical applications of NMR.

   (Chemical analysis; v. 48)
   "A Wiley-Interscience publication."
   Includes bibliographical references and index.
   1. Nuclear magnetic resonance spectroscopy.
I. Cox, Richard Havey, 1943–   joint author.
II. Title.   III. Series.

QD96.N8L49   543'.08   77-1229
ISBN 0-471-53403-X

Printed in the United States of America

10 9 8 7 6 5 4 3 2 1

Vol. 33. **Masking and Demasking of Chemical Reactions.** By D. D. Perrin
Vol. 34. **Neutron Activation Analysis.** By D. De Soete, R. Gijbels, and J. Hoste
Vol. 35. **Laser Raman Spectroscopy.** By Marvin C. Tobin
Vol. 36. **Emission Spectrochemical Analysis.** By Morris Slavin
Vol. 37. **Analytical Chemistry of Phosphorus Compounds.** Edited by M. Halmann
Vol. 38. **Luminescence Spectrometry in Analytical Chemistry.** By J. D. Winefordner, S. G. Schulman, and T. C. O'Haver
Vol. 39. **Activation Analysis with Neutron Generators.** By Sam S. Nargolwalla and Edwin P. Przybylowicz
Vol. 40. **Determination of Gaseous Elements in Metals.** Edited by Lynn L. Lewis, Laben M. Melnick, and Ben D. Holt
Vol. 41. **Analysis of Silicones.** Edited by A. Lee Smith
Vol. 42. **Foundations of Ultracentrifugal Analysis.** By H. Fujita
Vol. 43. **Chemical Infrared Fourier Transform Spectroscopy.** By Peter R. Griffiths
Vol. 44. **Microscale Manipulations in Chemistry.** By T. S. Ma and V. Horak
Vol. 45. **Thermometric Titrations.** By J. Barthel
Vol. 46. **Trace Analysis: Spectroscopic Methods for Elements.** Edited by J. D. Winefordner
Vol. 47. **Contamination Control in Trace Element Analysis.** By Morris Zief and James W. Mitchell
Vol. 48. **Analytical Applications of NMR.** By D. E. Leyden and R. H. Cox

*to Alice and Marilyn for their patience
and to our colleagues for their help*

PREFACE

Nuclear magnetic resonance has perhaps had more impact on a diverse variety of chemical investigations than any other single instrumental development. Provided that analytical chemistry is broadly defined as the practice of characterizing the chemical and physical properties of compounds or mixtures, nuclear magnetic resonance is an important tool for that practice. We have adopted this broad definition for the purpose of this book. It was our intent to prepare a monograph that would be useful to students and researchers who wish to gain an introduction to the principles, practices, instrumentation, and applications of nuclear magnetic resonance in analytical chemistry. We hope that the book will provide the reader with an overview of the power and potential of nuclear magnetic resonance in his field of research.

We recognize that the majority of readers will not read the book from beginning to end. For this reason we have attempted to make each chapter stand alone. This has been done by providing extensive references so that the reader may easily go to the original literature for further details. We have provided examples of applications throughout the text so that the concepts discussed can be related at once to chemical problems. Because of this effort the reader will find occasional repetition of statements in the book. We have done this for emphasis and for the benefit of those who will read only isolated chapters or sections.

The examples of applications of nuclear magnetic resonance selected for presentation in Chapter 7 are arbitrary and probably biased by the experience and interests of the authors. Many fine works are not mentioned. We hope that through our arbitrary selections we have been able to demonstrate the broad scope of the types of information that may be obtained by the use of nuclear magnetic resonance.

We wish to thank all copyright holders who kindly granted permission to use figures and tables from their publications. We especially thank CRC Press, Inc., for the use of material that appeared in *Critical Reviews of Analytical Chemistry*.

Our deep appreciation is extended to Becky McRorie and Martha Dove for preparation of the manuscript and to P. MacCarthy for his suggestions.

D. E. LEYDEN
R. H. COX

Athens, Georgia
November 1976

CONTENTS

CHAPTER 1	Introduction	1
CHAPTER 2	Theory of Nuclear Magnetic Resonance	9
CHAPTER 3	Experimental and Instrumental Aspects of NMR	49
CHAPTER 4	Analysis of Spectra and Determination of Structure	95
CHAPTER 5	Carbon-13 NMR	194
CHAPTER 6	Methods of Quantitative Measurements	256
CHAPTER 7	Examples of Analytical Applications	298
APPENDIX		425
INDEX		445

Analytical Applications of NMR

CHAPTER

1

INTRODUCTION

1.1	Historical	2
1.2	Applications and Limitations	3
1.3	Scope	6
1.4	Convention and Units	7

Before preparing a monograph entitled *Analytical Applications of NMR*, the scope of these analytical applications must be determined. What is analytical chemistry? For some the discipline conjures up memories of rather dull and tedious college courses in quantitative analysis in which an archaic method was used to determine an element in an uninteresting sample. However, to others analytical chemistry is a science practiced on occasion by all chemists in an attempt to characterize the properties and composition of a compound or material. The tools of this science have expanded dramatically in the past 30 years, and they have been combined effectively with the art of the discipline. Undergraduates now characterize materials in detail during a few laboratory periods. Such efforts would have been major research projects a few decades ago. It was decided, then, that the purposes of this monograph would be to provide an introduction to the fundamental principles of nuclear magnetic resonance (NMR) spectroscopy, to provide insight into operational procedures, and to demonstrate by selected examples the tremendous scope that nature has provided for the analytical applications of NMR.

NMR spectroscopy is probably the single most important physical tool available to the chemist today. Since the only requirement for a nucleus to give an NMR spectrum is that the nucleus possess a nonzero magnetic moment, there is hardly an organic or inorganic compound that is not amenable to study by NMR. It is probably safe to say that at least one, if not more, NMR spectrum is obtained of each new compound isolated today. Scientists have come to rely on the information provided by NMR to such an extent that it is not uncommon to find several NMR spectrometers within a given laboratory. These spectrometers range from the routine-type instrument to the more sophisticated research-type instrument. The developments in NMR instrumentation have continued to decrease the sample size requirement for an NMR spectrum and hence increase the areas for application of NMR spectroscopy.

1.1 HISTORICAL

Like most spectroscopic techniques, NMR had its beginning in the earlier experiments of various physicists. The concept that certain nuclei possess a magnetic moment was proposed by Pauli in 1924 (1) to account for the hyperfine structure observed in atomic spectral lines. Molecular beam experiments by Stern and Gerlach (2,3) demonstrated that the component of electron magnetic moments is quantized in a magnetic field. Latter refinements of these experiments permitted the measurement of the proton magnetic moment (4,5). Guoy-balance-type experiments by Lasarew and Schubnikow in 1937 demonstrated nuclear paramagnetism and showed that the thermal equilibrium of nuclear spins in a magnetic field is rapidly established even at low temperatures (6).

In the strictest sense, the first NMR experiment was made in 1939 by Rabi and associates (7). While passing a beam of hydrogen molecules through a magnetic field, radio-frequency electromagnetic energy was applied to the molecules in the magnetic field. The frequency of the electromagnetic irradiation was varied until at a certain point energy was absorbed by the beam of hydrogen molecules causing a slight deflection of the beam and a decrease in the number of molecules reaching the detector.

Although the early experiments were restricted to molecular beams, it had been pointed out that it should be possible to observe the absorption of energy in a magnetic field using samples of the solid, liquid, and gaseous states (8). Several attempts were made in the early 1940s to detect NMR absorption in solids. However it was not until late 1945 that two groups first observed NMR absorption in experiments using bulk materials. Purcell, Torrey, and Pound (9) at Harvard observed proton NMR absorption in solid paraffin wax and Bloch, Hansen, and Packard (10) at Stanford observed the proton absorption in liquid water. The Nobel Prize was awarded to these workers in 1952 for their discoveries.

The potential for applications of NMR in chemistry dates to 1951 with the experiments of Arnold, Dharmatti, and Packard (11). In an examination of the NMR spectra of some alcohols they showed that different signals could be observed for the chemically different types of protons. Since 1951 developments in both the theory and experimental aspects of NMR have been rapid. Major advances have been made in the quality of NMR spectra, in the ease of obtaining NMR spectra, and in improving the sensitivity of the NMR technique. One of the major improvements has been the development of magnets capable of operation at high field strengths with a uniform homogeneous field. This has led to improved sensitivity and has eased the problem of spectral interpretation. The development of low-priced minicomputers in recent years had led to the present-day pulsed Fourier transform NMR spectrometers that are becoming routine instruments in many laboratories.

1.2 APPLICATIONS AND LIMITATIONS

NMR spectra have been obtained from samples of the solid, liquid, and gaseous states. Although data on molecular motions within a crystal and internuclear distances have been obtained from the spectra of solid samples, NMR spectra of the solid state is usually considered under the topic of wide-line NMR, and will not be discussed here. Those interested in this aspect of NMR are referred to the general references at the end of this chapter. We shall be concerned here primarily with high-resolution NMR spectroscopy where the observed spectral line widths approach the limits of detection as a result of inhomogeneties in the magnetic field.

Unlike other forms of spectroscopy, where one either obtains too little information or more information than can readily be interpreted, the interpretation of NMR spectra in terms of the fundamental parameters is straightforward in most cases. The parameters obtained from NMR spectra (chemical shifts, coupling constants, intensities, and relaxation times) provide the necessary information for solving a wide variety of problems in the chemical and biological sciences.

Much of the early work in NMR was devoted to proton NMR spectroscopy of organic compounds. The major reason for this is that of all the nuclei that give an NMR spectrum the proton is the most sensitive to NMR detection. Subsequently there has been a great deal of activity in ^{19}F, ^{31}P, ^{11}B, and ^{14}N NMR spectroscopy (12). During the past few years the techniques for obtaining natural abundance ^{13}C NMR spectra in a reasonable length of time have been developed such that ^{13}C NMR is rapidly approaching proton NMR in terms of importance (13,14). The recent focus on biochemically related research areas has created some activity in ^{15}N NMR (15). While the previously stated nuclei have received the most attention to date, there are increasing reports of the NMR spectra of the heavier metal nuclei (16).

The major application of NMR spectroscopy has been in structure elucidations. Using the chemical shifts and coupling constants obtained from a spectrum, considerable progress can usually be made towards determining the structure of an unknown compound through the use of empirical correlations established with known compounds. In many instances it has been found that certain groupings of nuclei always give rise to a characteristic absorption pattern. This often allows one to immediately distinguish isomers. Two examples are shown in Figures 1.1 and 1.2. A *para*-disubstituted benzene will in most cases give an absorption pattern similar to that in Figure 1.1C, which appears on first inspection to contain two doublets in the low field, aromatic region of the spectrum. Similarly the large doublet and smaller septet in Figure 1.2B are characteristic of the isopropyl group.

Other applications of NMR include the quantitative determination of the percentage composition of a complex mixture. Typical examples might be the

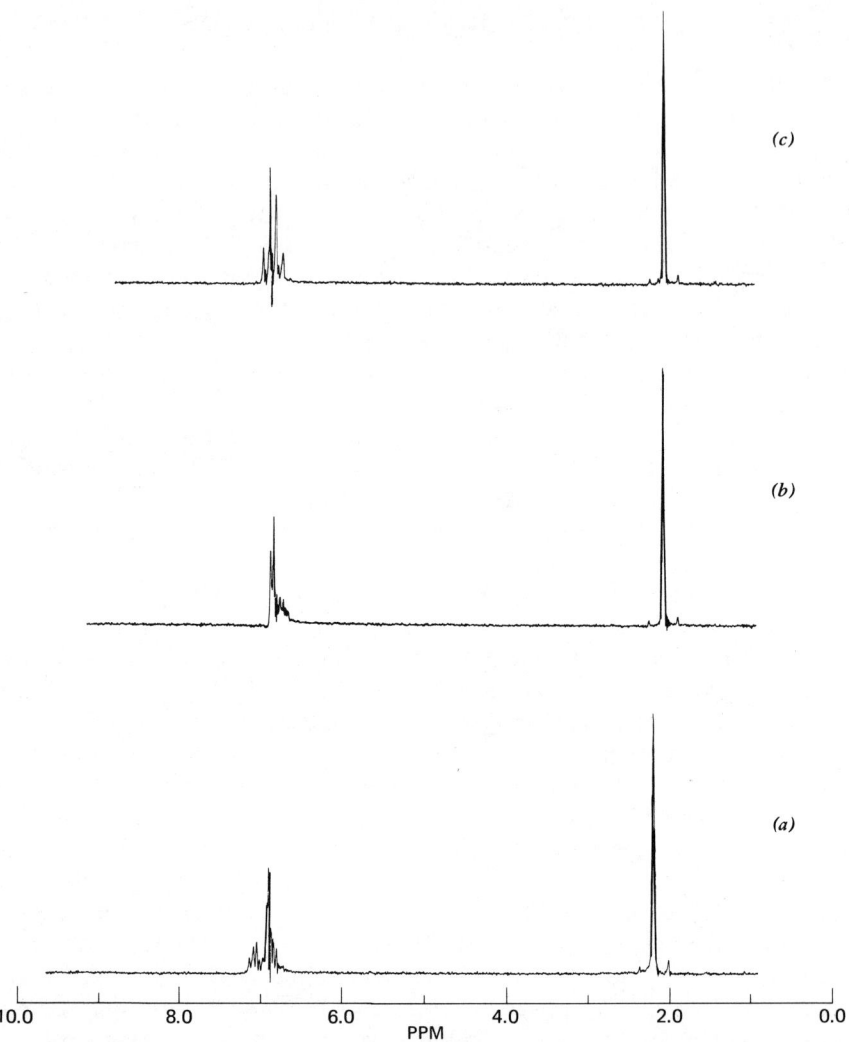

Fig. 1.1 The 100-MHz proton NMR spectrum of the isomeric chlorotoluenes: (*a*) *o*-chlorotoluene; (*b*) *m*-chlorotoluene; and (*c*) *p*-chlorotoluene.

determination of the percent deuteration at a particular site in a molecule, the product distribution in a reaction mixture, or the analysis of a complex drug mixture. In order for this technique to be successful one must first assign a peak in the spectrum to each individual component. From the known number of protons giving rise to each peak and the relative area under the peak the percentage composition of the mixture may be calculated. Another

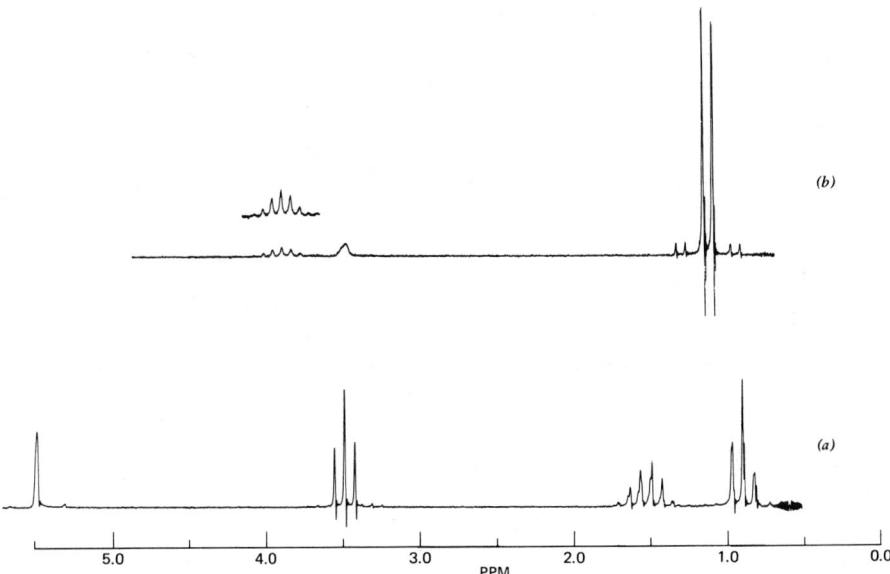

Fig. 1.2 The 100-MHz proton NMR spectrum of the propyl alcohols: (*a*) *n*-propyl alcohol; and (*b*) *iso*-propyl alcohol.

application of quantitative analysis using NMR is the determination of the percent hydrogen of an unknown compound. A known amount of a reference compound is added to the solution and from the relative areas under the peaks, the percent hydrogen of the unknown compound may be calculated.

During recent years, the application of NMR to problems involving time dependent phenomena has been increasing. In some cases, is possible to obtain reaction rates using NMR by carrying out the reaction in the probe of the spectrometer. This requires that one follow either the dissappearance of a peak in the spectrum due to a reactant or the appearance of a peak due to a product as a function of time. One recent application of this technique has been the study of chemically induced dynamic nuclear polarization (CIDNP). Other examples of NMR studies of time-dependent phenomena include conformational analysis, rotational isomerism, restricted rotation, and fast chemical exchange; all of which may be investigated while the chemical system is at equilibrium. Examples of studies of this type are given in Chapter 6.

One of the major limitations of NMR for analytical applications is the inherent insensitivity compared with other spectroscopic techniques. For normal continuous wave spectrometers, somewhere between 5 and 50 mg of sample (depending on the molecular weight) are required to obtain a proton spectrum in a single scan. Thus restrictions may be imposed by the solubility and/or the availability of the sample. Part of these difficulties may be overcome by using time averaging techniques, smaller sample volumes, and spectrometers operating at higher magnetic fields. The availability of Fourier transform accessories has been of tremendous value in obtaining spectra on small amounts of sample. In those cases where a spectrum cannot be obtained in a single scan time considerations may become important due either to the long time required to obtain the spectrum or to problems with decomposition of the sample.

Another problem often encountered in the application of NMR to structural elucidations of complex molecules is the overlap of absorptions in the spectrum. If a molecule contains several chemically similar types of protons, the absorptions of each of the protons may overlap to such an extent that it is not possible to obtain useful information from that region of the spectrum. In some cases these problems may be alleviated by obtaining the spectrum at higher magnetic fields, by obtaining the spectrum in various solvents, and through the use of lanthanide shift reagents (Chapter 4).

1.3 SCOPE

The topic of NMR spectroscopy has already been treated in several excellent texts. However, the emphasis has been on the theory of NMR or the application of NMR for organic structure determinations. With the recent advances in the sensitivity enhancement of NMR spectra there are a growing number of analytical applications of NMR that are becoming more practical. It is this area of NMR to which this text is devoted.

Our approach in this text has been to provide a brief background into the theory of NMR and then discuss possible analytical applications of proton NMR through the use of examples from the literature. The results of the theory of NMR are discussed in Chapter 2. The emphasis is concentrated on providing a physical picture for the NMR process rather than developing the theory in rigorous mathematical detail.

In Chapter 3 the experimental and instrumental aspects of proton NMR are discussed. Attention is focused on the various parameters that affect the quality of the spectrum, the manner in which the NMR spectrum is obtained, and on the differences in the various types of NMR spectrometers.

The analysis of NMR spectra to extract the fundamental parameters is discussed in Chapter 4 along with the various aids used to simplify the analysis of complex spectra. The use of the NMR parameters for structure determination through the aid of correlation tables and empirical relationships is discussed.

Because of the recent advances in Fourier transform NMR (FT-NMR) techniques it is clear that applications of carbon-13 NMR (CMR) will continue to increase in the future. There are some practical problems with analytical applications of CMR that are not encountered in proton NMR. Nevertheless future analytical applications of CMR seem likely to increase. Therefore a brief background into CMR is presented in Chapter 5.

The remaining portion of this text is devoted to the use of NMR parameters for quantitative measurements. The methods used in making quantitative measurements are discussed in Chapter 6. In Chapter 7 various literature examples are discussed to illustrate the advantages and limitations of applications of NMR to analytical chemistry.

1.4 CONVENTION AND UNITS

Two methods have been used previously for reporting proton chemical shifts, the tau (τ) and delta (δ) scales. In order to be consistent we shall use only the δ scale in this text. Furthermore low field shifts will be reported as positive and high field shifts as negative values with respect to the reference at $\delta 0.0$. The unit for frequency will be the hertz (Hz). As is now standard practice spectra will be presented using the convention of increasing field from left to right.

NMR spectroscopists and chemists have become accustomed to writing $\bar{H}$ for the magnetic field vector. Actually $\bar{H}$ is the magnetic field intensity vector and $\bar{B}$ is the magnetic induction field vector. As long as electromagnetic units are used, $\bar{B}$ and $\bar{H}$ may be interchanged, although the observable magnetic properties depend on $\bar{B}$ rather than $\bar{H}$. Equations presented in this text will use $\bar{B}$ for consistency in referring to the magnetic field. The electromagnetic unit, the gauss, is used for the magnetic induction field, $\bar{B}$.

REFERENCES

1. W. Pauli, *Naturwissenschaften*, **12**, 741 (1924).
2. O. Stern, *Z. Phys.*, **7**, 249 (1921).
3. W. Gerlach and O. Stern, *Ann. Phys. Lpz.*, **74**, 673 (1924).
4. I. Eastermann and O. Stern, *Z. Phys.*, **85**, 17 (1933).
5. R. Frisch and O. Stern, *Z. Phys.*, **85**, 4 (1933).
6. B. G. Lasarew and L. W. Schubnikow, *Phys. Z. Sowjet.*, **11**, 445 (1937).
7. I. I. Rabi, S. Millman, P. Kusch, and J. R. Zacharias, *Phys. Rev.*, **55**, 526 (1939).
8. C. J. Gorter, *Physica*, **3**, 995 (1936).
9. E. M. Purcell, H. C. Torrey, and R. V. Pound, *Phys. Rev.*, **69**, 37 (1946).
10. F. Bloch, W. W. Hansen, and M. E. Packard, *Phys. Rev.*, **69**, 127 (1946).
11. J. T. Arnold, S. S. Dharmatti, and M. E. Packard, *J. Chem. Phys.*, **19**, 507 (1951).

12. For reviews of these topics see J. W. Emsley, J. Feeney, L. H. Sutcliffe, Eds., *Progress in Nuclear Magnetic Resonance Spectroscopy*, Pergamon, New York; E. F. Mooney, Ed., *Annual Reports On NMR Spectroscopy*, Academic, New York.
13. G. C. Levey and G. L. Nelson, *Carbon-13 NMR For Organic Chemists*, Wiley-Interscience, New York, 1972.
14. J. B. Stothers, *Carbon-13 NMR Spectroscopy*, Academic, New York, 1972.
15. R. L. Lichter, in *Determination of Organic Structures by Physical Methods*, F. C. Nachod and J. J. Zuckerman, Eds., Vol. 4, Academic, New York, 1971, p. 195.
16. T. Axenrod and G. A. Webb, Eds., *Nuclear Magnetic Resonance Spectroscopy of Nuclei Other Than Protons*, Wiley, New York, 1974.

General References for NMR

J. D. Roberts, *Nuclear Magnetic Resonance*, McGraw-Hill, New York, 1959.

J. A. Pople, W. G. Schneider, and H. J. Bernstein, *High-resolution Nuclear Magnetic Resonance*, McGraw-Hill, New York, 1959.

J. W. Emsley, J. Feeney, and L. H. Sutcliffe, *High Resolution Nuclear Magnetic Resonance Spectroscopy*, Vols. 1 and 2, Pergamon, New York, 1965.

L. M. Jackman and S. Sternhell, *Applications of Nuclear Magnetic Resonance Spectroscopy in Organic Chemistry*, Pergamon, New York, 1969.

E. D. Becker, *High Resolution NMR, Theory and Chemical Applications*, Academic, New York, 1969.

A. Carrington and A. D. McLachlan, *Introduction to Magnetic Resonance*, Harper & Row, New York, 1967.

J. D. Roberts, *An Introduction to Spin–Spin Splitting in High-Resolution NMR Spectra*, Benjamin, New York, 1961.

P. L. Corio, *Structure Of High-Resolution NMR Spectra*, Academic, New York, 1967.

J. D. Memory, *Quantum Theory Of Magnetic Resonance Parameters*, McGraw-Hill, New York, 1968.

CHAPTER

2

THEORY OF NUCLEAR MAGNETIC RESONANCE

2.1	Magnetic Properties of Nuclei			10
2.2	The NMR Experiment			11
2.3	Intensity			17
2.4	Linewidths			18
2.5	Bloch Formalism			20
2.6	The Chemical Shift			22
2.7	Theory of Chemical Shifts			25
	2.7.1	Local Diamagnetic Screening		26
	2.7.2	Local Paramagnetic Screening		27
	2.7.3	Screening from Other Atoms in the Molecule		27
		2.7.3.1	Shielding Due to Carbon–Carbon and Carbon–Hydrogen Bonds	29
		2.7.3.2	Shielding Due to Carbon–Carbon Double and Triple Bonds	29
		2.7.3.3	Shielding Due to the Carbonyl Group	30
		2.7.3.4	Shielding Due to Miscellaneous Groups	31
	2.7.4	Ring Currents		31
2.8	Solvent Effects			33
2.9	Theory of Spin–Spin Splitting			35
2.10	Coupling Constants–Practical Considerations			38
2.11	Relaxation Mechanisms			39
	2.11.1	Dipole–Dipole Relaxation		40
	2.11.2	Chemical Shift Anisotropy		41
	2.11.3	Scalar Coupling		42
	2.11.4	Spin Rotation		43
	2.11.5	Quadrupole Relaxation		44
	2.11.6	Interaction with Paramagnetic Species		44
2.12	Effect of Exchange Processes on NMR Spectra			45

In this chapter we shall discuss the basic principles involved in the NMR experiment. The theoretical treatment will be limited in that we shall present the results of the theory and concentrate on providing a physical picture of the NMR phenomenon. Several excellent accounts of the theory of NMR are available and those interested in a more detailed account of the theory are referred to these texts (1–4).

2.1 MAGNETIC PROPERTIES OF NUCLEI

Certain nuclei, when placed in a magnetic field, behave as if they were spinning charged particles. Nuclei that possess this property have angular momentum p. From quantum mechanics the maximum observable component of the angular momentum is quantized and must be an integral or half-integral multiple of $\hbar$ (Planck's constant h divided by 2π). Furthermore only certain states of p are allowed. Defining I as the spin quantum number, the maximum observable component of p is I. The angular momentum thus has $2I + 1$ states $(-I, -I + 1, \ldots, I - 1, I)$ along which the component of p may have values.

The spinning nucleus generates a magnetic moment, μ, which is parallel to and proportional to p. This relationship is given by equation 2.1 where γ is the

$$\mu = \gamma p \tag{2.1}$$

magnetogyric ratio and has different values for different nuclei. The magnetic moment is also quantized. The maximum observable component of μ has values of $m\mu/I$, where m is the magnetic quantum number and may have the values

$$m = -I, -I + 1, \ldots, I - 1, I \tag{2.2}$$

A third magnetic property of nuclei related to the spin number I is the electric quadrupole moment Q. The electric quadrupole moment is a measure of the nonsphericity of the electric charge distribution within the nucleus. Nuclei with $I = \frac{1}{2}$ do not have electric quadrupole moments. However nuclei with $I \geq 1$ often show line broadening along with a loss of certain information from the interaction of the nuclear magnetic moment with the electric quadrupole moment. This will be discussed further in Section 2.11.5.

The only requirement for a nucleus to give an NMR absorption is that the nucleus possess a magnetic moment. All experiments to date indicate that the magnetic moment is zero if $I = 0$. Thus nuclei with $I \neq 0$ give rise to NMR absorption. Rules for nuclear spins have been summarized in terms of the mass number A and the atomic number Z as follows (1):

1. The nuclear spin is half-integral if the mass number A is odd.
2. The nuclear spin is integral if the mass number A is even and the atomic number Z is odd.
3. The nuclear spin is zero if both the mass number A and the atomic number Z are even.

Included in the first category with $I = \frac{1}{2}$ are the important nuclei ^{1}H, ^{13}C, ^{19}F, and ^{31}P. Some nuclei with $I = \frac{3}{2}$ are ^{7}Li, ^{11}B, ^{23}Na, and ^{35}Cl. The two major nuclei in the second category with $I = 1$ are ^{2}H (deuterium, D) and ^{14}N. Examples of nuclei with $I = 0$ are ^{12}C, ^{16}O, and ^{32}S.

The results of rule 3 are important from the standpoint of the spectra of organic molecules since a compound containing only C, H, and O will give a proton spectrum without any complications due to interactions of the proton spins with the spins of the other nuclei in the molecule. However, in some cases the proton spectrum either does not provide the desired information or the information is ambiguous and a carbon spectrum is needed. Severe restrictions in terms of time and instrumental requirements are placed on obtaining a carbon spectrum due to the low natural abundance of the only magnetic isotope of carbon, carbon-13 (1.1%). It is this area that has benefited most from developments in NMR instrumentation in recent years.

2.2 THE NMR EXPERIMENT

Both classical and quantum mechanics have been used to theoretically treat the NMR experiment (1). Although both treatments give identical results, it is convenient to use parts of each method in describing the NMR experiment. The classical approach is better suited to derive a physical picture of the NMR experiment, whereas the quantum mechanics approach relates the NMR experiment to the absorption of energy during a transition between two energy states similar to other forms of spectroscopy. We shall discuss the classical mechanics approach first.

When a nucleus of spin $I \neq 0$ is placed in a magnetic field B_0 (typically 14, 092, or 23,487 gauss), the magnetic moment takes up the allowed orientations along which the component of the magnetic moment may have values. For a spin $\frac{1}{2}$ nucleus, the possible orientations in terms of the magnetic quantum number are $m = +\frac{1}{2}$ and $-\frac{1}{2}$ as shown in Figure 2.1. The interaction between μ and B_0 results in a torque acting on μ which tends to tip it towards B_0. However, since the nucleus is spinning, instead of tipping μ

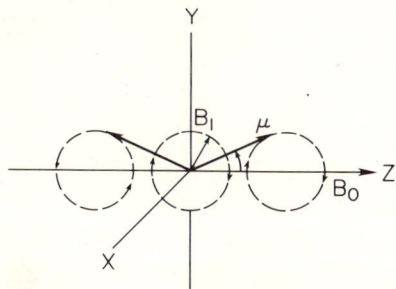

Fig. 2.1 The two orientations of the magnetic moments of a spin $I = \frac{1}{2}$ nucleus.

towards B_0, the torque causes μ to precess about the magnetic field B_0, analogous to the way in which a gyroscope precesses about the earth's magnetic field. (It should be pointed out that in the absence of an applied field B_0, the components of μ are randomly distributed since the earth's magnetic field is not strong enough to interact with μ.) The precessional frequency of μ about B_0 is given by the Larmor equation

$$v_0 = \frac{\gamma}{2\pi} B_0 \qquad (2.3)$$

where v_0 is the precessional frequency in cycles/sec. As can be seen from equation 2.3, the precessional frequency is dependent on both B_0 and γ and thus is different for each type of nucleus.

If a second, smaller magnetic field B_1 is applied in the x–y plane (Figure 2.1), and rotating in the same direction as μ, interactions between B_1 and μ occur. So long as B_1 is rotating at some frequency v other than the Larmor frequency v_0, the effect of B_1 on μ results in slight oscillations of the angle between μ and B_0. When B_1 is rotating at a frequency $v = v_0$, μ will feel the effects of both B_1 and B_0 and will exhibit large oscillations in the angle between μ and B_0 such that the direction of μ with respect to B_0 changes. Energy is absorbed by the nucleus and we speak of the magnetic moment μ as having "flipped" from one orientation in the magnetic field to the other. The absorption of energy when $v = v_0$ is the NMR phenomenon.

In an actual experiment the rotating field B_1 is obtained by passing a current through a coil at a frequency v. This generates a magnetic field that is linearly polarized along the x axis and may be thought of as resulting from two equal fields rotating in opposite directions in the x–y plane (Figure 2.2). The resultant field is therefore $2B_1$ along the x axis.

According to the Larmor equation 2.3, the resonant condition depends on the magnetogyric ratio, γ, the strength of the applied field, B_0, and the frequency of irradiation, v. Commercial NMR spectrometers usually employ a

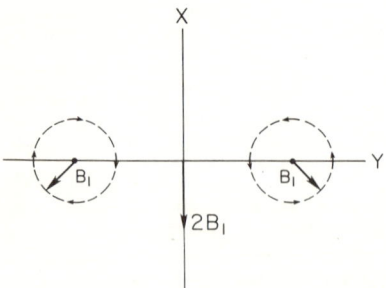

Fig. 2.2 A schematic illustration of the rotating external magnetic field B_1.

THE NMR EXPERIMENT

TABLE 2.1. NMR Properties of Selected Nuclei

Isotope	I	NMR Frequency in a 23,487 Gauss Field (MHz)	Natural Abundance (%)	Relative Sensitivity At Constant Field At Natural Isotopic Abundance
^{1}H	$\frac{1}{2}$	100.00	99.985	1.00
^{2}H	1	15.35	0.015	0.00000145
^{7}Li	$\frac{3}{2}$	38.86	92.58	0.27123
^{11}B	$\frac{3}{2}$	32.08	80.42	0.133
^{13}C	$\frac{1}{2}$	25.14	1.1	0.00018
^{14}N	1	7.22	99.63	0.001
^{15}N	$\frac{1}{2}$	10.13	0.37	0.000004
^{17}O	$\frac{5}{2}$	13.56	0.037	0.00001
^{19}F	$\frac{1}{2}$	94.08	100.00	0.833
^{23}Na	$\frac{3}{2}$	26.45	100.00	0.0925
^{27}Al	$\frac{5}{2}$	26.06	100.00	0.206
^{29}Si	$\frac{1}{2}$	19.86	4.70	0.00037
^{31}P	$\frac{1}{2}$	40.48	100.00	0.066
^{35}Cl	$\frac{3}{2}$	9.79	75.53	0.0035
^{119}Sn	$\frac{1}{2}$	37.27	8.58	0.0044
^{195}Pt	$\frac{1}{2}$	21.50	33.8	0.0034
^{199}Hg	$\frac{1}{2}$	17.83	16.84	0.00019
^{207}Pb	$\frac{1}{2}$	20.92	22.6	0.002

magnetic field of approximately either 14,092 or 23,487 gauss. The frequency of irradiation necessary for the resonant condition falls in the radio frequency range (4–100 MHz) and overlaps with the standard FM band. In Table 2.1 are given the frequencies of absorption with other properties important in observing an NMR spectrum for several nuclei.

In the quantum mechanical treatment of the NMR experiment, the interaction between the applied field and the magnetic moment appears in the Hamiltonian operator, $\mathcal{H}$, as

$$\mathcal{H} = -\gamma \hbar \bar{B}_0 \cdot \bar{I} \qquad (2.4)$$

Similarly the interaction between B_1 and μ appears as another term in the Hamiltonian as

$$\mathcal{H}' = 2\mu_x B_1 \cos 2\pi\nu t = 2\gamma\hbar B_1 I_x \cos 2\pi\nu t \qquad (2.5)$$

Solution of the Hamiltonian yields a discrete set of $2I + 1$ energy levels for the system

$$E_m = -\gamma\hbar m B_0 \qquad (2.6)$$

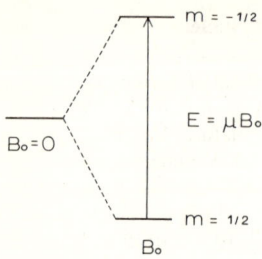

Fig. 2.3 The splitting of the energy levels for a spin $I = \frac{1}{2}$ nucleus when placed into the magnetic field B_0.

where m has values of $-I, -I+1, \ldots, I-1, I$. These energy levels correspond to the possible orientations of the magnetic moment with respect to B_0. For the case of a nucleus with $I = \frac{1}{2}$, there are two energy levels as shown in Figure 2.3, corresponding to orientation of μ either aligned or opposed to the applied field B_0. The separation between the energy levels is linearly related to the applied field B_0.

Transitions between the two energy levels occur when the energy of B_1 corresponds to the energy difference between the two states. The probability of a transition per unit time is given by time-dependent perturbation theory as

$$P_{mm'} = \gamma^2 B_1^2 I(m|I_x|m')I^2 \delta(v_{mm'} - v) \tag{2.7}$$

where $2B_1$ is the magnitude of the field applied in the x direction, $(m|I_x|m')I^2$ is the matrix element of the nuclear spin operator in the x direction, $\delta(v_{mm'} - v)$ is the Dirac δ function, and $v_{mm'}$ is the frequency corresponding to the energy difference between the two states m and m'. One important result of equation 2.7 is the selection rule that transitions occur only between energy states which differ in m by ± 1.

A second result of equation 2.7 is that the Dirac δ function vanishes for values of the frequency v unless v is exactly the frequency separation between the two states m and m'. When $v = v_{mm'}$, an infinitely sharp absorption line is predicted. In actual practice the NMR absorption usually approximates a Lorentzian line shape and the Dirac δ function is replaced with a line shape function $g(v)$, which is dependent on the frequency such that

$$\int_0^\infty g(v)dv = 1 \tag{2.8}$$

This substitution is necessary in order to reproduce the broadened lines normally found in NMR spectra.

On further consideration of the time-dependent perturbation theory and equation 2.7, one might raise the question as to why is it possible to observe an NMR transition. The probability of spontaneous absorption and emission is negligible. Furthermore the probabilities of induced absorption and induced emission are equal. Therefore the only way in which it would be possible to observe a transition would be if there exists a difference in the populations of the energy levels.

When a sample of nuclei is placed into a magnetic field B_0, the magnetic moments of the nuclei become orientated in the magnetic field according to the allowed energy levels. For a system of identical $I = \frac{1}{2}$ nuclei, there will be two allowed energy levels corresponding to the orientation of μ either aligned with or opposed to B_0. At 23,487 gauss, the separation between the energy levels of a sample of protons is only a few millicalories. The tendency for the nuclei to populate the lower energy level is opposed by thermal motions, which tends to equalize the populations. The equilibrium population of the two energy levels at some temperature T is given by the Boltzmann equation

$$\frac{n_+}{n_-} = \exp\left(\frac{-\Delta E}{kT}\right) \qquad (2.9)$$

where $n+$ and $n-$ are the number of nuclei in the upper and lower energy level, respectively, k is the Boltzmann constant, T is the absolute temperature, and ΔE is the energy separation between the two energy levels. Substituting for ΔE, equation 2.9 becomes

$$\frac{n_+}{n_-} = \exp\left(\frac{-2\mu B_0}{kT}\right) \qquad (2.10)$$

The solution of equation 2.10 for the excess of nuclei in the lower energy level may be approximated as

$$\frac{n_- - n_+}{n_-} \approx \frac{2\mu B_0}{kT} \qquad (2.11)$$

(For protons in a magnetic field of 23,487 gauss at 25°C, the excess population in the lower energy level is only 1.5×10^{-5} nuclei.) This slight excess of nuclei in the lower energy state leads to a net absorption of energy. Furthermore this leads to the important result that the net absorption will depend on the number of nuclei under consideration and also points out the fact that NMR is relatively insensitive compared to other spectroscopic techniques that operate at much higher frequencies. Since the excess population is directly proportional to B_0, sensitivity enhancement has been one of the major reasons for developing larger magnets during recent years.

Since the energy levels for a collection of identical nuclei are degenerate and therefore equally populated in the absence of a strong magnetic field, let us take a closer look at the process by which the Boltzmann distribution for the allowed energy levels becomes established. When a sample is placed into a strong magnetic field, a finite amount of time is required in order for the Boltzmann distribution to become established. Since the nuclei are not interacting with B_1 at this point and the probability for spontaneous emission is negligible, there must be some process for absorption and emission to occur in order to establish the equilibrium spin distribution.

The nuclei are undergoing thermal motions and are interacting with their surroundings (lattice). This interaction with the lattice provides a mechanism for energy transfer between the spin system and the lattice such that transitions between the energy levels occur. This process is called spin-lattice or longitudinal relaxation and is responsible for establishing the equilibrium spin distribution. It is a nonradiative first-order rate process, and is characterized by a spin-lattice relaxation time T_1. We shall discuss several mechanisms in Section 2.11 that can contribute to T_1.

It should be pointed out that this process is also responsible for the reestablishment of the equilibrium spin distribution after the sample has undergone transitions from the absorption of rf energy. Interactions of the spins with B_1 tend to equalize the population of the energy levels. The equilibrium spin distribution is reestablished by the process of spin-lattice relaxation. The magnitude of T_1 depends on the physical state of the sample, the temperature, and the type of nucleus under consideration. For liquids, typical T_1 values are in the range 10^{-2}–10^2 sec for nuclei with $I = \frac{1}{2}$.

In many cases, it is found that the width of NMR absorption signals is larger than can be accounted for in terms of spin-lattice relaxation alone. It is convenient to define an additional relaxation process, the spin-spin or transverse relaxation time T_2 such that

$$T_1 \geq T_2 \tag{2.12}$$

Spin-spin relaxation is a process in which neighboring nuclei exchange spin orientations by an interaction between their magnetic moments. This process results in no change in the total energy of the system.

With reference to Figure 2.1, consider the interaction between two $I = \frac{1}{2}$ nuclei, 1 and 2, which are precessing at the same frequency but with opposite orientations with respect to B_0. Since the rotating component of μ_1 is equal to that of μ_2 in the x–y plane, the rotating component of μ_1 in the x–y plane can be thought of as a second field B_1 acting on μ_2. The resulting interaction can cause μ_2 to flip to the opposite orientation with respect to B_0, while at the same time μ_2 causes μ_1 to change to the opposite orientation. The result of this process is that two transitions, one absorption and one emission, have occurred with no net change in the total energy of the system. Spin-spin relaxation does not affect the relative population of the energy levels but can affect the line width, as discussed in Section 2.4.

2.3 INTENSITY

One of the major advantages of NMR over other forms of conventional spectroscopy is that the integrated intensity (area) of an NMR absorption signal is directly related to the number of nuclei giving rise to the signal and is not related by an absorption coefficient. The area under an NMR signal (equation 2.13) depends on a number of factors including B_0, B_1, T_1, T_2, T, and the number and type of nuclei giving the signal.

$$A \propto \frac{NB_1\mu^2}{kT(1 + \gamma^2 B_1^2 T_1 T_2)^{\frac{1}{2}}} \quad (2.13)$$

The dependence of the area on B_0 results from the effect of B_0 on the separation of the energy levels between which the transitions are occuring and its effect on the relative population of the energy levels. Equation 2.13 is valid only for relatively low values of the rotating field B_1. The area will increase with an increase in B_1 until the point where $\gamma^2 B_1^2 T_1 T_2$ becomes significant with respect to 1. When this occurs the area starts to decrease with a further increase in B_1. When the area decreases with an increase in B_1, saturation is said to be occurring. A saturation factor Z_0 is defined as

$$Z_0 = \frac{1}{1 + \gamma^2 B_1^2 T_1 T_2} \quad (2.14)$$

To help understand this dependence of the area on B_1, consider the NMR process. When B_1 of correct frequency is applied to the sample, transitions of equal probability occur from the lower to the higher energy level due to absorption of energy and from the higher to the lower energy level due to induced emission and relaxation. Since the populations of the two energy levels are not equal, there will be a larger number of transitions from the lower to the higher energy level and a net absorption of energy (i.e., an NMR signal). As larger and larger values of B_1 are applied, the population of the two energy levels tend to become more equal such that a decrease in the net absorption of energy is observed, that is, a decrease in the area of the signal.

Since the area of the signal depends upon the relaxation times T_1 and T_2, and they are usually different for chemically different nuclei in a sample, each peak will saturate at a different value of B_1. Therefore care should be exercised in obtaining and integration of an NMR spectrum to insure that B_1 is low enough such that saturation is not occurring.

The peak height of an NMR signal is given by an equation similar to that for the area

$$In \propto \frac{NB_1\mu^2 T_2}{kT(1 + \gamma^2 B_1^2 T_1 T_2)} \quad (2.15)$$

Since the two equations are similar one might question why peak heights cannot be used as an indication of the relative number of nuclei giving rise to the signals rather than going to the trouble of integrating the signals to obtain the areas. One important difference in Equations 2.13 and 2.15 is the dependence of the peak height on T_2. Normally T_2 values are different for chemically different nuclei in the sample. Furthermore, as discussed in Section 2.4, the width of an NMR peak is also dependent on T_2 and can be different for chemically different nuclei. Due to this dependence on T_2, peak heights will usually be different for an equal number of chemically different nuclei and cannot be used in quantitative analysis as a measure of the relative number of nuclei giving rise to the signal.

2.4 LINEWIDTHS

Transition probability theory predicts an infinitely sharp absorption line when the frequency of the radiofrequency (rf) field B_1 exactly equals the frequency corresponding to the energy difference between the two energy levels for the transition in question. In actual practice, however, the lines are broadened by various factors and it appears as if absorption is occurring over a range of frequencies. Therefore it is convenient to introduce a line shape function $g(v)$ which is proportional to the absorption at frequency v. Experimentally the absorption usually has a Lorentzian shape and $g(v)$ obeys the equation

$$g(v) = \frac{a}{b^2 + (v - v_0)^2} \tag{2.16}$$

where a and b are constants, v_0 is the frequency of absorption and v is the frequency of the rotating rf field.

Several factors including inhomogeneity in the magnetic field, relaxation, dipole–dipole interaction, and electric quadrupole effects can contribute to the broadening of an NMR absorption peak. Normally the applied magnetic field B_0 is not constant throughout the entire sample volume such that molecules in different parts of the sample are experiencing slightly different magnetic fields

$$B = B_0 + \Delta B_0 \tag{2.17}$$

This results (Equation 2.3) in the absorption occurring over a range of frequencies, Δv, rather than at a discrete frequency, v_0, such that the observed peak is actually a superposition of absorptions by molecules in different parts of the sample. With the magnets used in commercial NMR spectrometers, the inhomogeneity in the magnetic field limits the observed line width to ~ 0.1 Hz.

Additional broadening due to relaxation is a result of the finite lifetime of a nucleus in a given energy level. As pointed out previously, thermal motions can induce transitions between energy levels. In general, the lifetime in a particular energy level will be on the order of the spin-lattice relaxation time T_1. The broadening due to T_1 can be estimated using the Heisenberg uncertainty principle

$$\Delta E \Delta t \approx \hbar \qquad (2.18)$$

Defining the line width at half-height as $v_{\frac{1}{2}}$, substitution of equation 2.18 into equation 2.3 and rearrangement gives the relationship

$$v_{\frac{1}{2}} \approx \frac{1}{T_1} \qquad (2.19)$$

Thus with $T_1 = 1$ sec, a line width of about 1 Hz would be expected.

In viscous liquids or solids, interactions between magnetic moments of adjacent nuclei lead to a greater broadening than predicted by spin lattice relaxation. Consider two nuclei, 1 and 2, located at a distance r from each other. The field experienced by nucleus 1 due to the presence of nucleus 2 is proportional to μ_2/r^3. Therefore the field at a given nucleus will depend upon the magnitude of this effect from neighboring nuclei and may either add to or subtract from the magnetic field experienced by the nucleus, depending on the orientation of the adjacent magnetic moments with respect to the applied magnetic field B_0. Since all nuclei in a sample will not experience the same dipole interaction, absorption will occur over a range of frequencies leading to a broadened line. This process is similar to the broadening caused by inhomogeneity in B_0, with the exception that the inhomogeneity arises from within the sample itself. This broadening is referred to as that due to spin–spin relaxation. Using the uncertainty principle one can derive a relationship similar to equation 2.19 between T_2 and $v_{\frac{1}{2}}$ as

$$v_{\frac{1}{2}} \approx \frac{1}{T_2} \qquad (2.20)$$

For a Lorentzian line shape, equation 2.20 becomes

$$v_{\frac{1}{2}} \approx \frac{1}{\pi T_2} \qquad (2.21)$$

Normally, with mobile liquid and gaseous samples, the interaction between adjacent magnetic moments averages out due to the rapid tumbling of the molecules, such that spin–spin relaxation is not an important source of broadening. Under these conditions T_2 and T_1 become approximately equal and the line width can be used to approximate T_1 using equation 2.21. In any event, the line width provides an estimate on the lower limit of T_1, since $T_2 \leq T_1$.

Another source of line broadening exists for nuclei with spin $I > \frac{1}{2}$. As pointed out in Section 2.2, nuclei with $I > \frac{1}{2}$ have electric quadrupole moments due to the nonspherical electron distribution. Molecular tumbling will lead to fluctuating electric field gradients around the nucleus. These gradients can cause transitions to occur among the nuclear quadrupole energy levels. The effect on the nuclear energy levels is the same as if relaxation were occurring by a magnetic interaction.

2.5 BLOCH FORMALISM

Before proceeding with further aspects of NMR, it is important to reconsider the NMR experiment in terms of the total sample. So far our approach has been to consider the properties of an isolated nucleus, and apply the results to a macroscopic sample of nuclei. Bloch (5–7) found that many of the NMR properties of nuclei could be predicted from classical mechanics by considering the total magnetization, $\overline{M}$, of the sample and the relaxation times T_1 and T_2. This approach has certain advantages in describing time-dependent effects on NMR spectra and also leads to alternative definitions for some of the concepts introduced earlier. We shall briefly consider this approach below. A more detailed account may be found in the references given at the end of this chapter (1–4).

Consider the NMR experiment in terms of a collection of identical spin $I = \frac{1}{2}$ nuclei. When the sample is placed into the magnetic field, the magnetic moments of the nuclei take one of the two orientations in the magnetic field according to the Boltzmann distribution and will precess around the z axis (Figure 2.4). The phase relationship between the various spins is such that they are randomly distributed around the precessional (z) axis. The equilibrium magnetization, M_0, will have a component, M_z, along the z axis due to the slight excess of spins aligned with the field B_0. The component of magnetization, M_{xy}, in the x–y plane is zero, due to cancellation resulting from random distribution around the precessional axis.

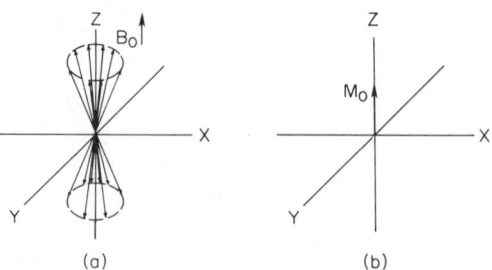

Fig. 2.4 (a) The orientations of the magnetic moments of a collection of identical nuclei when placed into a magnetic field B_0 and (b) the resultant magnetization M_0 along the z axis.

If the small rotating rf field, B_1, is applied in the x–y plane with a frequency equal to that of the precessional frequency, v_0, B_1 interacts with the magnetic moments and forces some of them to precess in phase with B_1. Simultaneously energy is absorbed by the spin system resulting in some of the magnetic moments changing their orientation with respect to B_0. In terms of the components of the magnetization, the absorption of energy leads to a decrease in M_z due to the spin populations becoming more equal and it creates a component in the x–y plane, M_{xy}, due to B_1 forcing some of the magnetic moments to precess in phase with B_1. The changes in the total magnetization of the sample when energy is absorbed may be detected by a receiver coil suitably placed around the sample and displayed in the form of the usual NMR spectrum. An absorption signal (v-mode) is observed for the component of M_{xy}, which is rotating out of phase with B_1, and a dispersion signal (u-mode) is observed for the component of M_{xy} rotating in phase with B_1.

After the absorption of energy the components of the magnetization decay back to their equilibrium values of zero for M_{xy} and M_0 for M_z. The inverse of the spin-lattice relaxation time T_1 is the first-order rate constant for the rate of decay of M_z, and the inverse of the spin–spin relaxation time T_2 is the first-order constant for the rate of decay of M_{xy}.

The Bloch equations for the time dependence of the magnetization are as follows:

$$\frac{dM_x}{dt} = \gamma(M_y B_0 + M_z B_1 \sin rt) - \frac{M_x}{T_2} \tag{2.22}$$

$$\frac{dM_y}{dt} = \gamma(M_z B_1 \cos rt - M_x B_0) - \frac{M_y}{T_2} \tag{2.23}$$

$$\frac{dM_z}{dt} = \gamma(-M_x B_1 \sin rt - M_y B_1 \cos rt) + \frac{(M_0 - M_z)}{T_1} \tag{2.24}$$

After transforming to a coordinate system that is rotating about the z axis at the precessional frequency (rotating frame reference), solution of the Bloch equations, assuming the magnetic field B_1 is swept slowly, yields (1)

$$u = M_0 \frac{\gamma B_1 T_2^2 (v_0 - v)}{1 + T_2^2(v_0 - v)^2 + \gamma^2 B_1^2 T_1 T_2} \tag{2.25}$$

$$v = -M_0 \frac{\gamma B_1 T_2}{1 + T_2^2(v_0 - v)^2 + \gamma^2 B_1^2 T_1 T_2} \tag{2.26}$$

$$M_z = M_0 \frac{1 + T_2^2(v_0 - v)^2}{1 + T_2^2(v_0 - v)^2 + \gamma^2 B_1^2 T_1 T_2} \tag{2.27}$$

The steady-state solution (equations 2.25, 2.26, and 2.27) assumes that equilibrium has been attained between the magnetization and the radiofrequency field, $dM/dt = 0$. If the magnetic field B_1 is scanned too rapidly, transient effects (2) occur that affect the shape of the resonance line. One of the more common effects is the appearance of wiggles or ringing at the tail of the absorption peak (Section 3.2.5).

2.6 THE CHEMICAL SHIFT

Our discussion to this point has focused on the interaction of a magnetic field with either a single nucleus or a collection of chemically equivalent nuclei. When a nucleus possesses a magnetic moment, an NMR absorption is observed for that nucleus when the Larmor equation (equation 2.3) is satisfied. Either the applied magnetic field or the frequency may be varied to achieve the resonance condition. Normally NMR spectrometers are set up to operate at an essentially constant magnetic field such that a wide range in the frequency (1–100 MHz at 23,487 gauss) is required to observe absorptions from different types of magnetic nuclei. The actual variation in the frequency employed is small (50–10,000 Hz) such that only one type of magnetic nucleus is under consideration in any one experiment (i.e., a proton spectrum). Therefore a different frequency range is scanned to observe the absorptions from different nuclei.

When an NMR spectrum is obtained on a sample that contains several chemically different types of the same nuclei, say protons, it is found that separate signals are observed for each type of chemically different proton. The separation between the signals for the different protons is called the chemical shift and is perhaps the most important parameter to be derived from an NMR spectrum. Different chemical shifts arise from the fact that the magnetic field experienced by the nucleus depends upon its environment and is not the same as the applied field. Thus each type of proton in an organic molecule will experience a slightly different magnetic field and separate absorptions will be observed for each group of chemically different protons.

The origin of the chemical shift arises from the screening or shielding of the nucleus by the electrons surrounding the nucleus and from electrons in other parts of the molecule. When a nucleus is placed into a magnetic field, the motion of the surrounding electrons induces a secondary magnetic field at the nucleus. The direction of the secondary magnetic field is usually such that it opposes the applied field. Thus the induced secondary magnetic field partially screens or shields the nucleus from the applied field, the magnitude of which depends on the local electron density and on the nature of neighboring groups within the molecule. Each different type of proton in the sample will experience a magnetic field that differs from the applied field by a small amount

$$B_{nucleus} = B_0(1 - \sigma) \quad (2.28)$$

where σ is the screening constant. Substitution of equation 2.28 into equation 2.3 gives the resonance condition as

$$v = \frac{\gamma}{2\pi} B_0(1 - \sigma) \quad (2.29)$$

The screening constant is usually small and, at constant frequency, determines the amount by which the applied field must be increased to satisfy the resonance condition.

In practice it is impossible to determine σ as this would require measuring the absorption of a bare nucleus stripped of its electrons. Instead a reference compound is employed and all chemical shifts are determined with respect to the absorption of the reference compound. Tetramethylsilane {TMS, $(CH_3)_4Si$} is employed as the reference compound for proton NMR since it gives a single absorption band that appears at a higher applied field than most other proton absorptions. Since the resonance condition may be satisfied by varying either the frequency or the applied field, chemical shifts may be expressed in frequency units (Hz) or field units (gauss). In order to eliminate the need to specify the magnetic field or the frequency used in measurements of chemical shifts, a method for reporting chemical shifts in nondimensional units is usually employed. For measurements made at constant frequency, the chemical shift δ is defined by

$$\delta = \frac{B_r - B_s}{B_r} \times 10^6 \text{ ppm} \quad (2.30)$$

where B_r is the applied field at constant frequency for absorption of the reference and B_s is the applied field for absorption of a given nucleus in the sample. Alternatively, for measurements made at constant applied field where the frequency is varied, the chemical shift δ is defined by

$$\delta = \frac{v_s - v_r}{v_r} \times 10^6 \text{ ppm} \quad (2.31)$$

where v_s is the frequency of absorption for a given nucleus in the sample and v_r is the frequency of absorption for the reference. Using this convention most proton chemical shifts will be expressed as positive numbers and approximately 95 % of all proton chemical shifts are in the range $\delta = 1 \sim 10$. Protons that are highly shielded will appear at a high applied field or low

TABLE 2.2. Some Typical Proton Chemical Shifts[a]

Compound	δ(ppm)
CH_4	0.22
cyclo-C_3H_6	0.22
CH_3CH_3	0.85
$(CH_3)_4C$	0.94
$HC\equiv C-H$	1.80
$CH_3\overset{\overset{O}{\|}}{C}CH_3$	2.04
$C_6H_5CH_3$	2.35
$CH_2=CH_2$	5.28
C_6H_6	7.27
H_2CO	9.57

[a] Data taken from various sources (19,28,40).

frequency and will have a small δ value, whereas protons that are less shielded will show absorptions at low field or high frequency and will have large δ values (Table 2.2).

A second convention for reporting chemical shifts was used in the early literature. Using this convention the chemical shift in parts per million increases with increasing shielding of the nucleus. The τ scale is defined as

$$\tau = 10.0 - \delta \tag{2.32}$$

such that values may be readily converted from one scale to the other. Since it has been recommended that the δ scale be adopted as the official scale, further discussions of chemical shifts in this book will refer to the δ scale only.

Occasionally, problems arise when reporting chemical shifts for nuclei other than protons. Several reference compounds have been used in the past for a given nucleus. In some cases the reference is less shielded than most compounds (CS_2 for ^{13}C NMR), and in others the reference is more shielded than the majority of compounds (TMS for ^{13}C NMR). Care should be exercised in referring to the literature and in reporting new chemical shift data for other nuclei to insure that the direction of the shift with respect to the reference is clearly stated. In the past, some workers have reported chemical shifts upfield from the reference as positive δ numbers whereas others have reported these as negative δ values. We shall adopt the convention in this book that a positive chemical shift is always downfield from the reference, irrespective of the nucleus under discussion.

2.7 THEORY OF CHEMICAL SHIFTS

As pointed out in Section 2.6, the chemical shift is determined in part by the interaction of the applied magnetic field with the electrons surrounding the nucleus in question. This interaction induces a secondary internal magnetic field that shields the nucleus from the applied magnetic field. The nucleus in question is also influenced by induced fields from other parts of the molecule, which may either add to or oppose the applied magnetic field. Several attempts have been made to theoretically account for the screening constant σ in terms of the electronic motion in a molecule. Using a second order perturbation approach, Ramsey has developed a theory which in principle accounts for the shielding in molecules (8). Ramsey considered the screening constant to be the sum of two contributions, a diamagnetic term, σ_d, and a paramagnetic term σ_p. The diamagnetic term σ_d is positive, leading to shifts to higher fields (shielding), and is approximated using the Lamb formula (9) for the shielding of atoms. The paramagnetic term σ_p is negative (deshielding) and results from the fact that the electrons in a molecule are not symmetrically disposed around the nucleus in question.

Application of the Ramsey formalism is difficult for anything except very small molecules (H_2) due to a number of unknown terms that must be approximated in evaluating σ_p. For larger molecules the two terms become approximately equal leading to their cancellation. Subsequent treatments of the screening constant have used the Ramsey theory as a starting point for further discussion. Instead of trying to explicitly calculate the screening constant, the approach has been to discuss the various factors that can contribute to σ and concentrate on their relative importance and magnitude. Saika and Slichter suggested that an additional term be included to account for contributions from other atoms within the molecule (10). Pople has shown that another term is needed to account for ring currents in aromatic compounds (11). In our discussion to follow, we shall consider the screening constant to be made up of five contributions

$$\sigma = \sigma_d^{loc} + \sigma_p^{loc} + \sigma_{other} + \sigma_{currents} + \sigma_{sol} \qquad (2.33)$$

where σ_d^{loc} is the local diamagnetic term, σ_p^{loc} is the local paramagnetic term, σ_{other} is the contribution from other atoms in the molecule, $\sigma_{current}$ is the contribution from ring currents, and σ_{sol} is an additional term to account for solvent effects. We shall discuss each of these terms in a qualitative fashion as to their relative importance. A more detailed account may be found in the references provided.

2.7.1 LOCAL DIAMAGNETIC SCREENING

Local diamagnetic screening arises from the induced magnetic field due to the circulation of electrons around a nucleus and results in the nucleus being shielded from the applied magnetic field (Figure 2.5). This contribution for the hydrogen atom may be approximated using the Lamb formula

$$\sigma_d^{loc} = \frac{4\pi e^2}{3mc^2} \int_0^\infty r p(r) dr \tag{2.34}$$

where p is the electron density associated with the hydrogen atom at a distance r from the nucleus. Depending on how the electron density is evaluated, for the hydrogen atom, equation 2.34 leads to

$$\sigma_d^{loc} = 20\lambda \times 10^{-6} \tag{2.35}$$

where λ is the effective number of electrons in the hydrogen 1s atomic orbital. Equation 2.35 shows that σ_d^{loc} is of the order of a few parts per million. Since most proton chemical shifts are within this same order of magnitude, it has been suggested that σ_d^{loc} is the primary factor controlling proton chemical shifts.

A second consideration that emerges from equation 2.35 is that σ_d^{loc} depends on the electron density associated with the atom. Several attempts have been made to correlate proton chemical shifts with substituent electronegativity assuming that the only influence of the substituent is through an inductive or resonance effect. For a related series of compounds such as CH_3X, a linear correlation is observed between the proton chemical shifts and the electronegativity of X (12). However significant variations are observed indicating the importance of other shielding mechanisms.

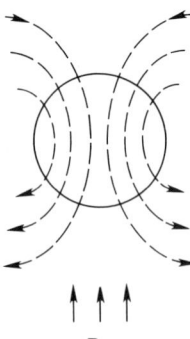

Fig. 2.5 The screening of an isolated nucleus arising from an induced magnetic field due to the circulation of the electrons around the nucleus.

2.7.2 LOCAL PARAMAGNETIC SCREENING

The Lamb formula is not applicable for calculating the screening constant in a polyatomic molecule since the electron density is not symmetrical about the nuclei. The paramagnetic term makes a negative contribution to the screening constant and arises from the interaction of ground state wavefunctions with excited state wavefunctions. Calculation of σ_p^{loc} therefore requires a knowledge of the excited state wavefunctions. The formula for σ_p^{loc} contains an expression for the difference in energy between the ground state and excited states in the denominator. Usually a knowledge of the excited state energies is not available. One approach to overcome this difficulty has been to simplify this term by using an average over the excited states of the molecule $\cdots$ the average energy approximation (8).

In the case of calculations of the screening in the hydrogen molecule, it is found that σ_p^{loc} contributes approximately 20% to the overall screening (8). For the most part the contribution from σ_p^{loc} to the screening of protons is considered to be negligible. However, it can become the dominant shielding mechanism for larger nuclei. For example, the chemical shift difference between F^- and F_2 is 200 ppm (13). Since F^- is symmetrical, the contribution from σ_p^{loc} is zero. In F_2 the bonding is primarily through the overlap of p orbitals such that the electron distribution is not symmetrical and the 200 ppm shift difference between F^- and F_2 is thought to be due to σ_p^{loc} in F_2 (10). Similar results are found for other nuclei. The relative importance of the paramagnetic term increases as the number of electrons around the atom increases. For lithium-7 it is thought that the diamagnetic and paramagnetic terms are approximately equal. For nuclei larger than ^7Li, the paramagnetic term is thought to be the dominant factor in the shielding mechanism and can lead to extremely large chemical shift differences from one compound to the next (14). For example, the normal range for proton shifts is ~ 12 ppm (1,2), for carbon-13 ~ 200 ppm (15), and for lead-207 ~ 3000 ppm (16).

2.7.3 SCREENING FROM OTHER ATOMS IN THE MOLECULE

The origin of the effect of other atoms or groups of atoms on the shielding of a nucleus lies in the magnetic anisotropy of these atoms (1). Since the electron density around a proton is relatively low, centers of higher electron density in other parts of the molecule may lead to induced fields which affect the shielding of the proton. This is especially true if the bonding orbitals over which the electrons are distributed are not spherically symmetrical.

In order to consider this effect in more detail, let us consider the model H–A, where A is a center of electron density due to an atom or group of

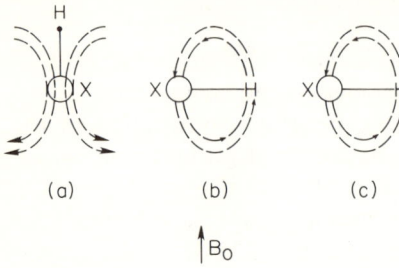

Fig. 2.6 The (a) shielding and (b and c) deshielding of a proton due to an atom or a group of atoms X located in a molecule.

atoms in another part of the molecule such that the proton is not necessarily bonded to A. When the sample is placed into a magnetic field, the circulation of the electrons around A generates a magnetic dipole moment at A. The magnitude of the magnetic dipole moment depends on the magnetic susceptibility χ_A of A and, for diamagnetic samples will oppose the applied magnetic field. The induced field experienced by H depends on the orientation of H–A in the magnetic field, and is illustrated in Figure 2.6. Thus the induced field at A results in deshielding at H if H–A is parallel to the applied magnetic field, and shielding if H–A is perpendicular to the applied magnetic field. Since the molecules are tumbling in solution, the induced field at H is the average of the contributions from the orientations shown in Figure 2.6, and is given (1) by

$$\Delta B_{(H)} = \frac{B_0}{3R^3} (2\chi^{\parallel} - \chi^{\perp} - \chi'^{\perp}) \tag{2.36}$$

where $\Delta B_{(H)}$ is the induced field at H, R is the distance from the proton to A, and $\chi^{\parallel}$ and $\chi^{\perp}$ are the magnetic susceptibilities of A when H–A is orientated parallel and perpendicular to the applied magnetic field, respectively. In general the induced fields do not cancel due to tumbling, except in the case when the electron distribution around A is symmetrical, such that either shielding or deshielding is observed at H, depending on the relative magnitudes of $\chi^{\parallel}$ and $\chi^{\perp}$.

Various attempts have been made to calculate the magnetic susceptibility of groups of atoms. The most successful approach has been to consider a point magnetic dipole centered at some position in A (17,18). Severe approximations are required for solution with the result that the reported magnitude of the susceptibility of various groups often covers a wide range. Nevertheless the results are useful for predicting the shielding and deshielding effects of these groups.

2.7.3.1 Shielding Due to Carbon–Carbon and Carbon–Hydrogen Bonds

As pointed out in Section 2.7.1, the local diamagnetic screening is the primary factor controlling proton chemical shifts as evidenced by the correlation of chemical shifts in methyl derivatives with the electronegativity of the substituent. A similar relationship holds for ethyl derivatives as well. However the α-proton chemical shift appears at lower field for an ethyl compound compared to the shift in the corresponding methyl compound. On the basis of inductive effects alone, one might have predicted the opposite result since a methyl group is known to be slightly electron donating. However the previously cited trend is completely general in that the chemical shift of a proton appears to lower field with an increasing number of C—C bonds attached to the carbon to which the proton is attached (i.e., CH_4, $\delta = 0.22$; CH_3CH_3, $\delta = 0.85$; $CH_3CH_2CH_3$, $\delta = 1.34$) (19).

Examination of the spectrum of cyclohexane at low temperatures reveals two signals separated by 0.47 ppm for the axial and equatorial protons (20). For monosubstituted cyclohexane derivatives at low temperature where the rate of conversion from one chair conformation to the other is slow, the proton on the carbon α to the substituent exhibits two resonances for the axial and equatorial conformations. It is now established that in general, an axial proton is more shielded than an equatorial proton (19). Yet the origin of the different chemical shifts for axial and equatorial protons in cyclohexanes remains largely unknown.

Attempts have been made to rationalize the previously cited shift differences in terms of the anisotropy of the magnetic susceptibility of the C—C and C—H bond (21–23). However calculated values of the anisotropies are far too large to be the sole causative factor. While there can be little doubt that there is some long range shielding effects of C—C and C—H bonds, the origin of these effects remains unknown. About all that can be said at this point is that the theory predicts that a nucleus lying along the axis of a single carbon bond will be deshielded whereas a nucleus lying above or below the bond will be shielded in accord with the experimental results.

2.7.3.2 Shielding Due to Carbon–Carbon Double and Triple Bonds

The long-range shielding effects of the carbon–carbon double bond are complicated. The induced magnetic field due to the circulation of the π electrons in the double bond depends on its orientation with respect to the magnetic field. Attempts to calculate this anisotropy using a point dipole approximation (24,25) have led to the following conclusions concerning the long-range shielding effect of a C=C bond: (1) protons located in the region above or below the plane of the double bond (x–y plane) between the two

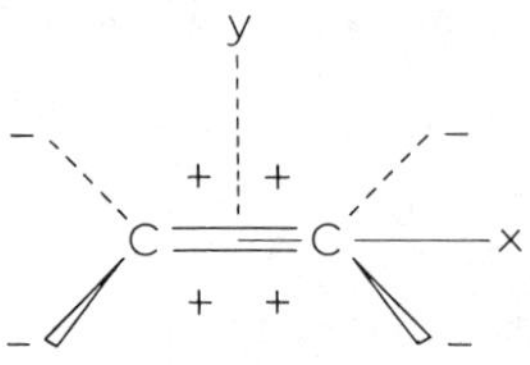

Fig. 2.7 The shielding (+) and deshielding (−) regions due to a carbon–carbon double bond.

carbons will experience shielding (Figure 2.7) and (2) protons located in the x–y plane will be deshielded.

The long-range shielding effect of a C≡C bond is even more dramatic than either a C=C or C—C bond. On the basis of electronegativity effects alone (local diamagnetic shielding) one would expect the chemical shift of acetylenic protons to be further downfield than the shift of olefinic protons. However the chemical shift of the protons in acetylene is between that of olefinic and aliphatic protons, and is only slightly downfield from the chemical shift of ethane.

Fig. 2.8 The shielding (+) and deshielding (−) regions due to a carbon–carbon triple bond.

Theory predicts a relatively large anisotropy along the axis of the molecules and a much smaller value perpendicular to this axis due to the axial symmetry of the C≡C bond (17). Therefore if the applied magnetic field is along the axis of the acetylene molecule, a relatively large induced field is generated due to the circulation of the π electrons around the carbon–carbon bond (Figure 2.8) and a much smaller induced field when the applied field is perpendicular to the axis of the molecule. Therefore nuclei that lie along the axis of the molecule will experience a shielding effect whereas nuclei in the region of the C≡C bond will experience deshielding (Figure 2.8). Verification of these predictions have been found experimentally (26,27).

2.7.3.3 Shielding Due to the Carbonyl Group

The chemical shift of formaldehyde $\left(\begin{array}{c}\text{O}\\\|\\\text{H—C—H, 9.57 ppm}\end{array}\right)$ (28) is considerably downfield from the proton shift of methane (0.22 ppm) (19). While part of this difference is due to the inductive effect of the carbonyl group, the shift difference is too large to be accounted for in terms of electron withdrawing effects on the local diamagnetic shielding alone. Theoretical calculations of the anisotropy of the carbonyl group predict that a nucleus will

THEORY OF CHEMICAL SHIFTS

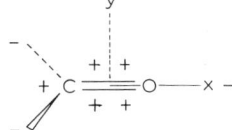

Fig. 2.9 The shielding (+) and deshielding (−) regions due to a carbon–oxygen double bond.

experience deshielding if it is located in the plane of the carbonyl group ($x-y$) (Figure 2.9) and shielding if it is located above and below the plane ($x-y$) of the bond. Furthermore shielding is predicted for a small region in the $x-y$ plane near the carbon end of the carbonyl group (25).

Experimental support for these predictions has come from a variety of sources. For example, it is well known that in α,β-unsaturated carbonyl compounds, the *cis-β*-protons are deshielded with respect to the *trans-β*-protons (29). Additional examples can be found in the NMR spectra of various steroids (30).

2.7.3.4 Shielding Due to Miscellaneous Groups

Long-range shielding due to a cyano group, (C≡N), is predicted to be similar to that of the carbon–carbon triple bond (31,32). However, it has been suggested that the shift observed in cyano compounds can be rationalized in terms of the electrostatic effect on the diamagnetic anisotropy alone (33).

The long-range shielding effect of an sp^2-hybridized nitrogen atom is thought to be due to the long pair of electrons such that protons located in the plane of the lone pair will experience deshielding (34). The change in the chemical shifts of pyridine upon protonation have been explained in this manner.

Several attempts have been made to account for the long-range shielding effects by halogens (35). It has been suggested that these effects result from the anisotropy due to the lone pair of electrons. However, the magnitude of this effect remains in question.

2.7.4 RING CURRENTS

The chemical shift of the protons in benzene are deshielded by approximately 2 ppm from those in ethylene, 7.27 versus 5.30 ppm. Since the proton in question is bonded to an sp^2-hybridized carbon in both cases, and the electron density on each carbon is equal, the local diamagnetic screening should be approximately equal in both cases. Therefore an additional screening contribution is needed to account for the difference in chemical shift.

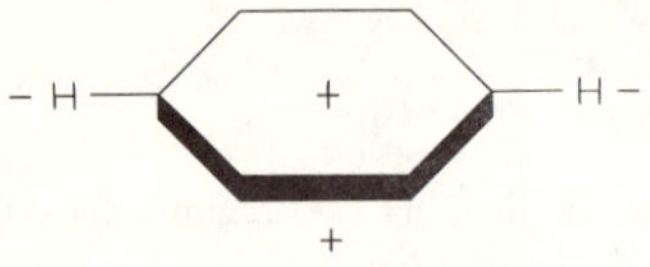

Fig. 2.10 The shielding (+) and deshielding (−) regions due to a benzene ring.

It is well known that an aromatic molecule possesses excess magnetic susceptibility in the direction perpendicular to the plane of the ring over that parallel to the plane (36). It was suggested that this results from a ring current due to the circulation of the π electrons around the orbitals of the ring induced by the externally applied magnetic field. Pople (11) first suggested that this would lead to deshielding of aromatic protons relative to ethylene and, using a point dipole approximation, calculated the deshielding to be 1.75 ppm. Refinements of this theory by Waugh and Fessenden (37) and Johnson and Bovey (38) by assuming the ring current to be located above and below the ring lead to a deshielding value of ~ 2.2 ppm.

The result of the theories just stated show that the secondary magnetic field due to the ring current is opposed to the externally applied field such that protons located inside the region of the ring will experience shielding whereas protons located outside the region of the ring will be deshielded (Figure 2.10). Further refinements of the theory by Pople and Untch (39) have suggested that the shielding associated with an annulene ring will depend upon the number of π electrons in the ring. Rings containing $4n\ \pi$ electrons will exhibit long-range shielding effects (paramagnetic ring current) such that protons located outside the area of the ring will be shielded whereas protons located inside the ring will be deshielded.

Many examples other than benzene exist to support the previously stated conclusions. These include various aromatic hydrocarbons, aromatic heterocycles, annulenes, paracyclophanes, and porphyrins (40). One of the most interesting examples is that provided by the 15,16-dialkyldihydropyrenes (**1**) (41). By varying the size of the alkyl group, the distance dependence of the effect of the ring current is illustrated. Furthermore, this ring system may be easily reduced by two electrons to a 16 π-electron dianion illustrating the

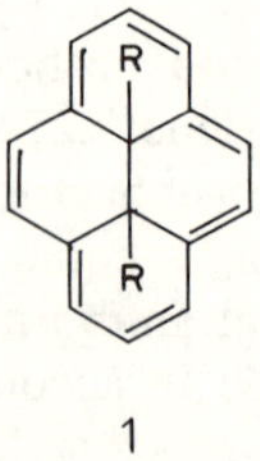

1

TABLE 2.3. Chemical Shifts of Some 15,16-dialkyldihydropyrenes[a,b]

R	Alkyl Proton Shifts			Exterior Proton Shifts
	α	β	γ	
Neutral hydrocarbons				
$\overset{\alpha}{C}H_3$	−4.25			7.95–8.67
$\overset{\alpha}{C}H_2\overset{\beta}{C}H_3$	−3.96	1.86		7.95–8.67
$\overset{\alpha}{C}H_2\overset{\beta}{C}H_2\overset{\gamma}{C}H_3$	−3.95	1.87	0.65	7.95–8.67
Dianions				
$\overset{\alpha}{C}H_3$	21.00	11.70		−(3.19–3.96)
$\overset{\alpha}{C}H_2\overset{\beta}{C}H_3$	21.15	11.70		−(2.50–3.14)
$\overset{\alpha}{C}H_2\overset{\beta}{C}H_2\overset{\gamma}{C}H_3$	21.24	12.59	5.51	−(2.56–3.14)

[a] Data taken from Mitchell, Klopfenstein, and Boekelhedie (41).
[b] In ppm from TMS.

reversal in the direction of the shielding effects with a paramagnetic ring current (Table 2.3).

The question of whether similar ring currents exist in nonaromatic rings has been discussed by a number of workers (42,43). Cyclopropane possesses a diamagnetic susceptibility (44) and it has been suggested that the high field shift of cyclopropane protons ($\delta = 0.22$) is evidence for a ring current. Theoretical treatments using the models previously mentioned tend to support this view. Anet and Schenck (45) have considered the possibility of ring currents in homoaromatic and antiaromatic systems. Using a method based on solvent effects on chemical shifts, they conclude that a diamagnetic ring current exists in 1,3,5-cycloheptatriene, cyclopentadiene and norbornadiene and a paramagnetic ring current exists in cyclooctatetraene.

2.8 SOLVENT EFFECTS

The effect of solvent on the screening of a particular proton may be divided into five contributions: bulk susceptibility of the medium, van der Waals interactions, anisotropy of the susceptibilities of the surrounding molecules, reaction field of the medium, and specific solute–solute interactions (46,47). The need for considering the bulk susceptibility term has been largely eliminated by the use of an internal reference (TMS) and will not be discussed further.

The contribution due to van der Waals interactions is thought to be due to distortions of the electronic environment of the nucleus by the solvent. This leads to a decrease in the diamagnetic screening of the nucleus and results in a downfield shift. In practice it has been found that polyhalogenated solvents produce the largest shifts which can be on the order of 0.5 ppm.

In most cases the change from a nonpolar nonaromatic solvent to an aromatic solvent, such as benzene, will result in an upfield shift of the solute protons. It has been suggested that this upfield shift is due to the solute molecules orienting themselves closer to the face of the aromatic ring, on the average, than to the edge of the aromatic ring. Allowing for an average over all solute–solvent orientations, this results in an upfield shift due to the anisotropy of the aromatic ring. Similar considerations applied to rod-like solvents such as carbon disulfide suggest that the solute lies closer, on the average, to the axis of the solvent molecule resulting in downfield shifts of the solute protons.

The reaction field contribution is due to the polarization of the surrounding medium by a polar solute which creates an electric field, the reaction field at the solute. Using the Onsager model (48) for the reaction field, Buckingham (49) has developed the theory for the reaction field effect on chemical shifts for a spherical molecule which predicts that the solvent shift should be linear in $(\varepsilon - 1)/(2\varepsilon - n)$ where ε is the dielectric constant of the medium and n is the refractive index of the solute. This relationship has been found to correctly predict the solvent shifts in various systems. However, in other cases it does not hold and the discrepancy is thought to be due to deviations from the spherical shape. Another theoretical approach has assumed an ellipsoidal shape and arrives at a linear relationship between the solvent shift and $(\varepsilon - 1)/(\varepsilon - \beta)$. Regardless of the model chosen, the practical results are that as the polarity of the solvent is increased, the chemical shifts move downfield, and protons closer to the polar site in the molecule experience larger shifts than those protons further removed from the polar site in the molecule.

In addition to the solvent shift with aromatic solvents due to anisotropy effects, specific shifts have been observed for polar solutes in aromatic solvents. These shifts are referred to as aromatic solvent-induced shifts (ASIS) and are thought to be due to some type of specific solute–solvent interaction. The results have generally been interpreted in terms of a collision complex brought about by dipole-induced dipole interactions or some other weak association between the electron donor aromatic solvent and some positively polarized part of the solute molecule. The energy of the interaction is on the order of 1 kcal/mole. For the most part these "complexes" are envisioned as a 1:1 association.

The major application of ASIS to date has been to differentiate between various angular methyl groups. Other applications are covered in the two excellent reviews of solvent effects in NMR (46,47). The reader is also referred to the reviews cited for a more detailed treatment of general solvent effects.

2.9 THEORY OF SPIN–SPIN SPLITTING

In 1951 it was reported that the spectra of several liquids contained more lines (multiplets) than could be accounted for on the basis of the number of chemically different nuclei in the sample (50–52). This multiplet splitting of resonance lines results from the interaction of a nuclear magnetic moment with the magnetic moments of neighboring nuclei in the same molecule which causes splitting of the energy levels and hence multiple transitions. The energy of this interaction is expressed as

$$E_{NN'} = hJ_{NN'} \bar{I}_N \cdot \bar{I}_{N'} \tag{2.37}$$

when $J_{NN'}$ is the coupling constant between the two nuclei N and N', and $\bar{I}_N \cdot \bar{I}_{N'}$ are the nuclear spin angular moments.

The mechanism proposed for this coupling interaction (53) assumes that the nuclear spins interact through polarization of the spins by the bonding electrons. As an example, consider the interaction between two bonded A and B nuclei of spin $I = \frac{1}{2}$. If the nuclear spin of A is oriented parallel to the applied magnetic field B_0, the electron near the vicinity of A will tend to orient its spin antiparallel to the nuclear spin due to the pairing of magnetic moments. Since A and B are bonded, the electron in the vicinity of B must have its spin antiparallel to that of A, due to the Pauli exclusion principle. The nuclear spin on B will in turn be antiparallel to the electron spin on B with the overall result that the two nuclear spins are oriented in an antiparallel fashion (Figure 2.11). The opposite orientation with the nuclear spins of A and B being parallel represents a slightly different energy arrangement. Both orientations, parallel and antiparallel, occur to approximately equal extents (Boltzmann distribution) in a bulk sample since the energy difference between the two orientations is small. When nucleus A undergoes a transition from one orientation with the magnetic field B_0 to the opposite orientation, two transitions occur corresponding to the two different orientations of nucleus B with respect to A. The frequency separation between the two transitions is proportional to the energy of interaction between the nuclear spins.

The magnitude of the spin–spin coupling constant J_{AB} is expressed in hertz (cycles/sec in the older literature) and may be either positive or negative. If the antiparallel orientation of the nuclear spins is the lower energy arrangement, the coupling constant J_{AB} is positive, whereas if the parallel orientation is lower in energy, the coupling constant is negative.

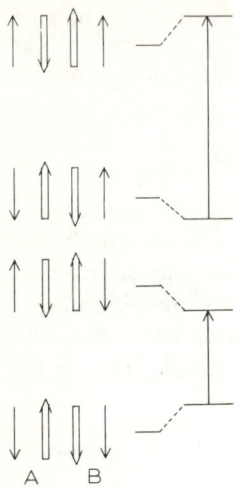

Fig. 2.11 The possible orientations of the magnetic moments (→) and electrons (⇒) of two bonded nuclei A and B, their relative energies, and the two nuclear transitions of nucleus A.

The *absolute* sign of a coupling constant cannot be readily determined from a normal high-resolution NMR spectrum. *Relative* signs of coupling constants, however, may be determined from double resonance experiments (Chapter 4) and, in some cases, from the analysis of non-first-order spectra. The absolute sign of the coupling constant between the *ortho* protons in *p*-nitrotoluene has been found to be positive from an experiment in which the molecule was oriented in the magnetic field by a strong electric field (54). A number of relative signs of coupling constants have now been determined with respect to this vicinal coupling. Some typical signs of coupling constants are given in Table 2.4. The relative signs of coupling constants can be of importance in structure elucidation.

A theory for the electron–nuclear spin–spin interactions has been proposed by Ramsey (53). The complete Hamiltonian for the motion of electrons in the presence of nuclei which possess magnetic moments is divided into three parts.

$$\mathcal{H}' = \mathcal{H}_1 + \mathcal{H}_2 + \mathcal{H}_3 \tag{2.38}$$

The first part represents the magnetic shielding of the interactions of the nuclear spins by the electron orbital motion. The second term is included to account for the dipole–dipole interaction between the electron magnetic moments and the nuclear magnetic moments. The final term arises from relativistic effects and is often referred to as the Fermi contact term. This term represents the interaction between electrons in s orbitals and nuclear magnetic moments.

TABLE 2.4. Signs of Some Proton–Proton Coupling Constants[a]

Fragment	Sign[b]	Fragment	Sign[b]
 H₂C (geminal)	−	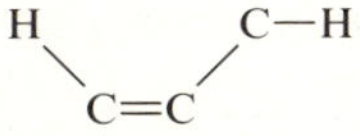 H\C=C/C−H (cis-like with C−H)	−
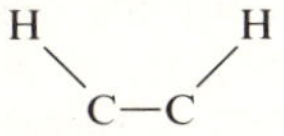 H−C−C−H	+	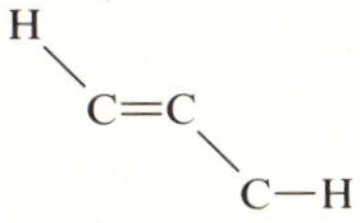 H\C=C\C−H	−
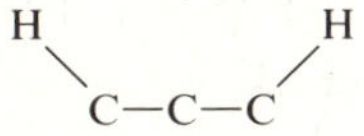 H−C−C−C−H	−	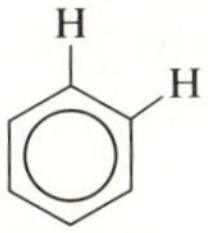 ortho benzene H,H	+
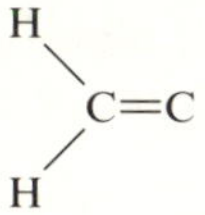 H\C=C/H (geminal vinyl)	+ or −	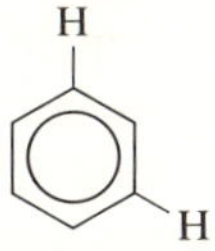 meta benzene H,H	+
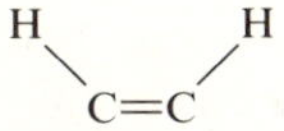 H\C=C/H (cis)	+	 para benzene H,H	+
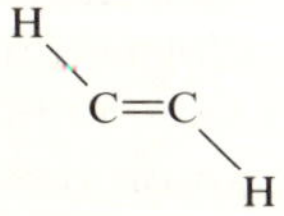 H\C=C\H (trans)	+		

[a] Data taken from Emsley, Feeney, and Sutcliffe (19).
[b] Relative signs with respect to J_{C-H} taken as positive.

Calculation of coupling constants from first principles is far too complex as it requires a knowledge of the total wave functions for the electronic ground and excited states. Several attempts have been made to calculate the coupling constant in the hydrogen molecule with reasonable success. The results suggest that the Fermi contact term accounts for greater than 90% of spin–spin interaction (55).

Several important conclusions concerning coupling constants result from the Fermi contact term. The magnitude of the coupling constant between two directly bonded nuclei depends upon the product of the magnetogyric ratios of the two nuclei $J \alpha \gamma_N \gamma_{N'}$. Thus if the coupling constant between two nuclei has been determined, the coupling constant between N and an isotope of N' may be calculated provided the magnetogyric ratio for the isotope of N' is known. Second, the magnitude of the coupling is directly proportional to the s electron densities in the bonding orbitals of N and N' used in forming the bond. This has led to correlations of coupling constants with the percent s character in the bond (56).

2.10 COUPLING CONSTANTS—PRACTICAL CONSIDERATIONS

In the analysis of NMR spectra (the extraction of the chemical shifts and coupling constants of the spectrum), the chemist is immediately faced with the problem of determining which peaks are due to spin–spin coupling and which are due to the absorptions of uncoupled nuclei. This task is not as formidable as it may appear on first glance. One soon learns the characteristic absorption patterns given by certain groupings of nuclei. From the intensities and number of lines in these patterns, one is able to pick out these multiplets in the spectrum with practice, In more complicated cases where there is considerable overlap of the peaks in a spectrum, one can make use of the fact that the relative intensities of the multiplets are in direct proportion to the number of nuclei giving rise to the peaks.

One important difference between coupling constants and chemical shifts is that coupling constants are independent of the applied magnetic field. When there is some doubt as to whether a particular multiplet is due to spin–spin coupling, or to the overlap of resonances from slightly different nuclei, one can obtain the spectrum at a higher magnetic field. If the multiplet is due to spin–spin coupling, only minor intensity changes will be observed in the multiplet, whereas if the multiplet is due to different nuclei, the absorptions will be separated more at higher magnetic field. In addition, one can carry out spin-decoupling experiments to determine which multiplets are due to nuclei that are mutually spin coupled. Obtaining the spectrum in another solvent may also change the spectrum such that characteristic multiplets may be recognized. These, along with other aids in the analysis of NMR spectra, are discussed in Chapter 4.

Since the spin–spin interaction is transmitted through the bonding electrons, coupling constants are characteristic of the arrangement of the nuclei in a molecule. Coupling between two nuclei may occur even when the two nuclei are separated by several bonds. For aliphatic organic compounds, proton–proton coupling constants are normally observed for protons separated by one, two, three, and sometimes four bonds. Spin coupling between nuclei separated by a larger number of bonds is observable if some of the intervening bonds are π bonds. The magnitude of the coupling constant decreases, in general, as the number of bonds separating the coupled nuclei increases. Typical values of coupling constants are discussed in Chapter 4.

Coupling constants are just as useful, if not more so, as chemical shifts in analytical applications of NMR. Since it is usually not possible to calculate a coupling constant from theory, the use of coupling constants relies very heavily on empirical correlations established through the examination of compounds of known structure. In this respect, it has been found that coupling constants depend on (1) the geometrical relationship between the two coupled nuclei; (2) the electronegativity of substituents; (3) the hydridization of the coupled nuclei; and (4) solvent effects. Thus coupling constants provide invaluable information in applications such as structure determinations, conformational analysis, and bonding. Applications of coupling constants will be discussed in Chapter 4, after we have discussed the analysis of NMR spectra.

2.11 RELAXATION MECHANISMS

Previously (Section 2.2) we have seen that when a spin system is placed into a magnetic field, or when a spin system is disturbed from its equilibrium state, it relaxes back to its equilibrium state by first-order processes characterized by two relaxation times T_1 and T_2. We have already discussed some processes by which spin–spin relaxation can occur, and we shall restrict ourselves here to a brief discussion of spin-lattice relaxation processes. Furthermore we shall restrict our discussion to relaxation in liquids. Excellent accounts of relaxation are available for those readers interested in more detail (57,58).

Any process that gives rise to fluctuating local magnetic fields can produce spin-lattice relaxation through the interaction of the local magnetic fields with the spin system. The fluctuating local magnetic fields invariably arise from thermal motions in the sample. The thermal motions of the molecules in the sample are occurring over a wide range of frequencies such that the fluctuating local magnetic fields have frequency components covering a wide range..In many ways the frequency components of the local magnetic fields behave similar to the frequency of the rotating field B_1 such that frequency components near the precessional frequency can induce transitions among the nuclear spins. The extent to which thermal motions are effective in producing spin-lattice relaxation depends on the magnitude of the local magnetic fields and the rate at which the fluctuations occur. It is convenient to define a *correlation time*, τ_c, which is a measure of the time required for the local field to acquire a new value. For translational motion, the correlation time may be thought of as the time required for the molecule to move one molecular diameter. In the case of rotational motion the correlation time is the time required for a molecule to rotate through an angle of one radian.

We shall briefly discuss the following types of processes that give rise to spin-lattice relaxation:

1. Nuclear dipole–dipole interactions
2. Chemical shift anisotropy
3. Scalar coupling
4. Spin rotation
5. Quadrupole interactions
6. Interactions with paramagnetic species.

2.11.1 DIPOLE–DIPOLE RELAXATION

Dipole–dipole relaxation results from fluctuating local magnetic fields arising from the interaction of the magnetic moment of another nucleus with the magnetic moment of the nucleus being relaxed. As the molecule tumbles in solution due to Brownian motion, the local field due to another magnetic moment will fluctuate in both magnitude and direction. The interacting magnetic moments may be in the same molecule (intramolecular) or in different molecules (intermolecular). The strength of the interaction depends on the orientation of the magnetic moments, the strength of the magnetic moments, and the distance separating the magnetic moments. Since other magnetic nuclei are always present in the sample, this is a general relaxation process and is often the dominant relaxation mechanism.

Consider a spin system containing two magnetic nuclei. The intramolecular interaction arises from rotational motion of the molecule. The distance separating the two nuclei remains fixed, as does the magnitude of the magnetic moments, while the orientation of the magnetic moments fluctuate with time. The intermolecular interaction arises from translational motion which is a function of the self-diffusion of the molecules. In general the correlation times for these two processes will not be equal such that $1/T_1$ is the sum of both terms

$$\frac{1}{T_1} = \frac{1}{T_{1_{\text{inter}}}} + \frac{1}{T_{1_{\text{intra}}}} \qquad (2.39)$$

Since the interaction depends on the motion of the molecules, both relaxation processes are dependent on the viscosity of the medium and the size and shape of the molecules.

Using the theory of Bloembergen, Purcell, and Pound (59), the dipole–dipole relaxation may be expressed as

$$\frac{1}{T_1} \propto \frac{\tau_c}{1 + 4\pi^2 v_0^2 \tau_c^2} \qquad (2.40)$$

RELAXATION MECHANISMS

TABLE 2.5. Some T_1 Values for Protons in Organic Molecules[a]

Molecule	T_1 (sec)	Temperature °C
Water	3.6	
Acetic acid	2.4	20
Methyl iodide	3.8	29
Benzene	19.3	25
11% benzene in CS_2	60	25
Toluene-CH_3	9	25
Toluene-aromatic	16	25

[a] Data taken from Pople, Schneider, and Bernstein (1).

For nonviscous liquids where molecular motion is rapid, τ_c is of the order $10^{-11} - 10^{-12}$ such that $1/\tau_c \gg 2\pi\nu_0$ and $1/T_1$ is proportional to τ_c. In this region of correlation times, both T_1 and T_2 are equal. As the liquid becomes more viscous, the correlation time increases and T_1 continues to decrease to the point where $\tau_c(2\pi\nu_0) = 1$. Further increases in the correlation time from this point result in an increase in T_1, and in the limit of slow motion where $1/\tau_0 \ll 2\pi\nu_0$, $1/T_1$ is proportional to $1/\tau_c$. In this region however, $1/T_2$ continues to increase linearly with the correlation time such that the line width continues to broaden (59).

The range of T_1 values for the dipole–dipole interaction is from 0.1 to 100 sec. For small organic molecules, T_1 is typically on the order of 10–20 sec and since molecular motion is rapid, $T_1 = T_2$. However for polymers the correlation times are such that T_1 is near the minimum and the contribution from incomplete averaging to T_2 controls the relaxation. The viscosity of the solution may be changed by raising the temperature which in turn increases τ_c and results in narrower lines.

Some values of T_1 for small organic molecules are given in Table 2.5. It should be kept in mind that T_1 will not be the same for all nuclei in the sample if they are magnetically different.

2.11.2 CHEMICAL SHIFT ANISOTROPY

As pointed out in Section 2.7, the magnetic field experienced by a nucleus is determined in part by shielding and deshielding effects of other nuclei in the molecule. The magnitude of the shielding depends on the orientation of the molecule with respect to the applied magnetic field. On the average, the nucleus experiences an average shielding due to rapid tumbling of the molecule in solution. On a much smaller time scale, however, the shielding may be anisotropic giving rise to fluctuating local magnetic fields that vary as the molecule rotates in solution. Thus the local fields provide a relaxation mechanism.

The magnitude of this relaxation mechanism is dependent on the externally applied magnetic field B_0 since it is related to the chemical shift interaction. In the case of nonviscous liquids where the extreme narrowing limit applies ($1/\tau_c \gg 2\pi\nu_0$), the relaxation due to the chemical shift anisotropy is given by

$$\frac{1}{T_1} = \tfrac{2}{15} \gamma^2 B_0^2 (\sigma_\| - \sigma_\perp)^2 \tau_c \tag{2.41}$$

where $\sigma_\|$ and $\sigma_\perp$ are the shielding tensors parallel and perpendicular to the magnetic field. The ratio of T_2/T_1 is $\tfrac{6}{7}$.

The range of T_1 values for this relaxation mechanism is of the order of from 10 to 100 sec. This mechanism is thought to be important only at higher magnetic field strengths. At field strengths of the order of 60 k gauss, this mechanism can become comparable with that due to dipole–dipole interaction. One example of relaxation via chemical shift anisotropy has been reported (60). The spin-lattice relaxation time T_1 from $CH_3{}^{13}CO_2H$ varies linearly with the square of the frequency over the range of 9–60 MHz.

2.11.3 SCALAR COUPLING

Spin–spin coupling arises from the influence of the magnetic moment of one nucleus on another magnetic nucleus and is due to the difference in the local magnetic field experienced by one nucleus due to the other. If the local magnetic field fluctuates with time it can provide a relaxation mechanism called scalar coupling. Suppose one nucleus A has a spin $I > \tfrac{1}{2}$. If this nucleus has a short relaxation time compared to that provided by scalar coupling, then the local field experienced by a coupled nucleus B will be an average and, instead of observing the expected multiplet, one observes a single line.

The relaxation due to scalar coupling is given by

$$\frac{1}{T_1^B} = \frac{2A_2}{3} I_s(I_s + 1) \frac{\tau_A}{1} + (\nu_B - \nu_A)^2 \tau_A^2 \tag{2.42}$$

where A is the spin–spin coupling constant in units of radians per sec and τ_A is the relaxation time of nucleus A. An example of this type of relaxation is provided by molecules with protons bonded to nitrogen. Nitrogen-14 has a spin $I = 1$ and is relaxed primarily by a quadrupole relaxation mechanism.

As a result spin–spin coupling is not observed with ^{14}N and one observes only a broadened resonance for protons attached to ^{14}N. In the event that the electron distribution surrounding nitrogen is symmetric, relaxation due to the quadrupole interaction becomes negligible, and one can observe spin coupling between nitrogen and protons (ammonium salts and isonitriles). Similar effects are also observed for protons attached to boron-11 ($I = \frac{3}{2}$).

Scalar relaxation can also occur when chemical exchange is present. If the exchange rate of nucleus A is much larger than the coupling constant between nuclei A and B, and is larger than $1/T_1$ for both nuclei A and B, then only a single line will be observed provided the time the nuclei are uncoupled is short compared to the time the nuclei are coupled. An example of this type of relaxation is given by the catalyzed exchange of the hydroxyl proton in alcohols where a single line is observed for the hydroxyl proton. The analysis of the line shape of single lines due to this type of relaxation can provide useful chemical information.

The range of T_1 values due to scalar relaxation is of the order 1–100 sec. This relaxation process can become comparable to or even greater than that provided by the dipole–dipole interaction in cases where the exchange rate is relatively fast.

2.11.4 SPIN ROTATION

When a molecule undergoes rotation, the motions of the electrons in the molecule generates a molecular magnetic moment at the nuclei. This gives fluctuating magnetic fields that provide a relaxation mechanism. The magnetic fields are proportional to the rotational angular momentum that is undergoing changes in both magnitude and direction due to Brownian motion.

The rate of this relaxation process is proportional to the angular momentum correlation time, which is the time a molecule spends in any given angular momentum state. This serves to distinguish this relaxation process from the others since τ_c decreases as the temperature increases whereas the angular momentum correlation time increases with increasing temperature. As a result T_1 becomes longer as the temperature decreases if this is the dominant relaxation mechanism.

The range of T_1 values for spin rotation relaxation is of the order 10^{-2}–100 sec. For protons, in most organic molecules, it is not a dominant relaxation mechanism. However, for small symmetric molecules, it can become a dominant relaxation mechanism for nuclei that have large chemical shift ranges. Some examples have been found in fluorine-19 relaxation studies (61).

2.11.5 QUADRUPOLE RELAXATION

The quadrupole relaxation arises from interactions between the nuclear spins of nuclei with $I > \frac{1}{2}$ with the electric field gradient at the nucleus. Reorientations of the molecule results in random fluctuations of the components of the quadrupole coupling tensor as a function of time, thereby providing a relaxation mechanism. In the narrowing limit where $v_0 \tau_c \ll 1$, $1/T_1$ due to this mechanism is proportional to the quadrupole coupling constant $(e^2 Qq/\hbar)$ and the correlation time, and is given by

$$\frac{1}{T_1} = \frac{3}{40} \frac{2I+3}{4I^2(2I-1)} \left(1 + \frac{\eta^2}{3}\right)\left(\frac{e^2 Qq}{\hbar}\right)^2 \tau_c \qquad (2.43)$$

where η is the asymmetry parameter (57).

In nonviscous liquids, the molecular correlation time is of the order 10^{-11}–10^{-12} sec and the relaxation time is determined by the quadrupole coupling constant. The quadrupole interaction is the dominant relaxation mechanism for nuclei with $I > \frac{1}{2}$ and T_1 values from 10^{-7}–10^2 sec. In cases where the electron distribution about spin $I > \frac{1}{2}$ nuclei is symmetrical, the quadrupole coupling constant is zero, and relaxation by this mechanism vanishes. Several examples of predominantly quadrupole relaxation may be found in alkyl lithium compounds (62), nitrogen compounds (63), and boron compounds (64). If the quadrupole coupling constant is known from other studies, $1/T_1$ provides a convenient way of determining the molecular correlation time.

2.11.6 INTERACTION WITH PARAMAGNETIC SPECIES

The presence of paramagnetic species in the sample provides a very effective relaxation mechanism. The magnetic moments of the unpaired electrons on the paramagnetic species provide the fluctuating local magnetic field. Since the electron magnetic moment is much larger than that of the nuclear magnetic moment only small concentrations of paramagnetic species are necessary in order for this relaxation mechanism to become the dominant mechanism. The theory for this process has been considered by Bloembergen, Purcell, and Pound (59) and is given by

$$\frac{1}{T_1} \propto \frac{4\pi^2 \gamma^2 n N_p \mu_{\text{eff}}^2}{kT} \qquad (2.44)$$

where N_p is the number of paramagnetic ions or molecules per cm^3, n is the viscosity of the solution, and μ_{eff} is the effective magnetic moment of the paramagnetic species (59).

In order for an ion to be effective in producing relaxation by this mechanism, the electron spin-relaxation time must be long compared to the molecular correlation time. That is to say, the spin of the electron magnetic moment remains fixed while the nuclei are undergoing reorientation. The most effective ions for this type of relaxation are Fe^{3+}, Cr^{3+}, Mn^{2+}, Eu^{2+}, Gd^{3+}, and Cu^{2+}. Less effective ions are Fe^{2+}, Co^{2+}, Ni^{2+}, and the remaining lanthanide ions. Oxygen is one paramagnetic molecule that is fairly effective in this type of relaxation.

In most cases paramagnetic species produces unwanted line broadening and their presence in NMR samples should be avoided. It is not uncommon for lines to be broadened into the baseline such that only a very broad line is observed if it is detectable at all. However there are applications where the presence of paramagnetic species can be used to probe the hydration shell of a nucleus and to obtain information on the binding sites of metal ions in biological systems. The addition of a paramagnetic species to a sample has proven valuable in order to shorten the relaxation times of carbonyl carbons when running ^{13}C NMR spectra (65).

2.12 EFFECT OF EXCHANGE PROCESSES ON NMR SPECTRA

Any process that results in the exchange of nuclei among two or more magnetic environments may have a profound effect on the shape of NMR absorptions. In the

$$ROH^* + HA \rightleftharpoons ROH + H^*A \qquad (2.45)$$

absence of coupling, if a nucleus (i.e., H*) is exchanging among two or more sites at a rate faster than the frequency separation between the absorption frequencies of the exchange sites, the nucleus will experience an average of the magnetic field at the individual exchange sites and will exhibit a single, sharp absorption. The frequency of the absorption is the average of the absorptions of the individual exchange sites

$$\delta_{obs} = \sum \delta_i p_i \qquad (2.46)$$

where p_i is the population of site i. If, on the other hand, the rate of exchange is slow, one may observe separate absorptions characteristic of each individual exchange site. At intermediate rates of exchange, one may observe spectra ranging from a single, broadened peak to broadened peaks slightly offset in frequency from the absorptions of the individual exchange sites.

If the exchanging nucleus is spin coupled with other nuclei, one may observe a collapse of the multiplet if the rate of exchange is fast. For example, in the case

$$CH_3OH^* + HA \rightleftharpoons CH_3OH + H^*A \qquad (2.47)$$

if the rate of exchange is slow, the methyl protons will be able to distinguish the two spin orientations of the OH proton and will appear as a doublet. If the rate of exchange is fast, the methyl protons will observe an average of the OH proton (i.e., exchange may change the spin orientation of the OH proton from aligned with the field to opposed to the field and vice versa) and a single line will be observed for the methyl protons. Intermediate rates of exchange will lead to absorptions ranging from a broadened multiplet to a broadened single absorption.

Provided one can alter the rate of exchange by changing the concentration, pH, or temperature, etc., analysis of the lineshapes in the region of intermediate exchange may provide useful information concerning the rate process. Typical examples of the application of lineshape analysis of NMR spectra for examining rate processes include the areas of conformational analysis, hindered rotation, protolysis, nitrogen inversion, and intermolecular rearrangements. The analysis of NMR lineshapes is discussed in more detail in Section 6.7.

REFERENCES

1. J. A. Pople, W. G. Schneider, and H. J. Bernstein, *High-Resolution Nuclear Magnetic Resonance*, McGraw-Hill, New York, 1959.
2. J. W. Emsley, J. Feeney, and L. H. Sutcliffe, *High-Resolution Nuclear Magnetic Resonance Spectroscopy*, Vol. 1, Pergamon, New York, 1965.
3. A. Carrington and A. D. McLachlan, *Introduction to Magnetic Resonance*, Harper & Row, New York, 1967.
4. J. D. Memory, *Quantum Theory of Magnetic Resonance Parameters*, McGraw-Hill, New York, 1968.
5. F. Bloch, *Phys. Rev.*, **70**, 460 (1946).
6. R. K. Wangness and F. Bloch, *Phys. Rev.*, **89**, 728 (1953).
7. F. Bloch, *Phys. Rev.*, **102**, 104 (1956).
8. N. F. Ramsey, *Phys. Rev.*, **78**, 699 (1950).
9. W. E. Lamb, Jr., *Phys. Rev.*, **60**, 817 (1941).
10. A. Saika and C. P. Slichter, *J. Chem. Phys.*, **22**, 26 (1954).
11. J. A. Pople, *J. Chem. Phys.*, **24**, 1111 (1956).
12. J. N. Shoolery and B. P. Dailey, *J. Am. Chem. Soc.*, **77**, 3977 (1955).
13. H. S. Gutowsky and C. J. Hoffman, *J. Chem. Phys.*, **19**, 1259 (1951).
14. C. J. Jameson and H. S. Gutowsky, *J. Chem. Phys.*, **40**, 1714 (1964).
15. J. B. Stothers, *Carbon-13 NMR Spectroscopy*, Academic, New York, 1972.
16. P. R. Wells, in *Determination of Organic Structures By Physical Methods*, F. C. Nachod and J. J. Zuckerman, Eds., Vol. 4, Academic, New York, 1971, p. 233.
17. J. A. Pople, *Proc. Roy. Soc.* (London), **A239**, 550 (1957).
18. H. M. McConnell, *J. Chem. Phys.*, **27**, 226 (1957).

REFERENCES

19. J. W. Emsley, J. Feeney, and L. H. Sutcliffe, *High-Resolution Nuclear Magnetic Resonance Spectroscopy*, Vol. 2, Pergamon, New York, 1965.
20. F. R. Jensen, D. S. Noyce, C. H. Sederholm, and A. J. Berlin, *J. Am. Chem. Soc.*, **82**, 1256 (1960).
21. A. G. Moritz and N. Sheppard, *Mol. Phys.*, **5**, 361 (1962).
22. J. I. Musher, *J. Chem. Phys.*, **35**, 1159 (1961).
23. A. A. Bothner-By and C. Naar-Colin, *J. Am. Chem. Soc.*, **80**, 1728 (1958).
24. J. Tillieu, *Ann. Phys.*, **2**, 471, 631 (1957).
25. J. A. Pople, *J. Chem. Phys.*, **37**, 60 (1962).
26. H. Heel and W. Zeil, *Z. Electrochem.*, **64**, 962 (1960).
27. W. Zeil and H. Buchert, *Z. Phys. Chem.*, **38**, 47 (1963).
28. B. L. Shapiro, R. M. Kopchik, and S. J. Ebersole, *J. Chem. Phys.*, **39**, 3154 (1963).
29. L. M. Jackman and R. H. Wiley, *J. Chem. Soc.*, 288 (1960).
30. N. S. Bhacca and D. H. Williams, *Application of NMR Spectroscopy in Organic Chemistry*, Holden-Day, San Francisco, 1965.
31. G. S. Reddy and J. H. Goldstein, *J. Chem. Phys.*, **39**, 3509 (1963).
32. R. F. Zurcher, in *Nuclear Magnetic Resonance in Chemistry*, B. Pesce, Ed., Academic, New York, 1965, p. 45.
33. A. D. Cross and I. T. Harrison, *J. Am. Chem. Soc.*, **85**, 3223 (1963).
34. V. M. S. Gil and J. N. Murrell, *Trans. Farad. Soc.*, **60**, 248 (1964).
35. R. F. Zurcher, in *Progress in Nuclear Magnetic Resonance Spectroscopy*, J. W. Emsley, J. Feeney, and L. H. Sutcliffe, Eds., Vol. 2, Pergamon, New York, 1967, p. 205.
36. L. Pauling, *J. Chem. Phys.*, **4**, 673 (1936).
37. J. S. Waugh and R. W. Fessenden, *J. Am. Chem. Soc.*, **79**, 846 (1957).
38. C. E. Johnson, Jr., and F. A. Bovey, *J. Chem. Phys.*, **29**, 1012 (1958).
39. J. A. Pople and K. G. Untch, *J. Am. Chem. Soc.*, **88**, 4811 (1966).
40. L. M. Jackman and S. Sternhell, *Applications of Nuclear Magnetic Resonance Spectroscopy in Organic Chemistry*, Pergamon, New York, 1969, pp. 94–98.
41. R. N. Mitchell, C. E. Klopfenstein, and V. Boekelheide, *J. Am. Chem. Soc.*, **91**, 4931 (1969).
42. D. J. Patel, M. E. H. Howden, and J. D. Roberts, *J. Am. Chem. Soc.*, **85**, 3218 (1963).
43. J. J. Burke and P. C. Lauterbur, *J. Am. Chem. Soc.*, **86**, 1870 (1964).
44. J. R. Lacher, J. W. Pollock, and J. D. Park, *J. Chem. Phys.*, **20**, 1047 (1952).
45. F. A. L. Anet and G. E. Schenck, *J. Am. Chem. Soc.*, **93**, 556, 3310 (1971).
46. P. Laszlo, in *Progress in Nuclear Magnetic Resonance Spectroscopy*, J. W. Emsley, J. Feeney, and L. H. Sutcliffe, Eds., Vol. 3, Pergamon, New York, 1967, p. 231.
47. J. Ronayne and D. H. Williams, in *Annual Reports of NMR Spectroscopy*, E. F. Mooney, Ed., Vol. 2, Academic, New York, 1969, p. 83.
48. L. Onsager, *J. Am. Chem. Soc.*, **58**, 1486 (1936).

49. A. D. Buckingham, *Can. J. Chem.*, **38**, 300 (1960).
50. H. S. Gutowsky and D. W. McCall, *Phys. Rev.*, **82**, 748 (1951).
51. E. L. Hahn and D. E. Maxwell, *Phys. Rev.*, **84**, 1246 (1951).
52. W. G. Proctor and F. C. Yu, *Phys. Rev.*, **78**, 471 (1950).
53. N. F. Ramsey, *Phys. Rev.*, **91**, 303 (1953).
54. A. D. Buckingham and K. A. McLauchlan, *Proc. Chem. Soc.*, 144 (1963).
55. M. Barfield and D. M. Grant, in *Advances in Magnetic Resonance*, J. S. Waugh, Ed., Vol. 1, Academic, New York, 1965, p. 149.
56. N. Muller and D. E. Pritchard, *J. Chem. Phys.*, **31**, 768 (1959).
57. A. Abragam, *The Principles of Nuclear Magnetism*, Oxford Univ. Press, London, 1961.
58. T. C. Farrar and E. D. Becker, *Pulse and Fourier Transform NMR*, Academic, New York, 1971, pp. 46–65.
59. N. Bloembergen, E. M. Purcell, and R. V. Pound, *Phys. Rev.*, **73**, 679 (1948).
60. T. C. Farrar, S. J. Druck, R. R. Shoup, and E. D. Becker, unpublished manuscript reported in reference 59, p. 60.
61. A. A. Maryott, T. C. Farrar, and M. S. Malmberg, *J. Chem. Phys.*, **54**, 64 (1971).
62. G. E. Hartwell and A. Allerhand, *J. Am. Chem. Soc.*, **93**, 4415 (1971).
63. W. B. Moniz and H. S. Gutowsky, *J. Chem. Phys.*, **38**, 1155 (1963).
64. T. Tsang and T. C. Farrar, *J. Chem. Phys.*, **50**, 3498 (1969).
65. O. A. Gansow, A. Burke, and G. N. LaMar, *Chem. Commun.*, 456 (1972).

CHAPTER

3

EXPERIMENTAL AND INSTRUMENTAL ASPECTS OF NMR

3.1	The Spectrometer	50
	3.1.1 Magnet	50
	3.1.2 Transmitter	52
	3.1.3 Probe	53
	3.1.4 Radiofrequency Detection	55
	3.1.5 Integrator	56
	3.1.6 Sweep Units	57
	3.1.7 Field/Frequency Stabilization	57
3.2	Factors Influencing the Resolution and Line Shape	59
	3.2.1 Homogeneity of the Magnetic Field	59
	3.2.2 Radiofrequency Power Level	61
	3.2.3 Radiofrequency Phase	62
	3.2.4 Spinning Rate	62
	3.2.5 Sweep Rate	63
	3.2.6 Signal Filtering	63
3.3	Sample Preparation	64
3.4	Chemical Shift Measurements	66
	3.4.1 Reference Compounds	66
	3.4.2 Frequency Calibration	68
3.5	Intensity Measurements	69
3.6	Factors Influencing Sensitivity	70
3.7	Sensitivity Enhancement	73
3.8	FT Techniques	73
3.9	Pulsed FT-NMR	75
	3.9.1 Data Acquisition	80
	3.9.2 Data Reduction	82
3.10	Quadrature FT-NMR	87
3.11	Stochastic Resonance	87
3.12	Rapid-Scan or Correlation FT-NMR	88
3.13	Commercial NMR Spectrometers	89
	3.13.1 Routine Proton Spectrometers	89
	3.13.2 CW Research Spectrometers	90
	3.13.3 FT Spectrometers	90
	3.13.4 Superconducting Spectrometers	91
	3.13.5 Spectrometers for Process Monitoring	92

Since the first NMR spectra on liquid samples were obtained in 1945 (1,2), the complexity of NMR spectrometers has increased considerably. Fortunately the difficulty in obtaining NMR spectra has decreased. In this chapter we discuss NMR spectrometers in relation to the instrumentation necessary for obtaining a spectrum. In addition, the various factors that influence the appearance of a spectrum and special techniques for obtaining spectra on small samples are also discussed. Finally, the various types of commercially available spectrometers are briefly reviewed.

3.1 THE SPECTROMETER

Present-day commercial NMR spectrometers may be divided into two basic types: (1) the routine spectrometer and (2) the more sophisticated research-type spectrometer. Both types of instruments contain complex electronic components and differ primarily in the design and number of components, and in the types of experiments they are capable of performing. Regardless of the type of instrument under consideration, or whether it is an older instrument with basically tube-type electronics or a more modern instrument with solid-state electronics, there are several components that are common to all types of high-resolution spectrometers. These include a magnet, a means of magnetic field stabilization, a sample probe, a radio-frequency oscillator, and a recorder for display of the spectrum. These components will be discussed in more detail in the following.

3.1.1 MAGNET

The magnet of an NMR spectrometer is required to produce the condition for the absorption of radiofrequency energy and is designed to produce as homogeneous a magnetic field as possible. One of three types of magnet is currently employed in NMR spectrometers: (1) a permanent magnet; (2) an electromagnet; or (3) a superconducting solenoid. Regardless of the type of magnet used, it should be capable of providing a highly uniform, stable magnetic field with a homogeneity of the order of three parts in 10^9. In order to achieve this homogeneity with permanent and electromagnets, parallel pole caps are used and their design is such as to take into account the relationship between the ratio of the diameter of the pole caps and the gap width, and its effect on the homogeneity.

In addition, electric shim coils (homogeneity coils) are incorporated on the pole faces to allow for additional corrections of the homogeneity (3,4). After the magnet is mechanically shimmed for the best possible homogeneity, the current in the homogeneity coils is varied to improve the homogeneity further in the region of the sample. Controls are usually provided for ad-

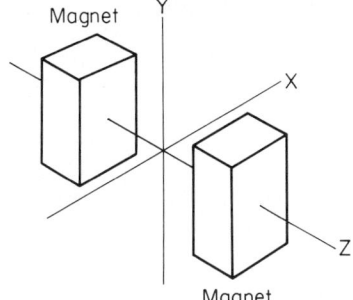

Fig. 3.1 Relative orientation of the x, y, and z axis with respect to the laboratory magnet.

justing a vertical gradient (y axis), a horizontal gradient in the plane parallel to the pole caps (x axis), and a horizontal gradient across the magnet (z axis) (Figure 3.1). In addition, it is desirable to have the contour of the magnetic field as flat as possible in the region of the sample. Curvature shim coils are provided to bend the magnetic field about the z axis to make corrections in the contour of the field. Additional shim coils to correct the x gradients along the y axis (x–y), the z gradients along the y axis (y–z), and second- and fourth-order corrections are provided on most instruments. The homogeneity is very sensitive to changes in the y gradients and many instruments provide a means for automatically adjusting the y gradient for maximum homogeneity.

The majority of permanent magnets are designed to produce a magnetic field of ~14,000 gauss (60 MHz for protons), although one has appeared recently with a field of 22,100 gauss (90 MHz). Permanent magnets have the advantage that a magnet power supply and a means for cooling the magnet are not required. Furthermore they usually have excellent long-term field stability. A disadvantage of permanent magnets is, of course, that the magnetic field is fixed and cannot be varied over a wide range. Furthermore, permanent magnets are subject to temperature-dependent variations so they are usually thermostated to operate at constant temperature. Permanent magnets are also subject to field variations induced by the placement of magnetic materials in the vicinity of the magnet. Nevertheless the stability and homogeneity of the magnetic field from a permanent are entirely satisfactory for use in NMR spectrometers.

The upper limit of the field strength of conventional electromagnets is ~25,000 gauss as a result of homogeneity and stability considerations. A very stable power supply is required to produce the high currents needed. In addition, a means for cooling the magnet and maintaining a constant magnet temperature ($\pm 0.1\,°C$) is required. An obvious advantage of an electromagnet over a permanent magnet is the ability to vary the magnetic field strength by varying the current passing through the coils. The stability of a conventional electromagnet is improved by using some type of stabilizing device (5). This usually consists of an additional set of coils on the pole pieces that detects changes in flux across the magnet gap. These coils are connected to some type of galvanometer that detects the current induced in the coils by the changes in flux and applies a correction voltage to another set of coils to compensate for the change in the magnetic field strength. As an additional means of improving the stability and homogeneity of an electromagnet, the room temperature should be thermostated to $\pm 1°$C.

The magnetic field strength used in NMR spectrometers has increased considerably in recent years through the use of superconducting solenoids (6). Fields up to $\sim 86{,}000$ gauss (360 MHz for ^{1}H) have been achieved recently. The major problem with superconducting solenoids is obtaining the field homogeneity necessary for NMR applications. Homogeneity coils are used extensively. In addition, the solenoids presently in use operate at liquid helium temperature. Closed-looped cryostats are used to reduce the helium loss due to evaporation. In spite of the difficulties involved, the use of superconducting solenoids will most likely increase in the future. The advantage in terms of sensitivity and chemical shift separation far outweigh the difficulties involved in operating a superconducting solenoid.

3.1.2 TRANSMITTER

The radiofrequency (rf) transmitter supplies the necessary rf power for NMR absorption at a particular magnetic field strength. The range of frequencies covered is ~ 4 to 360 MHz for the NMR absorption of all magnetic nuclei with today's spectrometers. The stability requirements for the transmitter are essentially identical to those of the magnetic field. The frequency is usually derived by the multiplication of the frequency of a thermostated, quartz crystal. A means is provided for controlling the B_1 power. The electronic circuits are so highly tuned that separate transmitters are used for different nuclei in many spectrometers. In others, different frequencies are derived by the multiplication of the frequency of a master crystal oscillator or by substitution of different crystals.

There are a growing number of spectrometers utilizing a frequency synthesizer as the frequency source. The frequency of most synthesizers can be controlled to 0.1 Hz or better. These systems have the advantage that several frequencies may be derived from the multiplication, addition, etc. of one frequency source. In this manner, all frequencies are "locked" together such that drifts in the frequencies are controlled by variations in the master frequency source. Provision is made for amplification of the frequency and for controlling the level of the rf power delivered to the probe.

3.1.3 PROBE

The probe or sample holder is mounted between the pole caps of the magnet such that its position may be adjusted to place the sample in the region of optimum field homogeneity. In addition, the probe contains an air turbine for spinning the sample, sweep coils, a coil(s) for transmitting the rf signals and detecting the resonance absorption, and a means for controlling the temperature in the region of the sample. The probe body is maintained at constant temperature either by circulating water at constant temperature through coils in the probe, by placing the probe in the thermostated housing surrounding the magnet, from air circulating in the room, or by using a combination of these methods (Figure 3.2).

Molecules in different parts of the sample will experience slightly different magnetic fields due to residual field gradients in the region occupied by the sample. As a result, the absorption will occur over a range of frequencies such that extremely broad lines are observed. The effective homogeneity of the magnetic field is improved by spinning the sample about the y axis. This has the effect of averaging the magnetic field in the x–z plane and reducing the line width due to inhomogeneities in the magnetic field. Spinning rates of 20–40 rev sec^{-1} are normally used. Linewidths at half-peak height (resolution) of 0.5 Hz or less are readily obtained, provided the homogeneity coils are properly adjusted. Higher spinning rates can lead to an increase in the noise level due to vibrations as the sample tube rotates. It is essential that uniform sample

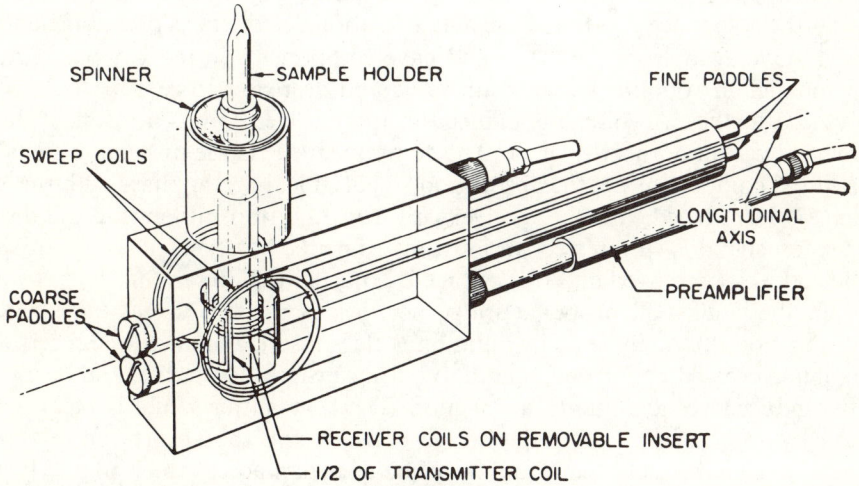

Fig. 3.2 A drawing of an NMR probe showing the basic components. (Courtesy of Varian Associates, Palo Alto, Calif.).

tubes be used to prevent this unwanted noise. Spinning of the sample also produces unwanted "sidebands" (reproductions of the spectral peaks) symmetrically disposed about the spectral peaks at frequencies separated from the spectral peaks by multiples of the spinning frequency. These peaks are easily recognized by their frequency dependence on the spinning rate. Normally they can be effectively removed by proper adjustment of the homogeneity coils and careful selection of samples tubes.

Two basic designs in the detection system of probes are used: (1) the crossed-coil probe or nuclear induction probe (6) and (2) the single-coil probe (2). In the crossed-coil probe, separate transmitter and receiver coils are used. The transmitter coil for producing the B_1 rotating field is wound in two sections such that it surrounds the receiver coil and has its magnetic axis parallel to the x axis. The receiver coil is wound on a glass insert at a right angle to the transmitter coil so that its magnetic axis is also parallel to the x axis. Leakage between the two coils is reduced by placing a Faraday shield between the two coils, as it is impossible in practice to obtain perfect 90° orientation between the two coils. Further reduction in the leakage is achieved by incorporating two paddles (inductors) which are coupled to the transmitter and receiver coils (7). In actual practice some leakage between the two coils is desirable to serve as a reference for the voltage induced by absorption.

The introduction of some leakage between the transmitter and receiver coils also serves to suppress the dispersion mode signal. When B_0 is located away from its resonance value and small amplitudes of B_1 are applied, the magnetization vector M_0 will be along the z axis (Figure 2.4). As B_0 is brought near the resonance condition, the magnetization vector starts precessing and will move away from the z axis and have components in the x–y plane. At resonance M_y obtains a maximum value and decreases to zero as B_0 is increased further. A voltage is induced in the receiver coil as a result of the magnetization of the x–y plane. The component M_x is the dispersion signal and the component M_y the absorption signal. If leakage in phase with the x axis is introduced into the receiver coil from the transmitter coil, the dispersion signal is suppressed. In most instruments, however, phase-sensitive detection is employed and the leakage is adjusted to a minimum.

In the single-coil probe, a bridge network is used to detect the NMR absorption (8). A single coil wound on a glass insert, similar to the receiver coil in a crossed-coil probe, is used to both transmit the rf signal and receive the induced voltage due to absorption. The transmitter signal is balanced out by carefully balancing the bridge network and the absorption or dispersion signal is detected as an out of balance emf across the bridge. Most spectrometers today using this type of probe utilize a twin-T bridge (9). The induced emf across the bridge is a mixture of both the absorption and dispersion mode signals. The dispersion signal can be suppressed by the introduction of a slight imbalance in the bridge tuning network. Adjustments are provided for balancing the phase and amplitude of the bridge network such that only the absorption mode is detected.

The electronic circuit employed in an NMR probe are usually tuned such that they will transmit and receive only frequencies over a very narrow range. Normally, separate probes tuned for one particular frequency are used for the individual nuclei.

The majority of probes have some means for controlling the temperature in the region occupied by the sample such that variable temperature experiments are possible over the range -100-to-$200°C$. The range is limited by the expansion and contraction characteristics of the glass insert. The insert is usually contained in a vacuum-jacketed dewar to prevent heat loss to the magnet which would result in a loss of resolution and field drifts. Generally, a stream of nitrogen gas at a controlled temperature is passed over the sample. The probe usually contains a heating coil and a sensing device to control the current to a heating coil located below the sample area. For high-temperature operation, the stream of nitrogen gas is heated to the desired temperature before it passes over the sample. Several different systems have been used for low-temperature operation. In one method, the nitrogen gas is precooled by passing it through a heat exchanger filled with liquid nitrogen and then heating the gas stream in the probe to the desired temperature. A method similar to this uses the boiloff from a liquid-nitrogen reservoir as the source of the cold nitrogen gas and then the gas is heated in the probe to reach a particular temperature. In another method, the gas stream is split with one part being cooled by passing it through a heat exchanger and then remixing it with the uncooled gas stream. The temperature is controlled by varying the fraction of gas passing through the heat exchanger. In yet another method, gas at high pressure is precooled and then allowed to expand through a Joule–Thompson device. Control of the temperature is achieved by using a sensor in connection with a heating coil. Regardless of the method used, the control of the temperature should be $\pm 1°C$ or better. Temperature gradients within the sample area may present some problems. However, probe design and/or mixing the sample usually eliminates problems due to temperature gradients.

3.1.4 RADIOFREQUENCY DETECTION

The signal induced in the receiver coil of the probe due to NMR absorption is extremely weak (~ 1 mv), and must be amplified before it can be displayed. The amplification is limited by the noise level due to any inbalance in the tuned electric circuits and due to sample spinning. The signal from the receiver coil is usually fed into a preamplifier mounted directly on the probe to eliminate as much electrical noise as possible. The signal from the preamplifier is further amplified in the receiver and mixed with the output of a local oscillator frequency to produce a beat frequency at an IF frequency that is further amplified and fed into a phase detector (10). The local oscillator frequency is also mixed with the transmitter frequency, to provide a reference frequency which is also amplified and fed into the phase detector. The absorption or dispersion signal is selected in the phase detector, the signal is rectified, filtered, and the d.c. signal is fed into either an oscilloscope or recorder for display of the signal.

3.1.5 INTEGRATOR

Most instruments today employ some type of modulation (11) along with phase sensitive detection to eliminate baseline drift as a result of improper balance of the circuits. Either field or frequency modulation may be used with identical results. If the output from an audiooscillator is applied to the sweep coils of the probe, the magnetic field at the sample is $B_0 + B_m \cos \omega_m t$, where B_m is the amplitude of the field modulation and ω_m is the angular frequency of the field modulation. If the magnetic field is moved through resonance, signals (sidebands) appear not only at resonance due to B_0 (centerband) but also on either side of the main peak at frequencies separated from the main peak by $\pm n\omega_m$, where n is an integer. Usually only the first sideband is important. Similar results may be obtained by modulating the transmitter frequency. The frequencies of all the resonances from the receiver coil are mixed with the transmitter frequency. After phase detection, the signals contain both d.c. and audio components. Any inbalance in the probe will affect only the d.c. signals. Therefore one can detect the audio components of the signal, convert to d.c. signals, and display the signals. This minimizes the baseline drifts in the spectrum.

The integrator not only contains the components necessary for modulation and phase sensitive detection, but also for the electronic integration of the area under an absorption signal. As pointed out previously, the area under an absorption signal is directly proportional to the number of nuclei giving rise to the signal and this provides the basis for quantitative analysis applications of NMR. Most spectrometers today are equipped with electronic integration circuits which permit the integration without interference with normal operation of the spectrometer. In one type of circuit, a Philbrick chopper stabilizer d.c. amplifier is used in conjunction with a Miller integrator (9).

3.1.6 SWEEP UNITS

As pointed out previously, the NMR absorption condition can be obtained by keeping the magnetic field constant while varying the frequency (frequency sweep) or by holding the frequency constant and varying the magnetic field (field sweep). Most spectrometers provide two types of magnetic field sweep, a slow sweep and a fast, recurrent sweep. Both sweeps are linear sweeps. That is to say, the rate of change of the field is constant with time. For the fast sweep, the output from a sawtooth generator is amplified and applied to two small Helmholtz coils usually located on the sides of the probe such that their magnetic axis in the same direction as the main magnetic field. The sweep time can usually be varied from seconds to minutes. The absorption signals are usually displayed on an oscilloscope. The return of the magnetic field to its original value for the start of another sweep is so rapid that absorption is not detected on the return. This mode of operation is useful when searching the magnetic field for either the resonance condition or a particular peak. It is also used to monitor the shape of a peak when adjusting the homogeneity coils for the maximum resolution. When the sweep rate is fast with respect to the relaxation times of the nucleus under consideration, the absorption peak will be distorted and ringing will be observed on the trailing edge of the absorption peak. The homogeneity coils are adjusted to produce the maximum ringing.

The slow sweep unit is used to sweep the magnetic field when a recorder presentation of the signals is being used. A d.c. voltage is applied to the flux stabilizer. The flux stabilizer senses this voltage as an error signal (a field change in the flux of the magnetic field) and supplies a correction voltage to the coils located on the pole faces resulting in a sweep of the magnetic field. The sweep rate may be varied by varying the applied d.c. voltage. Slow sweep rates may be obtained such that little distortion of the absorption signal occurs and the maximum resolution may be realized. In spectrometers employing some type of field/frequency stabilization, the travel of the recorder arm in the x direction can be synchronized with the slow sweep unit such that the spectra may be recorded on precalibrated charts.

3.1.7 FIELD/FREQUENCY STABILIZATION

A high degree of stability (one part in 10^6 or better) of an NMR spectrometer system is achieved by proper control of the magnetic field and the irradiating frequency as mentioned previously. While this is adequate for most purposes requiring short term stability, drifts in either magnetic field or frequency may occur over a longer period of time such that broadened signals are observed. Since it is really the ratio of the magnetic field to the frequency which is important for long-term stability, two methods using modulation techniques have been devised such that drifts in the magnetic field are automatically compensated for by corresponding changes in the frequency and vice versa. Depending on the spectrometer, the modulation frequency varies from 2 to 5 KHz and normally, the first sideband is utilized.

One method of field/frequency stabilization referred to as an external lock system, utilizes two separate samples in the probe (12,13). A small, stationary sample tube of water (the control sample) is placed as close as possible to the region occupied by the sample under consideration (analytical sample) so that both samples experience as close to the same magnetic field as possible. A separate set of homogeneity coils are usually provided for the control sample. The centerband frequency is determined by a stable, constant frequency crystal oscillator while the modulation frequency is varied by a feedback loop such that resonance is maintained at the upper sideband frequency of the sample. The control sample is detected as the dispersion mode signal such that small drifts in the magnetic field are detected as plus or minus voltages. These voltages are applied to the feedback loop as corrections such that the control sample is always maintained at resonance. The analytical sample is then field-swept by applying a d.c. voltage in the normal manner. The stability is maintained to better than one part and 10^8/hr with this method. Since the analytical sample and the control sample are in different regions of the magnetic field, a control is provided for correcting the magnetic field in order that precalibrated charts may be used to record the spectra. An external locked system has the advantage that the field/frequency ratio is maintained constant at all times. A similar arrangement may be used so that frequency sweeps of the analytical sample are obtained.

A second method of field/frequency stabilization referred to an internal locked system is also based on sideband modulation and offers certain advantages over an external locked system (14–16). The control sample is derived from a sharp peak in the sample and is usually the absorption from the reference material tetramethysilane (TMS). Thus both the analytical and control samples are subjected to the same magnetic field. Separate sideband frequencies and phase sensitive detectors are used for the control and analytical samples. The control sample is detected as the dispersion mode signal and the error voltages are used in a feedback loop to maintain the magnetic field at resonance. In the frequency sweep mode of operation, the control sample is maintained at resonance with a fixed frequency modulation. The analytical sample is swept in a linear manner by varying the modulation frequency of the analytical oscillator. The frequency of the analytical channel is mechanically linked to the recorder arm, such that the spectra can be recorded on precalibrated chart paper. In the field sweep mode of operation, the fixed frequency modulation oscillator is used for the analytical channel while the variable frequency modulation oscillator is used to linearly sweep the control sample. As the control sample modulation is swept, error signals are fed through the feedback loop to sweep the magnetic field through the resonances of the analytical sample. The stability of an internal lock is on the order of one part in 10^9/hr. With an internal lock system, the lock must be broken each time a sample is changed. This usually presents no problems and the gain in stability and the added advantage of frequency sweep operation for decoupling experiments (Section 4.11.4) more than compensates for this inconvenience.

Many commercial spectrometers today offer both external and internal lock capabilities. Furthermore, it is not necessary for the analytical and control channels to detect the same nucleus. Many nuclei have low sensitivity and broad signals which prevents their use in the control sample. By using a probe that is tuned to two frequencies and two radiofrequency oscillators, it is possible for the analytical channel to detect one nucleus (e.g., ^{13}C) and the control channel to detect another nucleus (e.g., ^{1}H, ^{2}D, and ^{19}F).

3.2 FACTORS INFLUENCING THE RESOLUTION AND LINE SHAPE

Assuming that the NMR sample has been prepared to the best of one's ability (Section 3.3), there are a number of instrumental factors, under operator control, that have a profound effect on the quality of the spectrum obtained. For the majority of samples, the natural line width of the absorption signal is less than the line width due to inhomogeneities in the magnetic field. Thus the resolution in most cases depends on one's ability to adjust the current in the homogeneity coils. In addition, the line shape depends on the rf power level, the phase of the detector, the spinning rate, the sweep rate, and the electronic filtering used before recording the spectrum. The operator should be aware of these factors and make the proper adjustments in order to obtain the best spectrum possible.

3.2.1 HOMOGENEITY OF THE MAGNETIC FIELD

When the spectrometer system is set up in the laboratory, the magnet is adjusted (shimmed) mechanically to obtain the best possible homogeneity of the magnetic field. The remaining gradients are corrected as far as possible by adjusting the current in the homogeneity coils located on the pole faces of the magnet. Before adjusting the homogeneity coils, it is important that the sample be located in the "center" of the magnetic field. Usually the position of the probe along the z axis is fixed by the probe holder and controls on the probe holder permit one to adjust the position of the probe along the x and y axis. After obtaining a peak on the oscilloscope using the fast sweep, the position of the probe along the x and y axis is adjusted so that the position of the peak does not vary (assuming no drift in the magnetic field) when the current in the x and y homogeneity coils is varied between the two limits. This adjustment should be made each time the probe is removed from the magnet. Some probe holders are equipped with stops such that, after initial set-up, the probe can be returned to the "center" of the magnetic field without the need to readjust the position along the x and y axis each time the probe is changed.

In order to adjust the current in the homogeneity coils, it is desirable to display the signal from a sample giving a single, narrow, absorption signal. Usually tetramethylsilane or chloroform are used for proton spectra. The substance used to supply the lock on internal locked spectrometers can be used. With the signal displayed on an oscilloscope using the fast sweep, the current in each of the homogeneity coils is adjusted with the sample non-spinning. One criteria for obtaining the maximum homogeneity is to make the adjustment on each coil for the maximum peak height. If ringing of the peak is observed, the current in each coil can be adjusted for the maximum ringing. Afterwards spinning of the sample is started and the adjustments of the y and curvature are repeated. Spinning of the sample about the y axis improves the effective homogeneity in the x and z directions. However, spinning does not effectively improve the homogeneity along the y axis or the curvature, both of which are very critical in determining the linewidth. At this stage it is usually more convenient to observe the signal on a recorder. The current in the curvature coil is adjusted such that the line shape is

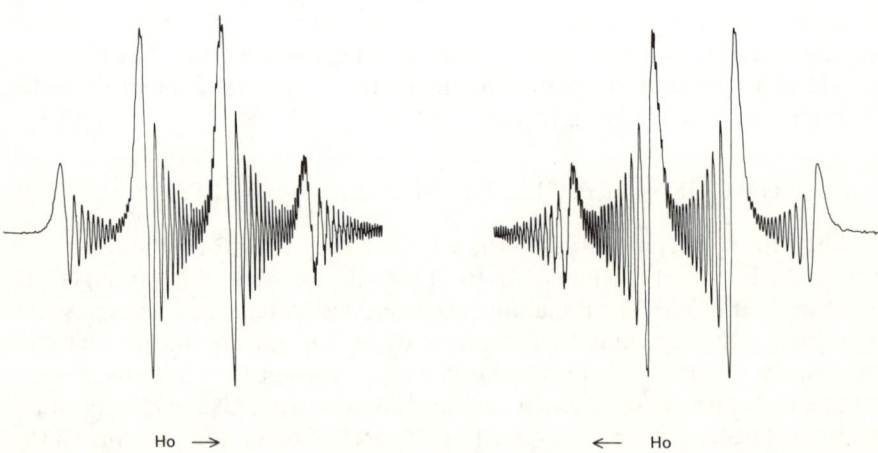

Fig. 3.3 An expansion of the aldehyde quartet in the 60-MHz proton NMR spectrum of acetaldehyde.

FACTORS INFLUENCING THE RESOLUTION AND LINE SHAPE 61

symmetrical about the center of absorption. One way to do this is to adjust the curvature such that identical ringing is observed when the field is swept in both directions (Figure 3.3). The current in the y coil is then adjusted to produce the maximum amount of ringing and/or the maximum peak height.

For normal, routine, day-to-day operation, the homogeneity coils should be adjusted at the beginning of each day. Unless subjected to some external influence, such as a change in the room temperature or a change in the temperature of the magnet cooling water, it will only be necessary to adjust the y and curvature controls when different samples are run throughout the day. For extremely accurate work requiring the maximum resolution, however, it is a good idea to adjust all the homogeneity controls for each sample.

3.2.2 RADIOFREQUENCY POWER LEVEL

The rf power level should be adjusted to avoid saturation effects. Saturation results in a broadening of the absorption signal and will be most noticeable at the center of the absorption signal (Figure 3.4). Saturation also depends on the time spent on resonance (the sweep rate). One procedure for determining whether saturation is occurring is to adjust the rf power level

(a)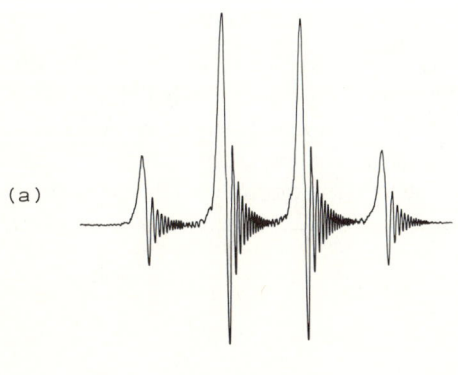

(b)

(c)

Fig. 3.4 The effect of increasing the rf power level B_1 on a typical NMR spectrum: (a) normal spectrum; (b) increased B_1; and (c) saturation due to a further increase in B_1.

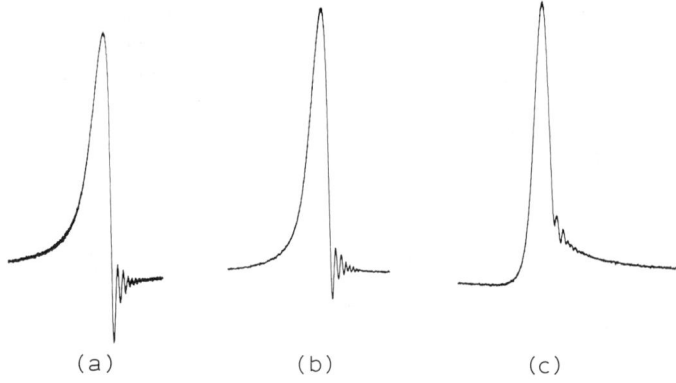

Fig. 3.5 The effect of improper phase adjustment (a and c) on the appearance of an NMR signal and the corrected (b) signal.

to give the desired signal intensity at a particular sweep rate. If the signal intensity does not decrease when the sweep rate is decreased then saturation is not occurring.

3.2.3 RADIOFREQUENCY PHASE

Incorrect adjustment of the phase-sensitive detector results in signals that are a mixture of the absorption and dispersion modes and will lead to distortion of the absorption signal (Figure 3.5). The normal procedure for setting the phase is to record the strongest signal in the spectrum at high gain and adjust the phase control so that a flat baseline is obtained. This adjustment should be repeated for each sample since the phase varies with the magnetic susceptibility of the sample.

3.2.4 SPINNING RATE

The sample is spun at the rate of 20–40 rev sec^{-1} about the y axis in order to average the inhomogeneties in the x and z directions (Section 3.1.3). If lower spinning rates are used, the band may be broadened due to incomplete averaging of the magnetic field. In addition, spinning sidebands appear symmetrically disposed about the main peak due to modulation effects. The sidebands can be effectively removed by careful selection of sample tubes, spinning at a faster rate, and proper adjustment of the homogeneity coils. Very high spinning rates should be avoided due to the problems of vortexing of the sample and the introduction of additional noise (Section 3.1.3).

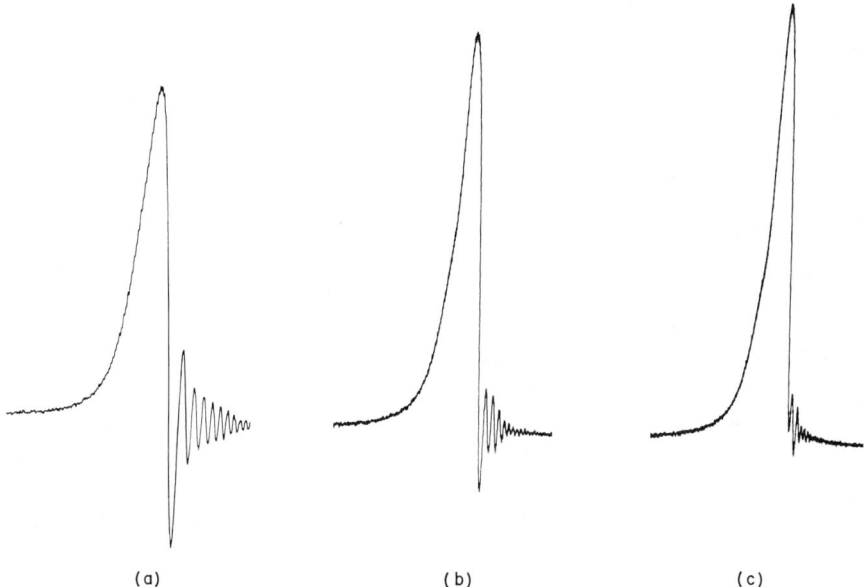

Fig. 3.6 The effect of sweep rate on the appearance of an NMR signal; (a) 0.5 Hz/sec; (b) 0.2 Hz/sec; and (c) 0.1 Hz/sec.

3.2.5 SWEEP RATE

The sweep rate used to record the spectrum can influence the shape of the absorption signal (Figure 3.6). As mentioned previously (Section 3.1.6), if the field is swept through resonance at a rate faster than the relaxation times T_1 and T_2 of the nucleus giving rise to the signal, distortions in the form of ringing or wiggles will appear on the trailing edge of the absorption. The ringing decays exponentially with time. When the rf field B_1 is rapidly changed to some position away from resonance, the rotating component of the magnetization M_{xy} (Section 2.5) is alternately in and out of phase with the rf field and the signals induced in the receiver coil are alternately plus and minus. Although ringing is desirable for tuning the homogeneity coils, the distortion can lead to errors when accurate peak frequencies are being determined and one should use slower sweep rates to reduce their effects.

3.2.6 SIGNAL FILTERING

Before the signal is displayed after it has come from the phase sensitive detector, it is filtered to suppress the background noise due to short-term fluctuations. Usually time-constants on the order of 0.01–to–10 sec are

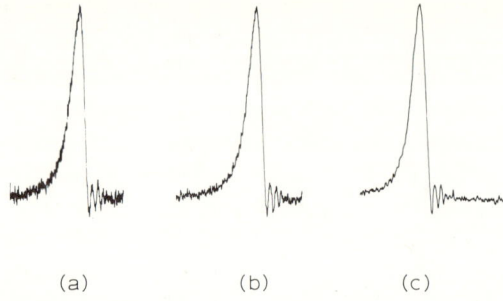

Fig. 3.7 The effect of increasing filtering on the appearance of an NMR signal: (a) smallest filter; and (c) largest filter.

available. The time constant used depends upon the sweep rate used. If longer time constants are used, then slower sweep rates must be used to eliminate distortions in the signal (Figure 3.7).

3.3 SAMPLE PREPARATION

Several factors including the state of the sample, the concentration, the solvent, and the presence of impurities should be considered in the preparation of a sample for NMR analysis. For high-resolution NMR spectra, a liquid compound may be examined as the pure liquid or may be dissolved in a suitable solvent. The latter is usually preferred, especially for viscous liquids, since high viscosity usually leads to signal broadening due to restricted molecular tumbling. In order to eliminate the dipole–dipole broadening in the solid state, solid samples must be dissolved in a suitable solvent. If the solid has a reasonable low melting point, it may be examined as a pure liquid at high temperatures. The spectra of gases may be obtained, but pressures of several atmospheres may be required in order to obtain reasonable sensitivity. Sample tubes are available which are safe up to about 20 atm pressure (17).

The concentration of the sample should be sufficient to obtain a reasonable signal-to-noise ratio (S/N). For 5-mm O.D. sample tubes, normally from 0.3 to 0.5 ml of volume is required. Depending on the molecular weight of the sample, from 5 to 50 mg of sample is usually sufficient for obtaining the spectrum in a single scan. Methods are available for obtaining spectra on smaller sizes (Section 3.7).

Several factors should be considered in the selection of a suitable solvent. Besides dissolving the sample, the solvent should not interact chemically with the sample and preferably, should not contain any hydrogen atoms whose absorptions would interfere with those of the sample. Carbon tetrachloride probably comes closest to being an "ideal" solvent. However its use is limited by the fact that numerous compounds are not sufficiently soluble in CCl_4. Several relatively inexpensive deuterated solvents of high isotopic purity are commercially available. The most common of these is chloroform-d. However, hydrogen bonding between $CDCl_3$ and the sample may result in substantial shifts from the resonance position in an inert solvent. Solvent shifts are also observed with aromatic solvents. A list of common solvents used in proton NMR spectroscopy is given in Table 3.1.

Considerable line broadening may be observed in an NMR spectrum if care is not exercised to ensure that paramagnetic or ferromagnetic impurities are eliminated. Any source of unpaired electrons will give rise to line broadening due to a reduction of the relaxation times. Molecular oxygen

TABLE 3.1. Common Solvents Used for Proton NMR[a]

Solvent	δ[b]
CCl_4	—
CS_2	—
SO_2	—
Chloroform-d[c]	7.28
Acetone-d_6	2.07
Dimethylsulfoxide-d_6	2.50
p-Dioxane-d_8	3.56
Acetonitrile-d_3	1.96
Methylene chloride-d_2	5.28
Nitromethane-d_3	4.29
Cyclohexane-d_{12}	1.42
Methanol-d_4[c]	3.34, 4.11
Tetrahydrofuran-d_8	1.79, 3.60
D_2O[c]	4.61
Benzene-d_6	7.24
Pyridine-d_5	7.18, 7.57, 8.57
Toluene-d_8	2.31, 7.10
Acetic acid-d_4[c]	2.06, 11.97
Trifluoroacetic acid[c]	11.34

[a] Data taken from the Sadtler Research Laboratories, Inc. (19).
[b] Chemical shift downfield from internal TMS of residual proton signals.
[c] Peak positions are variable depending on hydrogen bonding with the solute and temperature.

dissolved in most solvents causes some broadening and should be eliminated by purging the sample with oxygen-free nitrogen or by evacuation using several freeze-pump-thaw cycles.

Ferromagnetic particles, probably coming from a steel spatula or from dust particles in the atmosphere, give rise to considerable line broadening. In addition, solid particles in undissolved samples are another source of line broadening. The ferromagnetic particles may often be removed by using a fairly strong permanent magnet to retain the particle while the solution is withdrawn for transfer to another sample tube. Solid particles may be removed by filtration. Several devices are commercially available for filtering NMR samples (17,18) and it is a good idea to routinely filter each sample.

3.4 CHEMICAL SHIFT MEASUREMENTS

NMR spectra are normally displayed by using an x–y recorder in which the amplitude of the signal output from the detector is fed into the y axis. The x axis input is usually either a function of time or a function of the sweep frequency. The usual convention is to record the spectrum such that the magnetic field, B_0, increases from left to right across the chart. The absolute frequency of a particular absorption signal cannot be conveniently determined. Therefore the normal convention is to determine relative frequencies of absorptions with respect to some standard reference material.

3.4.1 REFERENCE COMPOUNDS

The ideal reference compound should not interact with either the solute or solvent and preferably should exhibit a single absorption peak in a region of the spectrum separated from normal absorptions. The standard reference compound for proton spectra in nonaqueous solvents is tetramethylsilane (TMS). Tetramethylsilane, $Si(CH_3)_4$, is a volatile liquid (b.p. 27°C) which facilitates its removal from the sample and the recovery of the solute under investigation. The single peak from the equivalent protons of TMS appears at higher field than most proton absorptions and thus does not interfere with absorptions from the solute. Normally a concentration of 1% TMS is sufficient for external locked spectrometers whereas from 3–5% is required for internal locked spectrometers if the TMS signal is used as the lock signal source.

Since TMS is insoluble in water, it cannot be used as a reference for aqueous solutions. The sodium salt of the compound 2,2-dimethyl-2-silapentane-5-sulphonic acid (DSS), $(CH_3)_3SiCH_2CH_2CH_2SO_3^-Na^+$, has been most widely used as a reference for aqueous solutions. Absorption frequencies are determined relative to the sharp peak given by the equivalent protons of the $(CH_3)_3Si$-group. The position of this absorption is usually not affected by the pH of the solution. The use of DSS as a reference has the disadvantage that additional absorptions are present in the spectrum due to the three methylene groups. At low concentrations (1%) of DSS, these absorptions will usually not interfere with the absorptions of the solute. However, if low concentrations of the solute are used, the absorptions due to the methylene peaks may interfere with the solute spectrum. An additional disadvantage of DSS is the difficulty of separating it from the solute after the spectrum is obtained.

Two methods are employed in NMR spectroscopy for referencing absorptions. The reference compound may either be dissolved directly in the sample under investigation (internal reference), or may be placed in a separate container from that of the sample (external reference). When an internal reference is used, both the sample and reference experience the same magnetic field. This is the most common type of referencing and usually presents no problems unless the reference interacts with either the solute or solvent, or with the solvent–solute interactions. Interactions of this type would influence the relative positions of the reference and sample absorptions. Usually these interactions are at a minimum when TMS is used as the internal reference.

When an external reference is used, the influence of the reference on the solvent–solute interactions is absent. The usual method is to place the reference in either an especially designed coaxial tube or a capillary tube (a melting point capillary tube) and then place this inside the normal NMR sample tube. The big disadvantage of an external reference is that the sample and reference do not experience the same magnetic field. The magnetization per unit volume induced in the sample (20) depends on the volume magnetic susceptibility K (equation 3.1)

$$M_0 = KB_0 \qquad (3.1)$$

For sufficiently dilute samples the volume magnetic susceptibility of the sample, K_s, is essentially that of the solvent. Since the reference is separated from the sample by a glass interface, the volume magnetic susceptibility of the reference, K_r, will be slightly different. If spherical cells are used, no susceptibility corrections are needed. For the normal cylindrical NMR sample tubes, the magnetic fields at the sample and reference are given by equation 3.2. Values for K may be obtained from the literature

$$(B_0)_s = B_0(1 - \tfrac{2}{3}\pi K_s) \qquad (3.2)$$

$$(B_0)_r = B_0(1 - \tfrac{2}{3}\pi K_r) \qquad (3.3)$$

(21) or may be determined by NMR methods (22–24). For quantitative proton NMR work, the relative frequency between the sample and reference should be corrected to take into account the difference in magnetic field. For

other nuclei, where larger frequency ranges are observed for the absorptions, these corrections are not as important.

In many of the earlier NMR studies, different reference compounds such as the solvents themselves were used. It is almost impossible to convert from one reference to another due to lack of information concerning the concentration of the solution and the extent of intermolecular association effects. For accurate work, conversion from one internal reference to another is possible only if the absorption frequencies are extrapolated to infinite dilution in the same solvent. Accurate frequencies from the conversion of data obtained with an external reference to an internal reference and vice versa is impossible.

3.4.2 FREQUENCY CALIBRATION

The relative positions of the absorptions due to the sample and reference compound are determined in frequency units. However, the standard for reporting chemical shifts (Section 2.6) is in the dimensionless units of ppm which are not dependent on the operating frequency of the spectrometer. Coupling constants are determined and reported in frequency units (hertz). Several procedures have been used to calibrate the absorption frequencies in NMR and the method of choice depends upon the complexity of the spectrum, the accuracy desired, and to a certain extent, on the type of spectrometer on which the spectrum is obtained.

For spectrometers without some means of field/frequency stabilization, two method of calibration are used. The superposition of band method can only be applied to spectra where the absorptions are well separated and depend upon one's ability to superimpose a signal from a sideband of the reference compound on the absorption in question from the sample. The normal procedure is to observe the signal from the sample on an oscilloscope using the fast sweep or on a recorder having a rapid response. The sweep time and filtering are adjusted so that a ringing pattern is observed. A sideband (usually the first sideband) of the reference is generated using an audiooscillator and its position is adjusted to exactly coincide with the signal from the sample. As the sideband signal is adjusted close to the signal of the sample, distortions in the ringing pattern of the sample signal will be observed. When the two frequencies exactly coincide, the ringing pattern returns to normal. At this point, the frequency of the audiooscillator, and hence the sample signal, can be read with the aid of a frequency counter. The frequency of the sample signal can usually be determined with an accuracy of $\pm 0.1\%$ using this method since it does not depend on the linearity of the sweep. This method is difficult to apply, however, to samples where the signals from the sample are overlapping such that clear ringing patterns are not observed for each peak.

The second method, the audio-sideband technique, uses the interpolation between two sidebands to obtain the frequencies of the sample absorptions with respect to the reference. Sidebands (usually of the reference) of known frequency are introduced by means of an audiooscillator and the relative positions of the signals are determined by interpolation. For routine work, one sideband placed such that the reference and sideband signals encompass all the sample signals is usually sufficient. This method suffers from errors due to the nonlinearity of the sweep. For more accurate work, two sidebands of the reference which encompass only a small region of the spectrum are used and this process is repeated in order to calibrate each region of the spectrum. By taking the average of several such determinations, the frequencies of the sample signals can usually be determined to an accuracy of $\pm 0.5\%$.

For routine spectra on instruments employing some means of field/frequency stabilization, it is possible to record the spectra on precalibrated chart paper. The accuracy of this method is limited by the calibration of the sweep widths and the accuracy of reading the signal's position from the chart paper. The sweep width calibration should be checked periodically to ensure that it is within the manufacturer's specifications. For more accurate work, sidebands of the reference can be introduced and the interpolation method used. Normally, one would use the smallest sweep width available. With an internal locked spectrometer, the recorder position is directly related to the sweep frequency. By using the smallest sweep width available, the recorder pen can be stopped at the signal of interest and its frequency determined by determining the frequency difference between the control and analytical channels.

3.5 INTENSITY MEASUREMENTS

The determination of the area under each peak or multiplet in an NMR spectrum provides information for the solution of structural and analytical chemical problems. Provided saturation is not occurring, the area under a peak is directly proportional to the number of nuclei giving rise to the absorption signal (Section 2.3). Thus, from the determination of the relative areas of two or more absorptions, the relative number of nuclei giving rise to each signal may be determined. This data is invaluable for such applications as structure determination, hydrogen analysis, analysis of mixtures, and the extent of isotopic substitution.

Most commercial NMR spectrometers incorporate an electronic integrator (Section 3.1.5) for the determination of the area under the absorption signals. Modulation techniques are usually employed to facilitate the determination of the integral and to eliminate errors which are introduced by

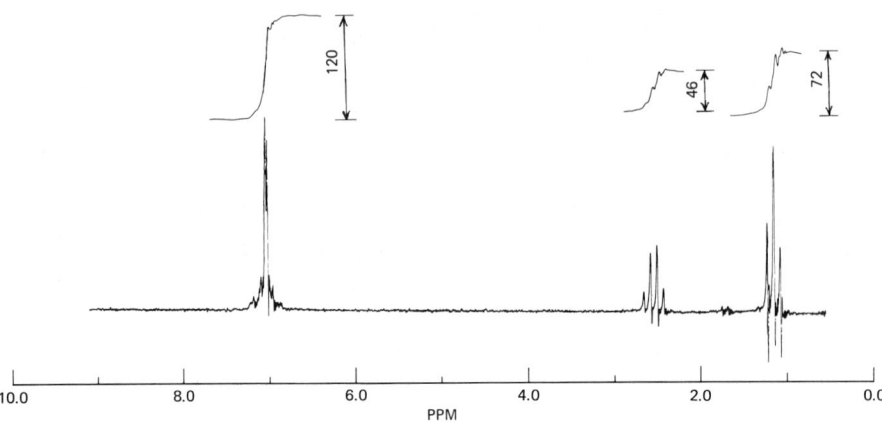

Fig. 3.8 The 100-MHz proton NMR spectrum of ethylbenzene and the corresponding integration traces.

fluctuations in the balance of the probe. The integral is obtained as a step function with the height of the steps being proportional to the area of the corresponding absorption signals. Normally, the integral is recorded on the same chart as the absorption spectrum (Figure 3.8) and the heights of the steps determined by direct measurement. Alternatively, the height of the integral steps can be read off directly by using a digital voltmeter. With a suitable signal-to-noise ratio, integration of NMR spectra with an accuracy within 1% can be obtained. In order to obtain this precision, however, there are a number of instrumental parameters, under operator control, that must be carefully adjusted. These parameters are discussed in detail in Section 6.4, along with some typical analytical applications of NMR integration.

3.6 FACTORS INFLUENCING SENSITIVITY

The major disadvantage with using NMR spectroscopy for quantitative analysis, compared to other spectroscopic techniques, is the inherent low sensitivity. The signal from the probe that is detected by the receiver is extremely weak and requires a great deal of amplification of the signal before it is finally presented at the recorder. The degree of amplification of the signal is limited by rf noise background which is amplified along with the signal. The electronic circuits employed in commercial NMR spectrometers are designed to minimize the background noise and the components are designed and tuned to yield the maximum sensitivity.

For purposes of discussion of the sensitivity of NMR spectrometers, it is convenient to introduce the concept of signal-to-noise ratio (S/N). The

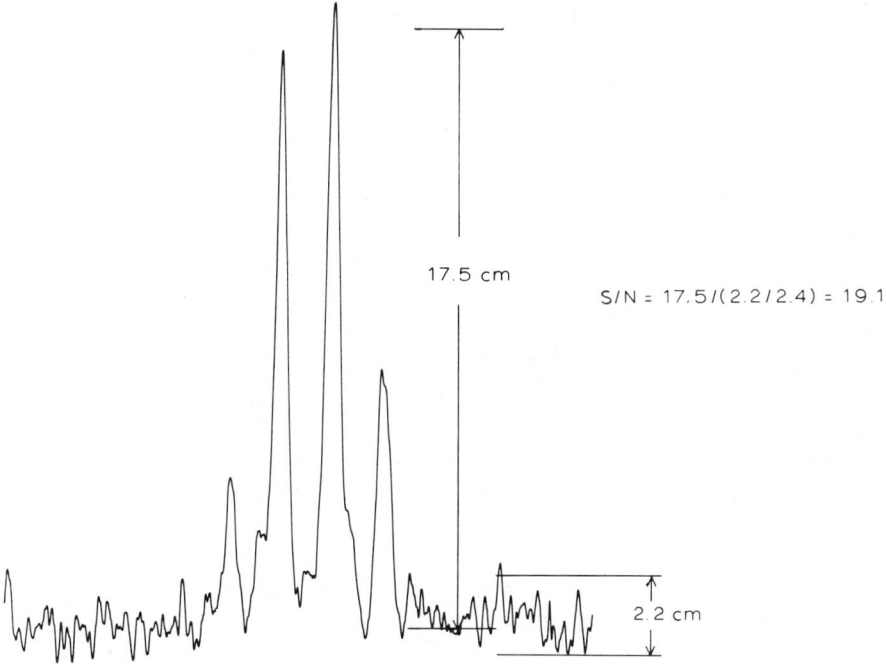

Fig. 3.9 A 60-MHz expansion of the methylene quartet of 1% ethylbenzene recorded under optimum conditions and the corresponding calculation of the signal-to-noise (S/N) ratio.

standard sample recommended by most manufacturers for determining the S/N is a 1% v/v solution of ethylbenzene in carbon tetrachloride. The spectrum is obtained under optimum conditions and several recordings are made of the methylene quartet (Figure 3.9). The S/N is then calculated (equation 3.5) from the average peak height of the tallest peak (average signal amplitude)

$$\text{RMS Noise} = \frac{\text{Average P–P Noise}}{2.5} \quad (3.4)$$

$$\text{S/N} = \frac{\text{Average Signal Amplitude}}{\text{RMS Noise}} \quad (3.5)$$

and the average peak-to-peak baseline noise level (Figure 3.9). For most 60 MHz spectrometers, a S/N of 5:1 to 15:1 is usually quoted, whereas for most 100 MHz spectrometers, a S/N of 25:1 or greater is common. In actual practice, the S/N is usually exceeded by a factor of 2–4, depending on the particular spectrometer.

An analysis of the S/N in terms of the factors that are important in the determination of sensitivity gives the S/N as (25)

$$\text{S/N} = 5.78 \times 10^{-32} I(I+1) \gamma^{11/4} \frac{1}{T^{3/2}} K \frac{1}{(F\Delta f)^{1/2}} B_0^{7/4} \frac{N_T}{l^{1/2} D^{1/2}} \quad (3.6)$$

where K is a factor that depends on the sweep rate, B_1 field, relaxation times, and field homogeneity, F is the noise figure of the reciever, Δf is the filter bandwidth in Hz, l is the reciever coil length in cm, D is the receiver coil diameter in cm, and N_T is the total number of nuclei within the active region of the coil.

The factor K includes many of the instrumental factors that are important in determining the line shape (Section 3.2). The signal amplitude increases with an increase in B_1 to the point where saturation occurs and then decreases. The maximum B_1 level is dependent on the relaxation times T_1 and T_2 and on the sweep rate used (Section 2.3). Better sensitivity is obtained with slower sweep times. However, a longer time is spent on resonance and hence, the greater the chance for saturation to occur. For dilute samples, a compromise is usually reached between the B_1 level and sweep rate to achieve the best S/N while insuring that saturation is not occurring.

Part of the baseline noise may be eliminated by filtering the signal before recording (Δf) (Section 3.2.6). The time constant (frequency response) of the filter is determined by the sweep rate since distortions in the line shape of the signal can occur if the time constant is set too long. In practice, the time constant should be greater than the sweep rate in hertz.

The signal intensity is determined in part by the excess of nuclei (Boltzmann distribution) in the lower energy spin state. Since the excess of nuclei in the lower energy spin state is directly proportional to the strength of the applied magnetic field B_0 (Equation 2.10), an increase in sensitivity can be obtained by running the spectrum with as large a magnetic field as possible. In theory, the S/N increases with magnetic field B_0 by the factor of $\frac{7}{4}$. In actual practice, the dependence is somewhat less than this value. For example, a spectrum run at 23.5 kgauss (100 MHz) compared to a spectrum run at 14 kgauss (60 MHz) shows an increase in the S/N of about 1.6. Further increases in the S/N could be realized by running the spectrum on a spectrometer utilizing a high-field superconducting solenoid.

The dependence of the S/N on the temperature (Equation 3.6) is a result of the temperature dependence of the Boltzmann distribution of excess nuclei in the lower energy state (Section 2.2). Although an increase in the S/N could be obtained by lowering the temperature, this has not been put into practice due to the added inconvenience of cooling the sample and to the high melting points of most solvents used in NMR spectroscopy.

As equation 3.6 shows, the S/N depends not only on the Boltzmann distribution of the nuclei, but also on the total number of nuclei N_T within the region of the coil. This is usually expressed in terms of a filling factor, which is the ratio of the sample volume to the coil volume. For maximum S/N, the coil dimensions (equation 3.6) should be as small as possible. The usual practice is to use thin-walled glass sample tubes with an outside diameter of 5 mm and have the receiver coil wound on a thin glass insert in order to maintain a high filling factor. The S/N does not increase extensively with larger coil diameters due to the problems with maintaining a homogeneous magnetic field over the receiver coil region of the probe. For nuclei other than protons where homogeneity requirements are not as rigid, the use of 10–15 mm sample tubes is common.

The most convenient way of increasing the S/N is to increase the concentration of the sample and, hence, increase the number of nuclei in the receiver coil area of the sample. Assuming that a signal can be identified with a S/N of 2 : 1, a concentration of 0.005–0.02 M is needed to detect protons that give a single, sharp peak. If the signal from a proton is split into a multiplet due to spin–spin coupling, higher concentrations are needed. For routine samples, concentrations of $\sim 0.2\,M$ will normally give a satisfactory S/N if the molecular weight of the sample does not exceed 300. The volume of sample required depends on the particular design of the probe but is usually somewhere in the neighborhood of 0.3–0.5 ml for a 5-mm O.D. sample tube.

3.7 SENSITIVITY ENHANCEMENT

In many cases, a sample is either not available in sufficient quantities, or is not soluble in suitable NMR solvents to an extent that a satisfactory S/N can be obtained. Methods have been devised to enhance the sensitivity such that spectra can be obtained from smaller sizes. Some of the more common methods include the use of microcells to concentrate the sample within the region of the receiver and transmitter coil (17,26–31) and time-averaging multiple scans (28) to improve the S/N. Excellent reviews of these methods are available (26–28,32). These and other methods are discussed in greater detail in Section 6.5.1.

3.8 FT TECHNIQUES

With conventional NMR spectrometers such as those described in Section 3.1, the spectra are obtained by sweeping either the frequency or the field through the region of NMR absorption. The irradiating field or frequency remains constant throughout the entire sweep range. This type of experiment is called the continuous-wave (cw) NMR experiment. While this type of experiment can be used with time averaging to improve the S/N, as discussed in the previous section, it is very time consuming since a considerable amount of time (100–1000 sec) must be spent in sweeping through each spectrum in order to prevent major line distortions. If many scans of the spectrum are required to obtain a usable S/N, the total time required to complete the experiment can be extremely long (overnight or even over a weekend). For example, to improve the S/N by a factor of 10, would require approximately 7 hr when a 100-sec sweep time is used. Furthermore, much of the total experiment time is wasted as far as obtaining the data is concerned since the majority of time spent on any one scan is used in scanning the baseline noise between peaks.

In theory, the total time required to achieve a desired S/N could be reduced drastically by using a multichannel spectrometer. For example, if instead of using one rf oscillator and receiver and scanning the spectral region of interest, one were to use a spectrometer that had a rf oscillator and receiver tuned for each data point (say 1 Hz) to be stored in the time averaging device, the entire spectrum could be obtained at one time. The total time required to complete one sweep could be reduced by a factor approaching 1/(number of data points X the length of the cw sweep). Although this type of experiment is clearly not practical from the standpoint of the cost and construction for such a multichannel spectrometer, it is just this type of experiment which Fourier transform techniques accomplish in a much more satisfactory manner (33,34).

Most Fourier transform NMR (FT-NMR) techniques use a method that simultaneously excites all the resonances of a given type of nucleus. The signal induced in the receiver coil is a complex waveform that is a superposition of the resonances of all the nuclei resonating within the frequency range of the exciting device. The complex waveform is sampled as a function of time and is referred to as the time domain spectrum. This process can be repeated a given number of times with each response being coherently added and stored in a digital computer. After the desired enhancement in the S/N has been achieved, the frequency spectrum (called the frequency domain spectrum) can be obtained by Fourier transformation of the time averaged time domain spectrum. This frequency spectrum is identical in all respects to the spectrum that would have been obtained using the cw technique with time averaging.

It is well known that the frequency response function $S(\omega)$ and the time response function $S(t)$ of a linear system form a FT pair

$$S(\omega) = \int_{-\infty}^{\infty} S(t) e^{-i\omega t} \, dt \tag{3.7}$$

$$S(t) = \frac{1}{2\pi} \int_{-\infty}^{\infty} S(\omega) e^{i\omega t} \, d\omega \tag{3.8}$$

where t represents the time response and ω is an angular frequency ($\omega = 2\pi\nu$) representing the corresponding frequency response. Under suitable conditions, an NMR spectrum can be considered as a linear system such that Fourier transformation of the time domain spectrum yields the frequency domain spectrum and vice versa. Since the time domain spectrum is a complex waveform representing the response of all nuclei of a particular type in the sample, Fourier transformation of the time domain spectrum simply separates the complex waveform into the various frequency and intensity components.

The major advantage of FT-NMR techniques over the conventional cw technique for improving the S/N is in the overall observation time required to achieve a given S/N. In general a reduction in the total time by a factor of $10 \sim 100$ can easily be realized using FT-NMR. As a result the proton NMR spectra of very dilute samples and the spectra of nuclei less abundant or less sensitive to NMR detection than protons can be obtained on a more or less routine basis. Furthermore, since a spectrum can be obtained in a few seconds, FT-NMR techniques allow one to examine unstable species with a short lifetime and to obtain information on chemical and molecular dynamics. Relaxation times for the individual nuclei in a sample may also be determined using FT-NMR techniques (35–39).

The collection of data, Fourier transformation, and data reduction is best accomplished using a small, on-line, digital computer. In fact, the developments in computer technology and the reduction in price of computers has paved the way for the development of FT-NMR techniques. Several methods have been developed to excite the resonances of all the nuclei in a sample and obtain the response of the spin system as a function of time. Some of the more common methods are discussed in the following, along with their advantages and disadvantages.

3.9 PULSED FT-NMR

The most widely used FT method to date for improving sensitivity is a pulsed FT-NMR technique (28,33,34,40). Basically, the pulsed FT technique uses a short, intense burst (pulse) of rf energy to excite the nuclei in the sample within a given frequency range. The signal induced in the receiver coil after the pulse is turned off is in the form of oscillating currents which are proportional to the resonance frequencies and intensities of all the resonating nuclei. A plot of this signal versus time as the nuclei return to equilibrium after the pulse is called the free induction decay (FID). This is the time domain spectrum. The FID is usually obtained within a few seconds or less after the pulse is turned off. Fourier transformation of the FID yields the frequency domain spectrum.

The instrumental and experimental procedures for pulsed NMR are consederably different from those encountered when using normal cw techniques. The conditions of field/frequency stabilization are common to both cw and pulse spectrometers. One of the major differences between cw and pulsed NMR experiments is in the magnitude of the irradiating field B_1. For cw experiments, B_1 is typically 0.1 mG whereas for pulsed experiments, B_1 may be as large as 100 G. This places severe restrictions on the amplification of the rf pulse. The timing of the pulse is also very important. A detailed treatment of pulsed NMR is beyond the scope of this text. Several review articles are available for those interested in the more detailed aspects of pulsed NMR (41). An excellent monograph on pulse and FT-NMR has been published (34). In our discussion to follow, we shall go through a typical pulsed FT-NMR experiments and discuss the various factors that should be considered when obtaining a spectrum using this technique.

As a starting point for our discussion of pulsed NMR, consider the total magnetization vector M of the sample in terms of the rotating frame of reference (Section 2.5). At equilibrium, the magnetization M of the sample lies parallel to B_0 along the z' axis. If an rf field B_1, with a frequency of $\omega = \omega_0$ is applied along the x' axis for a time t_p, M will begin to precess around B_1 during the time B_1 is turned on (Figure 3.10). The angle ϕ, through which M precesses around B_1 depends on the time the pulse is turned on, t_p, and the magnitude of the pulse B_1 (equation 3.9) (33).

$$\phi = t_p \gamma B_1 \text{(radians)} \tag{3.9}$$

During a pulse experiment, the magnitude of B_1 remains constant and the angle is controlled by varying the time the pulse is applied. If t_p is chosen such that $\phi = \pi/2$, M will precess from along the z' axis to the y' axis. This pulse is called a 90° pulse. Similarly, if t_p is chosen such that $\phi = \pi$, M will be along the negative z' axis as a result of the pulse (Figure 3.10). This is the 180° pulse.

Two major restrictions are placed on the pulse method for obtaining NMR spectra in order to insure that a true representation of the spectra are

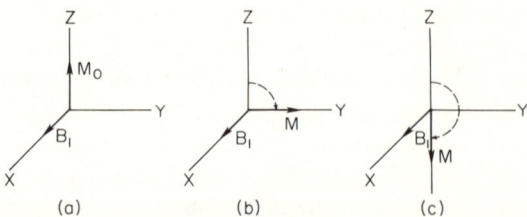

Fig. 3.10 A vectoral representation of the magnetization during a pulse experiment: (a) the magnetization at equilibrium; (b) after a 90° pulse; and (c) after a 180° pulse.

obtained: (1) the magnitude of B_1 must be sufficient to cause the magnetization of all nuclei of the same nuclear species to precess the same angle about B_1, and (2) the time t_p for which the pulse is applied should be short compared to the relaxation times T_1 and T_2 of the nuclei (33). In the rotating frame, the magnetization precesses about an effective field where ω_i is the angular frequency of precession of the nuclei in the

$$|B_{\text{eff}}| = \left(\frac{1}{\gamma}\right)[(\omega_i - \omega)^2 + (\gamma B_1)^2]^{\frac{1}{2}} \qquad (3.10)$$

sample. If B_1 is chosen such that

$$(\omega_i - \omega) \ll \gamma B_1 \qquad (3.11)$$

the term $\omega_i - \omega$ can be neglected in Equation 3.10 and B eff $\approx B_1$. Thus, if B_1 is chosen large enough, the magnetization of all the nuclei within the frequency range $(\omega_i - \omega)$ will precess about the x' axis when B_1 is applied to the sample. Stated another way, suppose we wish to examine a sample containing several nuclei whose precession frequencies cover a range of chemical shift, Δv Hz. If B_1 is such that

$$2\pi\Delta v \ll \gamma B_1 \qquad (3.12)$$

the magnetization of all the nuclei which resonate within the range Δv will precess around x' as a result of a pulse of rf frequency B_1 along the x' axis.

The time t_p for which the pulse of rf frequency is applied must be short compared to the relaxation times of the nuclei in order to ensure that the magnetization of all the spins is rotated by exactly the same angle. For a 90° pulse ($\phi = \pi/2$), substitution of equation 3.9 into 3.12 shows that

$$t_p \ll \frac{1}{4\Delta v} \text{ s.} \qquad (3.13)$$

For a sample containing nuclei that have chemical shifts over a frequency range Δv, the normal method of operation is to adjust the spectrometer frequency (carrier frequency) such that it lies at one end of the frequency range Δv. Due to the heterodyne processes in the receiver, the carrier frequency is subtracted from the absorption frequencies of the nuclei. If the carrier frequency were set in the middle of the chemical shift range Δv, the signal detected by the receiver would contain both positive and negative frequencies. If only one phase detector is used (the normal condition) the receiver makes no distinction between positive and negative frequencies when reading out the data. As a result, the final transformed frequency mode spectrum would appear as if it were folded about the carrier frequency and the positive and negative signal would appear 90° out of phase with each other. Although the peaks could be distinguished by their different phase relationship, overlap of the peaks could occur in a complex spectrum which would make it difficult to distinguish between the peaks. This problem is avoided altogether by setting the carrier frequency at one end of the frequency range Δ such that the frequency differences are all of the same sign. As a result, only one-half of the rf field B_1 is utilized in exciting the nuclei in the sample. Therefore the time for which the pulse is applied (equation 3.13) should be modified in actual practice to

$$t_p \ll \frac{1}{8\Delta\nu} \text{ s.} \tag{3.14}$$

Following a pulse of width t_p, the $x'y'$ component of the total magnetization is (33)

$$M_{xy} = M_0 \sin \omega_i t_p \tag{3.15}$$

Thus the signal induced in the receiver coil will increase with t_p, becoming a maximum for a 90° pulse. For pulse widths larger than 90°, the induced signal will decrease and become zero for a 180° pulse. (In actual practice, a sample with a strong signal is used and the pulse width is adjusted such that no signal is detected. One-half of this value is then taken to be the 90° pulse width t_p).

Initially following a 90° pulse, the magnetization lies along the y' axis and the magnetic moments are all in phase. When the pulse is turned off, the magnetic moments begin to precess around B_0 and will begin to lose their phase relationship due to inhomogeneties in the magnetic field and to relaxation processes. As a result of the relaxation and loss of phase coherence, the magnetization decays back to its equilibrium value along B_0. As the magnetic moments precess about B_0, a flux is induced in the receiver coil which alternates in sign and decays with time to zero when equilibrium is reestablished.

Since the carrier frequency is slightly off resonance, the signal detected in the receiver coil is in the form of a beat pattern corresponding to the difference in frequencies between the carrier frequency and the frequency of absorption of the nuclei. For a single type of nucleus, the pattern will appear as a sine wave whose amplitude decays with time (Figure 3.11). The frequency of oscillation of the sine wave corresponds to the frequency difference between the carrier frequency and the absorption frequency of the nuclei. If the nuclei under consideration are spin coupled to another type of nuclear species, the FID will appear as a beat pattern which is modulated with a frequency of J Hz, where J is the coupling constant between the two nuclei. Both the difference frequency and J may be extracted from the FID. For a sample containing several nuclei with slightly different chemical shifts, the beat

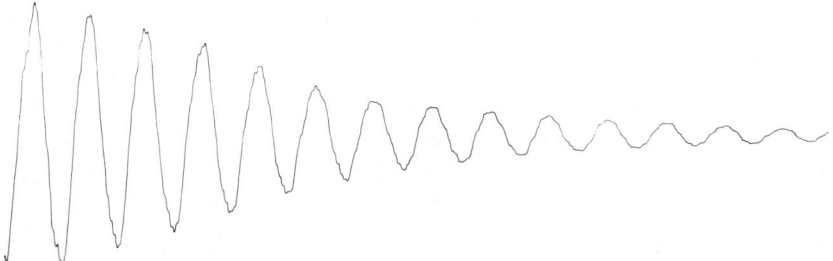

Fig. 3.11 The proton-free induction decay of tetramethylsilane following a 90° pulse.

pattern is very complex (Figure 3.12). The FID must be Fourier transformed in order to obtain the chemical shifts (frequency differences) and coupling constants (modulation frequencies).

For time averaging purposes, the advantage of the pulse method is that the FID is obtained rapidly. Pulses may be applied at a rapid rate with each FID being added and stored in the time averaging device. The rate at which the pulses may be repeated depends on the relaxation times of the nuclei in the sample. If $T_1 \approx T_2^*$ (T_2^* is usually ~ 1 sec due to inhomogeneities in the magnetic field) the pulse can be repeated after approximately $3T_2^*$ with no loss in signal intensity. However, if $T_1 \gg T_2^*$, repeating the pulse after $3T_2^*$

Fig. 3.12 The carbon-13-free induction decay of 7-azaindole.

will result in an attenuated signal. Under these conditions, there will have been little decay of the magnetization M_z back to its equilibrium value M_0 after a 90° pulse before the next pulse is applied. Since the signal depends on M_{xy}, which in turn depends on M_z, the amplitude of the signal will be greatly reduced unless the delay between pulses is sufficient to restore the magnetization along the z' axis.

Ernst and Anderson (33) have shown that the FT relationship between the time domain spectrum and the frequency domain spectrum holds for any pulse angle, t_p. If the pulse time is shortened, the angle through which the magnetization rotates towards the y' axis will be decreased and the amplitude of the signal will be decreased by a factor of $\sin \theta$ (equation 3.15). Simultaneously, the magnitude of M_z will decrease by a factor of $(1 - \cos \theta)$. Therefore, for a pulse angle smaller than 90°, the loss in signal amplitude is much less than the corresponding gain in time required to restore the magnetization along the z' axis. For example, for a pulse angle of 30°, the signal amplitude will be decreased by a factor of $\frac{1}{2}$ compared to a 90° pulse whereas the time required to restore equilibrium along z' will decrease by a factor of 0.14. Therefore the pulse may be repeated at a faster rate with a decrease in the total experiment time required to achieve a desired S/N. Since θ depends upon T_1 and T_2^*, the best value of θ in terms of the total time required to achieve a given S/N is chosen such that the maximum signal is obtained.

3.9.1 DATA ACQUISITION

The intense rf pulse used to rotate the magnetization towards the y' axis tends to saturate the receiver of the spectrometer (34,40). Therefore the start of data acquisition is usually delayed for a short time to allow the receiver to recover from the pulse. If the delay time is not adjusted correctly, phase errors are introduced in the transformed spectrum. A small error in the delay time can result in a spectrum that shows absorption peaks at the end near the carrier frequency and dispersion peaks at the other end of the spectrum. However, in practice, errors in the delay time are not all that critical. Several additional factors also lead to phase differences in the peaks. All of the phase shifts are corrected after the FID has been transformed to the frequency domain spectrum.

The sampling of the FID and storage for sensitivity enhancement is best accomplished using a digital computer. This requires that the FID be converted from analog into digital form before it is added to the memory. In order to obtain all the necessary information from the FID, it must be sampled at a rate sufficient to define the frequency of the FID properly. The sampling rate in turn determines the frequency range of the stored spectrum and hence the resolution of the frequency domain spectrum.

In practice the sampling rate is usually set to be somewhat greater than 2Δ. If there were noise present at a frequency higher than 2Δ, it would be folded back and added to the noise present at lower frequencies. This would obviously decrease the S/N and should be avoided. Most instruments also use low-pass filters (28) to attenuate the high frequency noise so that, when it is folded back into the spectrum, it will be too weak to decrease the S/N significantly. The cutoff frequency is usually chosen to match the frequency range, Δ, of interest. Usually the filters do not have such a sharp cutoff that a small frequency range can be selected without some interference from foldover of strong peaks just outside the frequency range.

For proton spectra at 100 MHz (1000 Hz spectral width), sampling rates of at least 2000 points sec^{-1} are required. Similarly, carbon-13 spectra at 25 MHz (5000 Hz spectral width) require sampling rates of at least 10,000 points sec^{-1}. With spectrometers operating at higher frequencies, or for other nuclei with a larger spectral width, even faster sampling rates are required.

The choice of the sampling time will also determine the resolution of the transformed frequency domain spectrum. The line width at half-peak-height from a cw experiment is $1/\pi T_2^*$ Hz, where T_2^* is the apparent spin–spin relaxation time that includes both relaxation due to T_2 processes and relaxation due to magnetic field inhomogeneity. Of course, we cannot improve the resolution beyond this point. However, the peak may be artificially broadened if the sampling rate is not sufficient. In order to achieve a resolution of R Hz, the FID must be sampled for at least $1/R$ sec. If the sampling time is too long however, the S/N decreases due to the fact that the noise content of the FID increases towards the tail-end of the FID. Usually a compromise is reached in setting the sampling rate such that the sensitivity is maximized at the expense of some loss in resolution.

Unless there are extremely narrow lines in the spectrum, the resolution will ultimately be determined by the available data storage capacity (memory) of the digital computer. For a proton spectrum at 100 MHz with a resolution of 0.5 Hz, at least 4000 (4K) data points are used to avoid folding back the high-frequency noise. A memory of 8K data points would not be uncommon. For a resolution of 1 Hz for carbon-13 spectra at 25 MHz, a memory of 16K is required taking into account the larger sampling rates for the noise.

An additional limitation of the S/N obtainable with pulse FT-NMR experiments is due to the dynamic range of the analog-to-digital (A–D) converter. If there is a large peak in the spectrum, perhaps due to solvent, and several smaller peaks, the larger peak will "fill-up" the memory storage at a fast rate while the storage for the smaller peaks remains only partially filled. This is referred to as overflow of the dynamic range of the A–D converter. The overflow limits the number of experiments that can be stored in the computer and hence, the attainable sensitivity enhancement. For this reason, large peaks in the spectrum should be avoided in a pulsed FT-NMR experiment if at all possible.

3.9.2 DATA REDUCTION

One of the major advantages of using an on-line digital computer to collect the FID is that additional computations may be carried out on the data to improve the appearance of the spectrum. These computations involve additional subroutines in the computer program which may be applied to either the original FID or to the transformed frequency domain spectrum. In general, it is usually more convenient to work with the FID (34,40).

Often a large peak with no physical meaning appears in the transformed spectrum near the carrier frequency. This peak is due to leakage of a small residual d.c. component of the initial pulse into the receiver, which prevents the average value of the data points in the FID from being zero. The first step in data reduction is usually to correct the FID such that the average value of the data points becomes zero. This can be accomplished by summing all the data points, determining the average value, and subtracting the average value from all the data points.

In most cases, the quality of the frequency domain spectrum can be improved by applying various smoothing routines to the FID. These routines have the same effect as further filtering of the data. The advantage, however, of carrying them out at this state of obtaining a spectrum, is that they can be carried out with the digital computer. A change in the type of filter used simply involves a change in the computer program. Normally, several smoothing programs may be stored in memory. The filtering process is a linear, time-independent transformation which can be represented in the form of a convolution integral (28)

$$f_0(t) = \int_{-\infty}^{\infty} h(\tau) f_i(t - \tau) d\tau \tag{3.16}$$

where $f_0(t)$ is the final time function, $f_i(t)$ is the input time function composed of both signal and noise, and $h(\tau)$ is the weighting function of the filter.

One of the more common smoothing routines (filter) applied to the data is a routine to improve the S/N. The FID contains both information about the signals of interest and random noise. Most of the signal information is contained in the initial part of the FID. The noise content of the FID increases towards the tail-end of the FID. If one could force the FID towards zero at the tail-end of the FID, then the noise content would be reduced. This can be accomplished by multiplication of each data point of the FID by an exponential function such as (28,34,40)

$$\exp\left(\frac{iC}{N}\right) \tag{3.17}$$

where i is the index of the data point (0 to $N - 1$), N is the total number of data points, and C is a constant that determines the multiplication factor. If negative values of C are used (-3 is usually used in practice), multiplication of the FID by Equation 3.17 will force the FID towards zero at the tail-end of the FID. As a result of this multiplication, the amount of noise that is transformed into the frequency domain spectrum is reduced and hence, the S/N is increased.

This procedure has the disadvantage that it introduces some broadening of the lines in the frequency domain spectrum (Figure 3.13). As mentioned

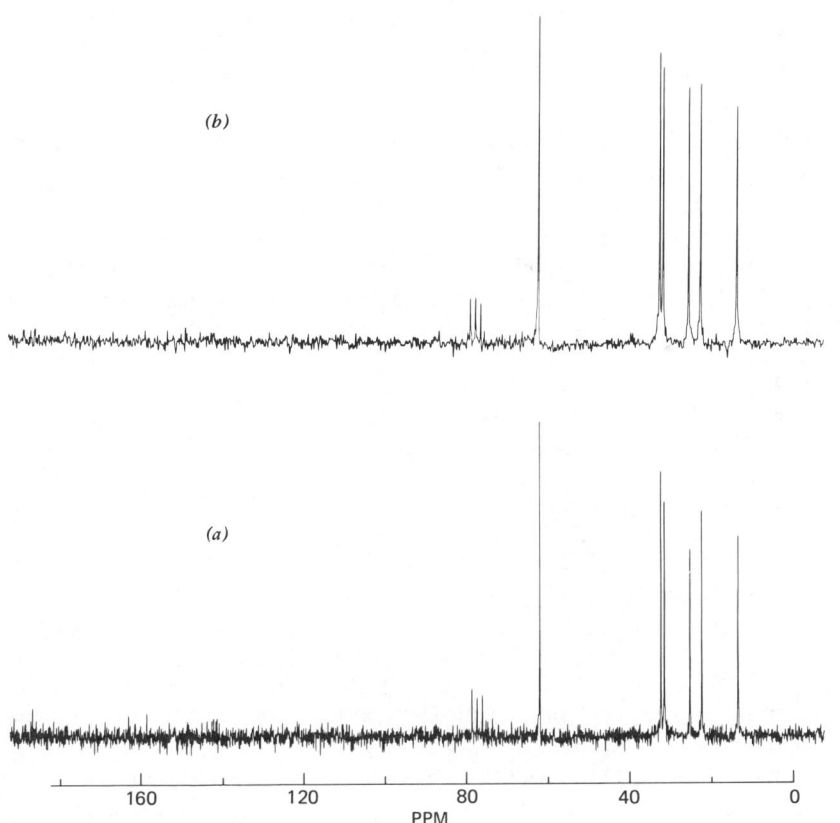

Fig. 3.13 The natural abundance, proton-decoupled carbon-13 spectrum of hexanol: (a) no filter; and (b) a filter of $C = -3$.

previously, the resolution and sensitivity of a FT spectrum depend on each other in an inverse manner such that increasing one usually decreases the other. The line broadening that is introduced is not severe however, such that for carbon-13 spectra, where improving the S/N is the main concern, this procedure results in a saving of time in obtaining a spectrum with a given S/N.

If the S/N is sufficient and one wished to improve the resolution, this can be accomplished in a similar manner by multiplying the FID by a positive exponential function. Of course, the S/N would decrease as a result of this type of filtering. As a general rule, one usually wishes to improve the resolution with proton spectra and the sensitivity with carbon-13 spectra. In addition to the previously stated smoothing routine, there are several other types of filters that can be applied to the FID to correct for such things as baseline ripple due to pulse feed-through.

After the application of the smoothing routines to the FID, the next step in the data reduction is Fourier transformation of the FID to yield the frequency domain spectrum. The Fourier transformation is carried out by the digital computer over the N data points in the FID. However, instead of carrying out the FT, as indicated in equation 3.7 (the discrete FT), a modification (34) of the algorithm of Cooley and Tukey (42) is used. This procedure results in a more efficient utilization of computer memory and also a reduction in the time required to do the Fourier transformation. For example, the procedure is such that as the Fourier transformation is carried out, the new data is written over the old data. Furthermore, the discrete FT requires N^2 multiplications whereas this procedure requires only $N \log N$ multiplications (34).

The actual evaluation of the Fourier transformation is performed by summing a series over the N data points rather than by evaluating the integral of equation 3.7 (34).

$$F(w) = \sum_{t=0}^{n-1} f(t)e^{-2\pi i f t/N} \qquad f = 0, 1, i - n - 1 \qquad (3.18)$$

From the trigonometric identity

$$e^{-iy} = \cos y - i \sin y \qquad (3.19)$$

equation 3.18 can be expressed as the complex sum of the cosine and sine transforms.

Fourier transformation of the FID yields both a real and imaginary part, each of which contains $N/2$ data points. The real part of the transform (cosine transform) corresponds to the absorption mode (v mode) spectrum whereas the imaginary part (sine transform) yields the normal dispersion mode

(u mode) spectrum. Normally, only the absorption mode spectrum is utilized.

Before the resulting frequency domain spectrum is displayed on a plotter, the transformed spectrum will usually require some phase corrections. The phase detector of the spectrometer is normally not adjusted before each experiment. This produces a spectrum that is a mixture of both the absorption and dispersion mode spectra similar to a cw experiment. The phase error due to the dispersion mode will be identical for all peaks in the spectrum (frequency independent) and is referred to as a zero-order phase error. The delay time after the pulse, before the start of data acquisition, introduces a phase error that varies linearly with frequency (first-order phase correction). At zero frequency (carrier frequency) this phase error is zero, but can become very large at higher frequency. Additional frequency dependent phase errors are also introduced by the filtering of the FID to eliminate higher frequency noise.

The zero-order phase error may be corrected (34) by generating the pure absorption mode spectrum $A(w)$ from a linear combination of the sine $S(w)$ and cosine $C(w)$ transforms

$$A(w) = pC(w) + (1 - p)^{\frac{1}{2}}S(w) \qquad (3.20)$$

where p is the correction factor. First-order phase errors may be corrected by a process such as

$$p = p_0 + p_i(\omega - \omega_0) \qquad (3.21)$$

which varies linearly with frequency. Although these phase corrections may be adjusted independently, the normal procedure is to combine the two equations and carry out both phase adjustments simultaneously. The phase correction may be accomplished with an interactive program which allows one to adjust both the zero- and first-order corrections while observing the effect of changing p on an oscilliscope. Alternatively, the phase correction can be accomplished automatically by using an iterative program which maximizes the area of the peak above the baseline while minimizing the area under the baseline.

If a digital computer is not available, Fourier transformation of the FID can be accomplished by analog techniques using a spectrum analyzer. The resulting spectrum, however, appears as the power spectrum

$$(v^2 + u^2)^{\frac{1}{2}} \qquad (3.22)$$

which represents the absolute value of M_{xy} with no phase information. Although this method avoids the problems of phase adjustment, it has the disadvantage that it produces some broadening at the base of the peaks (Figure 3.14). As a result, the intensities, line shapes, and even the frequencies

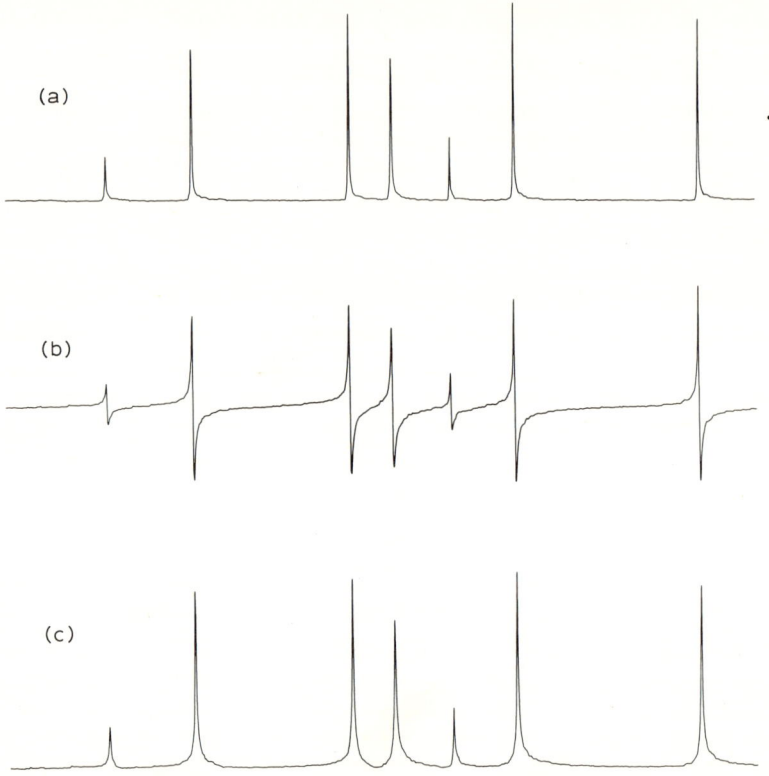

Fig. 3.14 An expansion of the natural abundance, proton-decoupled carbon-13 spectrum of 7-azaindole: (a) the real or v mode spectrum; (b) the imaginary or u mode spectrum; and (c) the power spectrum.

of the peaks may be distorted if one peak overlaps with another peak. The broadening at the base of a large peak could overlap and hide a smaller peak. For this reason, this mode of display is not recommended if it can be avoided.

When a digital computer is used for Fourier transformation, additional programs should be available to allow the user to carry out other data reduction procedures. The most common form of data reduction produces a printout of the memory location of the peak, the frequency of the peak with respect to any predetermined peak (usually the reference), the ppm value of the peak with respect to the predetermined peak, and some indication of the relative peak heights or areas. In addition, programs are usually available to obtain an integration of the entire spectrum.

3.10 QUADRATURE FT-NMR

The quadrature NMR experiment is very similar to the pulsed FT experiment. The major difference is that two phase detectors are used in quadrature such that one can simultaneously detect both the absorption and dispersion components of the magnetization (43). As a result, the carrier frequency can be placed in the center of the spectral region of interest such that use is made of both halves of the pulse power. This leads to an improvement in the sensitivity by a factor of ~ 1.5 over normal pulsed FT-NMR (44). Instruments employing this detection scheme are just coming into use.

3.11 STOCHASTIC RESONANCE

Another method that may be used to improve the S/N is stochastic resonance (45,46). Although the theory of stochastic resonance has been worked out in detail, it has not been applied as extensively as pulsed FT-NMR. Stochastic resonance is similar in many respects to pulsed FT-NMR in that all the resonances within a given frequency range are excited at one time and the response of the entire spin system is observed. However, the method of excitation is quite different.

In stochastic resonance, the sample is subjected to a random noise function $s(t)$ that contains frequency components covering the spectral region of interest. In this manner, all the nuclei within the spectral region will be excited at the same time. The response of the system $v(t)$ to this stochastic excitation is sampled as a function of time. The cross-correlation function $R_{sv}(\tau)$ is then calculated using equation 3.23

$$R_{sv}(\tau) = \lim_{T \to \infty} (2T)^{-2} \int_{-T}^{T} s(t)v(t + \tau)d\tau \qquad (3.23)$$

where T is the pulse period. Fourier transformation of the cross-correlation function $R_{sv}(\tau)$ yields the frequency domain spectrum.

In practice the noise excitation with a given width and power is applied for a specific time and the system response is obtained. This process can be repeated with the results being stored in a signal-averaging device. The FT of $S(t)$ and $V(t)$ are then taken separately and multiplied together to yield the frequency domain spectrum.

The S/N attainable with this method is equivalent to that attainable using pulsed FT-NMR for a given total experiment time. Stochastic resonance has the advantage that both sensitivity and resolution may be optimized at the same time. Furthermore, the peak power needed for stochastic excitation is much less than for pulsed FT-NMR so that larger spectral widths may be examined more conveniently.

The computer requirements for stochastic resonance are the same as for pulsed FT-NMR. However, the instrumentation required for stochastic resonance appears to be less complex.

3.12 RAPID-SCAN OR CORRELATION FT-NMR

In conventional cw NMR, spectra are obtained with a low rf power to avoid saturation and a slow sweep rate (~ 1 Hz sec^{-1}) in order to approximate slow passage conditions and produce undistorted lines. The magnetization of the sample is perturbed only slightly and a relatively weak signal is produced. Under conditions of rapid passage (~ 100 Hz sec^{-1}), conditions may be chosen such that the magnetization is tipped almost into the x–y plane. As a result of rapid passage, the lines in the spectrum are severely distorted to the point where they are no longer recognizable. Each line produces severe ringing which persists for a period of approximately $3T_2^*$. However, the slow passage spectrum can be obtained from the rapid scan response by cross correlating the response with either the response of a single line, or with a theoretical line shape function. This forms the basis of rapid scan NMR spectroscopy which compares favorably with pulsed FT-NMR as a method for improving the S/N of NMR spectra (47,48).

With rapid scan FT-NMR spectroscopy, one obtains the response $y(w)$ of the spin system as a function of time after rapid passage through the spectrum. The response is then Fourier transformed to yield the time-dependent response $h(t)$. The response $Y(w)_{\text{ref}}$ of a single line is then obtained and Fourier transformed to obtain its time-dependent response $H_{\text{ref}}(t)$. The outputs of both reference and sample are then cross-correlated, yielding the equivalent of the free induction signal of the spin system. Fourier transformation of the FID yields the frequency domain spectrum, which is equivalent to the slow passage spectrum obtained by cw techniques. The process may be repeated with the results being added coherently in a time-averaging device to improve the S/N.

One disadvantage of rapid-scan FT-NMR is that the linewidths obtained in the final spectrum are the sum of the linewidths of the sample and the reference line. This artificial broadening of the lines can be eliminated by utilizing a theoretical line-shape function for cross-correlation with the time-response of the spin system.

The sensitivity of the rapid scan FT-NMR method is slightly less than that which can be obtained using pulsed FT-NMR for a given time. When $T_1 = 10$ sec, the sensitivities are nearly equal for a spectral range of 1000 Hz. For the case where $T_1 = T_2^* = 1$ sec, the rapid-scan FT-NMR sensitivity is approximately 35% lower than that obtained using the pulsed FT-NMR method.

The computer requirements for rapid-scan FT-NMR are identical with those for pulsed FT-NMR. However, the instrumentation is far less complex. In fact, other than a normal cw-NMR spectrometer, only a sweep generator is needed.

One of the major advantages of the rapid-scan FT-NMR method is that dynamic range problems due to a large solvent peak in the spectrum are eliminated. This problem is eliminated by not scanning through the spectral region where the large peak appears. Of course, if the large peak appears in the middle of the spectrum, two scans may be required to cover the spectral range of interest. Other advantages are that there are no foldover problems, and the T_1 of selective peaks in the spectrum can be obtained by scanning only over the spectral region of interest.

3.13. COMMERCIAL NMR SPECTROMETERS

The developments in NMR instrumentation have primarily focused on three areas: (1) improving the sensitivity and hence reducing the sample size required for a spectrum; (2) reducing the difficulties in the day-to-day operation of NMR spectrometers; and (3) reducing the cost of spectrometers. The first commercial NMR spectrometer was introduced in 1953 and operated at 30 MHz for protons (49). Since that time, improvements in magnet technology have led to instruments operating at 40 MHz in 1955, 60 MHz in 1958, 100 MHz in 1962, 220 MHz in 1966, and 300 MHz in 1971 for protons (49). The first routine-type 60-MHz NMR spectrometer for protons was introduced in 1961 (49). Today, it is no more difficult to obtain an NMR spectrum than it is to obtain an infrared or uv spectrum. Bible (50) has reviewed the spectrometers introduced during the period 1963–1969. In the next section we shall discuss the developments in NMR instrumentation since 1970.

3.13.1 ROUTINE PROTON SPECTROMETERS

Three routine proton spectrometers have been made available since 1970. The EM-300 is a 30-MHz spectrometer using a permanent magnet (49). The low cost (about one-eighth of the cost of the first 30-MHz instrument) and ease of operation of the EM-300 makes it particularly attractive as a teaching instrument. It should also find use in industrial laboratories as an instrument for screening samples and thus, freeing the research-type instrument for more complex problems. Some disadvantages of the EM-300 are the lack of decoupling and variable temperature accessories and the low frequency.

The EM-360(49) and R-24A (51) are 60 MHz proton spectrometers. Both instruments have a sensitivity of $\sim 25:1$ (S/N for the methylene resonance of

1 % v/v ethylbenzene) and a resolution of 0.6 Hz. Facilities for homonuclear decoupling are available, but not for variable temperature. The simplicity, economy, and ease of operation of these spectrometers make them attractive for a wide variety of routine and, perhaps, research applications.

3.13.2 CW RESEARCH SPECTROMETERS

The R32 spectrometer introduced in 1972 (51) is a 90-MHz proton spectrometer using a permanent magnet. A resolution of 0.5 Hz and sensitivity of 50:1 are quoted for this instrument. Accessories include variable temperature, field/frequency stabilization, wide sweep ranges, homonuclear decoupling, and nuclei other than protons. Another 90-MHz proton spectrometer using a permanent magnet, the EM-390 (49), has been introduced recently. A resolution of 0.5 Hz with a sensitivity of 50:1 are quoted for the EM-390 (49). Standard accessories include an internal lock with automatic adjustment of the y-gradient homogeneity and wide sweep ranges. Optional accessories include spin decoupling, variable temperature, oscilloscope display, and signal averager.

A number of companies other than the original manufacturers now offer FT accessories for updating the previously described spectrometers and older continuous-wave spectrometers (52,53). This accessory is almost a prerequisite for running natural-abundance carbon-13 spectra and greatly reduces the sample size required for proton spectra. In addition, a 1-mm probe insert is now available for certain instruments (49,54) which further reduces the sample size required. These inserts use melting point capillaries as the sample tube and require ~ 5 μl of sample volume. With this accessory proton spectra from μg quantities of sample may be obtained within a matter of minutes using FT-NMR.

3.13.3 FT SPECTROMETERS

The area of FT spectrometers has received considerable attention during the past few years. These spectrometers use solid-state electronics and are capable of performing all the functions of a cw spectrometer, plus the added advantage of pulse capabilities. Furthermore, the cost of these instruments is little more than what a routine multinuclear cw instrument was a few years ago. There can be no doubt that the area of carbon-13 NMR has benefited most from these instruments. However, most of the instruments are capable of running proton spectra and a few have multinuclear accessories. Wideband proton noise decoupling, deuterium internal lock, variable temperature, and gated and off resonance decoupling are standard on most instruments. In addition, software and pulse programmers are available for carrying out automatic computer controlled T_1 measurements. A variety of computer configurations are available depending on the particular needs of the purchaser.

The WH90 spectrometer (55) is a multinuclear FT only instrument which measures proton spectra at 90 MHz. Its counterpart at 60 MHz is the WP60 (55). Both instruments have high sensitivity with a resolution for protons equal to that of cw instruments. Probe inserts for 15-mm sample tubes are available. The CFT-20 spectrometer (49) was designed primarily for carbon-13 work at 20 MHz with 12-mm sample tubes. It offers excellent sensitivity and resolution. Accessories for proton spectra are available. The FX60 spectrometer (54) is capable of running both proton and carbon-13 spectra. It offers the added advantage that the same probe is used for both proton and carbon-13 spectra and only the changing of one switch is required to convert from one nucleus to the other. The probe insert is designed for 10-mm sample tubes. The TT-14 spectrometer (52) utilizes a HR- or HA-60 magnet (49) already in the customers laboratory. It is designed for multinuclear work and offers the added advantage that the probe used in this system will take 20-mm sample tubes. Where sufficient sample is available, the 20-mm probe will increase the sensitivity some two to three times that of spectrometers using 10-mm sample tubes.

The R-26 NMR spectrometer (51) is an FT only instrument designed for carbon-13 work. A permanent magnet is used and 10-mm sample tubes are standard.

Since the instruments cited are not capable of cw operation, they have been designed to optimize operation in the FT mode. These instruments are easy to operate and there can be no doubt but what they will become the routine instruments of the future if they are not already at that stage.

3.13.4 SUPERCONDUCTING SPECTROMETERS

Improvements in superconducting magnet technology have continued to improve the resolution and increase the magnetic field strength. Two companies now offer high-field NMR spectrometers using superconducting solenoids. The SC-300 spectrometer operates at 300 MHz for protons (49). A full line of accessories are available with the SC-300 including decoupling, variable temperature, and nuclei other than protons. Although the SC-300 is a cw instrument, FT accessories are available. The HX-270 spectrometer operates at 270 MHz for proton spectra (55). Accessories similar to those previously stated are available with this instrument. This instrument is also available as a FT only spectrometer as the WH270 (55).

3.13.5 SPECTROMETERS FOR PROCESS MONITORING

Two of the more recent spectrometers for process monitoring are the Mini Spec R20 (55) and the PR-103 (56). Both instruments are pulse NMR spectrometers and are capable of distinguishing signals from protons in different chemical or physical states due to their different relaxation times. Either T_1 or T_2 of the sample may be determined. Once the conditions for sample measurement have been determined, several samples per hour may be run as no preconditioning of the sample is required in most cases. In addition, these instruments are suitable for on-line process monitoring. Examples of the application of these instruments are discussed in Section 6.5.3.

REFERENCES

1. F. Bloch, W. W. Hansen, and M. Packard, *Phys. Rev.*, **69**, 127 (1946).
2. E. M. Purcell, H. C. Torrey, and R. V. Pound, *Phys. Rev.*, **69**, 37 (1946).
3. W. A. Anderson, *Rev. Sci. Instruments*, **32**, 241 (1961).
4. W. A. Anderson, in *NMR and EPR Spectroscopy*, Pergamon, New York, 1960, p. 176.
5. H. Primas and H. H. Gunthard, *Rev. Sci. Instr.*, **28**, 510 (1957).
6. W. Naegele, in *Determination of Organic Structures by Physical Methods*, F. C. Nachod and J. J. Zuckerman, Eds., Vol. 4, Academic, New York, 1971, p. 1.
7. F. Bloch, W. W. Hansen, and M. Packard, *Phys. Rev.*, **70**, 474 (1946).
8. N. Bloembergen, E. M. Purcell, and R. V. Pound, *Phys. Rev.*, **73**, 679 (1948).
9. H. S. Gutowsky, L. H. Meyer, and R. E. McClure, *Rev. Sci. Instr.*, **24**, 644 (1953).
10. L. F. Johnson, in *NMR and EPR Spectroscopy*, Pergamon, New York, 1960, p. 90.
11. For a review of modulation, see O. Haworth and R. E. Richards, in *Progress in N.M.R. Spectroscopy*, J. W. Emsley, J. Feeney, and L. H. Sutcliffe, Eds., Vol. 1, Pergamon, New York, 1966, p. 1.
12. E. B. Baker and L. W. Burd, *Rev. Sci. Instr.*, **28**, 313 (1957).
13. W. A. Anderson, *Rev. Sci. Instr.*, **33**, 1160 (1962).
14. H. Primas, 5th European Congress on Molecular Spectroscopy, Amsterdam (1961).
15. R. Freeman and D. H. Whiffen, *Mol. Phys.*, **4**, 321 (1961).
16. R. Freeman and W. A. Anderson, *J. Chem. Phys.*, **37**, 2053 (1963).
17. Wilmad Glass Company, Route 40 and Oak Road, Buena, N.J.
18. Hamilton Company, Whittier, Calif.
19. *25 N.M.R. Solvents*, The Sadtler Research Laboratories, Inc.. Philadelphia, 1966.
20. J. A. Pople, W. G. Schneider, and H. J. Bernstein, *High-resolution Nuclear Magnetic Resonance*, McGraw-Hill, New York, 1959.

21. J. W. Emsley, J. Feeney, and L. H. Sutcliffe, *High Resolution Nuclear Magnetic Resonance Spectroscopy*, Vol. 1, Pergamon, New York, 1965, p. 605.
22. H. J. Bernstein and K. Frei, *J. Chem. Phys.*, **37**, 1891 (1962).
23. N. C. Li, R. L. Scruggs, and E. D. Becker, *J. Am. Chem. Soc.*, **84**, 4650 (1962).
24. D. C. Douglass and A. Fratiello, *J. Chem. Phys.*, **39**, 3163 (1963).
25. R. Gabillard and M. Soutif, in *La Resonance Paramagnetique Nucleaire*, P. Grivet, Ed., Centre National de la Recherche Scientifique, Paris, 1955, p. 159.
26. R. E. Lundin, R. H. Elsken, R. A. Flath, and R. Teranishi, in E. G. Brame, Jr., Ed., *Appl. Spectroscopy Rev.*, Vol. 1, Marcel Dekker, New York, 1967, p. 131.
27. G. E. Hall, in *Annual Review of NMR Spectroscopy*, E. F. Mooney, Ed., Vol. 1, Academic Press, New York, 1968, p. 227.
28. R. R. Ernst, in *Advances in Magnetic Resonance*, J. S. Waugh, Ed., Vol. 2, Academic, New York, 1966, p. 1.
29. Varian Associates, Palo Alto, Calif.
30. JEOL, U.S.A., Inc., Cranford, N.J.
31. E. G. Brame, Jr., *Anal. Chem.*, **37**, 1183 (1965).
32. A. Savitsky and M. J. E. Golay, *Anal. Chem.*, **36**, 1627 (1964).
33. R. R. Ernst and W. A. Anderson, *Rev. Sci. Instr.*, **37**, 93 (1966).
34. T. C. Farrar and E. D. Becker, *Pulse and Fourier Transform NMR, Introduction to Theory and Methods*, Academic, New York, 1971.
35. R. L. Vold, J. S. Waugh, M. P. Klein, and D. E. Phelps, *J. Chem. Phys.*, **48**, 3831 (1968).
36. R. Freeman and H. D. W. Hill, *J. Chem. Phys.*, **51**, 3140 (1969).
37. R. Freeman and H. D. W. Hill, *J. Chem. Phys.*, **54**, 3367 (1971).
38. J. L. Markley, W. J. Horsley, and M. P. Kelin, *J. Chem. Phys.*, **55**, 3604 (1971).
39. G. G. McDonald and J. S. Leigh, Jr., *J. Magn. Res.*, **9**, 358 (1973).
40. H. D. W. Hill and R. Freeman, *Introduction to Fourier Transform NMR*, Varian Associates, Palo Alto, 1970.
41. N. Boden, in *Determination of Organic Structure by Physical Methods*, F. C. Nachod and J. J. Zuckerman, Eds., Vol. 4, Academic, New York, 1971, p. 91.
42. J. W. Cooley and J. W. Tukey, *Math. Comput.*, **19**, 297 (1965).
43. J. D. Ellet, M. G. Gibby, U. Haeberlen, L. M. Huber, M. Mehring, A. Pines, and J. S. Waugh, in *Advances in Magnetic Resonance*, J. S. Waugh, Ed., Vol. 5, Academic, New York, 1971, p. 117.
44. E. D. Stejskal and J. Schaefer, *J. Mag. Res.*, **13**, 249 (1974).
45. R. R. Ernst, *J. Mag. Res.*, **3**, 10 (1970).
46. R. Kaiser, *J. Mag. Res.*, **3**, 28 (1970).
47. J. Dadok and R. F. Sprecher, *J. Mag. Res.*, **13**, 243 (1974).
48. R. K. Gupta, J. A. Ferretti, and E. D. Becker, *J. Mag. Res.*, **13**, 275 (1974).
49. Varian Associates, Palo Alto, Calif., technical bulletin.
50. R. H. Bible, Jr., *Appl. Spectros.*, **24**, 326 (1970).

51. The Perkin–Elmer Corporation, Norwalk, Conn., technical bulletin.
52. Nicolet Technology Corporation, Mountain View, Calif., technical bulletin.
53. Digilab Inc., Cambridge, Mass., technical bulletin.
54. JEOL, U.S.A., Inc., Cranford, N.J., technical bulletin.
55. Bruker Scientific Inc., Elmsford, N.Y., technical bulletin.
56. The Praxis Corporation, San Antonio, Tex., technical bulletin.

CHAPTER

4

ANALYSIS OF SPECTRA AND DETERMINATION OF STRUCTURE

4.1	Introduction	96
4.2	Nomenclature	97
4.3	First-Order Analysis of Spectra	101
4.4	Analysis of Complex NMR Spectra	104
	4.4.1 Quantum Mechanical Formalism	104
	4.4.2 Solution for a Two-Spin System	108
	4.4.3 Summary of Procedure for Calculating Spectra	111
4.5	Analysis of Specific Spin Systems	112
4.6	Two-Spin Systems	113
	4.6.1 Two Equivalent Nuclei (A_2)	113
	4.6.2 Two Nuclei (AB)	114
4.7	Three-Spin Systems	117
	4.7.1 Three Nuclei (AB_2 or AX_2)	117
	4.7.2 Three Nuclei (AMX)	120
	4.7.3 Three Nuclei (ABX)	122
	4.7.4 Three Nuclei (ABC)	127
4.8	Four-Spin Systems	128
	4.8.1 Four Nuclei ($AA'XX'$)	128
	4.8.2 Four Nuclei ($AA'BB'$)	132
	4.8.3 Four Nuclei (A_2B_2)	137
4.9	Larger Spin Systems	139
4.10	Computer Analysis of NMR Spectra	139
4.11	Aids in the Analysis of Spectra	142
	4.11.1 Variations in Magnetic Field	142
	4.11.2 Isotopic Substitution	143
	4.11.3 Double Resonance	145
	4.11.4 Homonuclear Spin Decoupling	147
	4.11.5 Spin Tickling	149
	4.11.6 INDOR	152
	4.11.7 Intramolecular Nuclear Overhauser Effect	154
	4.11.8 Heteronuclear Double Resonance	155
	4.11.9 Lanthanide Shift Reagents	156

96 ANALYSIS OF SPECTRA AND DETERMINATION OF STRUCTURE

4.12	Correlation of NMR Parameters	161
	4.12.1 Chemical Shifts	162
	4.12.2 Coupling Constants	176
	4.12.2.1 Geminal Coupling Constants	177
	4.12.2.2 Vicinal Coupling Constants	179
	4.12.2.3 Long-Range Couplings	183
	4.12.2.4 Coupling in Aromatic and Heteroaromatic Compounds	185
	4.12.2.5 Proton Coupling Constants with Other Nuclei	189

4.1 INTRODUCTION

The successful use of NMR spectroscopy as an analytical tool depends on the extraction of the fundamental parameters from the spectrum. The analysis of an NMR spectrum in terms of the chemical shifts and coupling constants is the most important step in the application of NMR spectroscopy. In certain cases the analysis may amount to the determination of the separation between the peaks in a multiplet and finding the frequency of the center of the multiplet. In other cases a full mathematical analysis using one of the existing computer programs is required for complete interpretation of the spectrum. Various additional experiments such as spin-decoupling, obtaining the spectrum in a different solvent, obtaining the spectrum in the presence of a shift reagent, and deuterium substitution may be required for the full analysis of a complex spectrum.

Once the parameters have been obtained they must be assigned to the individual protons of the sample using correlation tables of chemical shifts and coupling constants derived from known compounds. After the assignment has been made, the parameters may be used in conjunction with the correlation tables and empirical relationships to derive additional information about the compound under study.

For unknown compounds the NMR spectrum alone is not sufficient to establish the structure. One usually has additional information such as a molecular weight, an elemental analysis, a uv spectrum, and an infrared spectrum which, when used in conjugation with the NMR spectrum, allows one to propose a structure consistent with the known data. This structure may then be confirmed by comparison with data on an authentic sample. In this chapter we shall introduce the nomenclature and discuss the rules for the analysis of NMR spectra. Aids in the analysis of spectra and the use of correlation tables and empirical relationships in the determination of the structure of unknown compounds will be discussed.

4.2 NOMENCLATURE

In the discussion of the analysis of NMR spectra, chemists consider only the magnetic nuclei in a compound. The magnetic nuclei in a compound are referred to as the *spin system*. The spin system includes all the interacting magnetic nuclei (those that are spin coupled with each other) in the compound and it is not necessary for each nucleus to be coupled to every other nucleus in the spin system. In many cases two or more nuclei may have identical chemical shifts. These nuclei are said to be equivalent. In the analysis of NMR spectra, two types of equivalency are considered; the difference being determined by the coupling constants of the nuclei in question. Nuclei which have identical chemical shifts and identical coupling constants to all other magnetic nuclei in the molecule are said to be *magnetically equivalent*. Nuclei with identical chemical shifts but with different coupling constants to other magnetic nuclei are said to be *chemical shift equivalent*.

Two or more nuclei are chemical shift equivalent if there exists some symmetry element of the molecule such as an axis, center, or plane, which interchanges the nuclei. In addition, nuclei that are interchanged through some exchange process with a rate faster than about once in 10^{-3} s. are also chemical shift equivalent. As an example of chemical shift equivalence, consider *p*-chloronitrobenzene (**1**). This molecule has a plane of symmetry perpendicular to the plane of the ring that bisects the two substituted carbons.

$$
\begin{array}{c}
\text{NO}_2 \\
\text{H}_1 \quad \text{H}_3 \\
\text{H}_2 \quad \text{H}_4 \\
\text{Cl}
\end{array}
$$

1

As a result, protons 1 and 3 have identical chemical shifts, as do protons 2 and 4. However, protons 1 and 3 do not have identical coupling constants to the other magnetic nuclei. The coupling constant between protons 1 and 4 is a *para* coupling whereas the coupling constant between protons 3 and 4 is an *ortho* coupling. These two coupling constants have different values and therefore this molecule contains two sets of chemical shift equivalent nuclei.

As another example, consider the spectrum given by a 1,2-disubstituted ethane, XCH_2CH_2Y. In considering compounds of this type, the authors have found it convenient to consider the compound in terms of the Newman

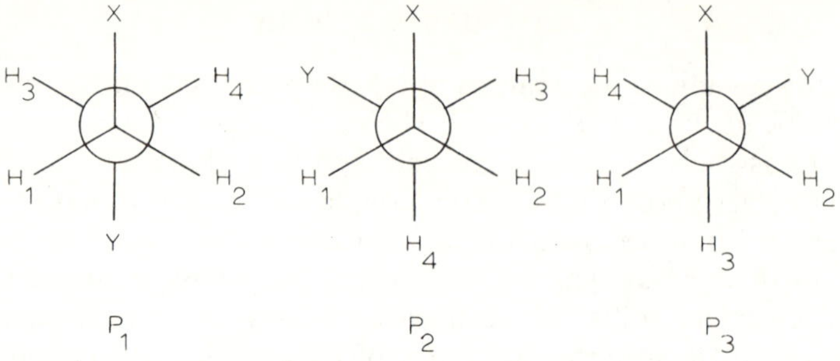

98   ANALYSIS OF SPECTRA AND DETERMINATION OF STRUCTURE

Fig. 4.1 Newman projections for the three most stable rotamers of a 1,2-disubstituted ethane $X\mathrm{CH}_2\mathrm{CH}_2Y$.

projections (Figure 4.1). Of the three staggered conformations, only the anti-conformation possesses a plane of symmetry. However, this is enough to make protons 1 and 2 have identical chemical shifts, as do protons 3 and 4. Perhaps this can be seen more easily if we consider the chemical shifts as an average of each of the Newman projections. The chemical shifts of protons 1 and 2 are given by equations 4.1 and 4.2 where p's refer to the populations of the conformations and the subscripts refer to the immediate environment of the protons. Since the populations of the two *gauche* conformations are equal

$$\delta H_1 = p_1 \delta_{Y,H} + p_2 \delta_{Y,H} + p_3 \delta_{H,H} \quad (4.1)$$

$$\delta H_2 = p_1 \delta_{Y,H} + p_2 \delta_{H,H} + p_3 \delta_{Y,H} \quad (4.2)$$

($p_2 = p_3$), protons 1 and 2 have identical chemical shifts. Similar results are obtained for protons 3 and 4. A similar treatment will show that protons 1 and 2 do not have identical couplings to protons 3 and 4. Considering the coupling constant to be an average of the coupling constants in the three conformations, the coupling constant J_{13} and J_{23} are given by equations 4.3 and 4.4 where J_t and J_g refer to *trans* and

$$J_{13} = p_1 J_g + p_2 J_t + p_3 J_g \quad (4.3)$$

$$J_{23} = p_1 J_t + p_2 J_g + p_3 J_g \quad (4.4)$$

gauche coupling constants, repectively. Since $J_t \neq J_g$, it can readily be seen that $J_{13} \neq J_{23}$.

This same treatment will also show why the two protons in a methylene group adjacent to a chiral center are not equivalent. Consider the Newman projections given in Figure 4.2 for a compound of the type $M\mathrm{CH}_2CXYZ$. The chemical shifts for protons 1 and 2 are given by equations 4.5 and 4.6.

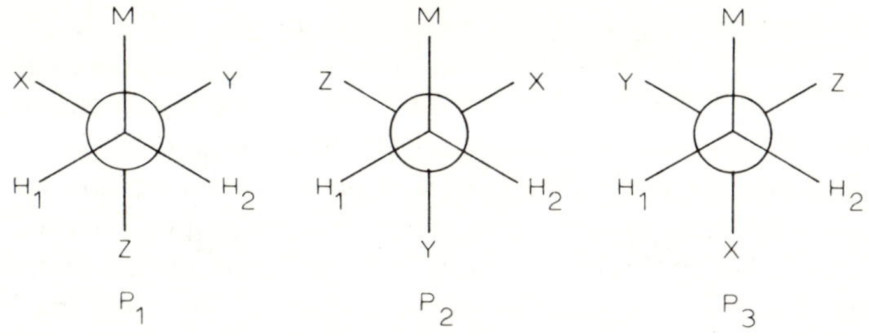

Fig. 4.2 Newman projections for the three most stable rotamers of a 1,1,1,2-tetrasubstituted ethane $XYZCHCH_2M$.

Since the populations of the three conformations are not equal, $p_1 \neq p_2 \neq p_3$, the chemical shifts of protons 1 and 2 are not equivalent

$$\delta H_1 = p_1 \delta_{X.Z} + p_2 \delta_{Y.Z} + p_3 \delta_{X.Y} \tag{4.5}$$

$$\delta H_2 = p_1 \delta_{Y.Z} + p_2 \delta_{X.Y} + p_3 \delta_{X.Z} \tag{4.6}$$

In order for nuclei to be magnetically equivalent, they must be chemical shift equivalent and have identical coupling constants to all other magnetic nuclei. This usually implies that the bond distances and bond angles with respect to the other nuclei be equal, or that they are equal with fast rotation about a single bond. As an example, consider the compound ethyl bromide, CH_3CH_2Br. If one draws the three staggered Newman projections similar to Figure 4.1 and considers that fast rotation is occurring about the carbon–carbon bond, writing equations for the chemical shifts and coupling constants for the methylene protons will show that the methylene protons have identical chemical shifts and identical coupling constants to each of the three methyl protons. The same is true for the three methyl protons. Therefore ethyl bromide contains two sets of magnetically equivalent nuclei. Methylene fluoride, CH_2F_2, provides another example. The two protons have identical chemical shifts and identical coupling constants with each of the fluorines. Similarly, the two fluorines have identical chemical shifts and identical coupling constants with each of the protons. The protons and fluorines in methylene fluoride therefore constitute two sets of magnetically equivalent nuclei.

The type of equivalence in a molecule has a large influence on the number of peaks observed in the spectrum of a spin system and the ease with which the parameters may be obtained from the spectrum. For molecules that contain magnetically equivalent nuclei, the coupling constant between the equivalent nuclei cannot be obtained from the spectrum (i.e., J_{H-H} in methylene fluoride). The spectrum given by a molecule that contains magnetically equivalent nuclei usually contains fewer absorptions and is normally easier to interpret. The spectrum given by a molecule containing chemically equivalent nuclei on the other hand is usually more complex and either a mathematical or computer analysis may be required to obtain the parameters.

A method for designating the magnetic nuclei in a sample in terms of the letters of the alphabet has developed over past years and will be used throughout this text (1). Using this convention, nonequivalent nuclei having chemical shifts that differ by magnitudes comparable to the coupling constants between the nuclei are referred to by the letters A, B, C, D, etc. If additional

TABLE 4.1. Examples of the Nomenclature of Spin Systems

Molecule	Notation
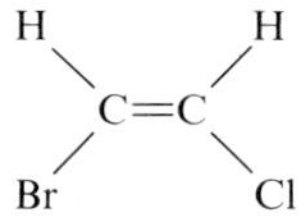 (H)(Br)C=C(H)(Cl)	AB
CH_2F_2	A_2X_2
CH_3CH_2Br	A_3B_2 or A_3X_2
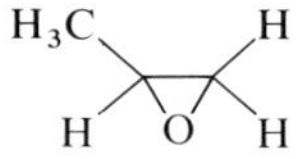 methyloxirane (H₃C, H, H, O, H)	$ABCX_3$
$BrCH_2CH_2Cl$	$AA'BB'$
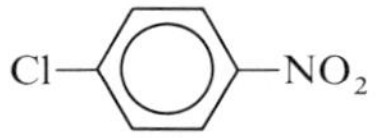 Cl–C₆H₄–NO₂	$AA'BB'$
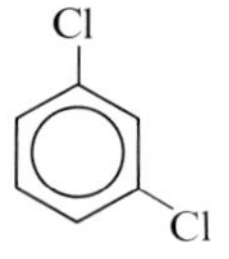 1,3-dichlorobenzene	AB_2C or AB_2X
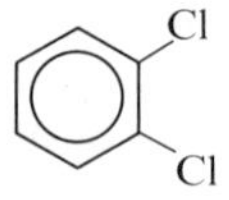 1,2-dichlorobenzene	$AA'BB'$
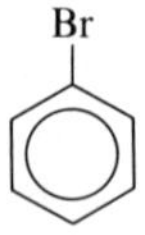 bromobenzene	$AA'BB'C$
$BrCH_2CH_2CH_2Cl$	$AA'BB'CC'$

magnetic nuclei are present in the molecule having chemical shifts separated from the other nuclei by large differences, these nuclei are referred to by letters at the end of the alphabet such as X, Y, Z. Equivalent nuclei are given the same letters of the alphabet.

Nuclei that are magnetically equivalent are given the same letter. For example, methyl iodide, CH_3I, is referred to as an A_3 spin system and methylene fluoride as an A_2X_2 spin system. The number in the subscript refers to the number of nuclei in the equivalent set.

Chemically equivalent nuclei are distinguished from each other by the use of primes as a superscript on the letter. Our example above of p-chloronitrobenzene is classified as an $AA'BB'$ spin system to indicate two sets of equivalent nuclei that are not magnetically equivalent. Additional examples of this notation for spin systems are given in Table 4.1.

This system of nomenclature is somewhat arbitrary in the choice of letters for the nuclei. For example the notation for CH_3CH_2Br could either be A_3X_2 or A_2X_3. No significance is attached to the relative order in which the nuclei are given. A useful convention might be to start at the high-field end of the spectrum and work towards the low-field end. About the only rule of thumb is that nuclei with different magnetogyric ratios should be distinguished by using letters at opposite ends of the alphabet (e.g., H—F is an AX spin system).

4.3 FIRST-ORDER ANALYSIS OF SPECTRA

Under certain conditions the coupling constants and chemical shifts may be extracted from the spectrum without carrying out a full mathematical analysis. However, it is important that one realizes the limitations of the first-order approach to analysis of an NMR spectrum. First-order analysis is applicable only to those spectra that may be classified as arising from magnetically equivalent nuclei or from cases where each nucleus has a different chemical shift. The second condition is that the chemical shift difference between two spin-coupled nuclei must be large compared to the coupling constant. As a general rule, the relationship $J/\delta < 0.05$ must be satisfied for a first-order analysis to be valid. When the previously stated conditions are met, first-order rules relating the number of peaks in the spectrum, the spacing of the peaks in a multiplet, and the intensities of the peaks in a multiplet are valid.

The number of peaks in a given multiplet is related to the number of nuclei coupled to the nuclei giving rise to the multiplet. Consider the case of two groups of magnetic equivalent nuclei A and B coupled to each other. The number of peaks in the multiplet due to A is given by $2nI + 1$ where n is the number of equivalent nuclei B and I is the spin number of B. Similarly, the number of peaks in the multiplet due to B is given by $2nI + 1$ where n is the number of equivalent nuclei A and I is the spin number of A. For those cases where $I = \frac{1}{2}$ (^{1}H, ^{19}F, ^{31}P), the number of peaks in A is $n + 1$ where n is the number of equivalent nuclei B. The spacing between the peaks in the A multiplet will be equal, and will be equal to the spacing between the peaks in the B multiplet. This spacing is the coupling constant J_{AB}. The chemical shift is the center of the multiplet. For nuclei with $I = \frac{1}{2}$, the intensities of the peaks in each multiplet are proportional to the coefficients of the binomial expansion $(X + 1)^n$, where n is the number of equivalent nuclei coupled to the nucleus in question (Table 4.2).

TABLE 4.2. Number of Peaks and Peak Intensities for First-Order Multiplets

Number of Adjacent Nuclei	Multiplet	Intensities
1	doublet	1 : 1
2	triplet	1 : 2 : 1
3	quartet	1 : 3 : 3 : 1
4	quintet	1 : 4 : 6 : 4 : 1
5	hextet	1 : 5 : 10 : 10 : 5 : 1
6	septet	1 : 6 : 15 : 20 : 15 : 6 : 1

As the chemical shifts become closer together, noticeable changes occur, first in the relative intensities of the peaks and, as the shifts become even closer, changes also occur in the number of peaks and in the spacing of the peaks. When $0.05 > J/\delta < 0.15$, the spectrum may still be interpreted as a first-order spectrum as far as the number of peaks in the multiplet and the spacing of the peaks is concerned. The intensities of the peaks will no longer be symmetrical about the center of the multiplet. The peaks on the sides between the two multiplets (inner peaks) will show increased intensities while the outer peaks will show decreased intensities. (Figure 4.3). For this type of spectrum the chemical shifts will no longer be at the center of the multiplets and the coupling constant may or may not be equal to the spacing between the peaks. One should always carry out a mathematical analysis for spectra of this type in order to obtain the true chemical shifts and coupling constants. When $J/\delta > 0.15$, the multiplets no longer appear at first-order spectra. The spacing of the peaks will most likely not be equal to the coupling constant. The intensities of the inner lines increase more as the outer lines decrease in intensity. Furthermore, additional peaks may appear in the spectrum. The complete interpretation of this type of spectrum always requires a full mathematical analysis.

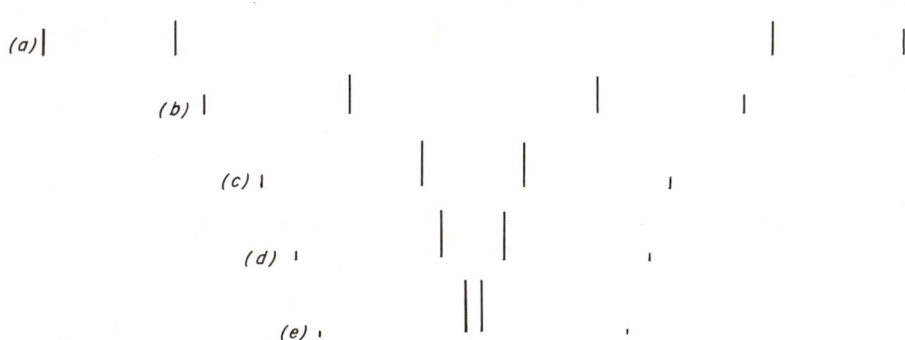

Fig. 4.3 A schematic illustration of the dependence of the intensities of an AB spectrum on the chemical shift difference δ_{AB}.

The spectra for spin systems involving nonequivalent nuclei may also be interpreted using first-order rules so long as the chemical shift differences are large compared to the coupling constants. Consider the case involving three spin $\frac{1}{2}$ nuclei AMX, where each nucleus is coupled to the other two. One way to construct the multiplets due to each nucleus is to use a graphical approach and consider the coupling to the other nuclei individually. For nucleus A, start by drawing a single line for the chemical shift of A in the absence of any coupling to the other nuclei (Figure 4.4). Next consider the coupling of M with A. Since M is a spin $\frac{1}{2}$ nucleus, coupling with A will split the line due to A into a doublet with spacing J_{AM}. Coupling of A with nucleus X will further split each peak of the doublet into doublets with spacing J_{AX}. The result is that the resonance of nucleus A will appear as a quartet with two different spacings, J_{AX}, J_{AM}, each repeated twice (Figure 4.4). Similarly, the resonance of M and X will also appear as quartets with spacings of J_{AM} and J_{MX} and J_{AX} and J_{MX}, respectively. The coupling constants are assigned on the basis that the repeated spacing (J) will appear in the resonances of both coupled nuclei.

The reverse of the previously outlined approach is used to analyze the first-order spectrum of an unknown compound. First identify the repeated spacings and construct a graph similar to that above in the reverse order.

Before quoting a spacing as a coupling constant, one must ensure that the spectrum is in fact first-order. If the intensities of the peaks are not exactly symmetrical in the multiplets, a full analysis should be carried out. In this case the spacings may be used as an approximation of the coupling constants in a computer analysis (Section 4.10). Finally, there are several situations where the spin system contains magnetically nonequivalent nuclei and yet, the spectrum approximates a first-order spectrum (deceptively simple

104   ANALYSIS OF SPECTRA AND DETERMINATION OF STRUCTURE

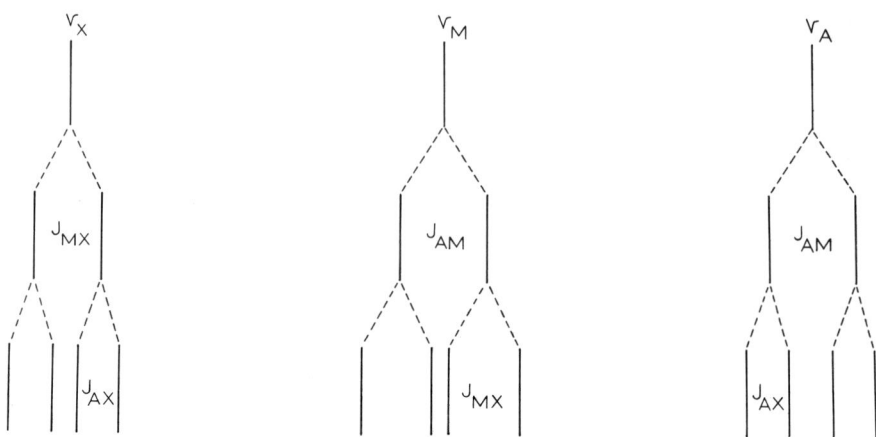

Fig. 4.4 A schematic representation of an AMX spectrum with the repeated spacings.

spectra, Section 4.7.3). The analysis of a spectrum of this type may prove difficult unless one can obtain the spectrum under conditions where the simplicity is removed.

4.4 ANALYSIS OF COMPLEX NMR SPECTRA

The first-order analysis of spectra previously outlined is limited in that it applies only to first-order spectra and not to more complex spectra. In this section we shall present a general method for the calculation of any type of NMR spectrum and for assigning each absorption in the spectrum to a specific transition based on a quantum mechanical formalism. If a set of chemical shifts and coupling constants can be found such that when they are used to calculate an NMR spectrum, they reproduce the experimental spectrum in transition frequencies and intensities, the set of parameters are the chemical shifts and coupling constants describing the experimental spectrum. We shall discuss certain types of spin systems where explicit equations have been derived that permit one to analyze the spectrum for values of the chemical shifts and coupling constants. A more detailed account of the analysis of NMR spectra may be found in a number of texts (1–5).

4.4.1 QUANTUM MECHANICAL FORMALISM

In order to calculate an NMR spectrum, it is necessary to first calculate the energy levels and stationary state wave functions for the spin system in the absence of the rf field. Transitions between the energy levels are induced by the applied rf field. The intensities of the transitions (transition probabilities) are usually calculated by perturbation methods (1). In our discussion to follow we shall consider the external magnetic field to be in the negative z direction.

The solution for the energy levels and the stationary state wave functions of the spin system requires the solution of the time-independent Schrödinger equation

$$\mathcal{H}\Psi = E\Psi \tag{4.7}$$

where $\mathcal{H}$ is the Hamiltonian operator, Ψ is the wave function, and E is the total energy of the spin system (1). By analogy with calculations of electronic energy levels in the HMO theory (6), we assume that Ψ can be expressed as a linear combination of functions ϕ_i such that

$$\Psi = \sum_i C_i \phi_i \tag{4.8}$$

where the C_i's are the weighting factors. If the ϕ_i's form an orthonormal and complete set, solution of equation 4.7 using the variational procedure yields a set of linear equations

$$C_N(\mathcal{H}_{mn} - E\delta_{mn}) = 0 \tag{4.9}$$

with

$$\mathcal{H}_{mn} = \int \phi_m \mathcal{H} \phi_n \, d\tau \tag{4.10}$$

where δ_{mn} is the Kronecker delta and is equal to 1 if $m = n$ and equal to zero if $m \neq n$.

For nontrivial solutions of equation 4.9, a necessary and sufficient condition is that the determinant of the coefficients, C_N, vanish

$$|\mathcal{H}_{mn} - E\delta_{mn}| = 0 \tag{4.11}$$

Equation 4.11 may be expanded in the form of a matrix (Hamiltonian matrix)

$$\begin{vmatrix} H_{11} - E_{11} & H_{12} & \cdots & H_{1n} \\ H_{21} & H_{22} - E_{22} & \cdots & H_{2n} \\ H_{n1} & H_{n2} & \cdots & H_{nn} - E_{nn} \end{vmatrix} = 0 \tag{4.12}$$

and the problem reduces to the evaluation of the matrix elements H_{mn} and the solution of the Hamiltonian matrix.

If it is assumed that the molecules are undergoing rapid rotation such that dipole–dipole interactions may be neglected, and the rf field is small such that saturation is avoided, a satisfactory Hamiltonian for the solution of equation 4.12 may be expressed as the sum of a field dependent term and a field independent term

$$\mathcal{H} = \mathcal{H}^\circ + \mathcal{H}' \tag{4.13}$$

The field-dependent term takes into account the interaction of the nuclear spins with the external magnetic field and is given by (1)

$$\mathcal{H}° = \sum_{i=1}^{n} v_i(\bar{I}_z)_i \qquad (4.14)$$

where v_i is the frequency of absorption of nucleus i expressed in hertz (chemical shift term) and $\bar{I}_z$ is the z component of the spin angular momentum. The field-independent term that represents the interaction between the nuclear spins is given by (1)

$$\mathcal{H}' = \sum_{i<j} J_{ij} \bar{I}_{(i)} \cdot \bar{I}_{(j)} \qquad (4.15)$$

where J_{ij} is the coupling constant between nuclei i and j expressed in hertz and $\bar{I}_i \cdot \bar{I}_j$ is the vector dot product of the spin angular momentum. The dot product is usually written in expanded form $\bar{I}_i \cdot \bar{I}_j = I_{ix}I_{jx} + I_{iy}I_{jy} + I_{iz}I_{jz}$ and the condition $i < j$ ensures that each coupling constant is counted only once.

Before proceeding with the evaluation of the Hamiltonian matrix it is convenient at this point to introduce a few additional concepts. For a spin system containing only spin $I = \frac{1}{2}$ nuclei, solution of the Schrödinger equation yields 2^N linear equations of the type given by equation 4.11 and therefore, a Hamiltonian matrix of order 2^N, where N is the total number of spin $I = \frac{1}{2}$ nuclei. Solution of equation 4.12 yields 2^N energy levels.

As pointed out in Chapter 2, for each spin $I = \frac{1}{2}$ nucleus, there are two states corresponding to the alignment of the nuclear spin angular momentum either with or opposed to the external magnetic field. The z component of the spin I_z has values of $\pm\frac{1}{2}$ in these states. We shall use the convention of writing these spin states as α and β when I_z is $+\frac{1}{2}$ and $-\frac{1}{2}$, respectively.

For a single nucleus, the two states α and β are eigenfunctions of equation 4.7. The simplest formulation of the functions ϕ_i for a spin system that yields a solution of equation 4.7 is that of the 2^N basic product functions (1)

$$\phi_i = \alpha_{(1)}\beta_{(2)}\alpha_{(3)}\beta_{(4)} \cdots \beta_{(N)} \qquad (4.16)$$

The nuclear designations are normally not included with the understanding that the ith spin state refers to the ith nucleus. The basic product functions represent the total number of ways in which the spin states of the nuclei may be arranged. It is convenient to arrange the basic products functions in terms of a total spin number F_z which represents the expectation value of the total spin component in the z direction (1).

$$F_z = \sum_i I_z(i) \qquad (4.17)$$

TABLE 4.3. Basic Product Functions for a Two-Spin System

F_z	Basic Product Function
+1	$\alpha\alpha$
0	$\alpha\beta$
0	$\beta\alpha$
−1	$\beta\beta$

The basic product functions for a spin system of two nonequivalent nuclei are given in Table 4.3.

In cases where two or more nuclei in a molecule are equivalent due to symmetry, the basic product functions as previously written may no longer belong to irreducible representations of the symmetry group. For example, in the case previously stated, interchanging the spins $\alpha_{(1)}\beta_{(2)} \rightarrow \alpha_{(2)}\beta_{(1)}$, where $F_z = 0$ results in a change in the basic product function such that it becomes equivalent with the remaining basic product function with $F_z = 0$. For cases where nuclei 1 and 2 are equivalent, it is necessary to construct a set of basic product functions that do belong to irreducible representations. These are called basic symmetry functions and are usually linear combinations of the basic product functions (1). They are further classified according to whether they are symmetrical or antisymmetrical with respect to interchange of the spins. The basic symmetry functions for a system of two magnetically equivalent nuclei are given in Table 4.4 (1). The basic symmetry functions for three magnetically equivalent nuclei, such as a freely rotating methyl group, are given in Table 4.5 (1).

TABLE 4.4. Basic Symmetry Functions for Two Equivalent Nuclei

F_z	ϕ	Symmetry
+1	$\alpha\alpha$	s
0	$(\alpha\beta + \beta\alpha)/\sqrt{2}$	s
0	$(\alpha\beta - \beta\alpha)/\sqrt{2}$	a
−1	$\beta\beta$	s

TABLE 4.5. Basic Symmetry Functions
for Three Equivalent Nuclei

F_z	ϕ	Symmetry
$\frac{3}{2}$	$\alpha\alpha\alpha$	s
$\frac{1}{2}$	$(\alpha\alpha\beta + \alpha\beta\alpha + \beta\alpha\alpha)/\sqrt{3}$	s
$\frac{1}{2}$	$(\alpha\alpha\beta + \alpha\beta\alpha - 2\beta\alpha\alpha)/\sqrt{6}$	a
$\frac{1}{2}$	$(\alpha\alpha\beta - \alpha\beta\alpha)/\sqrt{2}$	a
$-\frac{1}{2}$	$(\beta\beta\alpha + \beta\alpha\beta + \alpha\beta\beta)/\sqrt{3}$	s
$-\frac{1}{2}$	$(\beta\beta\alpha + \beta\alpha\beta - 2\alpha\beta\beta)/\sqrt{6}$	a
$-\frac{1}{2}$	$((\beta\beta) - \beta\alpha\beta)/\sqrt{2}$	a
$-\frac{3}{2}$	$\beta\beta\beta$	s

4.4.2 SOLUTION FOR A TWO-SPIN SYSTEM

Proceeding with the solution of the Hamiltonian matrix and using a two-spin system for illustrative purposes, the matrix may be written as

$$0 = \begin{vmatrix} \langle\alpha\alpha|\mathcal{H}|\alpha\alpha\rangle & \langle\alpha\alpha|\mathcal{H}|\alpha\beta\rangle & \langle\alpha\alpha|\mathcal{H}|\beta\alpha\rangle & \langle\alpha\alpha|\mathcal{H}|\beta\beta\rangle \\ \langle\alpha\beta|\mathcal{H}|\alpha\alpha\rangle & \langle\alpha\beta|\mathcal{H}|\alpha\beta\rangle & \langle\alpha\beta|\mathcal{H}|\beta\alpha\rangle & \langle\alpha\beta|\mathcal{H}|\beta\beta\rangle \\ \langle\beta\alpha|\mathcal{H}|\alpha\alpha\rangle & \langle\beta\alpha|\mathcal{H}|\alpha\beta\rangle & \langle\beta\alpha|\mathcal{H}|\beta\alpha\rangle & \langle\beta\alpha|\mathcal{H}|\beta\beta\rangle \\ \langle\beta\beta|\mathcal{H}|\alpha\alpha\rangle & \langle\beta\beta|\mathcal{H}|\alpha\beta\rangle & \langle\beta\beta|\mathcal{H}|\beta\alpha\rangle & \langle\beta\beta|\mathcal{H}|\beta\beta\rangle \end{vmatrix} \quad (4.18)$$

The order of the matrix may be reduced by using the selection rule that no mixing occurs between states that have different values of the total spin component F_z. With this rule the matrix becomes

$$\begin{vmatrix} \langle\alpha\alpha|\mathcal{H}|\alpha\alpha\rangle - E & 0 & 0 & 0 \\ 0 & \langle\alpha\beta|\mathcal{H}|\alpha\beta\rangle - E & \langle\alpha\beta|\mathcal{H}|\beta\alpha\rangle & 0 \\ 0 & \langle\beta\alpha|\mathcal{H}|\alpha\beta\rangle & \langle\beta\alpha|\mathcal{H}|\beta\alpha\rangle - E & 0 \\ 0 & 0 & 0 & \langle\beta\beta|\mathcal{H}|\beta\beta\rangle - E \end{vmatrix} = 0 \quad (4.19)$$

Using the complete Hamiltonian, the matrix elements are now evaluated. Each of the terms in equation 4.13 may be evaluated separately and summed together. The field dependent part of H_{11} becomes

$$\left\langle \alpha\alpha \middle| \sum_{i=1}^{N} v_i(I_z)_i \middle| \alpha\alpha \right\rangle \quad (4.20)$$

Since $I_{z(i)}$ equals $\frac{1}{2}$ if nucleus i has α spin and $-\frac{1}{2}$ if nucleus i has β spin, equation 4.20 yields $v_1/2 + v_2/2$ (1). The field-independent term (equation

4.13) is given by

$$\left\langle \alpha\alpha \left| \sum_{i=j} J_{ij} \bar{I}_i \cdot \bar{I}_j \right| \alpha\alpha \right\rangle \quad (4.21)$$

For diagonal elements, the solution of 4.21 yields $+\frac{1}{4} J_{ij}$ if nuclei i and j have parallel spins and $-\frac{1}{4} J_{ij}$ if nuclei i and j have antiparallel spins. Thus the $\mathscr{H}^1$ part of H_{11} yields $+\frac{1}{4} J_{12}$. The total matrix element H_{11} is (1)

$$H_{11} = \frac{v_1 + v_2}{2} + \frac{J_{12}}{4} \quad (4.22)$$

Similarly, the solution of the remaining diagonal matrix elements yields (1)

$$H_{22} = \frac{v_1 - v_2}{2} - \frac{J_{12}}{4} \quad (4.23)$$

$$H_{33} = \frac{-v_1 + v_2}{2} - \frac{J_{12}}{4} \quad (4.24)$$

$$H_{44} = \frac{-v_1 - v_2}{2} + \frac{J_{12}}{4} \quad (4.25)$$

The off-diagonal matrix elements do not contain a contribution from $\mathscr{H}^\circ$. The matrix element H_{23} becomes

$$H_{23} = \langle \alpha\beta | \sum_{i<j} J_{ij} \bar{I}_i \cdot \bar{I}_j | \beta\alpha \rangle \quad (4.26)$$

Solution of terms like equation 4.26 between different basic product functions yields $\frac{1}{2} J_{ij}$ if ϕ_m and ϕ_n differ only in the interchange of spins i and j and otherwise equals zero. Thus the solution of H_{23} yields

$$H_{23} = \tfrac{1}{2} J_{12} \quad (4.27)$$

Substitution of these values into equation 4.19 gives

$$\begin{vmatrix} \frac{v_1+v_2}{2}+\frac{J_{12}}{4}-E & 0 & 0 & 0 \\ 0 & \frac{v_1-v_2}{2}-\frac{J_{12}}{2}-E & \frac{J_{12}}{2} & 0 \\ 0 & \frac{J_{12}}{2} & \frac{-v_1+v_2}{2}-\frac{J_{12}}{2}-E & 0 \\ 0 & 0 & 0 & \frac{-v_1-v_2}{2}+\frac{J_{12}}{4}-E \end{vmatrix} = 0 \quad (4.28)$$

Two of the energy levels E_1 and E_2 are given by the terms H_{11} and H_{44}, respectively. The remaining two energy levels can be found from the solution of the 2 × 2 submatrix requiring the solution of the quadratic equation.

$$E^2 + \frac{J_{12}E}{2} - \frac{(v_1^2 + v_2^2)}{1} = 0 \tag{4.29}$$

Equation 4.29 may be readily solved to yield (1)

$$E_2 = \tfrac{1}{2}[(v_1 - v_2)^2 + J_{12}^2]^{\frac{1}{2}} - \tfrac{1}{4}J_{12}$$
$$E_3 = \tfrac{1}{2}[(v_1 - v_2)^2 + J_{12}^2]^{\frac{1}{2}} + \tfrac{1}{4}J_{12} \tag{4.30}$$

Once the energy levels have been found, they can be used to evaluate the coefficients C_N, and in turn, the corresponding wave functions. Using the selection rule that transitions occur only between states that have values of the total spin component F_z differing by ± 1, the frequencies of the allowed transitions may be calculated. The intensities of the allowed transitions (from state r to state s) are proportional to the square of the integral

$$M_{rs} = \int \phi_r \sum_{i=1}^{N} v_i(I_x)_i \phi_r \, d\tau \tag{4.31}$$

We shall not go into further details of the calculation of relative intensities. This subject has been treated in several texts (1,2) and the reader is referred to these for more detail.

Further reduction in the order of the submatrix to be evaluated may be made in certain cases. If the chemical shift difference between two nuclei is large compared to the coupling constant between the nuclei, the coupling constant between the two nuclei may be dropped in the off-diagonal elements (the X approximation). Furthermore, if the spin system contains different kinds of nuclei such as proton and a fluorine, the spin system can be considered in terms of the total spin component $F_z(H)$ and $F_z(F)$. To a high degree of approximation, no mixing occurs between functions that differ in the total spin components $F_z(H)$ and $F_z(F)$. Thus mixing occurs only when either $F_z(H) = \pm 1$ and $F_z(F)$ is constant or $F_z(H)$ is constant and $F_z(F) = \pm 1$ (1).

The procedure previously outlined may be used to derive the Hamiltonian matrix for any spin system. If the matrix can be diagonalized, the energy levels can be found and expressions for the transition frequencies derived. In many cases, however, reduction of the matrix leaves submatrices to be solved which are larger than 2 × 2. The solution of submatrices which are 3 × 3 or larger may prove to be too difficult to do by hand and furthermore, may not provide unique solutions. If this is the case, one can attempt to find solutions by trial and error using "guessed" values for the chemical shifts and coupling constants. Alternatively, computer programs are available to diagonalize a matrix using various approximations. The use of computer programs in the analysis of NMR spectra is discussed in Section 4.10.

4.4.3 SUMMARY OF PROCEDURE FOR CALCULATING SPECTRA

Before proceeding with the discussion of spin systems where explicit expressions have been derived for the energy levels, transition frequencies, and intensities, it is useful at this point to summarize the procedure used to calculate NMR spectra (1).

1. A complete set of basic product functions of the type $\phi_n = \alpha\beta\beta\alpha''''$ etc., is used to derive the matrix elements of the Hamiltonian matrix. Where appropriate, basic symmetry functions are used to take into account the symmetry of the spin system.

2. The part of the Hamiltonian $\mathscr{H}°$ representing the interaction of the external magnetic field with the spin system has values other than zero only for the diagonal matrix elements. This contribution is given by

$$\langle\phi_m|\mathscr{H}°|\phi_m\rangle = \sum_i v_i[(I_z)_i)] \tag{4.32}$$

where $(I_z)_i$ is $+\frac{1}{2}$ if nucleus i has α spin and $-\frac{1}{2}$ if nucleus i has β spin.

3. The part of the Hamiltonian $\mathscr{H}^1$ representing the interaction among the spins themselves may contribute to both diagonal and off-diagonal matrix elements. This contribution is given by

$$\langle\phi_m|\mathscr{H}'|\phi_m\rangle = \tfrac{1}{4}\sum_{i<j} J_{ij}T_{ij} \tag{4.33}$$

and

$$\langle\phi_m|\mathscr{H}'|\phi_n\rangle = \tfrac{1}{2}UJ_{ij} \tag{4.34}$$

where T_{ij} equals 1 if spins i and j are parallel in ϕ_m and equals -1 if spins i and j are antiparallel and U equals 1 if ϕ_m and ϕ_n differ only in the interchange of spin i and j and is zero otherwise. For basic product functions containing more than two spins, these matrix elements are evaluated by expansion.

4. The complete matrix may be reduced to several submatrices by using the rules that (1) no mixing occurs between basic product functions with different values of the total spin component, F_z, (2) no mixing occurs between basic product functions of different symmetry, and (3) no mixing occurs between basic product functions that differ in the value of the total spin components $F_z(A)$, $F_z(B)$ when the spin system contains two or more kinds of nuclei A and B. The last part also applies to spin systems containing the same type of nuclei if the chemical shift difference is large compared to the coupling constants between the nuclei.

5. The energies can be calculated by diagonalization of the submatrices of the total Hamiltonian matrix. Wavefunctions and coefficients may then be calculated.

6. Transitions are calculated as the frequency difference between the various energy levels using the selection rule that transitions occur only between states differing in F_z by ± 1. For symmetrical molecules, transitions occur between energy levels of states having the same symmetry.

7. For spin systems covered by part three of step 4, transitions occur between states in which only one of $F_z(A)$ or $F_z(B)$, differ by ± 1.

8. The relative intensities of the allowed transitions may be calculated using equation 4.31.

4.5 ANALYSIS OF SPECIFIC SPIN SYSTEMS

The method previously presented may be used to calculate any NMR spectrum in theory. From matching of a calculated spectrum with the experimental spectrum, the chemical shifts and coupling constants of the experimental spectrum may be found. Alternatively, a spectrum may be analyzed in certain cases from a knowledge of the transition frequencies. Since the transition frequencies represent the energy difference between two energy levels, and the energy levels are written in terms of the chemical shifts and coupling constants, it should be possible to derive expressions for the transition frequencies in terms of the chemical shifts and coupling constants. From a knowledge of the transition frequencies, it should therefore be possible to analyze a spectrum for the chemical shifts and coupling constants by taking the sum and differences of the expressions for the transition frequencies.

In actual practice, it may not prove possible to analyze every NMR spectrum. As pointed out previously, if the diagonalization of the Hamiltonian matrix involves submatrices larger than 2×2, it is not possible to obtain unique expressions for the energy levels and hence, unique expressions for the transition frequencies. For these cases, the only recourse is to use one of the exciting computer programs for the analysis of the spectrum (Section 4.10). Even here, the analysis may be difficult or impossible in practice if one is dealing with an extremely complex spectrum.

Fortunately, for many of the common spin systems encountered in NMR spectroscopy, unique expressions for the transition frequencies in terms of the chemical shifts and coupling constants have been derived thereby permitting the analysis of the spectrum (1–5,7). In the following sections, we shall present the expressions for several common spin systems and illustrate their use in the analysis of spectra.

4.6 TWO-SPIN SYSTEMS

4.6.1 TWO EQUIVALENT NUCLEI (A_2)

As pointed out previously, the spectrum of a group of magnetically equivalent nuclei consists of only a single line. The only parameter to be derived from the spectrum is the chemical shift. Since the spectrum does not depend upon the coupling constant between magnetically equivalent nuclei, many users of NMR spectroscopy have the misconception that there is no coupling between magnetically equivalent nuclei. In order to show that this is not the case, we present the energy levels and transition frequencies for the case of two magnetically equivalent nuclei. A method for obtaining the coupling constant between magnetically equivalent nuclei is discussed in Section 4.11.2.

The basic symmetry functions and diagonal matrix elements for two magnetically equivalent nuclei are given in Table 4.6 (1). Since there is no mixing between basic product functions of different symmetry, there are no off-diagonal matrix elements and the diagonal matrix elements are themselves eigenvalues. Using the selection rules that transitions are allowed when $\Delta F_z = \pm 1$ and that transitions are allowed only between basic product functions of the same symmetry, the allowed transitions and their energies are given in Table 4.7 (1). There are two allowed transitions, each with an energy of v_A. Therefore the spectrum consists of a single line centered at v_A, and does not depend on the coupling constant between the two nuclei due to the cancellation of the coupling constant term when the energy level differences are taken to find the energies of the transitions.

TABLE 4.6. Basic Symmetry Functions and Diagonal Matrix Elements for Two Nuclei (A_2)

Symmetry	Basic Symmetry Function	Diagonal Matrix Element H_{mm}
s_{+1}	$\alpha\alpha$	$v_A + \dfrac{J}{4}$
s_0	$\dfrac{\alpha\beta + \beta\alpha}{\sqrt{2}}$	$+\dfrac{J}{4}$
a_0	$\dfrac{\alpha\beta - \beta\alpha}{\sqrt{2}}$	0
s_{-1}	$\beta\beta$	$-v_A + \dfrac{J}{4}$

TABLE 4.7. Transition Energies for Two Nuclei (A_2)

Transition	Energy
$S_{-1} \to S_0$	ν_A
$S_0 \to S_{+1}$	ν_A

4.6.2 TWO NUCLEI (AB)

For the general two-spin ($I = \frac{1}{2}$) case, there are four possible transitions and each one is an allowed transition. Three parameters may be obtained from the spectrum ν_A, ν_B, and J_{AB}. The energy levels for a two-spin system were calculated in Section 4.4.2 and are tabulated in Table 4.8. To simplify the expressions we define δ as

$$\delta = \nu_A - \nu_B \qquad (4.35)$$

and C as

$$2C = [\delta^2 + J_{AB}^2]^{\frac{1}{2}} \qquad (4.36)$$

If we refer the transitions to $(\nu_A + \nu_B)/2$, allowed transitions and their energies and intensities are given in Table 4.9.

The AB spectrum consists of four lines where the outer lines are weaker in intensity than the inner lines (Figure 4.5). The separation between lines 1 and 3 and between lines 2 and 4 is equal to J_{AB}. The separation between lines 1 and 4 is equal to $2C + J_{AB}$ and between lines 3 and 2 to $2C - J_{AB}$. Thus, by working backwards, it is possible to obtain δ and J_{AB} from the line positions. It should be pointed out that the chemical shifts ν_A and ν_B are not, in general, equal to the midpoints of the doublets 1 and 3 and 2 and 4 but instead, are displaced towards lines 3 and 2 slightly.

TABLE 4.8. Energy Levels for Two Nuclei (AB)

n	E_r
1	$\dfrac{\nu_A + \nu_B}{2} + \dfrac{J_{AB}}{4}$
2	$\frac{1}{2}[(\nu_A - \nu_B)^2 + J_{AB}^2]^{\frac{1}{2}} - \dfrac{J_{AB}}{4}$
3	$-\frac{1}{2}[(\nu_A - \nu_B)^2 + J_{AB}^2]^{\frac{1}{2}} - \dfrac{J_{AB}}{4}$
4	$\dfrac{-(\nu_A + \nu_B)}{2} + \dfrac{J_{AB}}{4}$

TWO-SPIN SYSTEMS

TABLE 4.9. Transition Frequencies and Intensities for Two Nuclei (AB)

Line	Transition	Frequency	Relative Intensity
1	$E_3 \to E_1$	$C + \dfrac{J_{AB}}{2}$	$1 - \dfrac{J}{2C}$
2	$E_2 \to E_1$	$-C + \dfrac{J_{AB}}{2}$	$1 + \dfrac{J}{2C}$
3	$E_4 \to E_2$	$C - \dfrac{J_{AB}}{2}$	$1 + \dfrac{J}{2C}$
4	$E_4 \to E_3$	$-C - \dfrac{J_{AB}}{2}$	$1 - \dfrac{J}{2C}$

One cannot determine which lines are due to nucleus A and which are due to nucleus B from the spectrum. Furthermore, the sign of the coupling constant J_{AB} cannot be obtained from the spectrum. A reversal in the sign of J_{AB} reverses the labeling of the transitions 1 and 3 and 2 and 4. The appearance of the AB spectrum is determined only by the ratio of J and δ (Figure 4.3). As J becomes small with respect to δ, J_{AB}^2 becomes small and can be ignored, and the midpoints of the doublets 1 and 3 and 2 and 4 become the true chemical shifts. Furthermore, the intensities of all four lines become equal. This is the AX spin system, which is equivalent to dropping the off-diagonal matrix elements.

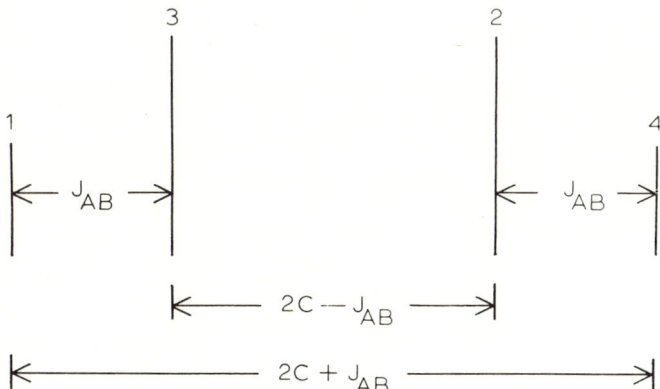

Fig. 4.5 A schematic illustration of an AB spectrum with the line numbering and spacings needed for the analysis.

ANALYSIS OF SPECTRA AND DETERMINATION OF STRUCTURE

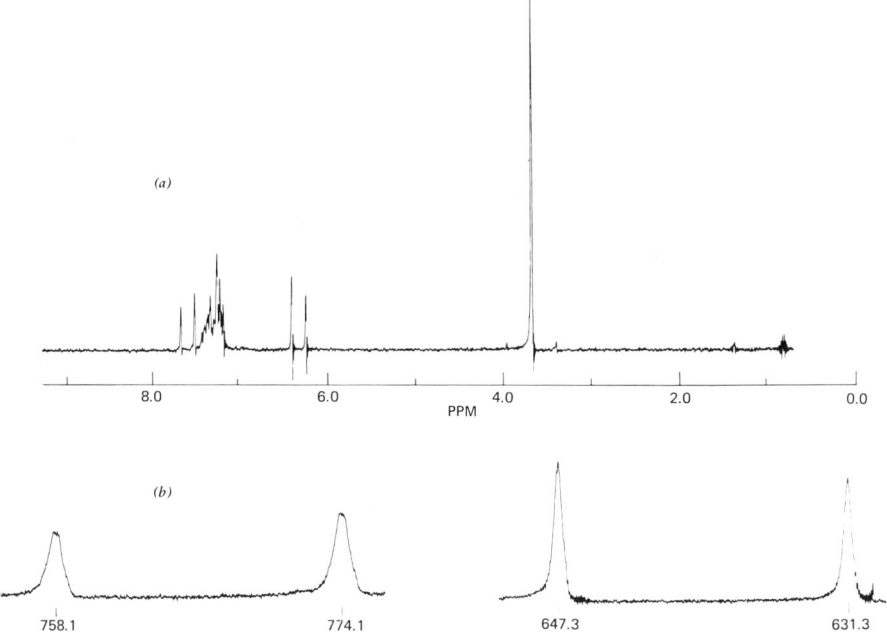

Fig. 4.6 (a) The 100-MHz NMR spectrum of *trans*-methylcinnamate. (b) An expansion of the AB region of the spectrum.

As an example of the analysis of a two-spin system, consider the spectrum in Figure 4.6.

$$|J| = 1\text{-}3 = 2\text{-}4$$
$$= 774.1 - 758.1 = 647.3 - 631.3 = 16.0 \text{ Hz}$$
$$\delta^2 = (\nu_A - \nu_B)^2 = 4C^2 - J^2$$
$$\delta = \nu_A - \nu_B = (4C^2 - J^2)^{\frac{1}{2}}$$
$$= [(2C - J)(2C + J)]^{\frac{1}{2}}$$
$$= [(3 - 2)(1 - 4)]$$
$$= [(758.1 - 647.3)(774.1 - 631.3)]^{\frac{1}{2}}$$
$$= [(110.8)(142.8)]^{\frac{1}{2}}$$
$$= 125.8 \text{ Hz}$$
$$\nu_A + \nu_B = (3 + 2) = (1 + 4)$$
$$= (758.1 + 647.3)$$
$$= 1405.4 \text{ Hz}$$

$$(v_A - v_B) + (v_A + v_B) = 125.8 + 1405.4$$

$$2v_A = 1530.2$$

$$v_A = 765.1 \text{ Hz}$$

$$v_B = 1405.4 - 765.1$$

$$= 640.3 \text{ Hz}$$

4.7 THREE-SPIN SYSTEMS

For the general three-spin system, the basic product functions are the various combinations of α and β for each nucleus

$$
\begin{array}{cccc}
 & & & F_z \\
 & \alpha\alpha\alpha & & +\tfrac{3}{2} \\
\alpha\alpha\beta & \alpha\beta\alpha & \beta\alpha\alpha & +\tfrac{1}{2} \\
\alpha\beta\beta & \beta\alpha\beta & \beta\beta\alpha & -\tfrac{1}{2} \\
 & \beta\beta\beta & & -\tfrac{3}{2}
\end{array}
\quad (4.37)
$$

A maximum of fifteen transitions are possible for the general three-spin case. Three of the transitions are combination transitions resulting from the simultaneous change in I_z of all three spins (i.e., $\beta\alpha\beta \rightarrow \alpha\beta\alpha$). These transitions are usually weak in intensity and are not observed unless the chemical shift difference between two of the nuclei is of the same order of magnitude as the coupling constant between them (strongly coupled case).

The diagonalization of the Hamiltonian matrix for a general three-spin system involves the solution of cubic equations for the 3×3 submatrices due to mixing of the basic product functions when $F_z = +\tfrac{1}{2}$ and $-\tfrac{1}{2}$. While methods have been presented for the analysis of a general three-spin system (8,9) the solution may not be unique and may lead to several sets of parameters. There are three types of three-spin systems where the Hamiltonian matrix may be reduced such that explicit expressions for the energy levels and hence explicit expressions for the transition frequencies have been derived. These cases result from the presence of symmetry or from ignoring certain off-diagonal elements due to the chemical shift differences being large with respect to the coupling constants (the X approximation). Each of these cases are discussed in the following, with emphasis on the parameters obtained from analysis of the spectra and the precautions that should be taken in the analysis.

4.7.1 THREE NUCLEI (AB_2 OR AX_2)

The AB_2 spin system results when two of the three nuclei are magnetically equivalent. Using the basic symmetry functions given in Table 4.4 for the B

TABLE 4.10. Basic Product Functions for Three Nuclei (AB_2)

Symmetry	F_z	Function
s	$\frac{3}{2}$	$\alpha\alpha\alpha$
s	$\frac{1}{2}$	$\beta\alpha\alpha$
s	$\frac{1}{2}$	$\dfrac{\alpha(\alpha\beta + \beta\alpha)}{\sqrt{2}}$
s	$-\frac{1}{2}$	$\dfrac{\beta(\alpha\beta + \beta\alpha)}{\sqrt{2}}$
s	$-\frac{1}{2}$	$\alpha\beta\beta$
s	$-\frac{3}{2}$	$\beta\beta\beta$
a	$\frac{1}{2}$	$\dfrac{\alpha(\alpha\beta - \beta\alpha)}{\sqrt{2}}$
a	$-\frac{1}{2}$	$\dfrac{\beta(\alpha\beta - \beta\alpha)}{\sqrt{2}}$

nuclei and α or β for the A nucleus, the basic product functions for the AB_2 system are given in Table 4.10. The presence of symmetry reduces the Hamiltonian matrix due to the rule that no mixing occurs between product functions of different symmetry. Four of the energy levels are given by the diagonal matrix elements for $F_z = \frac{3}{2}, -\frac{3}{2}, +\frac{1}{2}$ symmetric, and $-\frac{1}{2}$ antisymmetric. The remaining four energy levels may be obtained by solving the two 2×2 submatrices from the symmetric functions for $F_z = +\frac{1}{2}$ and $-\frac{1}{2}$. In order to simplify the expressions, the following are defined (1):

$$C_+ \cos 2\phi_+ = \frac{\nu_A - \nu_B}{2} + \frac{J_{AB}}{4}$$

$$C_+ \sin 2\phi_+ = \frac{J_{AB}}{\sqrt{2}}$$

$$C_- \cos 2\phi_- = \frac{\nu_A - \nu_B}{2} - \frac{J_{AB}}{4} \quad (4.38)$$

$$C_- \sin 2\phi_- = \frac{J_{AB}}{\sqrt{2}}$$

Since $\delta_{AB} = \nu_A - \nu_B$, the following expressions may be obtained

$$C_+ = \tfrac{1}{2}[\delta_{AB}^2 + \delta_{AB}J_{AB} + \tfrac{9}{4}J_{AB}^2]^{\frac{1}{2}}$$
$$C_- = \tfrac{1}{2}[\delta_{AB}^2 - \delta_{AB}J_{AB} + \tfrac{9}{4}J_{AB}^2]^{\frac{1}{2}} \quad (4.39)$$

TABLE 4.11. Transitions and Relative Intensities for Three Nuclei (AB_2)

Transition	Origin	Frequency	Relative Intensity
1	A	$v_A + \frac{v_B}{2} + \frac{3}{4}J_{AB} + C_+$	$(\sqrt{2} \sin \phi_+ - \cos \phi_+)^2$
2	A	$v_B + C_+ + C_-$	$[\sqrt{2} \sin (\phi_+ - \phi_-) + \cos \phi_+ \cos \phi_-]$
3	A	v_A	1
4	A	$v_A + \frac{v_B}{2} - \frac{3}{4}J_{AB} + C_-$	$(\sqrt{2} \sin \phi_- + \cos \phi_-)^2$
5	B	$v_B + C_+ - C_-$	$[\sqrt{2} \cos (\phi_+ - \phi_-) + \cos \phi_+ \sin \phi_-]^2$
6	B	$v_A + \frac{v_B}{2} + \frac{3}{4}J_{AB} - C_+$	$(\sqrt{2} \cos \phi_+ + \sin \phi_+)^2$
7	B	$v_B - C_+ + C_-$	$[\sqrt{2} \cos (\phi_+ - \phi_-) - \sin \phi_+ \cos \phi_-]^2$
8	B	$v_A + \frac{v_B}{2} - \frac{3}{4}J_{AB} - C_-$	$[\sqrt{2} \cos \phi_- \sin \phi_-]^2$
9	Comb.	$v_B - C_+ - C_-$	$[\sqrt{2} \sin (\phi_+ - \phi_-) + \sin \phi_+ \sin \phi_-]^2$

The allowed transitions for $v_B > v_A$ and their intensities are given in Table 4.11 (1). The transitions are labeled according to which nucleus is changing spin orientation in the limit of large chemical shift differences. A maximum of nine allowed transitions are possible for the AB_2 case. The transition 9 is a combination transition and is usually not observed due to its small intensity unless the chemical shift difference δ is of the same order of magnitude as the coupling constant J_{AB}. Line 3 is the transition between the antisymmetric energy levels. Like the A_2 case, the transition frequencies do not depend on J_{BB} and hence J_{BB} cannot be determined from the spectrum. The AB_2-spin system is similar to the AB-spin system in that the spectrum does not depend on the sign of J_{AB}. Furthermore, the appearance of the spectrum is strongly dependent on the ratio J_{AB}/δ_{AB}. Relationships for the intensities and frequencies of the lines as a function of J_{AB}/δ have been presented (3). The A_2B-spin system is just the mirror image of the AB_2 system and may be analyzed by reversing the line assignments. The intensities are identical to those given in Table 4.11 whereas the line frequencies are obtained by interchanging A and B and multiplying each term after the chemical shift term by -1 in Table 4.11.

Line assignments can usually be made without any difficulty. Inspection of Table 4.11 gives the following frequency relationships (7):

$$(v_1 - v_3) = (v_2 - v_4) = (v_5 - v_8)$$
$$(v_1 - v_2) = (v_3 - v_4) = (v_6 - v_7) \quad (4.40)$$
$$(v_1 - v_3) > (v_1 - v_2)$$

Transition 3 gives the chemical shift v_A directly. The chemical shift of v_B and J_{AB} are obtained from the following frequency relationships:

$$v_B = \frac{v_5 + v_7}{2}$$

$$J_{AB} = \frac{(v_1 - v_4) + (v_6 - v_8)}{3} \qquad (4.41)$$

As an example of the analysis of an A_2B spectrum, consider the spectrum of 2,6-dichlorotoluene, given in Figure 4.7. The analysis is as follows:

$$v_A = v_3 = 697.45 \text{ Hz}$$

$$v_B = \frac{v_5 + v_7}{2} = \frac{715.9 + 723.05}{2}$$

$$= 719.47 \text{ Hz}$$

$$|J_{AB}| = \frac{(v_4 - v_1) + (v_8 - v_6)}{3}$$

$$= \frac{(703.75 - 688.1) + (725.2 - 716.7)}{3}$$

$$= 8.38 \text{ Hz}$$

4.7.2 THREE NUCLEI (AMX)

The AMX notation applies to a system of three spins in which the coupling constants between the nuclei are small compared to the chemical shift differences between the nuclei (weakly coupled case). The basic product functions for the AMX case are those given in equation 4.37 for the general three-spin case. In setting up the Hamiltonian matrix for the AMX case, use is made of the X approximation for all three nuclei, All off-diagonal matrix elements are dropped and the energy levels are given by the diagonal elements. There are twelve transitions for the AMX case and their frequencies are given in Table 4.12.

The AMX-spin system is the simplest of all three-spin systems to analyze as it is considered a first-order spectrum. Before a spectrum is classified as an AMX system, care should be exercised to insure that it is in fact an AMX spectrum. The spectrum (Figure 4.4) consists of three multiplets of four lines each in which the intensity of each of the four lines is equal. The chemical shifts are given by the center of each multiplet. The coupling constants are

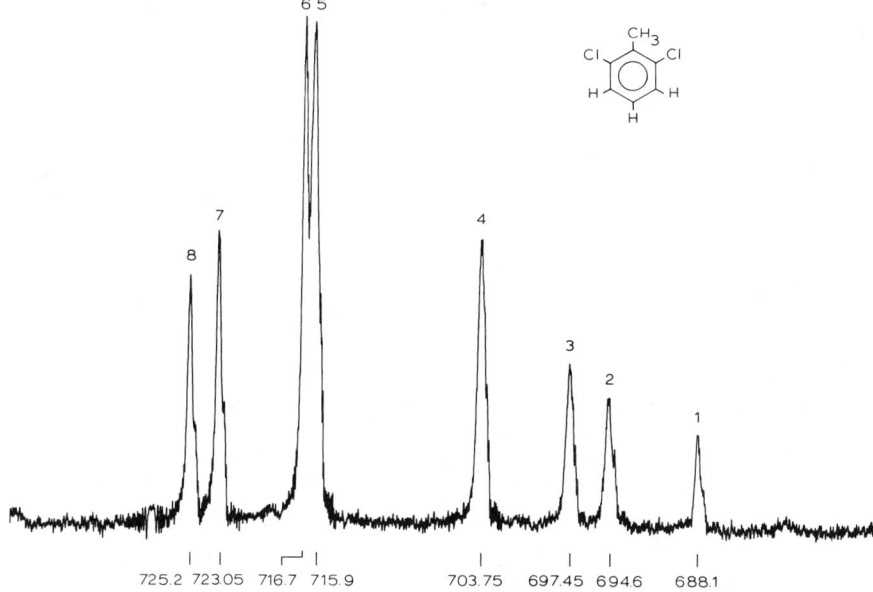

Fig. 4.7 An expansion of the A_2B portion of the 100-MHz NMR spectrum of 2,6-dichlorotoluene. This spectrum was obtained while decoupling the methyl protons.

TABLE 4.12. Transition Frequencies for Three Nuclei (AMX)

n	Origin	Frequency
1	A	$\nu_A - \frac{1}{2}(J_{AM} + J_{AX})$
2	A	$\nu_A - \frac{1}{2}(J_{AM} - J_{AX})$
3	A	$\nu_A + \frac{1}{2}(J_{AM} - J_{AX})$
4	A	$\nu_A + \frac{1}{2}(J_{AM} + J_{AX})$
5	M	$\nu_M - \frac{1}{2}(J_{AM} + J_{MX})$
6	M	$\nu_M - \frac{1}{2}(J_{AM} - J_{MX})$
7	M	$\nu_M + \frac{1}{2}(J_{AM} - J_{MX})$
8	M	$\nu_M + \frac{1}{2}(J_{AM} + J_{MX})$
9	X	$\nu_X - \frac{1}{2}(J_{AX} + J_{MX})$
10	X	$\nu_X - \frac{1}{2}(J_{AX} - J_{MX})$
11	X	$\nu_X + \frac{1}{2}(J_{AX} - J_{MX})$
12	X	$\nu_X + \frac{1}{2}(J_{AX} + J_{MX})$

obtained by finding the repeated spacings in each pair of multiplets. The relative signs of the coupling constants cannot be obtained from the spectrum.

$$v_A = \frac{v_3 + v_2}{2}$$

$$v_M = \frac{v_7 + v_6}{2}$$

$$v_X = \frac{v_{11} + v_{12}}{2} \qquad (4.42)$$

$$J_{AM} = (v_3 - v_1) = (v_4 - v_2) = (v_7 - v_5) = (v_8 - v_6)$$

$$J_{AX} = (v_2 - v_1) = (v_4 - v_3) = (v_{11} - v_9) = (v_{12} - v_{10})$$

$$J_{MX} = (v_6 - v_5) = (v_8 - v_7) = (v_{10} - v_9) = (v_{12} - v_{11})$$

In some cases, it may be tempting to classify a spectrum as AMX when the intensities of the lines in each multiplet are not exactly equal. If the chemical shifts of two of the nuclei move closer together (the ABX case in Figure 4.8) changes in the intensities and frequency relationships occur. Any attempt to analyze this spectrum as an AMX case will lead to erroneous results for both the chemical shifts and coupling constants since equation 4.22 is no longer valid. Similarly, when one four line pattern is observed but the other two patterns are obscured by overlap with other resonances, one should never attempt to analyze the observed four line pattern as the X part of an AMX spectrum. The errors associated with this are discussed along with the ABX-spin systems.

4.7.3 THREE NUCLEI (ABX)

The ABX classification is applied to a three-spin system where the resonance of one of the nuclei is well separated from the other two. The appearance of the ABX spectrum is similar to the AMX spectrum in that three multiplets are usually observed. However, the intensities of the lines in each multiplet are usually not equal (Figure 4.8). The ABX spectrum is one of the more common spin systems in NMR spectroscopy. As a result, it is the most common spin system in which errors are made in the analysis. The analysis of ABX systems and problems associated with their analysis have been reviewed (7,10). Usually, one can obtain the three chemical shifts v_A, v_B, and v_X and the three coupling constants J_{AB}, J_{AX}, and J_{BX} from analysis of the spectrum. In most cases, however, information about the relative signs of the coupling constants cannot be obtained without additional experiments.

The basic product functions given in equation 4.37 are used to set up the Hamiltonian matrix for the ABX spin system. Use is made of the X ap-

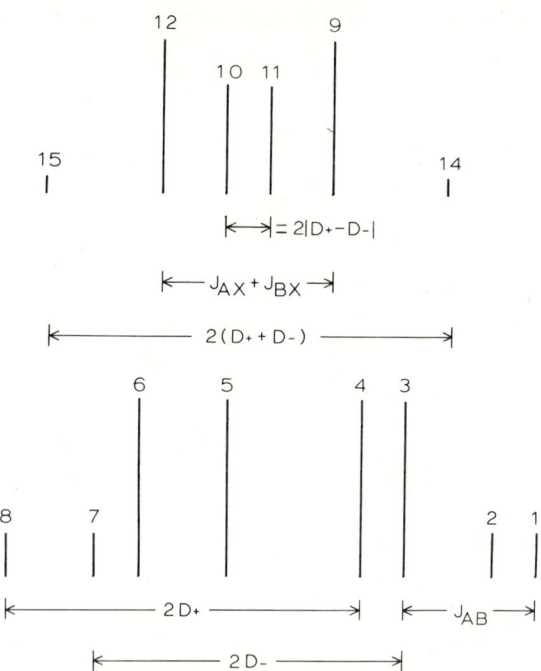

Fig. 4.8 A schematic drawing of an *ABX* spectrum with the repeated spacings

proximation to reduce the matrix. Thus mixing occurs only for functions which have the same $F_z(X)$. The function $\alpha\alpha\beta$ and $\beta\beta\alpha$ are not mixed with the other functions with $F_z = +\frac{1}{2}$ and $-\frac{1}{2}$, respectively, and they become wavefunctions for the *ABX* case. This reduces the problem to the solution of two 2×2 submatrices where the only off-diagonal element is $\frac{1}{2}J_{AB}$. Explicit expressions may be derived (1) for the energy levels yielding the transitions given in Table 4.13 for $v_A > v_B$, where the positive quantities D_+, D_- and angles ϕ_+ and ϕ_- are defined as (1)

$$D_+ \cos 2\phi_+ = \tfrac{1}{2}(v_A - v_B) + \tfrac{1}{4}(J_{AX} - J_{BX})$$
$$D_+ \sin 2\phi_+ = \tfrac{1}{2}J_{AB}$$
$$D_- \cos 2\phi_- = \tfrac{1}{2}(v_A - v_B) - \tfrac{1}{4}(J_{AX} - J_{BX}) \qquad (4.43)$$
$$D_- \sin 2\phi_- = \tfrac{1}{2}J_{AB}$$
$$D_\pm = \tfrac{1}{2}\{[v_A - v_B \pm \tfrac{1}{2}(J_{AX} - J_{BX})]^2 + J_{AB}^2\}^{\frac{1}{2}}$$
$$v_{AB} = \frac{v_A + v_B}{2}$$

124   ANALYSIS OF SPECTRA AND DETERMINATION OF STRUCTURE

TABLE 4.13. Transition Frequencies and Relative Intensities for Three Nuclei (ABX)

Transition	Origin	Frequency	Relative Intensity
1	B	$\nu_{AB} + \frac{1}{4}(-2J_{AB} - J_{AX} - J_{BX}) - D_-$	$1 - \sin 2\phi_-$
2	B	$\nu_{AB} + \frac{1}{4}(-2J_{AB} + J_{AX} + J_{BX}) - D_+$	$1 - \sin 2\phi_+$
3	B	$\nu_{AB} + \frac{1}{4}(2J_{AB} - J_{AX} - J_{BX}) - D_-$	$1 + \sin 2\phi_-$
4	B	$\nu_{AB} + \frac{1}{4}(2J_{AB} + J_{AX} + J_{BX}) - D_+$	$1 + \sin 2\phi_+$
5	A	$\nu_{AB} + \frac{1}{4}(-2J_{AB} - J_{AX} - J_{BX}) + D_-$	$1 + \sin 2\phi_-$
6	A	$\nu_{AB} + \frac{1}{4}(-2J_{AB} + J_{AX} + J_{BX}) + D_+$	$1 + \sin 2\phi_+$
7	A	$\nu_{AB} + \frac{1}{4}(2J_{AB} - J_{AX} - J_{BX}) + D_-$	$1 - \sin 2\phi_-$
8	A	$\nu_{AB} + \frac{1}{4}(2J_{AB} + J_{AX} + J_{BX}) + D_+$	$1 - \sin 2\phi_+$
9	X	$\nu_X - \frac{1}{2}(J_{AX} + J_{BX})$	1
10	X	$\nu_X + D_+ - D_-$	$\cos^2(\phi_+ - \phi_-)$
11	X	$\nu_X - D_+ + D_-$	$\cos^2(\phi_+ - \phi_-)$
12	X	$\nu_X + \frac{1}{2}(J_{AX} + J_{BX})$	1
13	Comb.	$2\nu_{AB} - \nu_X$	0
14	Comb.	$\nu_X - D_+ - D_-$	$\sin^2(\phi_+ - \phi_-)$
15	Comb.	$\nu_X + D_+ + D_-$	$\sin^2(\phi_+ - \phi_-)$

The ABX spectrum consists of a maximum of fourteen possible transitions since line 13 is never observed. The combination lines 14 and 15 appear in the X part of the spectrum as weak lines when $(\nu_A - \nu_B)$ is small. In many cases they are too weak in intensity to observe. In normal ABX spectra, the AB part of the spectrum will appear as eight lines which look as if they were two superimposed AB-spin systems (AB subspectra). This analogy also holds for the intensity of the lines. Close inspection of an ABX spectrum will show that there are three spacings which are repeated four times each in the spectrum. Use is made of the repeated spacings in the analysis of the spectrum.

Inspection of Table 4.13 shows that the coupling constant J_{AB} is repeated four times in the AB part of the spectrum. This is the only repeated spacing in the spectrum that is equal to a coupling constant. If the two AB subspectra can be recognized in the AB part of the spectrum, the separation between the centers of the outer AB pattern and the inner AB pattern yields $\frac{1}{2}|J_{AX} + J_{BX}|$. The quantities D_+ and D_- may also be obtained from line separations in the AB part of the spectrum. The center of the X part of the spectrum yields ν_X. The quantities $|J_{AX} + J_{BX}|$ and $2|D_+ - D_-|$ may be obtained from the four most intense lines in the X part of the spectrum. If the combination transitions 14 and 15 are observed, their separation yields $2(D_+ + D_-)$. The quantities obtained from the X part of the spectrum serve as a check on the values derived from the AB part of the spectrum and vice versa. The repeated spacings are summarized in equations 4.44 and graphically in Figure 4.8.

$$\begin{aligned} J_{AB} &= (v_4 - v_2) = (v_3 - v_1) = (v_7 - v_5) = (v_8 - v_6) \\ &(v_8 - v_7) = (v_6 - v_5) = (v_{12} - v_{11}) = (v_{10} - v_9) \\ &(v_3 - v_4) = (v_1 - v_2) = (v_{12} - v_{10}) = (v_{11} - v_9) \\ 2D_- &= (v_7 - v_3) \\ 2D_+ &= (v_8 - v_4) \\ 2|D_+ - D_-| &= v_{10} - v_{11} \\ 2|D_+ + D_-| &= v_{15} - v_{14} \\ v_X &= \frac{v_{12} + v_9}{2} = \frac{v_{10} + v_{11}}{2} \\ |J_{AX} + J_{BX}| &= (v_{12} - v_9) = 2\left[\frac{v_6 - v_4}{2} - \frac{v_5 - v_3}{2}\right] \end{aligned} \qquad (4.44)$$

The analysis of an ABX spectrum is usually straightforward provided either twelve or fourteen lines are observed in the spectrum. By making use of the repeated spacings, the correct line assignment can usually be made. The spectrum is not dependent on the sign of J_{AB} as a reversal in the sign of J_{AB} simply reverses the labeling of the transitions in the AB part of the spectrum (i.e., 1 and 3 and 4 and 2, etc.). The relative signs of J_{AX} and J_{BX} may be determined in favorable cases. However, the absolute signs of J_{AX} and J_{BX} cannot be determined.

As an example of the analysis of an ABX spectrum, consider the spectrum given in Figure 4.9. The analysis is outlined in the following (10).

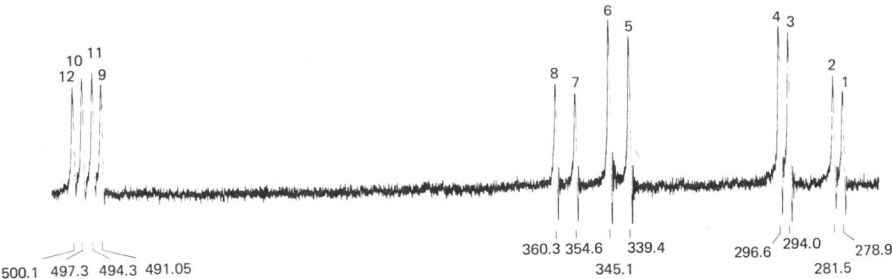

Fig. 4.9 An expansion of the ABX portion of the 100-MHz NMR spectrum of 1,4-diphenylazetidinone.

126   ANALYSIS OF SPECTRA AND DETERMINATION OF STRUCTURE

Step 1. Locate the two AB subspectra in the AB part of the spectrum. Reading from right to left there are two possibilities: lines 2, 4, 6, 8 and 1, 3, 5, 7 or lines 1, 3, 6, 8 and 2, 4, 5, 7. In order to decide on these two possibilities, it is necessary to calculate $\frac{1}{2}|J_{AB} + J_{BX}|$ from the centers of the two AB subspectra and compare these values with the value obtained from $|J_{AX} + J_{BX}|$ from the X part of the spectrum. Lines 9 and 12 yield a value of 8.1 Hz for $|J_{AX} + J_{BX}|$. The combination 2, 4, 6, 8 and 1, 3, 5, 7, yields a value of 1.55 Hz for $\frac{1}{2}|J_{AX} + J_{BX}|$ whereas the combination 1, 3, 6, 8 and 2, 4, 5, 7 yields 4.15 Hz. Clearly, the latter combination of lines for the two AB subspectra is the correct choice.

Step 2. Calculate the remaining quantities from the spectrum. The spacing that is repeated four times in the AB part of the spectrum (3-1, 4-2, 8-6, 7-5) yields a value of 15.1 Hz for J_{AB}. The frequency difference between the first and third lines in each of the two AB subspectra yields $2D_+$ and $2D_-$. Thus $(v_6 - v_1)/2$ yields a value of 31.85 Hz for D_+ $(v_5 - v_2)/2$ yields a value of 30.3 Hz for D_- (D_+ is assumed to be larger than D_-). The frequency separation between lines 9 and 10 $(v_9 - v_{10})$ yields $2|D_+ - D_-| = 3.5$ which serves as a check on the values determined in the AB part of the spectrum. The average $(v_{12} + v_9)/2$ gives v_X as 496.05 Hz. The value of $(v_A + v_B)/2$ may be found by taking the average of the centers of the two AB subspectra or by taking the average of all eight lines in the AB part of the spectrum. This yields a value of 318.75 for $(v_A + v_B)/2$.

Step 3. By squaring, summing, rearranging, and taking square roots of the equations 4.43 for $D_+ \cos 2\phi_+$, $D_+ \sin 2\phi_+$, $D_- \cos 2\phi_-$, and $D_- \sin 2\phi_-$, we arrive at the following equations:

$$\pm[4D_+^2 - J_{AB}^2]^{\frac{1}{2}} = (v_A - v_B) + \tfrac{1}{2}(J_{AX} - J_{BX})$$
$$\pm[4D_-^2 - J_{AB}^2]^{\frac{1}{2}} = (v_A - v_B) - \tfrac{1}{2}(J_{AX} - J_{BX}) \quad (4.45)$$

Equation 4.45 yields four possible solutions

$$\pm 61.88 = (v_A - v_B) + \tfrac{1}{2}(J_{AX} - J_{BX})$$
$$\pm 58.1 = (v_A - v_B) - \tfrac{1}{2}(J_{AX} - J_{BX}) \quad (4.46)$$

Two of the solutions may be discarded since both $(v_A - v_B)$ and $(J_{AX} - J_{BX})$ must be positive. Thus the combinations -61.88 and -58.1, and -61.88 and $+58.1$ are eliminated. Using $+61.88$ and $+58.1$ yields the following values:

$$v_A - v_B = 60 \text{ Hz}$$
$$J_{AX} - J_{BX} = 3.76 \text{ Hz} \quad (4.47)$$

The values $+61.88$ and -58.1 yield the second solution as

$$v_A - v_B = 1.89 \text{ Hz}$$
$$J_{AX} - J_{BX} = 60 \text{ Hz} \quad (4.48)$$

The second solution may be discarded at this point since J_{AX} and J_{BX} are both proton–proton coupling constants and their difference would never be 60 Hz. In cases where both solutions appear reasonable, one must find the value of $2\phi_+$ where $\phi \leq 2\phi_+ < 90°$ from $\sin 2\phi_+ = J_{AB}/2D_+$ and the two possible values of $2\phi_-$ where $\phi < 2\phi_- < 180°$ from $\sin 2\phi_- = J_{AB}/2D_-$. These are then used to calculate the two possible values of $\sin(\phi_+ - \phi_-)$ and $\cos(\phi_+ - \phi_-)$ which are used to calculate the intensities of the X lines. Choose the solution which gives X intensities consistent with the spectrum.

Step 4. From the correct value of $(v_A - v_B)$ and the previously calculated value of $(v_A + v_B)/2$, we can now calculate v_A and v_B as 384.8 and 288.8 Hz, respectively. Similarly, with the correct value of $J_{AX} - J_{BX}$, and the previously calculated value of $|J_{AX} + J_{BX}|$ we calculate J_{AX} and J_{BX} to be 5.93 and 2.17 Hz, respectively. Since we have calculated only $|J_{AX} + J_{BX}|$ and not the sign, the question arises as to whether this should be positive or negative. A change in sign does not affect the spectrum but simply exchanges D_+ and D_-. However, we have already chosen D_+ and D_- in the calculation of $(v_A - v_B)$ and $(J_{AX} - J_{BX})$. At this point, the incorrect sign of $|J_{AX} + J_{BX}|$ will interchange J_{AX} and J_{BX} and will not be consistent with the observed spectrum.

Our discussion to this point has focused on *ABX* spectra in which either twelve or fourteen lines are observed. There are many cases where lines overlap in the spectrum such that fewer than twelve or fourteen lines are observed. These situations have been described as *deceptively simple spectra* and usually arise as J_{AB} approaches $v_A - v_B$ or $(v_A + v_B)/2$ approaches $\frac{1}{2}(J_{AX} - J_{BX})$. Unless they are recognized as *ABX* spectra, they may be interpreted incorrectly to give the wrong set of parameters. Two of the more common types of deceptive *ABX* spectra exhibit four lines in the X part and five lines in the AB part, and three lines in the X part and two lines in the AB part of the spectrum. The five line AB pattern arises when either D_+ or $D_- \approx J_{AB}$, $J_{AB} \gg [v_{AB} + \frac{1}{2}(J_{AX} - J_{BX})]$. In this case, one of the AB subspectra becomes a single line. The two line AB pattern arises when $v_A = v_B$ and the spectrum appears as if it were an A_2X case. Other examples of spectra and combinations of parameters are discussed in a review (10) of the analysis of *ABX* spectra. When one encounters a deceptively simple *ABX* spectrum, it is often possible to remove some of the simplicity by obtaining the spectrum in another solvent or at a different magnetic field.

4.7.4 THREE NUCLEI (*ABC*)

The *ABC* three-spin system is encountered quite often in NMR spectroscopy. Yet there are no convenient methods for the analysis of an *ABC* spectrum by hand. Approximate values for the parameters may be obtained in favorable cases if one can analyze this type of spectrum as an *ABX* spectrum. However, it should be kept in mind that the parameters are approximate, and the more the spectrum deviates from a true *ABX* spectrum, the larger the errors in the parameters.

The analysis of an *ABC* spectrum is best carried out using one of the existing computer programs (Section 4.10). It should be remembered, however, that several solutions may be obtained from such an analysis. One can use the match between the calculated and observed intensities as a check on the "correctness" of a given set of parameters. The spectrum may be obtained at a different magnetic field and analyzed to check on the best parameters. In addition, one should have some idea of the parameters expected from analogy with the parameters for similar systems.

4.8 FOUR-SPIN SYSTEMS

The general four-spin system is characterized by four chemical shifts and six coupling constants. There are 16 basic product functions made up of the possible combinations of α and β for the four nuclei. A maximum of 56 transitions are allowed for the general four-spin system. The diagonalization of the Hamiltonian matrix for a four-spin system can involve the solution of two 4×4 and one 6×6 submatrices which is impossible to do by hand in any reasonable length of time. Therefore the analysis of a general four-spin system is best accomplished by using one of the computer programs discussed in Section 4.10. In certain cases, the size of the submatrices to be solved may be reduced due either to the presence of symmetry in the molecule or to making the *X* approximation. In these cases, explicit expressions for the transition frequencies in terms of the chemical shifts and coupling constants have been derived such that it is possible to analyze the spectrum by hand. It is these types of systems that we discuss here.

4.8.1 FOUR NUCLEI (*AA'XX'*)

The *AA'XX'* spin system arises from four magnetically nonequivalent nuclei in which there are two sets of two nuclei that are symmetrically equivalent and whose chemical shift difference is large compared to the coupling constants between the equivalent sets of nuclei. Typical examples of arrangements of nuclei giving this type of spin system are given in Table 4.1. Six parameters may be obtained from an *AA'XX'* spectrum: the two chemical shifts v_A and v_X and the four coupling constants $J_{AA'}$, $J_{XX'}$, J_{AX}, and $J_{AX'}$ ($J_{A'X}$ and $J_{A'X'}$ are equivalent to $J_{AX'}$ and J_{AX}, respectively).

The basic symmetry functions for two nuclei given in Table 4.4 are used for each set of nuclei *AA'* and *XX'* to construct the 16 basic product functions for the *AA'XX'* system. Use is made of the rule that no mixing occurs between basic product functions of different symmetry and the X approximation (no mixing occurs between basic product functions differing in both $F_z(a)$ and $F_z(s)$ i.e., off-diagonal elements of J_{AX} and $J_{AX'}$ vanish) to reduce the Hamiltonian matrix. Using the previously stated rules, the diagonalization of the matrix reduces to the solution of twelve 1×1 and two 2×2 submatrices. Twelve of the energy levels are given directly by the diagonal elements of the 1×1 matrices and the remaining four may be obtained from solution of the two 2×2 submatrices. Explicit expressions for the transition frequencies and their relative intensities may be calculated from the energy levels. These are given (1) in Table 4.14 with respect to v_A where

$$K = J_{AA'} + J_{XX'}$$
$$M = J_{AA'} - J_{XX'}$$
$$L = J_{AX} - J_{AX'} \quad (4.49)$$
$$N = J_{AX} + J_{AX'}$$

$$\cos 2\phi_s : \sin 2\phi_s : 1 = K : L : (K^2 + L^2)^{\frac{1}{2}}$$

$$\cos 2\phi_a : \sin 2\phi_a : 1 = M : L : (M^2 + L^2)^{\frac{1}{2}}$$

Only the A transitions are given as the X transitions are identical to the A transitions with the exception that they are centered about v_X.

Inspection of Table 4.14 shows that an $AA'XX'$ spectrum consists of a maximum of 24 allowed transitions. Each half of the spectrum contains 10

TABLE 4.14. Transition Frequencies for the AA' Part of Four Nuclei $(AA'XX')^a$

Transition	Transition Frequency	Relative Intensity
1	$\frac{1}{2}N$	1
2	$\frac{1}{2}N$	1
3	$-\frac{1}{2}N$	1
4	$-\frac{1}{2}N$	1
5	$\frac{1}{2}K + \frac{1}{2}(K^2 + L^2)^{\frac{1}{2}}$	$\sin^2 \phi_s$
6	$-\frac{1}{2}K + \frac{1}{2}(K^2 + L^2)^{\frac{1}{2}}$	$\cos^2 \phi_s$
7	$\frac{1}{2}K - \frac{1}{2}(K^2 + L^2)^{\frac{1}{2}}$	$\cos^2 \phi_s$
8	$-\frac{1}{2}K - \frac{1}{2}(K^2 + L^2)^{\frac{1}{2}}$	$\sin^2 \phi_s$
9	$\frac{1}{2}M + \frac{1}{2}(M^2 + L^2)^{\frac{1}{2}}$	$\sin^2 \phi_a$
10	$-\frac{1}{2}M + \frac{1}{2}(M^2 + L^2)^{\frac{1}{2}}$	$\cos^2 \phi_a$
11	$\frac{1}{2}M - \frac{1}{2}(M^2 + L^2)^{\frac{1}{2}}$	$\cos^2 \phi_a$
12	$-\frac{1}{2}M - \frac{1}{2}(M^2 + L^2)^{\frac{1}{2}}$	$\sin^2 \phi_a$

[a] Frequencies relative to v_A.

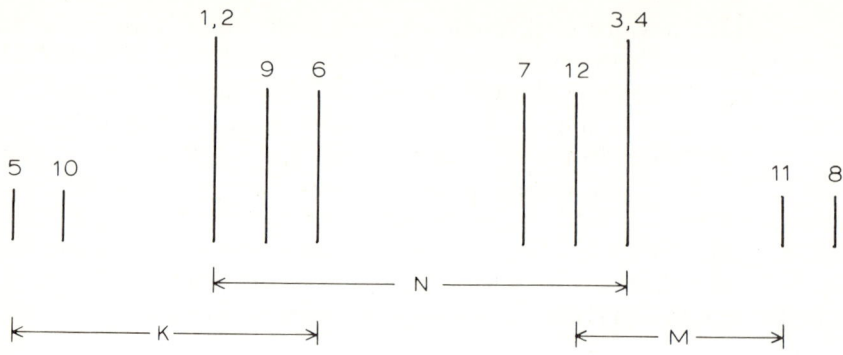

Fig. 4.10 A schematic illustration of one-half of an $AA'XX'$ spectrum.

lines due to the degeneracy of certain transitions. There are two intense lines separated by $|N|$ in each half of the spectrum (transitions 1, 2 and 3, 4) and two pairs of AB type quartets (transitions 5, 6, 7, 8 and transitions 9, 10, 11, 12) centered about v_A (Figure 4.10). The AB type quartets are actually AB subspectra in which L replaces δ_{AB} and K replaces J_{AB} in transitions 5, 6, 7, 8 and M replaces J_{AB} in transitions 9, 10, 11, 12 in Table 4.14.

The analysis of an $AA'XX'$ spectrum is straightforward provided all 10 transitions are observed in each half of the spectrum. The chemical shifts are obtained as the midpoints of each half-spectra and the parameters, K, M, L, and N are obtained from the line separations as follows:

$$v_A = \frac{v_1 + v_3}{2} = \frac{v_2 + v_4}{2}$$

$$v_X = \frac{v_3' + v_1'}{2} = \frac{v_4' + v_2'}{2}$$

$$K = v_5 - v_6 = v_7 - v_8$$

$$M = v_{10} - v_9 = v_{12} - v_{11} \qquad (4.50)$$

$$N = v_1 - v_3 = v_2 - v_4$$

$$(K^2 + L^2)^{\frac{1}{2}} = v_5 - v_7 = v_6 - v_8$$

$$(M^2 + L^2)^{\frac{1}{2}} = v_9 - v_{11} = v_{10} - v_{12}$$

The spectrum is invarient to the signs of N, L, K, and M as only their absolute values are obtained from the spectrum. A reversal in the sign of N merely changes the assignment of transitions 1, 2 and 3, 4. Furthermore, the labeling of the two AB subspectra changes if the signs of K, L, and M are reversed. If K and M are interchanged, the two AB subspectra are reversed, that is, the 5, 6, 7, 8 quartet becomes the 9, 10, 11, 12 quartet and vice versa. Thus the relative signs of J_{AX} and $J_{AX'}$ cannot be determined from the spectrum nor can the relative signs of $J_{AA'}$ and $J_{XX'}$ be determined. Furthermore, there is no way from the spectrum to decide which is J_{AX} and which is $J_{AX'}$ or which is $J_{AA'}$ and which is $J_{XX'}$. In order to properly assign the coupling constants in an $AA'XX'$ system, one must rely on other information such as analogy with other systems.

Several combinations of parameters give rise to $AA'XX$ spectra in which fewer than 10 lines are observed in each half of the spectrum (deceptively simple spectra). In these cases it may not be possible to analyze the spectrum for all of the parameters. For example, when $L = 0$ each half-spectrum appears as a 1:2:1 triplet with spacings of $N/2$. This is, of course, the A_2X_2 spectrum in which $J_{AX} = J_{AX'}$. If $L \neq 0$, but is small with respect to K and M, the spectrum may appear as a 1:2:1 triplet similar to the above in which only the value of N may be obtained. Therefore caution should be exercised before assigning a 1:2:1 triplet as an A_2X_2 spectrum, especially if by analogy with other systems there is no reason to believe that $J_{AX} = J_{AX'}$. An example of this type of deceptively simple $AA'XX'$ spectrum is given in Figure 4.11. Other examples have been discussed in the literature (7,11–14). Finally, other combinations of parameters have been discussed in which fewer than 10 lines are observed in each half-spectrum. It is important that one recognize that it may not always be possible to obtain all the parameters from the analysis of an $AA'XX'$ spectrum.

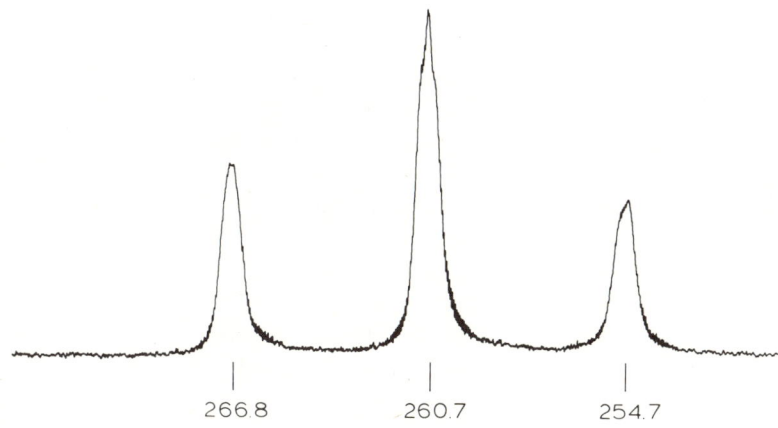

Fig. 4.11 An expansion of one-half of the 100-MHz NMR spectrum of $HOCH_2CH_2CN$.

4.8.2 FOUR NUCLEI ($AA'BB'$)

The $AA'BB'$ spin system arises from two sets of magnetically nonequivalent nuclei similar to the $AA'XX'$ system with the exception that the chemical shift difference ($v_A - v_B$) is smaller. The most noticeable difference in the spectrum compared to the $AA'XX'$ spectrum is that the intensity of the inner lines increases at the expense of the outer lines similar to the situation on going from an AX to an AB spectrum.

The basic product functions used for the $AA'XX'$ system are used to set up the Hamiltonian matrix for the $AA'BB'$ system. The matrix is reduced somewhat due to the rule that no mixing occurs between product functions with different symmetry. However, the off-diagonal matrix elements involving J_{AB} and $J_{AB'}$ are retained. The matrix can be factorized into two 1×1, five 2×2 and one 4×4 submatrices. The solution of the 2×2 submatrices is straightforward. However, the 4×4 submatrix cannot be readily solved and explicit expressions for the transition frequencies involving these four energy levels cannot be obtained. If one denotes the four energy levels as E_1, E_2, E_3, and E_4, they are solutions of the fourth power equation (equation 4.51) resulting from the expansion of the 4×4 determinant (2).

$$(K + E)E^3 + NE^2 - E[\tfrac{1}{4}N^2 + \delta_{AB}^2] - \tfrac{1}{4}N^3$$
$$-\tfrac{1}{4}L^2 \quad 3E^3 - [\tfrac{3}{4}N^2 + \delta_{AB}^2] = 0 \quad (4.51)$$

Using these values, expressions for the twelve transitions may be written as given in Table 4.15.

TABLE 4.15. Transition Frequencies for the AA' Part of Four Nuclei ($AA'BB'$)

Transition	Transition Frequency
1	$\tfrac{1}{2}N + \tfrac{1}{2}[\delta_{AB}^2 + N^2]^{\frac{1}{2}}$
2	$-\tfrac{1}{2}(\delta_{AB}^2 + N^2)^{\frac{1}{2}} - E_1$
3	$-\tfrac{1}{2}N + \tfrac{1}{2}(\delta_{AB}^2 + N^2)^{\frac{1}{2}}$
4	$E_2 - \tfrac{1}{2}(\delta_{AB}^2 + N^2)^{\frac{1}{2}}$
5	$\tfrac{1}{2}(\delta_{AB}^2 + N^2)^{\frac{1}{2}} - E_3$
6	$E_4 + \tfrac{1}{2}(\delta_{AB}^2 + N^2)^{\frac{1}{2}}$
7	$\tfrac{1}{2}(\delta_{AB}^2 + N^2)^{\frac{1}{2}} - E_4$
8	$E_3 + \tfrac{1}{2}(\delta_{AB}^2 + N^2)^{\frac{1}{2}}$
9	$\tfrac{1}{2}[(\delta_{AB} + M)^2 + L^2]^{\frac{1}{2}} + \tfrac{1}{2}(M^2 + L^2)^{\frac{1}{2}}$
10	$\tfrac{1}{2}[(\delta_{AB} - M)^2 + L^2]^{\frac{1}{2}} + \tfrac{1}{2}(M^2 + L^2)^{\frac{1}{2}}$
11	$\tfrac{1}{2}[(\delta_{AB} + M)^2 + L^2]^{\frac{1}{2}} - \tfrac{1}{2}(M^2 + L^2)^{\frac{1}{2}}$
12	$\tfrac{1}{2}[(\delta_{AB} - M)^2 + L^2]^{\frac{1}{2}} - \tfrac{1}{2}(M^2 + L^2)^{\frac{1}{2}}$

Frequencies relative to $(v_A + v_B)/2$.

FOUR-SPIN SYSTEMS

Defining the expressions in equation 4.52 for simplification, inspection of Table 4.15 reveals the following expressions relating the line separations with the quantities involving the chemical shifts and coupling constants:

$$C = [(\delta_{AB}^2 + N^2)]^{\frac{1}{2}}$$
$$D = [(\delta_{AB} + M)^2 + L^2]^{\frac{1}{2}}$$
$$F = [(\delta_{AB} - M)^2 + L^2]^{\frac{1}{2}} \quad (4.52)$$
$$G = (L^2 + M^2)^{\frac{1}{2}}$$

$$v_5 - v_6 = v_7 - v_8$$
$$v_5 - v_1 = v_3 - v_8$$
$$v_1 - v_6 = v_7 - v_3$$
$$v_9 - v_{10} = v_{11} - v_{12}$$
$$C = v_1 + v_3 = v_5 + v_8 = v_6 + v_7$$
$$N = v_1 - v_3$$
$$G = v_{10} - v_{12} = v_9 - v_{11}$$
$$\tfrac{1}{2}(D + F) = v_{10} + v_{11} = v_9 + v_{12}$$
$$\tfrac{1}{2}(D - F) = v_{10} - v_9 = v_{10} - v_{11} \quad (4.53)$$
$$E_2 - E_3 = v_5 - v_7 = v_6 - v_8$$
$$\tfrac{1}{2}K - E_2 - E_3 = v_5 - v_6 = v_7 - v_8$$
$$E_3 = +\frac{v_5 - v_8}{2}$$
$$E_4 = +\frac{v_6 - v_7}{2}$$
$$\tfrac{1}{2}K - E_1 - E_4 = v_2 - v_4$$
$$D = v_{10} + v_{12}$$
$$F = v_9 + v_{11}$$

From these expressions, the coupling constants may be determined from the expressions given in equation 4.54 for N, L, K, M, and δ_{AB}.

$$\delta_{AB} = (4 \cdot v_1 \cdot v_3)^{\frac{1}{2}} = [2(v_{10} \cdot v_{12} + v_9 \cdot v_{11})]^{\frac{1}{2}}$$
$$N = v_1 - v_3$$
$$K = v_5 - v_6 + v_2 - v_4 - N \quad (4.54)$$
$$M = \frac{v_9 \cdot v_{11} - v_{10} \cdot v_{12}}{\delta_{AB}}$$
$$L = (D^2 - M^2)^{\frac{1}{2}}$$

The analysis of an $AA'BB'$ spectrum depends, of course, on making the correct line assignment. For convenience, the 12 lines in one-half of an $AA'BB'$ spectrum are divided into four groups (2,14). The first group consists of lines 1 and 3, which are dependent only on δ_{AB} and N. Lines 1 and 3 comprise an AB subspectrum in both frequency and intensity. The intensity relationship often can be used to determine if additional transitions are overlapping with lines 1 and 3. The intensity of lines 1 and 3 contribute two out of a total of eight in each half-spectra.

The second group consists of lines 9, 10, 11, 12, which are due to the antisymmetric transitions. This group of lines is dependent on δ_{AB}, L and M, but is independent of the sign L and M. The intensity of this group of lines contributes a total of two to the overall intensity of eight.

The third group consists of lines 5, 6, 7, 8 and depends on δ_{AB}, K, L, and N. The appearance of this group of lines is dependent on the relative signs of K and N, but not on the relative sign of L. About the only useful rule involving this group of lines is that they are always symmetric about $\frac{1}{2}C$. They have a total intensity of two.

Lines 2 and 4 comprise the fourth group. They depend on δ_{AB}, K, L, and N and also on the relative signs of N and K. These lines may or may not overlap with lines 1 and 3 and contribute a total intensity of two out of the eight.

Due to the variety of combinations of N, L, K, and M and their effect on the appearance of an $AA'BB'$ spectrum, it is almost impossible to give a set of rules for the analysis of all $AA'BB'$ spectra. Although the relationships given in equation 4.53 are overdetermined such that internal checks are available for the correctness of a particular assignment and a unique analysis can be carried out, it is not uncommon to find $AA'BB'$ spectra containing fewer than twelve lines in each half-spectrum such that checks on the parameters may not be obtained. In fact, one may not be able to obtain all of the parameters, N, L, K, and M from every $AA'BB'$ spectrum. Usually, lines 9 and 12 and 5 and 8 decrease in intensity and vanish as L approaches zero or lines overlap such that it may not be possible to uniquely determine the parameters. The appearance of $AA'BB'$ spectra as a function of N, L, K, and M

has been discussed and the reader is referred to these references (11,12) for further details on the appearance of $AA'BB'$ spectra. Usually one can approximate the coupling constants by analogy with known systems and calculate a spectrum using one of the computer programs discussed in Section 4.10. The parameters may be varied until a good fit is obtained between the calculated and experimental spectrum.

There are four quite arbitrary types of $AA'BB'$ spectra where sufficient examples are available in the literature such that a unique assignment of the transitions can usually be made. These spectra are discussed in the following. It should be recognized that these are general examples and that it is possible to find spectra which may be borderline between two of the four types of spectra.

This first type of spectrum is that characterized by large values of J_{AB} and $J_{BB'}$ and much smaller values of $J_{AB'}$ and $J_{AA'}$. Typical examples of this type of spectrum are symmetrically *ortho*-disubstituted benzenes, symmetrical 1,4-disubstituted butadienes, or any molecule containing the symmetrical fragment —CH_A—CH_B—$CH_{B'}$—$CH_{A'}$—. A typical spectrum of this type is given in Figure 4.12 along with the correct line assignment. Various spectra of this type have been discussed (11).

A second class of $AA'BB'$ spectra are those that are characterized by large values of J_{AB} and small values for the remainder of the coupling constants. The most common example of this type of spectrum is that of the unsymmetrical *para*-disubstituted benzenes. Usually the analysis of this type of spectrum is relatively straightforward and a typical example is given in Figure 4.13 with the line assignment. Considerable overlap of the lines occurs in this type of spectrum due to $M \approx 0$.

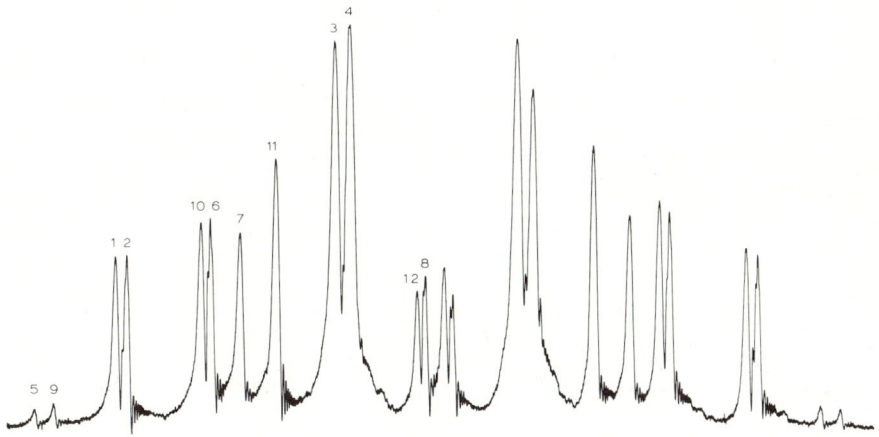

Fig. 4.12 An expansion of the 60-MHz NMR spectrum of *o*-dichlorobenzene.

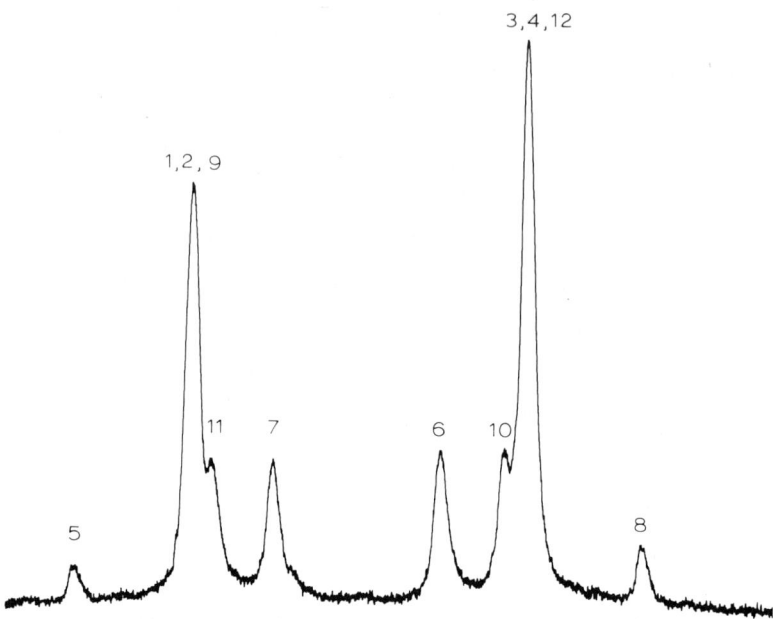

Fig. 4.13 An expansion of the low-field half of the 100-MHz NMR spectrum of *p*-bromoaniline.

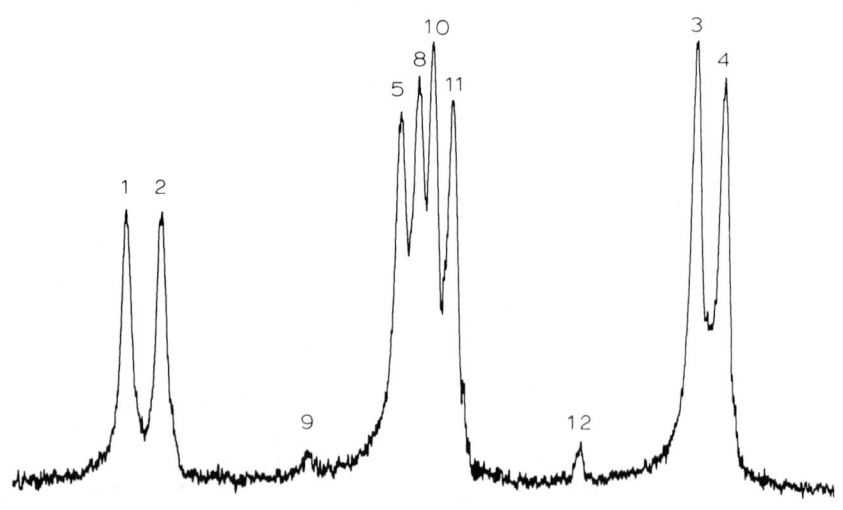

Fig. 4.14 An expansion of the low-field half of the 100-MHz NMR spectrum of 1-chloro-2-phenylethane.

The third type of $AA'BB'$ spectrum is one in which both $J_{AA'}$ and $J_{BB'}$ are large such that $K \gg N \gg L$ and M ranges in value from N to L. This is by far the most common type of $AA'BB'$ spectrum and is given by the $X-CH_2-CH_2-Y$ group. A typical example of this type of spectrum is given in Figure 4.14. With a large value of K, it is not uncommon to find spectra in which lines 6 and 7 are so weak in intensity that they are not observed. In such ten-line spectra, the value of K cannot be determined. An example of this type of spectrum is discussed latter. Should both K and M be large compared to L, it is possible for lines 9 and 12 and lines 6 and 7 to become so weak in intensity that they are not observed and for lines 10 and 11 to overlap. This results in a deceptively simple seven line spectrum similar to the true A_2B_2 spectrum.

A fourth type of $AA'BB'$ spectrum is given by systems where L is large. Typical examples of this type of spectrum are molecules containing the $-CH_2CH_2-$ fragment such that the protons are paired as

$$\begin{array}{c} H_A \;\; H_{A'} \\ | \;\;\; | \\ -C-C- \\ | \;\;\; | \\ H_B \;\; H_{B'} \end{array}.$$

In this type of spectrum, it is not uncommon for line 8 to appear in the B part of the spectrum as a strong line and for lines 2 and 4 to be relatively weak. A typical example of this type of spectrum is given in Figure 4.15.

In the analysis of an $AA'BB'$ spectrum, one can usually pick out lines 1 and 3 from which values of N and δ_{AB} may be obtained. The next step is the assignment of the two groups of four lines 5, 6, 7, 8 and 9, 10, 11, 12. This can usually be accomplished by using the relationships given in equation 4.53 and by analogy with other systems. Next, lines 2 and 4 are assigned. From the assignment of the remaining lines, values of K, M, and L may be obtained using the relationship in equation 4.54 to complete the analysis. It should be kept in mind that if fewer than 12 transitions are observed, it may not be possible to obtain all the parameters from the analysis of the spectrum. It may be possible in some cases to obtain the spectrum in another solvent or at a different magnetic field such that a complete analysis can be carried out. In any event, the analysis should not be considered to be complete without first calculating a spectrum with the parameters to insure that the parameters do in fact reproduce the experimental spectrum.

4.8.3 FOUR NUCLEI (A_2B_2)

When the two coupling constants J_{AB} and $J_{AB'}$ are equal ($L = 0$) a reduction in the number of allowed transitions occurs. Transitions 5 and 8 and 9

138   ANALYSIS OF SPECTRA AND DETERMINATION OF STRUCTURE

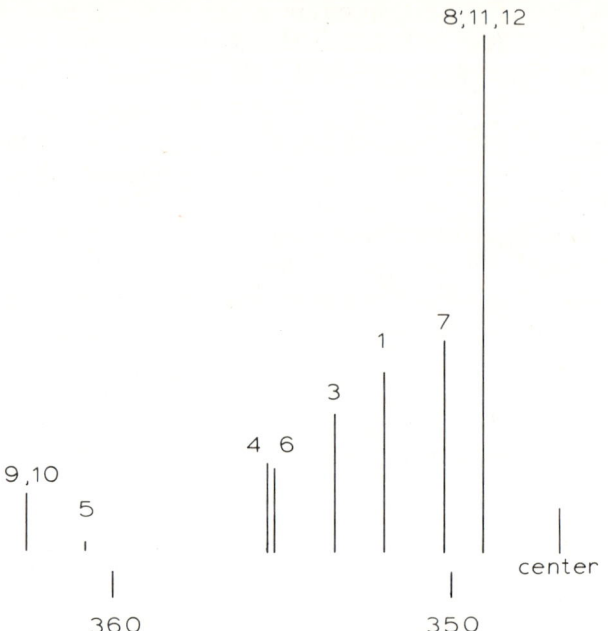

Fig. 4.15 A schematic representation of the low-field half of the proton spectrum given by compounds containing the

$$\begin{array}{cc} H_A & H_{A'} \\ | & | \\ -C-C- \\ | & | \\ H_B & H_{B'} \end{array}$$

fragment.

and 12 are forbidden (2). Furthermore, transitions 10 and 11 become degenerate such that a seven-line spectrum is observed (Figure 4.16). The only parameters that can be obtained from the analysis of an A_2B_2 spectrum are the chemical shifts and J_{AB}. The frequency of the degenerate transition is centered at the frequency v_A. Like the AB_2 spectrum, the A_2B_2 spectrum is not dependent on J_{AA} and J_{BB} and its appearance depends only on the ratio J/δ. As the chemical shift separation becomes larger, the seven-line spectrum reduces to the triplet typical of a first-order spectrum.

We have mentioned previously that, in an $AA'BB'$ spectrum, as L approaches zero, lines 9, 12, 5, and 8 become weak in intensity such that they may not be observed. Lines 10 and 11 merge together such that a similar seven line pattern may be observed for an $AA'BB'$ spectrum. When a seven-line pattern is observed for a four-spin system, caution should be exercised before labeling it as an A_2B_2 spectrum since it may be a deceptively simple $AA'BB'$ pattern. This is especially true if there is no good reason to suspect that the two couplings J_{AB} and $J_{AB'}$ should be equal.

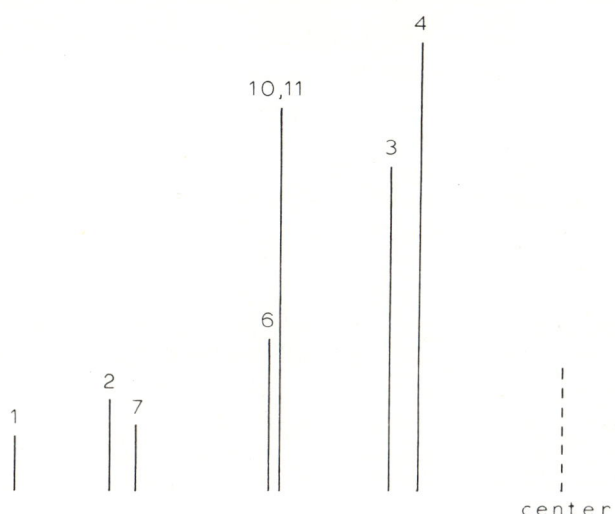

Fig. 4.16 A schematic drawing of one-half of an A_2B_2 spectrum.

4.9 LARGER SPIN SYSTEMS

The number of magnetic nuclei in a spin system does not have to be large (*ABC*) before the analysis of the spectrum becomes so complicated that analysis by hand is impossible. The reason for this is that the differences in the transition frequencies no longer depend upon only one or two of the parameters, but instead, depend on several of the parameters in a complicated way such that unique expressions for the individual parameters cannot be obtained. Several of the more common complex spin systems have been discussed by Emsley, Feeney, and Sutcliffe (2). In these cases, a computer analysis is required in order to obtain the parameters. Even with computer analysis, it may not be possible to obtain the parameters from all spectra.

4.10 COMPUTER ANALYSIS OF NMR SPECTRA

As can be seen from Section 4.4, the matrix of the spin Hamiltonian is written in terms of the chemical shifts and coupling constants. The solution of the matrix to yield the eigenvalues and eigenvectors from which the transitions and their intensities are calculated usually cannot be accomplished by direct calculation due to the complexity of the equations involved. However, this problem is ideally suited for solution using a computer. Subroutines exist for the diagonalization of the matrix using various approximations. Differences between the eigenvalues yield the transition frequencies while the eigenvectors are used to calculate the intensities. Computer programs have been written which perform all the necessary calculations from a given set of chemical shifts and coupling constants.

The analysis of an NMR spectrum requires the extraction of the chemical shifts and coupling constants from the spectrum. With complex spectra, differences in the eigenvalues which yield the transitions depend on two or more of the parameters such that the parameters cannot be found directly from the separation between transition frequencies. In theory, one could guess a set of parameters, carry out the calculation of the spectrum, and compare the calculated spectrum with the experimental spectrum. The "guessed" parameters could then be adjusted until the calculated spectrum is identical to the experimental spectrum. The "guessed" parameters are then assumed to be the correct parameters for the experimental spectrum.

Fortunately, computer programs are available in which the parameters are varied until the calculated spectrum is identical to the experimental spectrum in terms of transition frequencies (15–18). The comparison of the calculated intensities with the experimental intensities is left to the user. Two approaches have been taken for the variation of the parameters to match the experimental spectrum.

The procedure of Swalen and Reilly (15) consists of using the two computer programs NMRIT and NMREN in three separate stages. In the first stage, approximate chemical shifts and coupling constants are used with NMRIT to calculate a trial spectrum. The output consists of the energy levels and calculated transitions. Different parameters are tried until the calculated spectrum approximates the experimental spectrum.

The second stage consists of assigning the experimental transitions to the transitions calculated using NMRIT. The experimental frequencies and their assignments are used with the program NMREM to calculate the energy levels of the experimental spectrum and their errors. These energy levels are calculated by a least-squares technique utilizing the transition equations $E_i - E_j = v_{ij}$ and the equation which sets the trace of the matrix equal to zero. Errors in the energy levels are calculated from the inverse matrix of the normal equations and the sum of squares of the deviations between the observed and calculated frequencies. While it is not necessary to assign every transition, one must assign enough transitions to uniquely define each energy level.

Once the energy levels have been obtained, they are used in the third stage with NMRIT employing multiple iterations. The trial set of chemical shifts and coupling constants are used to compile and diagonalize the Hamiltonian matrix. The calculated energy levels are compared with the experimental energy levels from NMREN and a least-squares technique is used to determine adjustments to the trial parameters. The new adjusted parameters are then used to recompile and diagonalize the matrix. This iterative process is repeated until a preset number of interations have been performed. The output consists of a final set of parameters, calculated transition frequencies with their errors, and calculated intensities. This process can be repeated until the energy levels calculated from the derived parameter set agree with the experimental energy levels from NMREN to the desired accuracy.

The method of Castellano and Bothner-By (16) differs from the previously mentioned method in that the eigenvectors are used for obtaining the best values of the parameters. A set of trial parameters is used with the computer program LAOCN3 and the resultant spectrum is computed. The output includes all transitions within a preset range and with a preset minimum intensity. An origin number accompanies each transition allowing designation of the energy levels connected by the transition. The calculated spectrum is then matched with the experimental spectrum allowing assignment of the experimental transitions. Not all transitions need be assigned. The transitions are assigned a weighting factor of either 1 or 0, according to whether or not they are included in the calculation.

The assigned experimental transitions are used with the trial set of parameters in the second half of the calculation. Least-squares iterations are performed with the new eigenvectors being used to set up the matrix for the succeeding cycle. When the vector of corrections to the parameters is not equal to zero, corrections are made to the parameters in order to give the best least-squares fit. This procedure is repeated until the error in line fitting decreases insignificantly, increases, or a preset-number of iterations is reached. Finally a calculated spectrum is determined from the best set of eigenvalues and eigenvectors. The output includes the best values of the parameters, the error vectors and probable errors, and the calculated spectrum with its error in line fitting.

Modifications have been made in the above computer programs to allow for factoring of the matrix on the basis of symmetry (19,20). When the compound contains a set of magnetically equivalent nuclei, the calculations can be broken down into separate parts (submatrices) according to the total spin number F_z of the magnetically equivalent nuclei. Consequently, less computer storage is required (smaller matrices to diagonalize) and larger spin systems may be treated (19,20). The size of the spin system which can be treated by the programs previously mentioned is typically seven when there is no magnetic equivalence. Large spin systems can be treated if magnetic equivalence is present. The limitation in the size of the spin system that can be treated is in the user's computer as to the storage required for diagonalization of the matrix. Although the programs may be easily modified to handle larger spin systems, there are very few cases where a program that treats spin systems larger than seven is needed.

There is no substitute for experience in the computer analysis of NMR spectra. A spectrum must first be calculated in order for one to assign the experimental transitions. In systems approaching first order, this usually presents little problem since approximate parameters for the calculation of the spectrum may be obtained from a first-order analysis, even though the intensities are clearly not first order. In more complex systems where there is overlap of the transitions, several spectra may have to be calculated before the calculated spectrum matches the experimental spectrum sufficiently well such that the experimental transitions can be assigned. This task becomes easier with experience. The reader should consult the two reviews (20,21) on the computer analysis of NMR spectra for more insight into this area.

In each of the previously mentioned programs, the matching of the transition frequencies provides the criteria for the "correctness" of the computer fit of the spectrum. One program uses both the transition frequencies and intensities in fitting the spectrum (22). This method has not received widespread use due to the difficulties in determining the intensities accurately in a complex spectrum. Various programs are available (18) for taking the output from the above programs and plotting a spectrum. These programs may produce either a stick plot or a line plot assuming some line shape function. The use of a plot program allows one to compare the calculated with the experimental intensities as another check of the computer fit of the experimental spectrum.

4.11 AIDS IN THE ANALYSIS OF SPECTRA

As mentioned previously, it may not be possible to completely analyze every spectrum that is encountered. Spin systems containing equivalent nuclei are particularly difficult due to the fact that not all the allowed transitions are observed in many cases. In other spin systems, deceptively simple spectra may be observed, thereby limiting the parameters that can be obtained from the analysis. On the other hand, with complex spectra it may not be possible to obtain estimates of the parameters to use in the calculation of a trial spectrum for computer analysis. Several methods have been used in the past to aid in the analysis of spectra. Each of these methods is discussed briefly in the following sections.

4.11.1 VARIATIONS IN MAGNETIC FIELD

Often a complex spectrum may be simplified by obtaining the spectrum in a larger magnetic field. As pointed out previously, chemical shifts are field dependent whereas coupling constants are independent of the magnetic field. Therefore, the ratio J/δ may be changed and the spectrum made to approach first order by obtaining the spectrum in a stronger magnetic field. Commercial spectrometers are available for obtaining proton spectra at 30, 60, 90, 100, 220, 270, and 360 MHz, offering a wide range of possibilities.

The reverse of this process may be used to advantage in certain cases. As a general rule, certain information may not be obtained from a first-order spectrum (i.e., signs of coupling constants and certain coupling constants in cases where the intensities of certain absorptions are such that they are not observed). This information may become available in a more complex spectrum obtained by running the spectrum in a less intense magnetic field. The analysis of a spectrum at two magnetic fields can also be used as a check on the parameters obtained from the spectrum.

4.11.2 ISOTOPIC SUBSTITUTION

Isotopic substitution has been used as an aid in the analysis of NMR spectra in two principal ways: (1) by removing a spin from the system (replacing H with D) and (2) by removing the magnetic equivalence observed in certain cases. Using this technique, it is possible to obtain information from a spectrum which would otherwise prove difficult or impossible to obtain.

One of the most widely used applications of deuterium substitution is that of the identification of labile proton resonances. A typical example is the identification of the —O—H resonance in the spectra of alcohols. After obtaining a normal spectrum, the usual technique is to add a drop of D_2O to the NMR tube and shake for a few minutes to allow D to exchange for H. The spectrum is rerun and the peak due to the hydroxyl proton will not be present in this spectrum. This technique may be used to identify the resonance of any readily exchangeable proton.

A second related area where isotopic substitution has proven valuable in the analysis of spectra is in the simplification of a complex spectrum. Suppose one had a spectrum where it was not possible to assign the resonances to specific protons in the molecule. If one could synthesize the molecule with D at specific positions in the molecule, one could obtain the spectrum of the deuterated material and, from a comparison of the spectra, assign the resonances in the nondeuterated sample. This method can also be used in reverse; that is, if one can assign the spectrum of the nondeuterated material then one can use this technique to determine the site of incorporation of deuterium into a molecule. This has been used in various mechanistic studies.

The removal of resonances from complex spectra is only one aspect of deuterium substitution. Recall from Section 2.9 that the coupling constant between two nuclei is proportional to the product of the magnetogyric ratio

144   ANALYSIS OF SPECTRA AND DETERMINATION OF STRUCTURE

of the nuclei. Therefore, substitution of D for H will affect all coupling constants to the H. The magnetogyric ratio of deuterium, γ_D, is 6.55 times smaller than that of hydrogen. Consequently, substitution of D for H will reduce all coupling constants to the H be a factor of 6.55 ($J_{HD} = J_{HH}/6.55$). Thus, not only are resonances removed from the spectrum, but the remaining resonances are simplified by the reduction in the magnitude of the coupling constants. In many cases, the coupling constants can become negligibly small such that only line broadening is observed in the resonances of the protons that are spin coupled to deuterium.

The second area where isotopic substitution is used actually results in an increase in the complexity of a spectrum by removing magnetic equivalence. There are several instances where the coupling constant between magnetically equivalent nuclei is needed either for comparison with theoretical predictions or for a base for the discussion of substituent effects. Consider for example the early work on the theory of coupling constant mechanisms where the coupling constant in the hydrogen molecule H_2 was calculated.

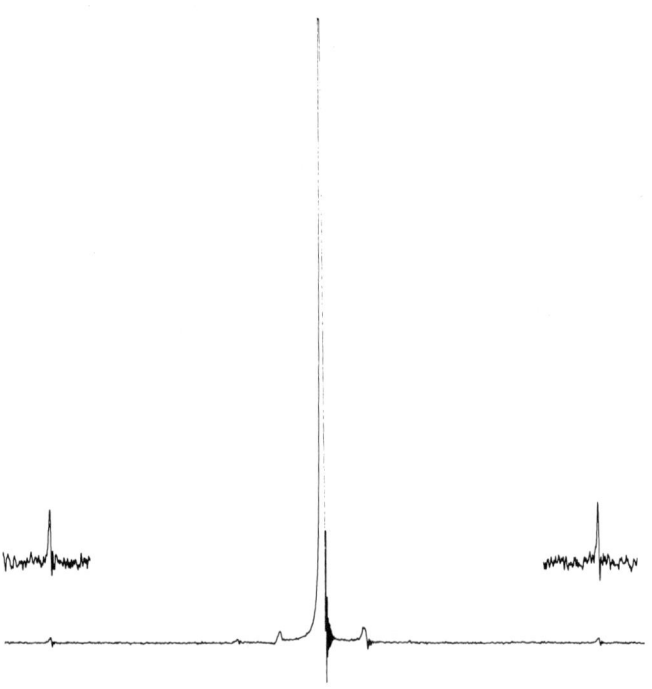

Fig. 4.17 An expansion of the 100-MHz NMR spectrum of $CHCl_3$ showing the carbon-13 satellites.

The nmr spectrum of H_2 consists of a single line due to the magnetic equivalence of the two hydrogens. This A_2 spin system can be converted into an AX spectrum by substituting one D for H (H—D). The proton spectrum of H—D consists of three equally intense lines due to coupling with deuterium ($I = 1$). From the ratio of J_{HH} to J_{HD} previously given, J_{HH} in H_2 can be calculated. Similarly, the substitution of D for H in methane would result in a three line spectrum from which J_{HH} in methane could be calculated. A variety of coupling constants have been obtained using this technique.

Another method for removing certain magnetic equivalence is to examine the ^{13}C satellite resonances. This does not necessarily require isotope substitution if sufficient sensitivity is available. The ^{13}C isotope ($I = \frac{1}{2}$) is present in 1.1% natural abundance. In a molecule like $CHCl_3$, 1.1% of the molecules contain ^{13}C, which is coupled to the proton. Therefore, the proton spectrum of $^{13}CHCl_3$ gives rise to a doublet that is symmetrically disposed about the single peak due to the $^{12}CHCl_3$ molecules (Figure 4.17). The ^{13}C satellite spectra can be used to advantage to remove the equivalence in symmetrical molecules. Consider the proton spectrum of benzene which consists of a single line. Approximately 6.6% of the molecules contain a ^{13}C carbon (the probability of having a molecule with two ^{13}C carbons is negligibly small).

2

The presence of the ^{13}C nucleus makes the spin system an $ABB'CC'D$ spin system (**2**). All of the interproton coupling constants can be obtained from the analysis of the ^{13}C satellite spectra on either side of the absorption due to the all ^{12}C molecules. This technique has been utilized to obtain the proton–proton coupling constants in a variety of molecules including ethylene (23), benzene (24), cyclopropane (25), various 1,2,-disubstituted ethanes (26), and the three membered ring heterocycles (27).

4.11.3 DOUBLE RESONANCE

Double resonance methods are often used in NMR spectroscopy to aid in the interpretation of complex NMR spectra and to extend the range of information that may be obtained from spectra. As the name implies, double resonance experiments are concerned with the simultaneous application of two rf fields to the sample under consideration. We shall continue to refer to B_1 as the observing field and use B_2 to refer to the second rf field used to perturb the system of nuclei. The notation commonly used for double resonance experiments is $X - \{Y\}$ where X is the nuclear species being observed and Y is the nuclear species being irradiated (28). Two classes of experiments are included under the general topic of double resonance. In homonuclear double resonance the second rf field B_2 is applied to the same type of nucleus as that of the observing field B_1, $H - \{H\}$. Heteronuclear double resonance differs in that the field B_2 is applied to a different type of nucleus than that observed by B_1 and it is denoted by $H - \{X\}$.

Several types of experiments are included under homonuclear double resonance and they differ primarily in the strength of the perturbing field B_2. Expressing field strengths in frequency units and using the conversion factor for proton of 1 mG equals 4.26 Hz, typical values of B_2 are from 1 to 8 mG compared to normal values of 0.1 to 0.3 for B_1. In principle, homonuclear double resonance experiments may be carried out by varying the magnetic field B_1, which is called field sweep, or by varying the frequency of B_1, which is called frequency sweep. We shall restrict ourselves here to frequency sweep experiments since not all types of double resonance experiments can be carried out using field sweep. Most commercial NMR spectrometers have, as standard equipment the accessories necessary for double resonance. Primarily, the basic instrumental requirement is that there be some means provided for field frequency stabilization. The second rf field B_2 is provided using sideband modulation.

The theory of double resonance techniques is beyond the scope of this book. Excellent reviews of this subject are available (28,29). Among the experiments included under homonuclear double resonance are spin decoupling, spin tickling, INDOR, and nuclear Overhauser effect (NOE). Each of these techniques is discussed in the following as to the experimental details and the type of information obtainable from their use.

During the course of the discussion, we shall refer to lines in the spectrum corresponding to transitions having an energy level in common as connected transitions (30). Connected transitions will be further classified depending on whether the energy of the common level is intermediate between those of the two terminal levels (progressive transition, Figure 4.18a) or whether the energy of the common level falls outside the range of the two other levels (regressive transition, Figure 4.18b).

We shall furthermore consider only frequency sweep spectra. In the construction of energy level diagrams, the external magnetic field is considered to lie along the negative z direction. When considering possible changes in a spectrum caused by the irradiating field B_2, a useful rule of thumb is that B_2 levels which are less than the linewidths will cause only population

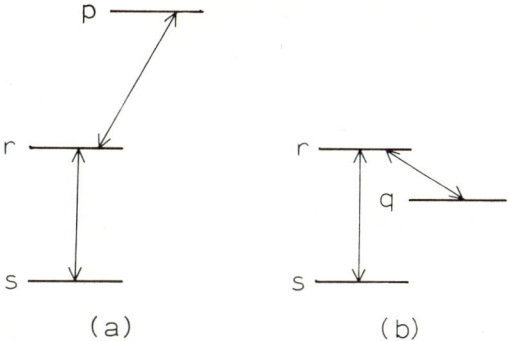

Fig. 4.18 An illustration of the two types of connected transitions in the NMR energy levels: (a) progressive transition; and (b) regressive transition.

changes whereas B_2 levels which are equal to the linewidth or larger will cause changes in the energy levels.

4.11.4 HOMONUCLEAR SPIN DECOUPLING

Spin decoupling implies that the spectra obtained are equivalent to those which would be observed in the absence of any coupling to the irradiated nuclei. Although this is not generally true, useful information to aid in the interpretation of complex NMR spectra can be obtained from spin-decoupled spectra. The most common application is to identify signals from nuclei which are spin coupled to the nucleus being irradiated. Usually the multiplet from one nucleus is irradiated with B_2 while observing the effects in the multiplets due to the other nuclei (Figure 4.19). Another application is the location of a "hidden" resonance in a complex spectrum with overlapping resonances by finding the irradiating frequency v_2 which causes collapse of splittings arising from coupling to the "hidden" nucleus.

Experimentally, in spin decoupling the irradiating field B_2 is adjusted such that the frequency of B_2, v_2, coincides with the chemical shift of the nucleus to be irradiated. The level of B_2 is adjusted to a level necessary to bring about spin decoupling, and the spectrum is obtained. Nuclei spin coupled to the nucleus being irradiated appear as if the irradiated nucleus was not present in the spin system, that is, a reduction in the number of lines in the resonances due to the spin coupled nuclei is observed. Although not strictly correct, the effect of B_2 can be thought of as causing rapid transitions among the possible spin states of the irradiated nucleus such that the remaining nuclei "see" only an average for the irradiated nucleus and not the individual spin states.

148   ANALYSIS OF SPECTRA AND DETERMINATION OF STRUCTURE

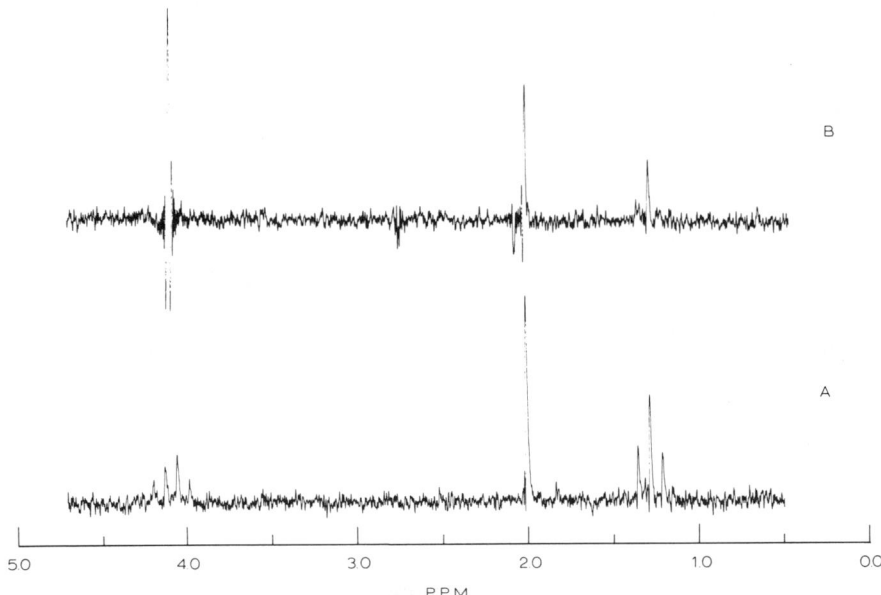

Fig. 4.19 (A) The 60-MHz NMR spectrum of ethylacetate. (B) The result of decoupling the methylene protons by irradiating the center of the methylene quartet with a strong rf field B_2.

True spin decoupling can be observed only when the chemical shift separation between the nuclei in question is much greater than the coupling constant. In the simplest cases in which a single nucleus is coupled to one other nucleus or a group of equivalent nuclei (AX, AX_n), the multiplet signal due to A is reduced to a single peak by irradiation at the resonance frequency, v_x, of X. In order to observe complete decoupling, the intensity of B_2 must be sufficiently high such that

$$\frac{\gamma B_2}{2\pi} \gg 2|J_{AX}| \tag{4.55}$$

Estimates of the level of B_2 required for decoupling may be obtained by experiments with one of the test samples (acetaldehyde or ethylbenzene) supplied by the spectrometer manufactor. At lower levels of B_2 than that required for decoupling, complex spectra are observed. For an AX spin system, these usually amount to spectra in which either four lines are observed for A or a two line pattern is observed in which the splitting is smaller than J_{AX}. Similar effects are also produced if the frequency of B_2 is incorrectly set from the correct value for decoupling, v_x (29).

Usually, decoupling cannot be obtained when the chemical shift separation is not much larger than the coupling constant between the nuclei. The B_2 level needed becomes comparable to the chemical shift difference such that irradiation of the signals from one nucleus perturbs the other nucleus also. If the chemical shift difference is not infinitely large compared to the coupling constant, the frequency of B_2 for optimum decoupling differs from the resonance frequency of the nucleus being irradiated. The correct frequency is displaced towards the signal being observed by an amount given by (29)

$$\left(\frac{\gamma B_2}{2\pi}\right)^2 \cdot \frac{1}{v_A - v_X} \qquad (4.56)$$

Caution should be exercised when chemical shifts are determined from decoupled spectra. The levels of B_2 used to produce decoupling are sufficiently high to produce significant changes in the resonance frequencies of the other nuclei. Since the field experienced by the nuclei is a vector sum of B_2 and the polarizing field B_0, all signals are shifted away from the point of irradiation, v_2, by an amount equal to

$$\left(\frac{\gamma B_2}{2\pi}\right)^2 \cdot \frac{1}{2(v_1 - v_2)} \qquad (4.57)$$

where v_1 is the frequency when B_2 is zero. For a B_2 level of 3 mG and a separation of 30 Hz between the observing and irradiating frequency, the shift is 2.7 Hz. This is known as the Bloch–Siegert shift and it affects all signals in the spectrum regardless of whether they are coupled to the nucleus being irradiated. An appropriate correction must be made to obtain the true chemical shift value. This shift is greatest for signals near the irradiating signal in a frequency swept spectrum; but in a field sweep spectrum, all signals experience the same shift.

Finally, one should not use decoupled spectra to obtain integration data for determining relative intensities. The high-irradiation level of B_2 can cause saturation of signals near the point of irradiation. The reduction of signal intensity is proportional to the square of the level of B_2 and decreases with increasing separation of the signal from the frequency of B_2. Although the decrease in signal intensity depends upon the relaxation times, it is not uncommon to find a reduction in intensity by 50% for a signal separated from the frequency of B_2 by a B_2 field of 3 mG.

4.11.5 SPIN TICKLING

Spin tickling differs primarily from spin decoupling in that a weaker irradiating field B_2 is used. Furthermore, only a single nondegenerate line in the spectrum is irradiated, rather than an entire multiplet. If the strength

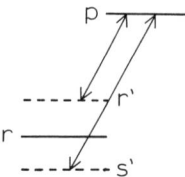

Fig. 4.20 A schematic illustration of the splitting of an energy level during a spin-tickling experiment.

of B_2 is about the same magnitude as that of the width of the line being irradiated (in Hz or milligauss), the result is that transitions which share a common energy level with the irradiated line will be split into doublets (31).

The splitting of the connected transitions is due to the mixing of the unperturbed states in the rotating frame. Irradiation of v_{rs} (Figure 4.20) will cause energy level r to be mixed into two new eigenstates r' and s'. The doublet arises from the transitions $r' \rightarrow p$ and $s' \rightarrow p$ in the progressive case or $q \rightarrow r'$ and $q \rightarrow s'$ in the regressive case. For a case such as that in Figure 4.20, where p and s differ by two units, the doublets will be broadened (31) and where s and q have the same spin quantum number, the doublets will be well resolved.

Consider the ABX spectrum given in Figure 4.21, with $|J_{AB}|J_{BX}| > |J_{AX}|$ and $v_X > v_B > v_A$. An energy level diagram for the three-spin system is given in Figure 4.22 with the correct labeling of the transitions for all positive coupling constants. If all coupling constants were positive, irradiation of line 12 in the X region with a tickling field should cause lines 7 and 8 in the B region and lines 3 and 4 in the A region of the spectrum to be split into doublets. Saying this another way, irradiating line 12 in the X region should cause those lines in the B region that have the same A spin state as line 12 to be split into doublets and those lines in the A region which have the same B spin state as line 12 to be split into doublets, that is, lines 3, 4, 7, and 8 should be split into doublets. Clearly from Figure 4.21 this is not the case. Since lines 1 and 2 in the A region and lines 5 and 6 in the B region are split into doublets, the energy level diagram should be relabeled accordingly. Irradiation of line 9 confirms this. Therefore, it is seen from the results previously stated that the coupling constant J_{AB} is opposite in sign from that of J_{AX} and J_{BX}. This method can be used to assign the energy level diagram of larger spin systems. However, the number of experiments increases considerably.

The spin tickling technique can be used to determine the relative signs of coupling constants and to trace the energy level diagram for a spin system.

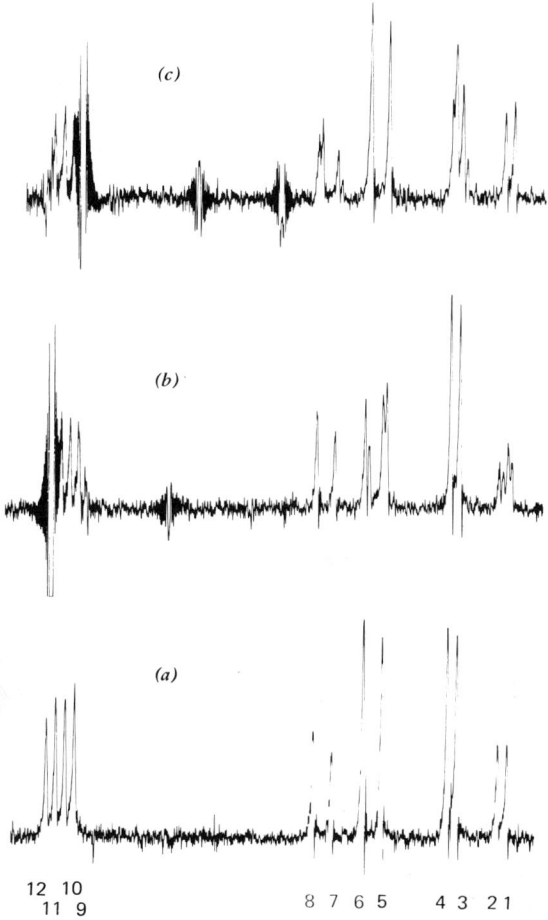

Fig. 4.21 (a) An expansion of the *ABX* part of the 60-MHz NMR spectrum of 1,4-diphenyl-azetidinone. (b) The result of irradiating line 12 with a spin-tickling rf field B_2. (c) The result of irradiating line 9 with a spin-tickling rf field.

152   ANALYSIS OF SPECTRA AND DETERMINATION OF STRUCTURE

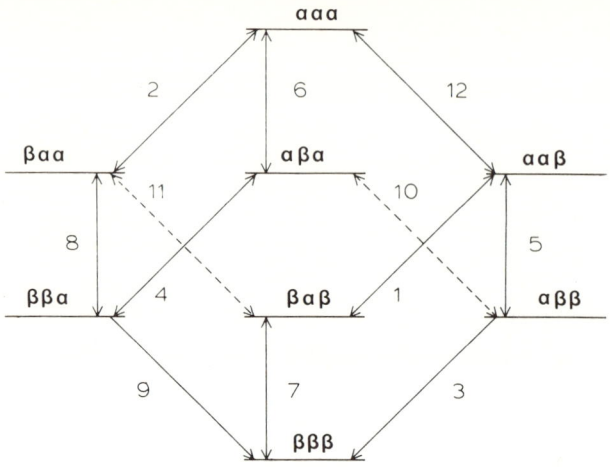

Fig. 4.22 An energy level diagram for a three-spin system. The transitions are labeled for a negative J_{AB}.

Furthermore, this technique can be used to locate transitions which are "hidden" due to overlap with other resonances by varying the frequency of B_2 until the doublet splitting is observed.

4.11.6 INDOR

The INDOR (INternuclear DOuble Resonance) (32) technique can be used to obtain information similar to that derived from spin tickling (33). However, the experimental conditions differ considerably. In INDOR experiments, a smaller B_2 is used compared to that used in tickling experiments. In addition, the intensity of a single line is monitored with the observing field B_1 while B_2 is swept through the spectrum. A change in intensity occurs when B_2 irradiates a line having a common energy level with the line being monitored. A positive signal is produced when the transition observed and that irradiated have a progressive relationship and a negative signal when the transition observed and that irradiated have a regressive relationship. No change in the intensity is observed if there is no coupling between the nuclei or if the two transitions have no common energy level. With reference to the energy level diagram for a three-spin system in Figure 4.22, if line X_{12} is monitored as B_2 is swept through the A and B portion of the spectrum, one observes positive peaks for A_4 and B_8 and negative peaks for

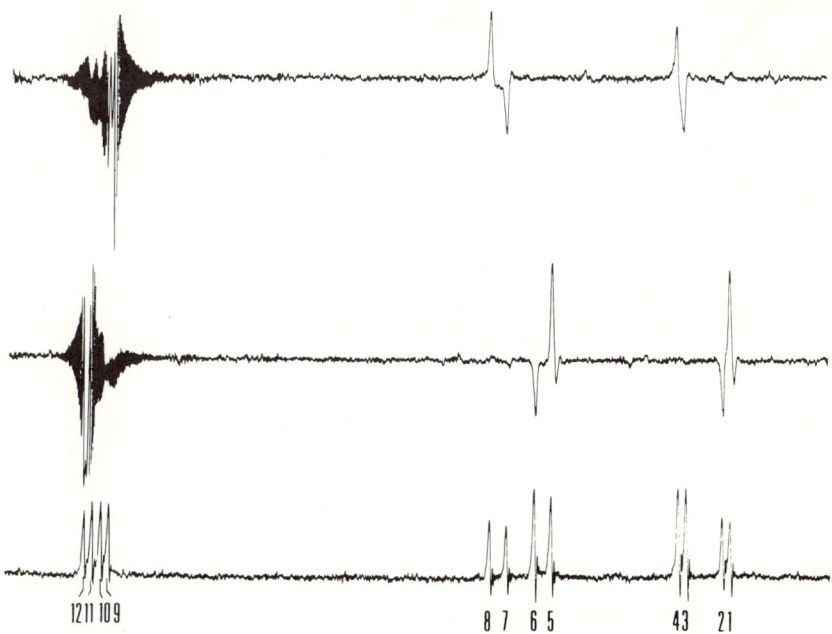

Fig. 4.23 An expansion of the ABX part of the 100-MHz NMR spectrum of 1,4-diphenyl-azetidinone and the results of INDOR spectra with B_1 monitoring lines 9 and 12.

A_3 and B_7 (Figure 4.23). If the spectrum is not first order, one can construct an energy level diagram and use the INDOR spectra to assign the transitions.

The experimental conditions for INDOR determination are rather rigid in terms of overall system stability. The observing field must be maintained at the center of a line and drifts in B_1 must be less than the line width in order to maintain an even baseline. Furthermore, there should be no short-term variations in the peak height due to changes in resolution or uneven spinning. Provision must be made for detecting changes in the intensity of the signal monitored by B_1 with a y–t recorder. The intensity of B_1 must be sufficiently low in order not to saturate the signal being monitored. Typical values of B_1 are 0.1–0.2 Hz. The irradiating field B_2 should be about 0.5 Hz, sufficient to cause population transfer between spin states without perturbing the energy levels.

Typical applications of the homonuclear INDOR technique are determination of relative signs of coupling constants and the location and measurement of the multiplicity of hidden signals. In general, if the accessories for INDOR are available, the INDOR method is preferred over spin tickling for determining relative signs of coupling constants.

4.11.7 INTRAMOLECULAR NUCLEAR OVERHAUSER EFFECT

The nuclear Overhauser effect (NOE) (34) differs from the previously mentioned decoupling techniques in that it does not depend on two nuclei being spin coupled. NOE experiments are used to provide information about molecular geometries under suitable conditions (35–38). Experimentally, the method involves the saturation of one signal in the spectrum and observation of changes in the intensities in the other signals. The magnitude of the intensity changes depend upon internuclear distances between the nuclei concerned. These intensity changes arise from perturbations of the relaxation processes which lead to thermal equilibrium between the spin states. The relaxation of a given nucleus is affected by all surrounding nuclei and depends upon the mean square value of the magnetic field produced by the surrounding nuclei. The relaxation from other nuclei is dominated by short range interactions and is proportional to r^{-6} where r is the distance separating the nuclei. For protons, an NOE effect is observed for nuclei usually separated by less than about 3.5 Å. The interactions leading to relaxation may be either intermolecular of intramolecular. For dilute solutions in proton-free solvents, only the intramolecular relaxation is significant.

The maximum intensity increase possible for only dipolar interactions is 50%. Any other type of relaxation will result in an intensity increase less than 50%. Furthermore, the intensity increase will be reduced if more than one type of nucleus is contributing to the relaxation of the nucleus in question. If a molecule is exchanging rapidly between different conformations, an enhancement may be observed if the internuclear distance is small in one of the conformations and will depend upon the wieghted average value of r^{-6}.

Care should be exercised in the preparation of a sample for an NOE experiment. Dissolved oxygen can cause relaxation thereby reducing the intensity enhancement. Therefore, the sample should be subjected to several freeze-thaw cycles and sealed under vacuum. Solvents with a low concentration of magnetic nuclei such as CS_2 should be used in order to reduce intermolecular relaxation. Deuterated solvents may be used to reduce the solvent contribution to the relaxation.

For a single line, B_2 levels of the order of 0.1 mG are required. Higher values are required to saturate a multiplet. This can lead to problems of saturation of the signal being observed if the irradiated signal is not well-separated from the observed signal. Experimentally, it is best to increase the level of B_2 slowly from zero until the maximum signal enhancement is observed. One should always use integration of peak areas for determining the enhancement as peak height often gives misleading results due to the removal of unresolved small couplings. Furthermore, one should compare the peak area obtained when B_2 is offset to a blank region of the spectrum.

4.11.8 HETERONUCLEAR DOUBLE RESONANCE

Although the types of experiments are basically the same, there are several important differences between homonuclear and heteronuclear double resonance as far as instrumentation is concerned (28,39). Many of the recent research NMR spectrometers have provisions for heteronuclear double resonance. However, the older instruments must be modified. First, a means must be provided for introducing the second rf to the sample. This has been accomplished by either placing a second transmitter coil in the probe or by double tuning the existing transmitter coil to accept the frequencies of both B_1 and B_2 (39). Second, since the frequency of B_2 usually differs considerably from that of B_1, some means should be provided for locking the frequency of B_2 to the frequency of B_1 in order to compensate for drifts in the magnetic field. The frequency of B_2 can be derived from a frequency synthesizer, from the master crystal oscillator of the spectrometer, or from a stable rf oscillator which is frequency modulated to give sidebands at the proper frequency for double resonance. Third, since proton–other nuclei coupling constants are usually much larger than proton–proton coupling constants, the strength of B_2 is considerably larger in heteronuclear double resonance experiments. The high-power levels required usually leads to heating of the probe and provision should be provided for maintaining the probe at constant temperature.

Heteronuclear spin decoupling $H - \{X\}$ has been used to remove the coupling constants to other nuclei in proton spectra. This provides a convenient means, not only for simplifying proton spectra for analysis, but also for identifying the proton–heteronuclei coupling constants. If the frequency relationship between B_1 and B_2 is known, heteronuclear spin-decoupling experiments also provide a means for determining the chemical shifts of other nuclei by determining the frequency of B_2 for maximum decoupling. This application is particularly useful for those nuclei which are not very sensitive to NMR detection.

Another major use of heteronuclear spin decoupling has been to remove the broadening effects due to quadrupole relaxation (28,39). The proton spectra of compounds containing ^2D, ^{11}B, and ^{14}N often show line broadening, due to the quadrupolar relaxation of these nuclei. Such broadening complicates the accurate determination of absorption frequencies and may be eliminated by irradiation at the absorption frequencies for these nuclei.

Heteronuclear spin tickling experiments provide a convenient means for determining the relative signs of proton–other nuclei coupling constants. The techniques are essentially the same as outlined in the previous discussion of homonuclear spin tickling. The coupling of protons to other magnetic nuclei, present in low natural abundance, is observed in proton spectra as satellite resonances displaced on both sides of the resonance due to species containing other nuclei that are nonmagnetic. The frequencies in the other nuclei spectrum and the relative signs of the coupling constants can be obtained by observing the changes in the satellite resonances while selectively irradiating the resonances due to the other nuclei. Similar techniques have also provided the signs and magnitudes of a variety of coupling constants between two hetero nuclei (39). Applications of the INDOR technique can also provide similar information.

One of the more important applications of heteronuclear double resonance has been in $X - \{H\}$ experiments. The NMR of heteronuclei is not as sensitive as that of proton NMR due to the smaller magnetic moments and the lower natural abundance of the heteronuclei. Some sensitivity enhancement is achieved by irradiation over all the proton frequency range using broadband decoupling which results in the collapse of the multiplet and the concentration of the intensity in a single line. Furthermore, when the protons are directly attached to the heteronuclei, an additional increase in sensitivity may be achieved due to a NOE. The broadband proton-decoupling technique is now standard practice in ^{13}C NMR studies where an increase in sensitivity by a factor of ~ 3 may be obtained in certain cases (40).

4.11.9 LANTHANIDE SHIFT REAGENTS

The discovery by Hinckley (41) in 1969 that certain paramagnetic lanthanide β-diketonates can be used to induce stereospecific shifts in the spectra of certain organic compounds has opened up a new area in NMR spectroscopy. Although large shifts induced by paramagnetic lanthanide ions had been observed previously, the ease of application, the small accompanying linebroadening, and the information obtainable from the use of lanthanide shift reagents (LSR) has greatly expanded previous applications. The initial application of LSR was to reduce the complexity of proton NMR spectra (in many cases the spectra are reduced to first order), (Figure 4.24) and additional applications have been rapidly forthcoming. Excellent reviews of LSR have appeared (42–44).

The usual approach has been to obtain a series of spectra in which increasing increments of a LSR have been added to the sample. A plot of the induced shifts, Δ_{obs} versus either the concentration of LSR or LSR/substrate, is then made. For small relative concentrations of LSR, these plots are usually linear. After assigning the resonances to specific protons and extracting the coupling constants from the "shifted" spectrum, the chemical shifts in the absence of the LSR are found by extrapolating the Δ_{obs} values back to zero concentration of LSR. It is assumed that the coupling constants obtained from the "shifted" spectrum are identical to the coupling constants in the absence of the LSR. There is some indication that LSRs may affect coupling

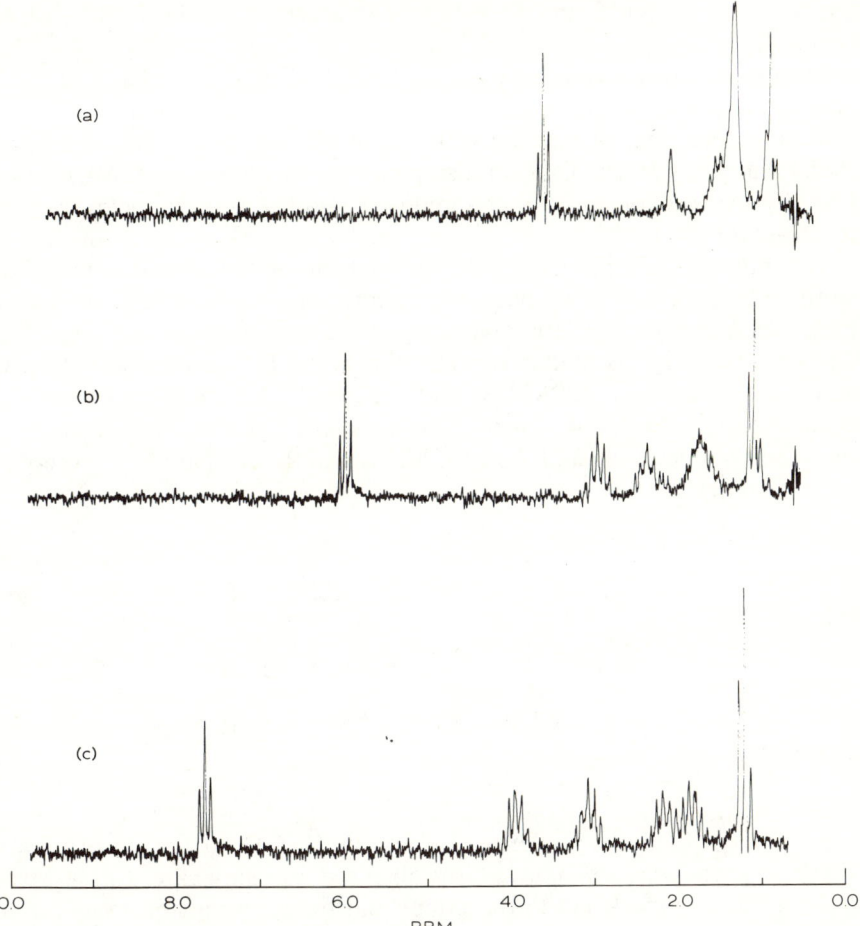

Fig. 4.24 (a) The 100-MHz NMR spectrum of hexanol. (b) The spectrum as a result of adding a small amount of Eu(DPM)$_3$ to the sample. (c) The spectrum as a result of adding a larger amount of Eu(DPM)$_3$ to the sample.

constants (45). However, using this technique, it is often possible to analyze a complex spectrum as a first order spectrum.

Another common practice when using LSRs has been to extrapolate the above plots to a 1:1 molar ratio of LSR to substrate and refer to these values of Δ_{obs} at the 1:1 ratio in order to assess the relative shifting power of various LSRs and the degree of complex formation for various substrates. It now appears that drawing such extrapolations should be avoided since the plots have been found to exhibit curvature at higher LSR/substrate ratios (42).

Several solvents have been used for LSR studies. On the basis of extrapolation of plots similar to that above to a 1:1 molar ratio of substrate to LSR, it appears that carbon tetrachloride is a better solvent (larger induced shifts) than chloroform. Carbon disulphide has also been used. Most of the LSR are hydroscopic and water should be avoided as it competes with the substrate for the LSR.

The induced shifts observed with LSR depend on the complexation of the lanthanide ion with a Lewis base site (usually a heteroatom) on the substrate molecule. Although a variety of paramagnetic complexes have been examined for their potential as shifts reagents, the best LSR to date are the tris-dipivaloylmethanates (DPM) (3) and the *tris*-1,1,1,2,2,3,3-heptafluoro-7,7-octanedionates (FOD) (4). Their utility as shift reagents is thought to arise from their ability to expand the coordination of the lanthanide by accepting additional ligands.

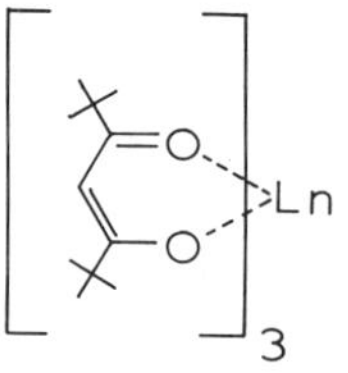

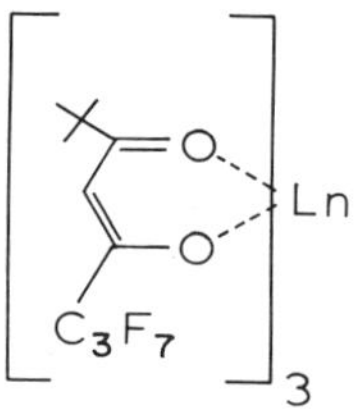

Several lanthanide complexes of DPM and FOD have been examined. While large induced shifts have been observed, in many cases the substrate resonances are broadened to the extent that the coupling constants cannot be obtained from the spectra. As a compromise between the magnitude of the induced shift and the accompanying line broadening, the best lanthanide ions appear to be Eu, Pr, and Yb. These LSRs also exhibit greater solubility in organic solvents normally used in NMR experiments. The induced shifts observed in proton spectra are to lower field with Eu and Yb LSR and to higher field with Pr LSR (42). Both $Eu(NO_3) \cdot 6H_2O$ and $Pr(NO_3)_3 \cdot 6H_2O$ are useful LSRs for aqueous solutions (46). Two mechanisms have been proposed to account for LSR-induced shifts Δ_{obs}; (1) the contact mechanism Δ_c and (2) the pseudocontact mechanism Δ_{pc}. The contact shift mechanism

$$\Delta_{obs} = \Delta_c + \Delta_{pc} \tag{4.58}$$

contributes to the induced shift only if there is transfer of unpaired spin density from the LSR to the substrate. This transfer of spin density may occur either through direct delocalization of the unpaired spin or by spin polarization. The induced shift due to the contact mechanism is given by (47)

$$\Delta_c = \frac{-2\pi\beta v AJ(J+1)gL(gL-1)}{3kTv} \qquad (4.59)$$

where β is the Bohr magneton, v is the nuclear Larmor frequency, A is the scalar coupling constant, J is the electronic-spin angular momentum and gL is the Landé g-factor. The contact-induced shift falls off rapidly with distance from the coordination site and, for proton spectra, is thought to be important only for protons in the immediate vicinity of the coordinating site in the substrate molecule. However, for nuclei other than protons, the contact shift mechanism may be the dominant shift mechanism (58).

The pseudocontact shift arises from a dipolar interaction between the nucleus and the electron spin magnetization of the paramagnetic lanthanide ions and results from the nonaveraging of the anisotropic electronic g-tensors in the complexes. Alternatively, the induced shift may be thought of as arising from the magnetic field generated by the unpaired electron spin. If it is assumed that the complex is axially symmetric, the pseudocontact shift is given by (48)

$$\Delta_{pc} = \frac{-vB^2 J(J+1)}{9kTr^3}(3\cos^2\theta - 1)(g_z - g_x)(g_z - g_y) \qquad (4.60)$$

where θ is the angle between the distance vector r joining the lanthanide ion and the nucleus in the complexed substrate and the crystal field axis of the complexed substrate and the g's are the g-tensor components. It is further assumed that the crystal field axis of the complex lies along a line joining the lanthanide ion and the coordinating site.

It is thought that the pseudocontact shift mechanism makes the major contribution to LSR-induced shifts of protons. The majority of data have been interpreted in terms of equation 4.61 where n is the constant

$$\Delta_{pc} = \frac{n(3\cos^2\theta - 1)}{Tr^3} \qquad (4.61)$$

of proportionality. Equation 4.61 predicts that LSR-induced pseudocontact shifts depend on the distance separating the lanthanide ion and the nucleus in question, the angle θ (the sign changes at $\theta = 54°44'$), and the temperature.

The $1/r^3$ distance dependence of LSR-induced shifts was recognized early and has been adequately demonstrated. The angular dependence of LSR-induced shifts was not considered until it was observed that some resonances are shifted downfield while others are shifted upfield in the same molecule (42–44). It now appears that both factors should be considered together. One problem with the angle dependence is the location of the lanthanide ion with respect to the substrate. Computer programs have been written for an iterative procedure whereby the location of the lanthanide ion is varied until a "best fit" is obtained for all shifted nuclei (49,50). The effects of temperature have not been investigated thoroughly and it is not clear at this time whether the observed effects are due to the temperature per se or to the effects of temperature on complex formation. It has been observed that the induced shift increases with decreasing temperature in some cases and decreases in others (42,44).

Considerable effort has been devoted to the determination of the stoichiometry of the LSR-substrate complex. Since only one set of resonances are observed for the substrate, there must be fast exchange on the NMR time scale between the substrate molecules. Various equilibria have been considered in order to derive expressions for the determination of the dissociation constant of the complex and the shift of the

$$\begin{array}{c} S + LSR \rightleftharpoons S\text{-}LSR \ 1:1 \\ S\text{-}LSR + S \rightleftharpoons S_2\text{-}LSR \ 1:2 \end{array} \quad (4.62)$$

pure complex. Most of the treatments have been patterned after the studies of charge transfer complex and hydrogen bonding. Caution should be exercised in such studies since interactions of the LSR with impurities or the solvent will affect the results.

Various functional groups have been examined with LSRs. Larger induced shifts are observed with the more basic functional groups. There appears to be a general correlation between the induced shift and the pKa of the substrate (51). Steric factors are also important. Several studies of the ability of various functional groups to coordinate with LSRs have established the following orders of effectiveness:

amine > hydroxyl > ketone > aldehyde > ether > ester > nitrile (52)
ether > thioether > ketone > ester (53)
phosphoryl > carbonyl > thiocarbonyl > thiophosphoryl (54)

Structural factors such as α–β unsaturation may change the order stated.

Chiral shift reagents have been prepared and found to be effective in resolving the spectra of enantiomeric mixtures (55,56). Most of these shift reagents use camphor to introduce the optically active center. When a chiral shift reagent is added to an enantiomeric mixture, it is found that the protons of one enantiomer appear at a lower field than the other enantiomer. These reagents have been used with amines, alcohols, esters, epoxides, and ketones. It appears as if the chiral shift reagent is forming complexes having different dissociation constants with the enantiomer. There is some evidence that the geometries of the complexes may also be different.

[Structure 5: a chiral camphor-derived β-diketonate complex with Ln, shown as [...]₃]

5

Although most of the work with LSRs has been with proton spectra, LSR-induced shifts have been observed in the spectra of other nuclei. These include ^{13}C (57), ^{14}N (58,59), and ^{31}P (46,60). Larger induced shifts are observed with these nuclei than with protons. This is attributed to a larger contribution from the contact shift mechanism which may in some cases become the dominate shift mechanism.

LSRs have found application in a variety of uses other than the analysis of complex spectra. LSR-induced shift data, in conjunction with the $1/r^3$ and θ dependence, are extremely valuable aids in assigning resonances to specific protons. They also provide information concerning the stereochemistry of the substrate (caution should be exercised in conformation studies since the formation of the complex may change the rotamer populations). Analysis of a mixture of similar compounds is facilitated by the use of LSR.

4.12 CORRELATION OF NMR PARAMETERS

When an NMR spectrum is used in addition to other spectral and chemical information, it is often possible to uniquely determine the structure of an unknown compound. In some cases, if a molecular formula is known, the NMR spectrum alone may be sufficient to determine the structure. In more complex cases, the NMR spectrum often provides information which would be difficult or impossible to obtain by other spectral methods.

Basically, there are three types of information available from an NMR spectrum: (1) the chemical shifts; (2) integrated intensity; and (3) coupling constants. Each of these pieces of information provides unique information for use in the determination of structure. The value of the chemical shifts allows one to determine the types of protons present in the sample, that is, protons bonded to saturated carbons, protons bonded to olefinic carbons, protons bonded to aromatic carbons, protons bonded to carbon in functional groups such as aldehydes, and protons bonded to heteroatoms. In favorable cases, one can further distinguish protons as to specific type such as methyl, methylene, and methine protons. Furthermore, the number of absorptions in the sample allow one to determine the number of chemically different types of protons. The integration provides a determination of the relative number of protons of each chemically different type. If one of the peaks in the spectrum can be assigned to a specific number of protons, a methyl group for example, the total number of protons in the sample can be determined from the relative areas of the remaining peaks. The coupling constants provide information on the geometrical relationship of the protons in the sample. The splitting pattern produced as a result of coupling not only allows one to determine which protons are adjacent in the sample, but also the number of adjacent protons. With experience, one will be able to recognize the splitting patterns in NMR spectra that are characteristic of particular groupings of nuclei or molecular fragments. From a knowledge of the chemical shifts and coupling constants, one can start putting together various fragments of an unknown structure much like one would go about solving a jigsaw puzzle.

The use of chemical shifts and coupling constants in structure determinations relies very heavily on the use of model compounds and the correlations between these parameters and structures derived from the examination of known compounds. Rather than try to cover each possible case that may arise (an impossible task), we shall discuss the most general aspects of chemical shifts and coupling constants in the following sections. There are several excellent compilations of proton NMR data available (61–66). These sources should be consulted for additional data on chemical shifts and coupling constants.

4.12.1 CHEMICAL SHIFTS

The examination of the NMR spectra of a wide variety of known compounds has shown that protons bonded to carbon appear in characteristic chemical shift ranges, depending upon the hybridization of the attached carbon and the substitution in the compound. Protons bonded to sp^3 hybridized carbons normally absorb within the range 0.5–5.0 ppm. These proton chemical shifts are influenced primarily by the electronegativity of adjacent substituents (diamagnetic screening) and long-range shielding effects of particular groupings of nuclei, as discussed in Section 2.7. Of the protons that appear in this range, it is usually easier to distinguish methyl protons from the others, since methyl protons can only appear in a limited number of splitting patterns (singlet, doublet, or triplet) and the signals of a methyl group are more intense than those of methylene and methine protons. In favorable cases where there is no overlap of the absorptions in the spectrum, one should be able to distinguish methylene from methine protons on the basis of the integration.

The chemical shifts of several substituted alkanes that can serve as model compounds are given in Table 4.16. The effect of a single substituent on proton chemical shifts in alkanes depends on the location of the substituent with respect to the proton and the type of proton in question. For protons directly bonded to the carbon to which the substituent is attached H–C–X, (α protons), the chemical shifts vary over a range of 3–4 ppm. Protons three bonds removed from the substituent (β protons vary over a range of ~ 1 ppm while protons four bonds from the substituent vary over a range of only ~ 0.3 ppm. Substituent effects on protons further than four bonds removed from the substituent are usually negligible.

TABLE 4.16. Chemical Shifts For Some Alkyl Derivatives With A Single Functional Group[a,b]

	Methyl	Ethyl		n-Propyl			iso-Propyl	
X	CH_3X	CH_3	CH_2X	CH_3	CH_2	CH_2X	CH_3	CHX
H	0.23	0.86	0.86	0.91	1.33	0.91	0.91	1.33
$CH=CR_2$	1.73	1.00	2.00					
$-C\equiv CR$	1.75	1.15	2.15	0.97	1.50	2.10	1.15	2.59
$-C_6H_5$	2.34	1.21	2.63	0.95	1.65	2.59	1.25	2.89
$-CHO$	2.18	1.13	2.46	0.98	1.65	2.35	1.13	2.39
$-COR$	2.10	1.05	2.47	0.93	1.56	2.32	1.08	2.54
$-CO_2H$	2.08	1.16	2.36	1.00	1.68	2.31	1.21	2.56
$-CO_2R$	2.01	1.12	2.28	0.98	1.65	2.22	1.15	2.48
$-CONR_2$	2.05	1.13	2.23	0.99	1.68	2.19	1.18	2.44
$-F$	4.27	1.24	4.36					
$-Cl$	3.06	1.33	3.47	1.06	1.81	3.47	1.55	4.14
$-Br$	2.69	1.66	3.37	1.06	1.89	3.35	1.73	4.21
$-I$	2.16	1.88	3.16	1.03	1.88	3.16	1.89	4.24
$-CN$	1.98	1.31	2.35	1.11	1.71	2.29	1.35	2.67
$-NO_2$	4.29	1.58	4.37	1.03	2.01	4.28	1.53	4.44
$-NH_2$	2.47	1.10	2.74	0.93	1.43	2.61	1.03	3.07
$-NHCOR$	2.71	1.12	3.21	0.96	1.55	3.18	1.13	4.01
$-OH$	3.39	1.18	3.59	0.93	1.53	3.49	1.16	3.94
$-OR$	3.24	1.15	3.37	0.93	1.55	3.27	1.08	3.55
$-OCOR$	3.67	1.21	4.05	0.97	1.56	3.98	1.22	4.94
$-SH$	2.00	1.31	2.44	1.02	1.57	2.46	1.34	3.16
$-SR$	2.09	1.25	2.49	0.98	1.59	2.43	1.25	2.93

[a] In ppm.
[b] Data taken from Emsley, Feeney, and Sutcliffe (65) and Jackson and Sternhell (66).

From the data in Table 4.16, it can be seen that one can easily distinguish several types of methyl protons. For example, the chemical shift separation between CH_3-CH_2-, $CH_3\overset{\overset{O}{\|}}{C}-$, and CH_3O- is large enough such that there should be no difficulty in assigning these peaks in an unknown struc-

TABLE 4.17. Substituent Effects on Aliphatic Proton Chemical Shifts[a]

Substituent	Type of Hydrogen[b]	Alpha Shift	Beta Shift
—Cl	CH_3	2.43	0.65
	CH_2	2.30	0.53
	CH	2.55	0.03
—Br	CH_3	1.80	0.83
	CH_2	2.18	0.60
	CH	2.68	0.25
—I	CH_3	1.28	1.23
	CH_2	1.95	0.58
	CH	2.75	0.00
—OH	CH_3	2.50	0.33
	CH_2	2.30	0.13
	CH	2.20	0.00
—OR (sat.)	CH_3	2.43	0.33
	CH_2	2.35	0.15
	CH	2.00	0.00
$-OC_6H_5, -O-\overset{\overset{O}{\|}}{C}$	CH_3	2.88	0.38
$-O\overset{\overset{O}{\|}}{C}OR$	CH_2	2.98	0.43
	CH	3.43	—
—C=C—	CH_3	0.78	—
	CH_2	0.75	0.10
	CH	—	—
$-\overset{\overset{O}{\|}}{C}-X$, X=OH, OR, H, N alkyl, or aryl	CH_3	1.23	0.18
	CH_2	1.05	0.31
	CH	1.05	—
—NRR′	CH_3	1.30	0.13
	CH_2	1.33	0.13
	CH	1.33	—

[a] Taken from Silverstein, Bassler, and Morrill (67).
[b] These values are to be added to the standard positions: CH_3, $\delta 0.87$; CH_2, $\delta 1.20$, CH, $\delta 1.55$.

ture. There is, however, considerable overlap of the various methyl resonances between 2.0 and 3.0 ppm such that it may be difficult to uniquely assign signals in this region to a particular functional group in the absence of further information.

Several attempts have been made to derive an additivity scheme for predicting the chemical shifts of aliphatic protons in compounds containing a single functional group. One such set of additivity parameters for calculating the chemical shifts of methyl, methylene, and methine protons is given in Table 4.17. It should be kept in mind that the use of the substituent parameters in Table 4.17 allows one to calculate approximate chemical shifts only. Normally, the deviations will be within 0.3 ppm. Furthermore, the data refer to dilute solutions (less than 10%) in either $CDCl_3$ or CCl_4. The use of another solvent may lead to significant deviations from the calculated value. The additivity parameters given in Table 4.17 should also prove useful for higher substituted alkanes provided the substituents are located at least six bonds from each other.

And additivity scheme has also been proposed for predicting the chemical shifts of methylene protons where the carbon is attached to two substituents (67). These additivity

$$\delta = 0.28 + \sum \sigma_{eff} \qquad (4.63)$$

constants σ_{eff}, known as Shoolery's constants, are given in Table 4.18. Again, chemical shifts calculated using equation 4.63 should be regarded as approximate values only. Deviations of up to 0.6 ppm have been observed in some cases. Although, the use of the substituent parameters in Tables 4.17 and 4.18 leads to significant deviations in some instances, they nevertheless have proven useful in assigning protons to specific absorptions in NMR spectra.

The chemical shifts of protons bonded to sp^2 hybridized carbons in olefins are dependent on the remaining substituents attached to the carbons. Data for a large number of compounds have been reported. Examination of this data has led to an additivity scheme which is very useful for predicting the chemical shifts of olefinic protons (69,70). The chemical shift of an olefinic proton can be calculated by taking into account the geometrical relationship of the proton with respect to the substituents by using equation 4.64 where 5.28

$$\begin{array}{c} H \diagdown \qquad \diagup cis \\ C{=}C \\ \diagup \qquad \diagdown \\ gem \qquad \qquad trans \end{array}$$

$$\delta = 5.28 + \sum \sigma_{eff} \qquad (4.64)$$

TABLE 4.18. Shoolery's Constants for Calculating the Chemical Shifts of Methylene Protons Bonded to Two Substituents $X-CH_2-Y$ [a,b]

X or Y	σ_{eff}	X or Y	σ_{eff}
—CH_3	0.47	—CN	1.70
—C=C	1.32	—COR	1.70
—C≡CR	1.44	—$COCRH_5$	1.84
—I	1.82	—$CONR_2$	1.59
—Br	2.33	—COOR	1.55
—Cl	2.53	—OCOR	3.13
—C_6H_5	1.85	—SR	1.64
—NR_2	1.57	—SCN	2.30
—NO_2	2.46	—CF_2	1.21
—OH	2.56	—CF_3	1.14
—OR	2.36	—N_3	1.97
—OC_6H_5	3.23	—NHCOR	2.27

[a] In ppm.
[b] Data taken from Silverstein, Bassler, and Morrill (67).

is the chemical shift of ethylene and σ_{eff} is the additivity parameter for the substituent. Substituent additivity parameters are given in Table 4.19. In most cases, chemical shifts calculated using equation 4.64 are within ±0.3 ppm of the experimental value. Larger deviations are observed in cases where there is extended conjugation, ring strain, or inhibition of resonance. Chemical shifts for various types of aliphatic and olefinic compounds are given in Table 4.20.

The proton chemical shifts of monosubstituted benzenes have been investigated by a number of workers. The chemical shifts of aromatic protons usually lie within the range of 6.0–8.5 ppm (Table 4.21). A ring current in the aromatic ring (Section 2.7) is thought to be responsible for the downfield shift of aromatic protons compared to olefinic protons. Spiesecke and Schneider have considered the effect of substituents on the protons shifts of monosubstituted benzenes (71). They suggested that the *para*-proton shifts are controlled primarily by the resonance effect of the substituent. In support of this, reasonable correlations were found between the *para*-proton shifts and the π-electron density of the attached carbon. Similar correlations were also found with Hammett σ-*para* substituent constants. *Ortho*-proton chemical shifts cannot be explained on the basis of resonance and inductive effects alone. For the halobenzenes, it appears as if there is a major contribu-

TABLE 4.19. Substituent Constants for Calculating the Chemical Shift of Olefinic Protons[a]

X	σ_{gem}	σ_{cis}	σ_{trans}
—CH_3	0.44	−0.32	−0.34
—Alkyl	0.44	−0.26	−0.29
—Alkyl–ring	0.71	−0.33	−0.30
—C≡C—	0.50	0.35	0.10
—C=C	0.98	−0.04	−0.21
—C=C (conj)	1.26	0.08	−0.01
—C_6H_5	1.43	0.39	0.06
—Aromatic	1.35	0.37	−0.10
—CH_2O–,	0.67	−0.02	−0.07
—CH_2S–	0.53	−0.15	−0.15
—CH_2Cl, –CH_2Br	0.72	0.12	0.07
—CH_2N	0.66	−0.05	−0.23
—C≡N	0.30	0.75	0.53
—C=O	1.10	1.13	0.81
—C=O (conj)	1.06	1.01	0.95
—CO_2H	1.00	1.35	0.74
—CO_2H (conj)	0.69	0.97	0.39
—CO_2R	0.84	1.15	0.56
—CO_2R (conj)	0.68	1.02	0.33
—CHO	1.03	0.97	1.21
—$CONR_2$	1.37	0.93	0.35
—COCl	1.10	1.41	0.99
—OR (R aliph)	1.18	−1.06	−1.28
—OR (R conj)	1.14	−0.65	−1.05
—Cl	1.05	0.14	0.09
—Br	1.02	0.33	0.53
—NR_2 (R aliph)	0.69	−1.19	−1.31
—NR_2 (R conj)	2.30	−0.73	−0.81
—SR	1.00	−0.24	−0.04
—SO_2—	1.58	1.15	0.95

[a] Data taken from Pascual, Meier, and Simon (69) and Tobey (70).

tion from the diamagnetic anisotropy of the substituent. *Meta*-proton chemical shifts do not appear to correlate with resonance, inductive, or diamagnetic anisotropic effects of the substituent. It appears as if all three factors are contributing to the *meta*-proton shifts.

Several di- and higher-substituted benzenes have also been examined. From the examination of several *para*-disubstituted benzenes, Diehl (72) has suggested that the effect of substituents on aromatic proton chemical

TABLE 4.20. Chemical Shifts of Protons Bonded To Carbon In Some Miscellaneous Compounds[a]

Compound	δ	Compound	δ
cyclopropane (△)	0.22	norcarane (H,H on cyclopropane)	0.02
cyclobutane (□)	1.96	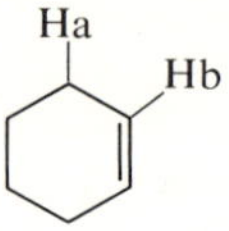 cyclohexene Ha, Hb	a 1.96 b 5.57
cyclopentane	1.51		
cyclohexane	1.44	cyclopropyl ketone	1.65
cycloheptane	1.54	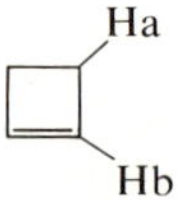 cyclobutene Ha, Hb	a 2.57 b 5.97
$H_2C=C=CH_2$	4.55	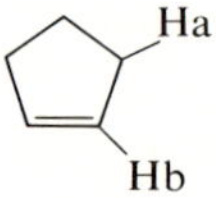 cyclopentene Ha, Hb	a 2.28 b 5.60
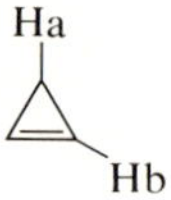 cyclopropene Ha, Hb	a 0.92 b 7.01	methylenecyclopropane CH_2a, Hb	a 5.38 b 0.99
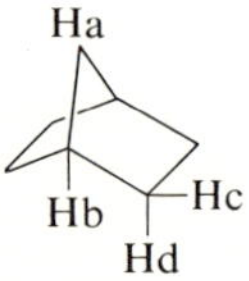 norbornane Ha, Hb, Hc, Hd	a 1.21 b 2.20 c 1.49 d 1.18	methylenecyclobutane H_2a, Hb, Hc	a 4.70 b 2.7 c 1.92
norbornene Ha, Hb, Hc, Hd, He, Hf	a 1.32 b 1.07 c 2.83 d 1.57 e 0.94 f 5.95	methylenecyclopentane CH_2a, Hb, Hc	a 4.82 b 2.7 c 1.92
		ethylene oxide	2.54
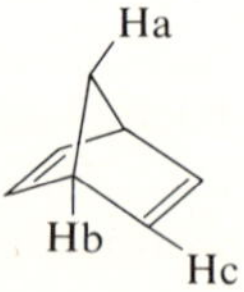 norbornadiene Ha, Hb, Hc	a 1.95 b 3.53 c 6.66	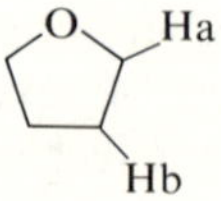 tetrahydrofuran Ha, Hb	a 3.63 b 1.79

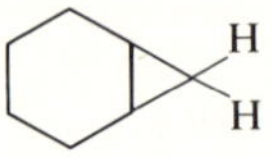

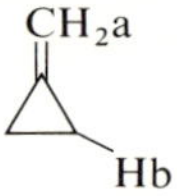

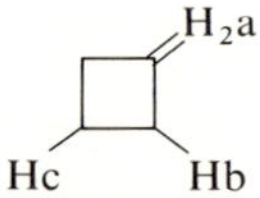

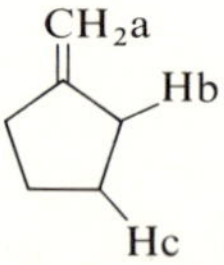

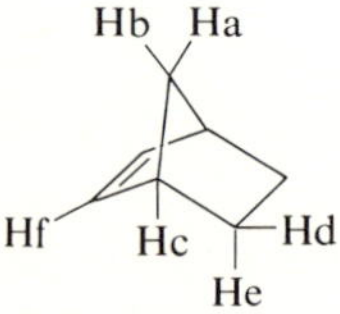

168

TABLE 4.20. (*Continued*)

Compound	δ	Compound	δ
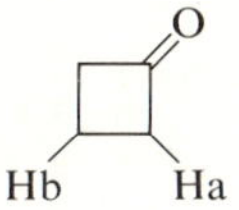 cyclobutanone (Hb, Ha)	a 3.03 b 1.96	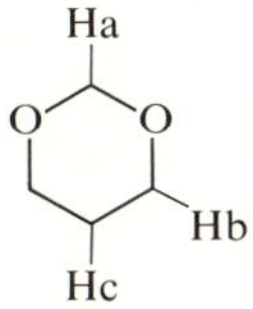 1,3-dioxane (Ha, Hb, Hc)	a 4.82 b 3.80 c 1.68
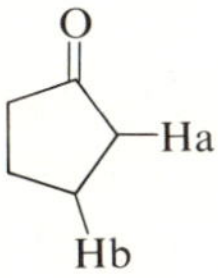 cyclopentanone (Ha, Hb)	a 2.06 b 2.02	 aziridine (Ha)	a 1.48
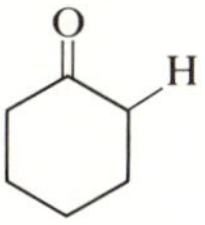 cyclohexanone (H)	2.25	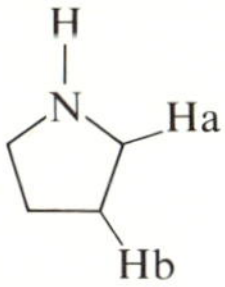 pyrrolidine (Ha, Hb)	a 2.74 b 1.62
methylenecyclohexane (CH₂a, Hb)	a 4.55 b 1.5	thiirane	2.27
		tetrahydrothiophene (Ha, Hb)	a 2.82 b 1.93
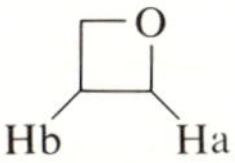 oxetane (Hb, Ha)	a 4.73 b 2.72	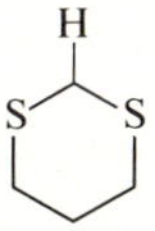 1,3-dithiane (H)	3.69
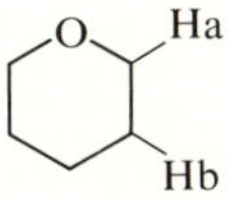 tetrahydropyran (Ha, Hb)	a 3.56 b 1.58	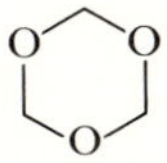 1,3,5-trioxane	5.00
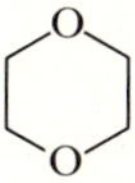 1,4-dioxane	3.59	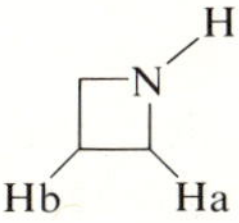 azetidine (Hb, Ha)	a 3.54 b 2.23
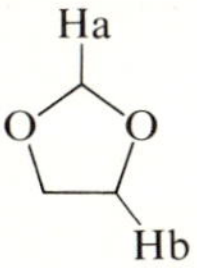 1,3-dioxolane (Ha, Hb)	a 4.77 b 3.77	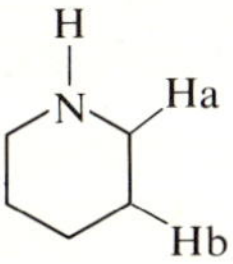 piperidine (Ha, Hb)	a 2.69 b 1.49

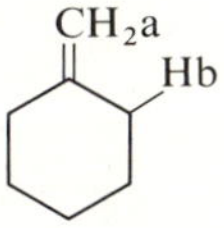

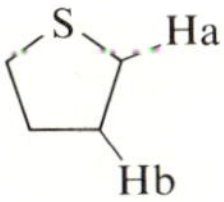

TABLE 4.20. (*Continued*)

Compound	δ	Compound	δ
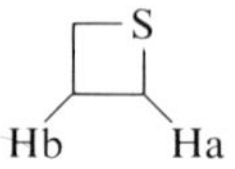 thietane (Hb, Ha with S)	a 2.82 b 1.93	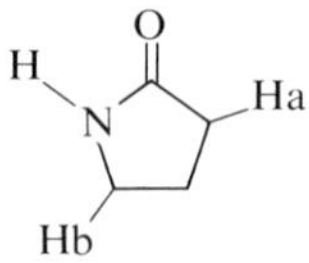 2-pyrrolidinone (Ha, Hb)	a 2.3 b 3.4
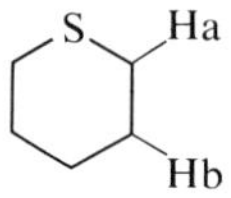 thiane (Ha, Hb)	a 2.57 b 1.6	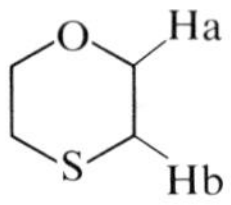 1,4-oxathiane (Ha, Hb)	a 3.88 b 2.57
1,3,5-trithiane	4.18	γ-butyrolactone (Ha, Hb, Hc)	a 2.31 b 2.08 c 4.28
morpholine (Ha, Hb)	a 3.57 b 2.83		
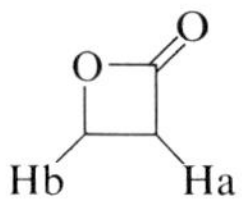 β-propiolactone (Hb, Ha)	a 3.48 b 4.22	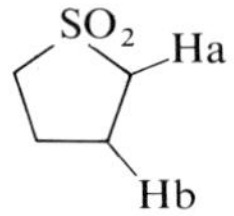 sulfolane (Ha, Hb)	a 2.92 b 2.16
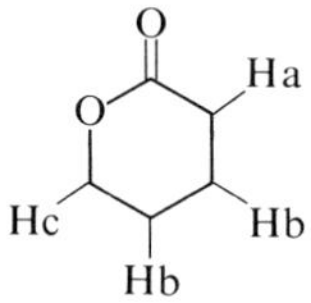 δ-valerolactone (Ha, Hb, Hc)	a 2.27 b 1.62 c 4.06	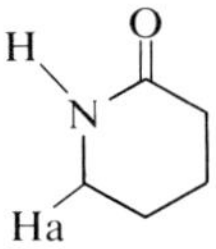 2-piperidinone (Ha)	3.17

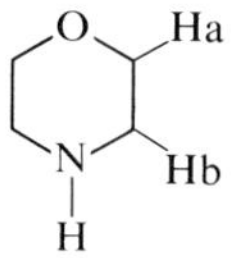

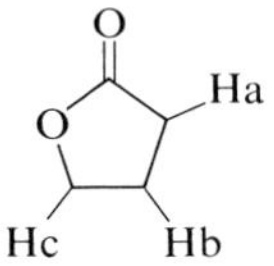

a Data taken from various sources.

shifts is additive. The effect of a single substituent X on the chemical shifts of a monosubstituted benzene with respect to benzene (Table 4.21) is denoted as $S_{o;x}$, $S_{m;x}$, and $S_{p;x}$. The calculation of the proton chemical shifts of a *para*-disubstituted benzene $p\text{-}C_6H_4XY$ is given as follows:

$$\delta^{xy}_{o;x} = S_{o;x} + S_{m;y} \tag{4.65}$$
$$\delta^{xy}_{m;y} = S_{m;x} + S_{o;y} \tag{4.66}$$

This additivity scheme can also be applied to *meta*-disubstituted benzenes. The respective chemical shifts are

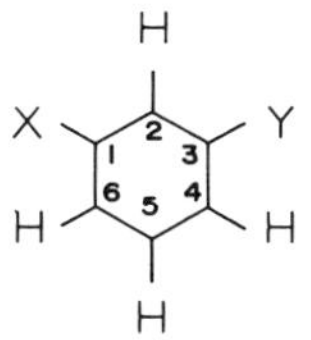

$$\delta_2 = S_{o;x} + S_{o;y} \tag{4.67}$$

$$\delta_4 = S_{p;x} + S_{o;y} \tag{4.68}$$

$$\delta_5 = S_{m;x} + S_{m;y} \tag{4.69}$$

$$\delta_6 = S_{o;x} + S_{p;y} \tag{4.70}$$

Similar treatment of an *ortho*-disubstituted benzene yields the chemical shifts

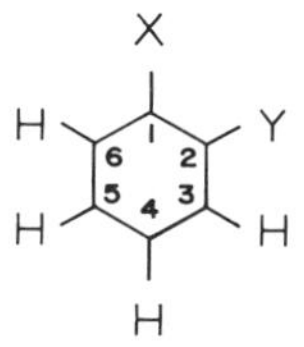

$$\delta_3 = S_{m;x} + S_{o;y} \tag{4.71}$$

$$\delta_4 = S_{p;x} + S_{m;y} \tag{4.72}$$

$$\delta_5 = S_{m;x} + S_{p;y} \tag{4.73}$$

$$\delta_6 = S_{o;x} + S_{m;y} \tag{4.74}$$

For the *para*- and *meta*-disubstituted benzenes, chemical shifts obtained from the equations previously stated agree with experimental values to within ±0.1 ppm in most cases. However, significant deviations are observed for some *ortho*-disubstituted benzenes (77). It has been suggested that the deviations result from steric interactions between the substituents which prevent the substituents from exerting their normal resonance and inductive

TABLE 4.21. Chemical Shifts for Some Monosubstituted Benzenes[a,b]

X	δ_{ortho}	δ_{meta}	δ_{para}
—CH$_3$	−0.28	−0.17	−0.25
—C(CH$_3$)$_3$	−0.01	−0.07	−0.13
—CH=CH$_2$	0.02	−0.06	−0.13
—C$_6$H$_5$	0.20	0.03	−0.06
—C≡CH	0.15	−0.05	−0.03
—C≡CC$_6$H$_5$	0.19	−0.02	−0.05
—C≡N	0.33	0.17	0.29
—CHO	0.53	0.18	0.27
—COCH$_3$	0.59	0.10	0.18
—COC$_6$H$_5$	0.43	0.09	0.18
—COCl	0.80	0.20	0.36
—CONH$_2$	0.69	0.17	0.24
—COOH	0.72	0.11	0.22
—COOCH$_3$	0.70	0.07	0.17
—COOC$_6$H$_5$	0.88	0.15	0.26
—F	−0.30	−0.03	−0.24
—Cl	0.00	−0.07	−0.13
—Br	0.16	−0.12	−0.07
—I	0.37	−0.25	−0.03
—NH$_2$	−0.81	−0.26	−0.65
—N(CH$_3$)$_2$	−0.69	−0.19	−0.67
—N$^+$(CH$_3$)$_3$I$^-$	0.71	0.39	0.32
—NO$_2$	0.92	0.25	0.38
—OH	−0.46	−0.15	−0.40
—OCH$_3$	−0.50	−0.10	−0.45
—OC$_6$H$_5$	−0.33	−0.05	−0.29
—Li	0.76	−0.21	−0.30
—MgBr	0.39	−0.20	−0.27
—SiCl(C$_6$H$_5$)$_2$	0.31	0.03	−0.14
—PbCl(C$_6$H$_5$)$_2$	0.66	0.27	0.10
—PO(OCH$_3$)$_2$	0.43	0.13	0.21
—SH	−0.06	−0.13	−0.21

[a] Data taken from references 73–76.
[b] With respect to benzene (7.27 ppm).

effects (78). Deviations are also observed when the additivity scheme is applied to 1,2,3-trisubstituted benzenes.

The chemical shifts of a wide variety of substituted heteroaromatic compounds and polycyclic aromatic compounds have been examined. Some representative values are given in Table 4.22. In general, the effect of substituents on these ring systems is similar to their effect on benzene chemical shifts. Of special interest are the chemical shifts of protons which interact sterically (peri interaction) such as the 1,8-protons on a naphthalene ring system or the 1,9-protons on a phenanthrene ring system. These protons are usually shifted downfield in comparison to proton chemical shifts in similar compounds where this interaction is absent. In polycyclic aromatic compounds, the ring current effects of the multiple rings appear to reinforce each other (79). Since this contribution to chemical shifts depends on the distance of the proton in question to the additional rings, protons closer to the attached ring(s) will be shifted downfield with respect to protons further removed from the additional ring(s) (for example the α- and β-proton shifts on naphthalene).

Protons bonded to atoms other than carbon absorb over a wider range than protons bonded to carbon. Protons bonded to oxygen, nitrogen, and sulfur are subject to hydrogen bonding. As a result, their chemical shifts are dependent on the solvent, concentration, and temperature. Furthermore, since these protons are acidic in nature, they readily undergo proton exchange when a trace of acid or water is present in the sample. As a result of this rapid exchange, adjacent protons see only an average for the spin states of these protons and hence coupling is usually not observed between protons bonded to carbon and adjacent protons bonded to heteroatoms. If a proton bonded to a heteroatom is suspected in the sample, the peak due to this proton can usually be identified by taking advantage of the exchange property. The usual procedure is to rerun the spectrum after adding a few drops deuterium oxide to the sample and shaking the solution. The peak due to the proton bonded to the heteroatom will either disappear or reduce in intensity due to exchange of the proton with deuterium. In addition, a new peak due to HOD will appear in the spectrum between 4.5 and 5.0 ppm.

Representative chemical shift ranges for protons bonded to heteroatoms are given in Table 4.23. The signal due to the hydroxyl proton of alcohols is usually a sharp singlet due to exchange. If the sample is purified to remove water and acid, it may be possible to slow the exchange and observe coupling with the hydroxyl proton. Furthermore, if two or more hydroxyl groups are present in the compound, it may be possible to observe a separate peak for each group if the exchange is slow. A method for determining the type of alcohol based on slowing the exchange has been presented (80). In highly purified dimethyl sulfoxide, the exchange of hydroxyl protons is slowed such that one observes coupling between the hydroxyl proton and protons on the carbon to which the hydroxyl group is attached. Thus the hydroxyl proton will appear as a singlet for a tertiary alcohol, a doublet for a secondary alcohol, and a triplet for a primary alcohol.

Since nitrogen-14 has a spin $I = 1$, one should observe coupling between amine protons and the nitrogen. However, there are two factors that usually prevent the coupling from being observed. If the proton on nitrogen is

TABLE 4.22. Representative Chemical Shifts in Aromatic Compounds[a]

Compound	Compound
pyrrole: 6.05, 6.62, NH 7.70	furan: 6.30, 7.40
thiophene: 7.04, 7.19	selenophene: 7.12, 7.70
pyrazole: 7.55, 6.25, 7.55, NH 13.7	imidazole: 7.14, 7.14, 7.70, N—H
benzofuran: 7.49, 6.66, 7.13, 7.52, 7.19, 7.42	indole: 7.55, 6.45, 6.99, 7.26, 7.09, 7.40, NH 9.80
benzothiophene: 7.71, 7.26, 7.27, 7.29, 7.30, 7.77	benzimidazole: 7.70, 7.26, NH
pyridine: 7.46, 7.06, 8.50	pyridazine: 7.46, 9.17
pyrimidine: 9.15, 7.09, 8.60	pyrazine: 8.5
quinoline: 7.68, 8.00, 7.43, 7.26, 7.61, 8.81, 8.05	isoquinoline: 7.70, 7.47, 7.57, 8.45, 7.49, 7.86, 9.13
quinoxaline: 8.04, 7.66, 8.73	phthalazine: 7.93, 9.44, 7.85
naphthalene: 7.81, 7.46	anthracene: 8.31, 7.91, 7.39

TABLE 4.22. (*Continued*)

Compound	Compound
Naphthalene-like fused ring: 7.71, 8.12, 7.82, 8.93, 7.88	Biphenylene: 6.60, 6.47
Azulene-like fused ring: 7.33, 8.26, 7.13, 7.82, 7.52	Cyclopentadienyl anion (−): 5.57
Cyclooctatetraene dianion (=): 5.69	Cyclooctatetraene dication (+): 9.28

a Data taken from various sources.

TABLE 4.23. Chemical Shifts for Protons Bonded to Heteroatoms[a]

Compound Type	δ[b]
Alcohols	0.5–5.0
Enols	15–19
Phenols	4.0–7.5
Phenols (intramolecular H-bonding)	10.0–12.0
Carboxylic acids	10–13
Oximes	7–10
Primary amines	1.1–1.8
Secondary amines	1.2–2.1
Anilines	3.3–4.0
Amides	5.0–6.5
N-Alkyl amides	6.0–8.2
N-Aryl amides	7.8–9.4
Ammonium salts	7.1–7.7
Thiols	1.0–2.0
Thiophenols	3.0–4.0

[a] Data taken from various sources.
[b] Dependent on concentration, solvent, and temperature.

exchanging rapidly, it will "see" an average of the spin states of nitrogen and hence will appear as if it were not coupled to nitrogen. Second, nitrogen has a quadrupole moment which induces relaxation, thereby decreasing the lifetime of the nitrogen spin states. Thus a proton attached to nitrogen will see an average for the spin states on nitrogen and will usually appear as a broadened peak. Normally, the N—H proton is exchanging at a rapid rate such that coupling is not observed between N—H and protons attached to the carbon to which nitrogen is attached. This coupling has been observed in cases where the sample was rigorously purified to remove all traces of water. Exchange is also slow in amine salts such that this coupling can be observed. Furthermore, in acid solution, the multiplicity of the signal due to the C—H protons on the carbon to which nitrogen is attached can be used to classify amines in certain cases (81). In addition, the N—H proton may appear as a broadened triplet due to coupling with nitrogen in favorable cases where the quadrupole moment on nitrogen is reduced.

Protons attached to sulfur usually exchange at a slow rate such that coupling is observed between the S—H proton and C—H protons on the carbon to which sulfur is attached. Although S—H protons can be exchanged with D_2O, S—H protons usually do not exchange with hydroxyl, amino, or carboxylic acid protons when these functional groups are present in the same compound.

4.12.2 COUPLING CONSTANTS

The magnitude of the proton coupling constants provide detailed information on the structure of unknown compounds. The examination of the spectra of a wide variety of known compounds has provided a number of empirical correlations between coupling constants and molecular structure. In most cases, a theoretical explanation of these trends has been presented. After one has carried out the spectral analysis to obtain the coupling constants, the next step is to utilize the coupling constants with the previously established correlations to confirm the presence or absence of particular molecular fragments in the unknown compound. In some cases, the presence or absence of particular molecular fragments may be confirmed by the presence or absence of certain splitting patterns in the spectrum without carrying out the detailed analysis.

Several compilations of coupling constants are available in review articles and texts (65,66,82,83). We shall not attempt to provide a theoretical treatment of coupling constants here. Rather, we will focus on the magnitude of coupling constants, the trends that have been established for coupling constants, and discuss some of the ways in which coupling constants can be used to provide information on the structure of unknown compounds.

In aliphatic compounds, coupling is normally observed between protons separated by two (H—C—H, ^2J, geminal coupling) and three bonds (H—C—C—H, ^3J, vicinal coupling). Coupling between protons separated by four or more bonds is usually not observed in aliphatic compounds except in some special molecular arrangements. When one of the intervening bonds is a π-bond, a small coupling between protons separated by four bond (H—C=C—C—H, allylic coupling) and five bonds (H—C—C=C—C—H, homoallylic coupling) may be observed. In aromatic compounds, coupling between protons separated by three bonds (*ortho* coupling), four bonds (*meta* coupling), and five bonds (*para* coupling) is normally observed. In addition, coupling may be observed between aromatic protons and protons on the benzylic carbon (benzylic coupling). In cases where the conjugation is extended, such as the cumulenes, coupling between protons separated by as many as nine bonds may be observed.

4.12.2.1 Geminal Coupling Constants

Geminal coupling constants for protons bonded to sp^3 hybridized carbons range in value from $+6$ to -30 Hz. Some values in representative compounds are given in Table 4.24. In general, electronegative substituents attached to the CH_2 group produce a decrease (more negative value) in the geminal coupling constant. Other factors which influence geminal coupling constants are the H—C—H bond angle and the orientation of lone electron pairs on adjacent heteroatoms. The orientation of π-substituents with respect to the H—C—H plane will also influence the magnitude of geminal coupling constants. In most cases, the geminal coupling constants in saturated systems are negative in sign. The exceptions are epoxides, aziridines, and 1,3-dioxalanes where the geminal coupling constant is positive. If a full spectral analysis is carried out and the signs of the coupling constants determined, a large negative coupling constant can usually be safely assigned to a geminal coupling since the majority of other large coupling constants are positive.

In contrast to saturated systems, the geminal-coupling constant for protons attached to an sp^2 hybridized carbon varies over a much larger range (Table 4.24). However, for most ethylene derivatives, the range is smaller than that found in saturated systems. Geminal couplings for protons attached to an sp^2 hybridized carbon are influenced by substituent electronegativity and orientation, and by ring strain for exocyclic methylene groups. A correlation between $^2J_{HH}$ and the electronegativity of the substituents has been proposed for ethylene derivatives $H_2C=CX,Y$

$$^2J_{H-H} = \frac{61.6}{E_x + E_y} = 12.9 \tag{4.75}$$

TABLE 4.24. Geminal Proton Coupling Constants[a]

Compound	^2J	Compound	^2J
CH_4	−12.4		
CH_3X	−9.2 to −16.9	norbornane CH$_2$	−5.4
$CH_2(CN)_2$	−20.4		
$K^+O_2^-CCH_2CHOHCO_2^-K^+$	−15.3		
cyclopropane	−0.5 to −9.9		
oxirane	+4.0 to +6.3	norbornane bridge CH$_2$	−9.5 to −13.0
thiirane	0 to −1.4	norbornene CH$_2$	−8.0 to −12.0
aziridine	0 to +1.5		
cyclobutane	−12.0 to −15.0	norbornene bridge	−10.4 to −13.7
cyclobutanone	−15.3 to −18.0	γ-butyrolactone α-CH$_2$	−17.0 to −18.9
cyclopentane	−12.0 to −15.0	γ-butyrolactone β-CH$_2$	−8.8 to −10.5
cyclopentanone	−19.0 to −19.5	$H_2C=O$	+41
		$H_2C=NR$	+8 to +16.5
cyclohexane	−11.6 to −15.0	$H_2C=CHX$	−3.2 to +7.4
cyclohexanone	−12.0 to −16.0	$H_2C=C=CR_2$	−9.0

[a] Data taken from various sources.

where E_x and E_y are the Pauling electronegativities of the substituents. In general, the magnitude of $^2J_{H-H}$ becomes more negative as the electronegativity of the substituents increase.

Geminal coupling constants in both saturated and unsaturated systems may be solvent-dependent (84). The general trend is towards a more negative value of J_{gem} as the polarity of the solvent increases, provided the negative end of the dipole moment of the molecule is directed away from the CH_2 group. This solvent-dependence may be used to determine the sign of the coupling constants in favorable cases.

4.12.2.2 Vicinal Coupling Constants

By far, the most informative coupling constant for structural determination is the vicinal coupling constant (H—C—C—H, $^3J_{H-H}$). Representative values of vicinal coupling constants are given in Table 4.25. In cases where there is a single bond joining the two carbons, rotation about the carbon–carbon is possible. In general, this rotation will be fast such that the observed vicinal coupling constant will be a population weighted average of the individual conformers. The magnitude of vicinal coupling constants where both carbons are sp^3 hybridized ranges from -0.3 to $+14$ Hz. The major influence on this type of vicinal coupling constant is the dihedral angle between the two C—H bonds (Figure 4.25). The electronegativity of substituents attached to the two carbons, the orientation of the substituents with respect to the two C—H bonds, bond angles, and bond lengths all influence the magnitude of the vicinal coupling constant. In general, the vicinal coupling decreases as the electronegativity of attached substituents increases.

On the basis of theoretical calculations, Karplus (85,86) suggested that the relationship between dihedral angle and vicinal coupling constant could be expressed as

$$J = J^0 \cos^2 \phi - C \quad \text{for} \quad 0° < \phi < 90° \tag{4.76}$$

$$J = J^{180} \cos^2 \phi - C \quad \text{for} \quad 90° < \phi < 180° \tag{4.77}$$

where J^0 (a standard coupling for a dihedral angle of 0°), J^{180} (a standard coupling for a dihedral angle of 180°), and C are constants. For an unsubstituted ethane H—C—C—H fragment, Karplus suggested the values $J^0 = 8.5$ Hz, $J^{180} = 9.5$ Hz, and $C = -0.3$. A plot of this relationship is given in Figure 4.25. While the value of C probably remains constant from one compound to another, it is clear that the values of J^0 and J^{180} are dependent on the substituents attached the H—C—C—H fragment. Examination of a variety of known compounds has shown that J^0 and J^{180}

TABLE 4.25. Vicinal Proton Coupling Constants[a]

Compound	^3J	Compound	^3J
CH_3CH_2X	7.0–9.0		
cyclopropane H,H	cis 2.2–12.5 trans 1.4–8.6	norbornane (exo,exo)	2.5–5.0
cyclobutane H,H	cis or 4.0–13.0 trans	norbornane (endo,endo)	9.0–10.0
cyclopentane H,H	cis or 4.0–13.0 trans	norbornane (exo,endo)	6.0–7.0
cyclohexane H,H	ax–ax 6–14 ax–eq 0–5 eq–eq 0–5	norbornane (bridgehead)	3.0–4.0
cyclopropene H,H	0.5–1.5	norbornane (H,H)	0.0–2.0
cyclobutene H,H	2.0–4.0	cyclopentene H,H	5.1–7.0
$CH_2=CH_2$	cis 11.5 trans 19.0	cyclohexene H,H	8.8–11.0
H,C=C,H (trans)	12.0–19.0	O=C–C=C H,H	5.0–8.0
H,C=C,H (cis)	6.0–12.0	C=C–C H,H	4.0–10.0
O=C(H)–C(H)	1.0–3.0	C=C–C=C H,H	9.0–11.0

[a] Data taken from various sources.

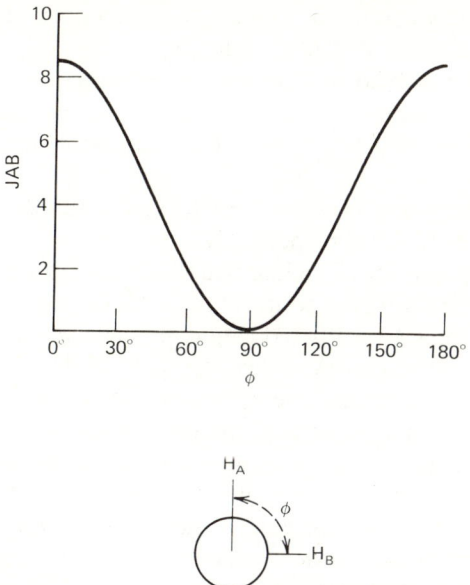

Fig. 4.25 A schematic illustration of the dihedral angle dependence of vicinal coupling constants (the Karplus relationship).

may vary from 8 to 16 Hz with J^{180} usually being the larger of the two couplings. This has led to the suggestion that it is probably better to consider the Karplus relationship in terms of a "family of curves" similar to the one shown in Figure 4.25 (83).

Provided reliable values of J^0 and J^{180} can be chosen, the application of equations 4.76 and 4.77 to derive stereochemical information is usually straightforward. After considering all reasonable conformations and measuring the respective dihedral angles with the aid of models, a series of simultaneous equations can be set up and solved such that they are internally consistent and an estimate of the magnitude of the dihedral angle obtained. With freely rotating systems, an additional unknown is the populations of the various conformers. In this case, one can assume reasonable conformers with fixed dihedral angles and set up the Karplus equations to solve for the relative populations. As a typical example, consider the 1,2-disubstituted ethane $X-CH_2CH_2-Y$. The most likely stable conformers and the corresponding equations are given in Figure 4.1 and equations 4.3 and 4.4. If J_t and J_g are known, or reasonable guesses of their values made, solution of equations 4.3 and 4.4 will yield the populations of the conformers.

Vicinal coupling constants in molecular fragments where one or both of the carbons is sp² hybridized (H—C(=)—C—H and H—C(=)—C(=)—H) also appear to follow a Karplus-like dependence on the dihedral angle (83). Rotational averaged values of these couplings are usually in the range 5–8 Hz. For the fragment H—C(=)—C—H, typical values of J_{gauche} and J_{trans} are 1.8–3.7 and 9.6–13.4 Hz, respectively.

The angular dependence of vicinal coupling constants has proven to be of great value in determining the conformations of six-membered rings. Where the chair form of the six-membered ring is the principal form, typical values of the vicinal coupling constants are J_{ax-ax} 8–13 Hz, J_{ax-eq} 2–6 Hz, and J_{eq-eq} 1–5 Hz. Even in cases where the spectrum cannot or has not been analyzed, the conformation of a particular proton on a cyclohexane ring can be determined provided its signals are separated from other signals. Due to the magnitude of the couplings, the width of the proton multiplet at half-peak height ($v_{\frac{1}{2}}$) is characteristic of the conformation of the proton (87). An equatorial proton will have $v_{\frac{1}{2}}$ usually smaller than 12 Hz whereas an axial proton will usually have $v_{\frac{1}{2}}$ larger than 15 Hz. Not only does this allow the conformation to be determined, but in many cases, the configuration of polysubstituted cyclohexanes may also be determined.

In smaller ring systems, the vicinal coupling constants may or may not provide as useful information concerning the conformation of the ring. As the ring becomes more planar, the dihedral angle between *cis* protons approaches 0° whereas the dihedral angle between *trans* protons decreases from 180° and may approach 90°. As a result, the *cis* coupling constant increases and the *trans* coupling constant decreases such that one can no longer assume that the *trans* coupling will be larger than the *cis* coupling. In three membered rings, J_{cis} is usually larger than J_{trans}. Typical ranges of the couplings are $J_{cis} = 2.2-12.5$ Hz, and $J_{trans} = 1.4-8.6$ Hz. For four- and five-membered rings, the vicinal couplings range from 4.0 to 13.0 Hz. There is no clear pattern as to whether J_{cis} will be larger or smaller than J_{trans}. It appears that each ring system must be considered separately (83).

The vicinal coupling constant in the molecular fragment H—C=C—H is just as informative as the previously discussed vicinal couplings in terms of their dependence on structure (Table 4.25). A large amount of data has been collected for this type of coupling and it is found that J_{trans} is always larger than J_{cis}. Values for J_{trans} in acyclic olefins are in the range 9.5–19.0 Hz whereas J_{cis} ranges from −2.0 to 11.7 Hz. Both J_{trans} and J_{cis} decrease as the electronegativity of the substituents attached to the H—C=C—H fragment increases. This relationship is approximately linear such that, if the substituents are known, structural assignments can usually be made when only one isomer is present by comparison of the changes in the coupling constants of the corresponding vinyl derivatives with ethylene. The vicinal coupling in H—C=C—H is also dependent on the ring size in cyclic systems (Table 4.25). This dependence on ring size is thought to be due to a dependence of the vicinal coupling on the angle H—C=C. It has also been suggested that this vicinal coupling is dependent on bond-order or bond-length of the double bond. However, sufficient data on bond lengths are not available to develop trends for use in the solution of structural problems (66).

4.12.2.3 Long-Range Couplings

Coupling constants between protons separated by four or more bonds are referred to as long-range couplings. Usually, in saturated systems, long-range coupling constants are not observed except in some special cases to be discussed later. The most common type of long range coupling is the allylic coupling (H—C—C=C—H) where both J_{cis} and J_{trans} may be observed. Some representative values are given in Table 4.26. The allylic coupling is dependent on stereochemistry similar to vicinal coupling constants (83). However, the allylic coupling is considerably smaller and may be either positive or negative in sign (-3 to $+2$ Hz), depending on the stereochemical arrangement of the protons. This relationship is shown approximately in Figure 4.26. In most cases, $|J_{trans}| > |J_{cis}|$. However, this

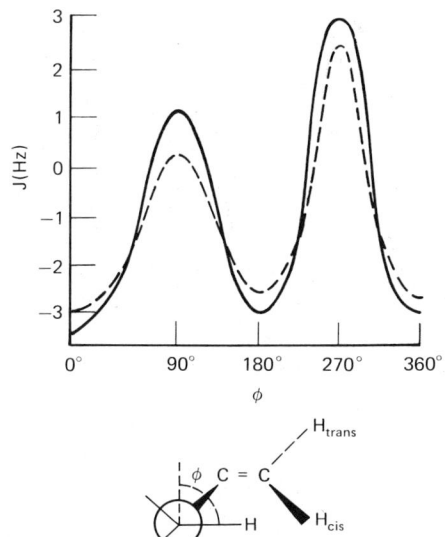

Fig. 4.26 A schematic illustration of the dihedral angle dependence of *cis* (– – –) and *trans* (———) allylic coupling constants (83).

TABLE 4.26. Long-Range Proton Coupling Constants[a]

Compound Type	J	Compound Type	J
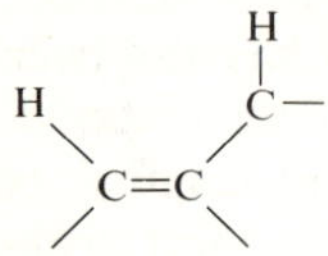 H-C=C-C-H (allylic)	−1.0 to −2.0	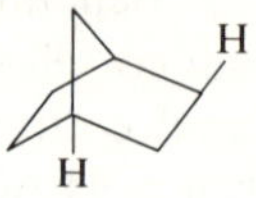 norbornane H-C-C-C-H	1.0
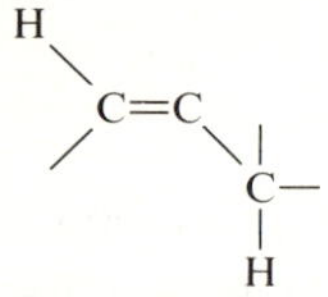 H-C=C-C-H (homoallylic, 4-bond)	−0.4 to −1.7	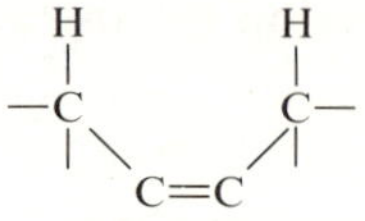 H-C-C=C-C-H	1.0–5.0
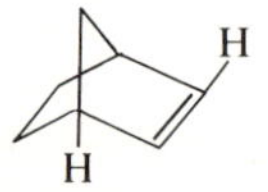 norbornene H-C-C=C-H	+0.5	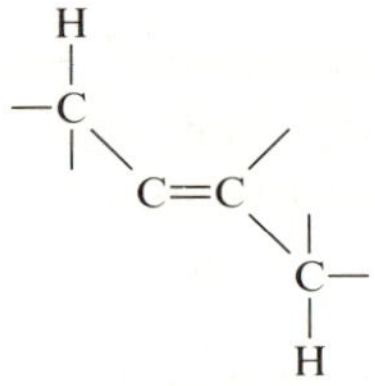 H-C-C=C-C-H	1.0–5.0
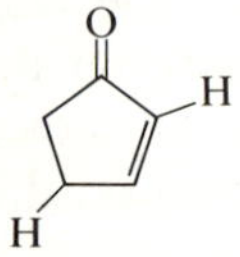 cyclopentenone	−2.1	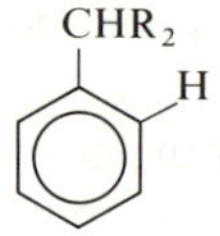 ortho CHR₂-C₆H₄-H	0.6–0.9
CH₂=C=CH₂	−7.37	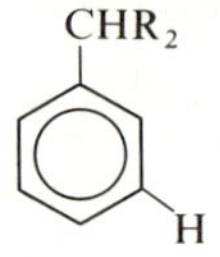 meta CHR₂-C₆H₄-H	0.2–0.8
CH₂=C=C=CH₂	7.01		
H−C≡C−C≡C−H	0.95	para CHR₂-C₆H₄-H	0.1–0.6
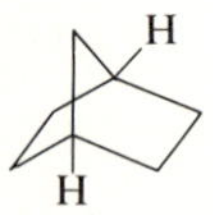 norbornane (exo,exo)	1.0–1.4	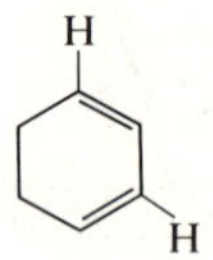 1,3-cyclohexadiene	1.04
norbornane (endo,endo)	1.0–1.4		
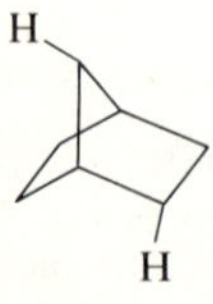 norbornane (bridge)	1.7–2.6	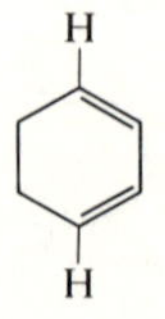 1,4-cyclohexadiene	1.11

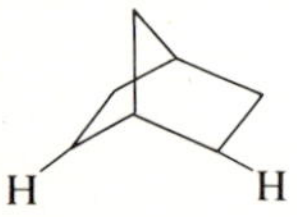

CORRELATION OF NMR PARAMETERS 185

TABLE 4.26. (*Continued*)

Compound Type	J	Compound Type	J
[norbornane with H exo and H endo]	6.7–8.1	[benzofuran-type with H and H, X, R]	0.6–1.0
[norbornane with two H]	8.0		

^{a} Data taken from various sources.

relationship is not firmly established to justify its use for structure determination.

Homoallylic long-range couplings (H—C—C=C—C—H) may also be observed (Table 4.26). These couplings are similar to allylic couplings in both magnitude and stereochemical dependence.

Long-range coupling is also observed in systems with extended conjugation separating the two protons such as in acetylenes, allenes and cumulenes (Table 4.26). The magnitude of this coupling is not attenuated rapidly and coupling between protons separated by as many as nine bonds has been observed. Usually, replacing a proton with a methyl group will not appreciably affect the magnitude of the coupling in conjugated systems.

Long-range coupling in saturated systems is usually observed when the protons are separated by four or five bonds and the protons are located in a planar, zig-zag arrangement. This type of coupling has come to be referred to as coupling-along-a W-path and appears to be independent of either the nature or hybridization of the intervening atoms. Some representative values are given in Table 4.26. The magnitude of this type of coupling falls off rapidly as the two protons lose coplanarity and is more commonly observed in unsaturated compounds.

4.12.2.4 Coupling in Aromatic and Heteroaromatic Compounds

The proton–proton coupling constants in aromatic and heteroaromatic compounds are of immense value in structural determination. In the case of di- and higher-substituted benzenes, one can easily distinguish between ^3J, ^4J, and ^5J and use these values to determine the substitution pattern.

If the heterocyclic system can be identified, the coupling constants can be used to readily determine the substitution pattern. Typical coupling constants in these systems are given in Table 4.27. These coupling constants vary regularly with the electronegativity of the substituents and ring size.

From the examination of a wide variety of monosubstituted benzenes, the trends in the coupling constants with substituent electronegativity have been defined (88). The *ortho* coupling, J_{12}, shows a pronounced increase with increasing substituent electronegativity. The *meta* coupling J_{15} shows

TABLE 4.27. Coupling Constants In Aromatic and Heteroaromatic Compounds[a]

Compound Type	J	Compound Type	J
benzene	$J_{12} = 6.0$–9.5 $J_{13} = 1.2$–3.3 $J_{14} = 0.0$–1.5	pyrazole	$J_{12} \sim 1.9$ $J_{23} \sim 2.0$
naphthalene	$J_{12} = 8.3$–9.1 $J_{23} = 6.1$–6.9 $J_{13} = 1.2$–1.6 $J_{14} = 0.0$–1.0	pyridazine	$J_{12} = 5.1$ $J_{23} = 8.0$–9.6 $J_{13} = 1.8$ $J_{14} = 3.5$
phenanthrene	$J_{12} = 8.0$–9.0 $J_{23} = 6.9$–7.3 $J_{34} = 8.0$–9.5 $J_{13} = 0.9$–1.6 $J_{24} = 1.2$–1.8 $J_{14} = 0.3$–0.7	pyrimidine	$J_{23} = 4.0$–6.0 $J_{12} = 0.0$–1.0 $J_{24} = 2.5$ $J_{13} = 1.0$–2.0
pyridine	$J_{12} = 4.0$–6.0 $J_{23} = 6.9$–9.1 $J_{13} = 0.0$–2.7 $J_{24} = 0.5$–1.8 $J_{15} = 0.0$–0.6 $J_{14} = 0.0$–2.3	pyrazine	$J_{12} = 1.8$–3.0 $J_{14} = 0.0$–0.5 $J_{13} = 1.3$–1.8
thiophene	$J_{12} = 4.9$–6.2 $J_{23} = 3.4$–5.0 $J_{13} = 1.2$–1.7 $J_{14} = 3.2$–3.7	furan	$J_{12} = 1.3$–2.0 $J_{23} = 3.1$–3.8 $J_{13} = 0.4$–1.0 $J_{14} = 1.0$–2.0
pyrrole	$J_{12} = 2.4$–3.1 $J_{23} = 3.4$–3.8 $J_{13} = 1.3$–1.5 $J_{14} = 1.9$–2.2	imidazole	$J_{23} \sim 1.6$ $J_{12} = 0.8$–1.5
		thiazole	$J_{23} = 3.2$ $J_{12} < 0.5$ $J_{13} = 1.9$

[a] Data taken from various sources.

a similar trend. The *meta* coupling J_{13} and the *para* coupling J_{15} show a decrease as the substituent electronegativity increases. The remaining *ortho* coupling, J_{23}, remains essentially constant with varying substituent electronegativity. It appears that the effect of substituents on the coupling constants of benzenes is primarily an inductive type effect and is rapidly attenuated with increasing distance from the substituent.

An additivity scheme for the coupling constants in disubstituted benzenes has been derived by using the coupling constants obtained from the monosubstituted benzenes. The additivity scheme is given in **8** where $J_o = 7.56$ Hz, $J_m = 1.38$ Hz, and $J_p = 0.69$ Hz are the coupling constants obtained from benzene itself. For *meta-* and *para*-disubstituted benzenes, the agreement between calculated and experimental coupling constants is usually within ± 0.1 Hz. Similar agreement is observed for other disubstituted benzenes, except for the case of strongly interacting *ortho*-substituents. Some examples where the additivity scheme fails are *ortho*-di-*t*-butylbenzene, *ortho*-dinitrobenzene (steric interactions between the two substituents) and *ortho*-nitrophenol (hydrogen-bonding between the nitro oxygen and phenolic hydrogen) (78). The breakdown of the additivity scheme for some *ortho*-disubstituted benzenes probably reflects geometrical distortion of the aromatic ring. If the interaction between *ortho*-substituents is taken into account, the additivity scheme

8

$$J_{12} = J_{12}^x + J_{23}^y - J_o \qquad J_{12} = J_{12}^x + J_{23}^y - J_o \qquad J_{12} = J_{12}^x + J_{12}^y - J_o$$
$$J_{23} = J_{23}^x + J_{23}^y - J_o \qquad J_{23} = J_{23}^x + J_{12}^y - J_o$$
$$J_{34} = J_{23}^x + J_{12}^y - J_o$$
$$J_{13} = J_{13}^x + J_{24}^y - J_m \qquad J_{13} = J_{13}^x + J_{13}^y - J_m \qquad J_{15} = J_{15}^x + J_{24}^y - J_m$$
$$J_{24} = J_{24}^x + J_{13}^y - J_m \qquad J_{15} = J_{15}^x + J_{13}^y - J_m \qquad J_{24} = J_{24}^x + J_{15}^y - J_m$$
$$J_{35} = J_{13}^x + J_{15}^y - J_m$$
$$J_{14} = J_{14}^x + J_{14}^y - J_p \qquad J_{25} = J_{14}^x + J_{14}^y - J_p \qquad J_{14} = J_{14}^x + J_{14}^y - J_p$$

should also be applicable to tri-substituted benzenes.

Substituent effects on the coupling constants in heteroaromatic compounds have been thoroughly investigated and are found to follow similar

TABLE 4.28. Proton–Fluorine and Proton–Phosphorus Coupling Constants[a]

Compound Type	J	Compound Type	J
—C(H)—C—F	47.5	>PH	180–225
—C(H)—C—F	25.7	>P(O)H	490–710
—C(F)—C(H)—F	57.2	>P(S)H	490–650
—C(F)—C(H)—F	20.8	—P$^+$—H	490–600
C(H)—CF$_3$	2.0–13.0	C—P(H)<	0–3.0
C(H)—C—CF$_3$	0.5–1.0	C—P(O), H	10–15
Fluorocyclohexane (F, H1, H2)	F_{ax}–H_1 49; F_{eq}–H_1 49; F_{ax}–2_{ax} 43.5; F_{ax}–2_{eq} 3; F_{eq}–2_{ax} <3; F_{eq}–2_{eq} <3	C—P(S), H	10–15
H$_2$C=CHF	gem 84.7; cis 20.1; trans 52.4	C—P$^+$—, H	10–15
CH$_3$(H)C=CF(H)	cis 2.6; trans 2.3	C—C—P<, H	13.7
fluorobenzene	6.2–10.1	C—C—P(O)<, H	18
		C—O—P<, H	0.5–12
		C—O—P(O)<, H	3.0–15

TABLE 4.28. (Continued)

Compound Type	J	Compound Type	J
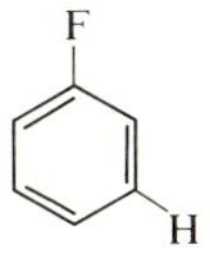 3-F-C$_6$H$_4$-H (meta)	6.2–8.3	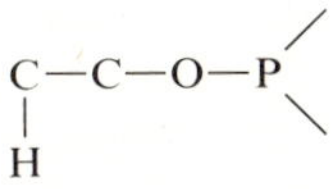 C–C(H)–O–P⟨	0–3.0
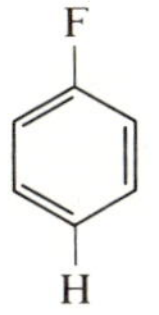 4-F-C$_6$H$_4$-H (para)	2.1–2.3	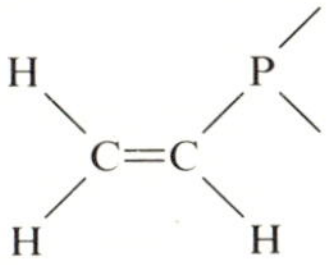 H_C=C_/P⟨ with H's	gem 11.7 cis 13.6 trans 30.2
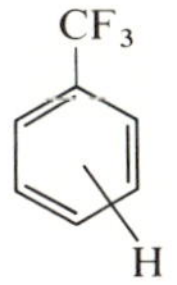 3-CF$_3$-C$_6$H$_4$-H	0.5–1.0		

a Data taken from various sources.

trends to those found for monosubstituted benzenes. The introduction of a heteroatom (compare substituted benzenes with pyridines in Table 4.27) in an aromatic ring results in a decrease in the *ortho* coupling constant. There is also a decrease in the *ortho* coupling constant on going from a 6- to a 5-membered ring. This trend is not observed for the *meta* coupling constants. One of the more interesting trends is the effect of protonation on the coupling constants of heteroaromatic compounds (89). Protonation of a nitrogen heterocycle results in an increase in the *ortho* coupling constant adjacent to nitrogen (i.e., J_{12} in pyridine) by from 0.5 to 1.5 Hz. This trend appears to be of general nature and may be used to identify the *ortho* coupling in nitrogen heterocycles. Removal of a proton (i.e., forming the nitranion of pyrrole) results in a corresponding decrease in the *ortho* coupling constant (90). These trends have been interpreted in terms of substituent effects similar to the effect of substituents on the coupling constants of monosubstituted benzenes (89).

4.12.2.5 Proton Coupling Constants with Other Nuclei

In principle, coupling can occur between any two magnetic nuclei. Fortunately, most organic compounds do not contain magnetic nuclei other than hydrogen. However, when other magnetic nuclei are present, coupling between the protons and other magnetic nuclei is detectable in the proton spectrum. The two most common "other nuclei" appearing in organic compounds are fluorine-19 and phosphorus-31. Both nuclei have spins of $I = \frac{1}{2}$ and thus couple with the protons in the sample just as if they were an additional proton; the difference being that the coupling constant between hydrogen and other nucleus is usually larger than a proton–proton coupling constant. Some typical values of proton–fluorine and proton–phosphorus coupling constants are given in Table 4.28. The magnitude of proton–fluorine coupling constants usually decreases with the number of bonds separating the two nuclei. For proton–phosphorus couplings, the magnitude of the couplings usually follow the trend $^1J > {}^3J > {}^2J$. This trend is also observed for the coupling constants between protons and other magnetic nuclei as well.

REFERENCES

1. J. A. Pople, W. G. Schneider, and H. J. Bernstein, *High-resolution Nuclear Magnetic Resonance*, McGraw-Hill, New York, 1959, Chap. 6.
2. J. W. Emsley, J. Feeney, and L. H. Sutcliffe, *High Resolution Nuclear Magnetic Resonance Spectroscopy*, Vol. 1, Pergamon, New York, 1965, Chap. 8.
3. P. L. Corio, *Structure of High-Resolution NMR Spectra*, Academic, New York, 1967.
4. E. D. Becker, *High Resolution NMR*, Academic, New York, 1969.
5. J. D. Roberts, *An Introduction to the Analysis of Spin–Spin Splitting in High-Resolution Nuclear Magnetic Resonance Spectra*, Benjamin, New York, 1961.
6. A. Streitwieser, Jr., *Molecular Orbital Theory for Organic Chemists*, Wiley, New York, 1961.
7. E. W. Garbish, Jr., *J. Chem. Ed.*, **45**, 311, 402, 480 (1968).
8. J. S. Waugh and S. Castellano, *J. Chem. Phys.*, **34**, 295 (1961); **35**, 1900 (1961).
9. R. W. Fessenden and J. S. Waugh, *J. Chem. Phys.*, **31**, 996 (1959).
10. G. Slomp, *Appl. Spectrosc. Rev.*, **2**, 263 (1969).
11. D. M. Grant, R. C. Hirst, and H. S. Gutowsky, *J. Chem. Phys.*, **38**, 470 (1963).
12. R. C. Hirst and D. M. Grant, *J. Chem. Phys.*, **40**, 1909 (1964).
13. B. Dischler and W. Maier, *Z. Naturforsch.*, **16A**, 318 (1961).
14. B. Dischler and G. Englert, *Z. Naturforsch.*, **16A**, 1180 (1961).
15. J. D. Swalen and C. A. Reilly, *J. Chem. Phys.*, **37**, 21 (1962).
16. S. Castellano and A. A. Bothner-By, *J. Chem. Phys.*, **41**, 3863 (1964).
17. Copies of these programs are available from the Quantum Chemistry Program Exchange, Department of Chemistry, Indiana University, Bloomington, Ind.

18. For a full listing, description and worked examples of these programs, see D. F. De Tar, Ed., *Computer Programs for Chemistry*, Vol. 1, Benjamin, New York, 1968, pp. 10 and 54.
19. R. C. Ferguson and D. W. Marquardt, *J. Chem. Phys.*, **41**, 2087 (1964).
20. C. W. Haigh, in *Annual Reports on NMR Spectroscopy*, E. F. Mooney, Ed., Vol. 4, Academic, New York, 1971, p. 311.
21. J. D. Swalen, in *Progress in N.M.R. Spectroscopy*, J. W. Emsley, J. Feeney, and L. H. Sutcliffe, Eds., Vol. 1, Pergamon, New York, 1966, p. 205.
22. T. Yamamoto and S. Fujiwara, *Bull. Chem. Soc. Japan*, **39**, 333 (1966).
23. N. Sheppard and R. M. Lynden-Bell, *Proc. Roy. Soc.*, **A269**, 385 (1962).
24. J. M. Read, R. E. Mayo, and J. H. Goldstein, *J. Mol. Spectrosc.*, **22**, 419 (1967).
25. V. S. Watts and J. H. Goldstein, *J. Chem. Phys.*, **46**, 4165 (1967).
26. N. Sheppard and J. J. Turner, *Proc. Roy. Soc.*, **A252**, 506 (1959).
27. F. S. Mortimer, *J. Mol. Spectrosc.*, **5**, 199 (1960).
28. J. D. Baldeschwieler and E. W. Randall, *Chem. Rev.*, **63**, 81 (1963).
29. R. A. Hoffman and S. Forsen, in *Progress in N.M.R. Spectroscopy*, J. W. Emsley, J. Feeney, and L. H. Sutcliffe, Eds., Vol. 1, Pergamon, New York, 1966, p. 15.
30. W. A. Anderson, R. Freeman, and C. A. Reilly, *J. Chem. Phys.*, **39**, 1518 (1963).
31. R. Freeman and W. A. Anderson, *J. Chem. Phys.*, **37**, 2053 (1962).
32. E. B. Baker, *J. Chem. Phys.*, **37**, 911 (1962).
33. For a review, see V. J. Kowalewski, in *Progress in N.M.R. Spectroscopy*, J. W. Emsley, J. Feeney and L. H. Sutcliffe, Eds., Vol. 5, Pergamon, New York, 1969, p. 1.
34. F. A. L. Anet and A. J. R. Bourn, *J. Am. Chem. Soc.*, **87**, 5250 (1965).
35. J. H. Noggle and R. E. Schirmer, *The Nuclear Overhauser Effect. Chemical Applications*, Academic, New York, 1971.
36. G. C. Bachers and T. Schaefer, *Chem. Rev.*, **71**, 617 (1971).
37. R. A. Bell and J. K. Saunders, in *Topics in Stereochemistry*, N. L. Allinger and E. L. Eliel, Eds., Vol. 7, Wiley-Interscience, New York, 1973, p. 1.
38. P. D. Kennewell, *J. Chem. Educ.*, **47**, 278 (1970).
39. W. McFarlane, in *Determination of Organic Structures by Physical Methods*, F. C. Nachod and J. J. Zuckerman, Eds., Vol. 4, Academic, New York, 1971, p. 139.
40. K. F. Kuhlmann and D. M. Grant, *J. Am. Chem. Soc.*, **90**, 7355 (1968).
41. C. C. Hinckley, *J. Am. Chem. Soc.*, **91**, 5160 (1969).
42. J. Reuben, in *Progress in Nuclear Magnetic Resonance Spectroscopy*, J. W. Emsley, J. Feeney and L. H. Sutcliffe, Eds., Vol. 9, Part 1, Pergamon, New York, 1973.
43. B. C. Mayo, *Chem. Soc. Rev.*, **2**, 49 (1973).
44. A. F. Cockerill, G. L. O. Davies, R. C. Harden, and D. M. Rackham, *Chem. Rev.*, **73**, 553 (1973).
45. B. L. Shapiro, M. D. Johnson, Jr., and R. L. R. Towns, *J. Am. Chem. Soc.*, **94**, 4381 (1972).

46. J. K. M. Sanders, S. W. Hanson, and D. H. Williams, *J. Am. Chem. Soc.*, **94**, 5325 (1972).
47. W. B. Lewis, J. A. Jackson, J. F. Lemons, and H. Taube, *J. Chem. Phys.*, **36**, 694 (1962).
48. H. M. McConnell and R. E. Robertson, *J. Chem. Phys.*, **29**, 1361 (1958).
49. M. R. Willcott III, R. E. Lenkinski, and R. E. Davis, *J. Am. Chem. Soc.*, **94**, 1742 (1972).
50. R. E. Davis and M. R. Willcott III, *J. Am. Chem. Soc.*, **94**, 1744 (1972).
51. L. Ernst and A. Mannschreck, *Tetrahedron Letts.*, 3023 (1971).
52. J. K. M. Sanders and D. H. Williams, *J. Am. Chem. Soc.*, **93**, 641 (1971).
53. H. Hart and G. M. Love, *Tetrahredon Letts.*, 625 (1971).
54. T. M. Ward, I. L. Allcox, and G. H. Wahl, Jr., *Tetrahedron Letts.*, 4421 (1971).
55. G. M. Whitesides and D. W. Lewis, *J. Am. Chem. Soc.*, **92**, 6979 (1970).
56. H. L. Goering, J. N. Eikenberry, and G. S. Koermer, *J. Am. Chem. Soc.*, **93**, 5913 (1971).
57. A. A. Chalmers and K. G. R. Pachler, *J. Chem. Soc.*, Perkin II, 748 (1974); for leading references.
58. M. Witanowski, L. Stefaniak, H. Januszewski, and Z. W. Wolkowski, *Tetrahedron Letts.*, 1653 (1971).
59. M. Witanowski, L. Stefaniak, H. Januszewski, and Z. W. Wolkowski, *Chem. Commun.*, 1573 (1971).
60. F. S. Mandel, R. H. Cox, and R. C. Taylor, *J. Magn. Reson.*, **14**, 235 (1974).
61. *Varian High Resolution NMR Spectra Catalog*, Vols. 1 and 2, Varian Associates, Palo Alto, Calif., 1963.
62. *Sadtler Standard NMR Spectra*, Sadtler Research Laboratories, Philadelphia, Pa.
63. W. Brugel, *NMR Spectra and Chemical Structure*, Academic, New York, 1967.
64. F. A. Bovey, *NMR Data Tables for Organic Compounds*, Wiley-Interscience, New York, 1967.
65. J. W. Emsley, J. Feeney, and L. H. Sutcliffe, *High Resolution Nuclear Magnetic Resonance Spectroscopy*, Vol. 2, Pergamon, New York, 1965.
66. L. M. Jackson and S. Sternhell, *Applications of Nuclear Magnetic Resonance Spectroscopy in Organic Chemistry*, Pergamon, New York, 1969.
67. R. M. Silverstein, G. C. Bassler, and T. C. Morrill, *Spectrometric Identification of Organic Compounds*, Wiley, New York, 1974.
68. J. N. Shoolery, Technical Information Bulletin, 2, No. 3, Varian Associates, Palo Alto, Calif., 1959.
69. C. Pascual, M. Meier, and W. Simon, *Helv. Chim. Acta.*, **49**, 164 (1966).
70. S. W. Tobey, *J. Org. Chem.*, **34**, 1281 (1969).
71. H. Spiesecke and W. G. Schneider, *J. Chem. Phys.*, **35**, 731 (1961).
72. P. Diehl, *Helv. Chim. Acta.*, **44**, 829 (1961).

73. S. Castellano, R. Kostelnik, and C. Sun, *Tetrahedron Letts.*, 4635 (1967).
74. S. Castellano, C. Sun, and R. Koselnik, *Tetrahedron Letts.*, 5205 (1967).
75. H. B. Evans, Jr., A. R. Tarpley, and J. H. Goldstein, *J. Phys. Chem.*, **72**, 2552 (1968).
76. S. Castellano, unpublished results.
77. R. H. Cox, *Spectrochimica Acta*, **25A**, 1189 (1969).
78. S. Castellano and R. Koselnik, *Tetrahedron Letts.*, 5211 (1967).
79. N. Jonathan, S. Gordon, and B. P. Dailey, *J. Chem. Phys.*, **36**, 2443 (1962).
80. O. L. Chapman and R. W. King, *J. Am. Chem. Soc.*, **86**, 1256 (1964).
81. W. R. Anderson, Jr. and R. M. Silverstein, *Anal. Chem.*, **37**, 1417 (1965).
82. A. A. Bothner-By in *Advances in Magnetic Resonance*, J. S. Waugh, Ed., Vol. 1, Academic, New York, 1965, p. 195.
83. S. Sternhell, *Quart. Rev.*, **23**, 236 (1969).
84. S. L. Smith, *Top. Cur. Chem.*, **27**, 117 (1972).
85. M. Karplus, *J. Chem. Phys.*, **30**, 11 (1959).
86. M. Karplus, *J. Am. Chem. Soc.*, **85**, 2870 (1963).
87. A. Hassner and C. Heathcock, *J. Org. Chem.*, **29**, 1350 (1964).
88. S. Castellano and C. Sun, *J. Am. Chem. Soc.*, **88**, 4741 (1966).
89. S. Castellano and R. Kostelnik, *J. Am. Chem. Soc.*, **90**, 141 (1968).
90. R. H. Cox, unpublished results.

CHAPTER

5

CARBON-13 NMR

5.1	Introduction	195
5.2	Experimental Techniques	197
	5.2.1 Observation of Spectra	197
	5.2.2 Calibration of Spectra	199
5.3	Chemical Shifts	200
5.4	sp^3 Carbons	202
	5.4.1 Alkanes	202
	5.4.2 Cycloalkanes	205
	5.4.3 Substituted Alkanes	206
	5.4.3.1 Deuterium	207
	5.4.3.2 Lithium	207
	5.4.3.3 Magnesium	209
	5.4.3.4 Halogens	209
	5.4.3.5 Alcohols	209
	5.4.3.6 Amines	210
	5.4.3.7 Carbonyl-containing Substituents	211
	5.4.3.8 Hydrocarbon Substituents	212
	5.4.3.9 Miscellaneous Substituents	212
	5.4.4 Substituted Cycloalkanes	213
	5.4.5 Saturated Heterocycles	214
5.5	sp^2 Hybridized Carbons	215
	5.5.1 Alkenes	215
	5.5.2 Cyclic Alkenes	217
	5.5.3 Substituted Alkenes	218
	5.5.4 Substituted Benzenes	218
	5.5.5 Aromatic Heterocycles	220
5.6	sp Hybridized Carbons	222
5.7	Functional Groups	224
5.8	Coupling Constants	225
	5.8.1 Carbon–Hydrogen Couplings	226
	5.8.2 Carbon–Carbon Couplings	228
	5.8.3 Carbon–Coupling with Other Nuclei	229
5.9	Assignment of CMR Spectra	230
	5.9.1 Chemical Shift	231

INTRODUCTION 195

5.9.2	Off-resonance and Selective Decoupling	232
5.9.3	Spectral Comparison	234
5.9.4	Specific Labeling	236
5.9.5	Undecoupled Spectra	236
5.9.6	Spin-lattice Relaxation	238
5.10	Quantitative Analysis by CMR	240
5.11	Structural Determinations Using CMR	241
5.12	Conformational Analysis	244
5.13	Mechanistic Studies	247
5.14	Biosynthetic Studies	247
5.15	Other Nuclei	249

5.1 INTRODUCTION

Although the first carbon-13 NMR (CMR) spectra were reported in 1957 (1), the development of CMR spectroscopy as a routine spectroscopic technique has progressed much more slowly than for other nuclei. The carbon-13 nucleus possesses a spin $I = \frac{1}{2}$ like the proton. However, the relatively low natural abundance (1.1%) and smaller magnetogyric ratio (Table 2.1) of carbon-13 result in it being approximately 5700 times less sensitive to NMR detection than the proton. An additional difficulty with CMR is the relaxation times of carbon-13 nuclei, which can be considerably longer than those for protons. This introduces problems with saturation and restricts the rate at which a scan of a CMR spectrum may be repeated in a time averaging experiment. Because of the sensitivity problems with CMR, the early studies of CMR spectra were necessarily limited to concentrated samples of relatively low molecular weight compounds, or to isotopically enriched samples, in order to obtain spectra with a reasonable S/N ratio. As more sophisticated NMR instrumentation was developed along with new techniques for improving the sensitivity, the applications of CMR continued to increase such that by the late 1960s, a considerable body of CMR data has been accumulated. However, it has only been with the development of FT-NMR techniques that CMR has become available as a routine, spectroscopic tool. With FT-NMR techniques, CMR spectra may now be routinely obtained within a matter of a few minutes to a few hours for typical organic compounds. The number of published papers reporting CMR data continues to increase each year such that it would not be surprising if applications of CMR surpassed those of proton NMR within the near future.

The early workers in CMR were quick to recognize the advantages of CMR over proton NMR for structural determination applications. Other than the obvious advantage that one is obtaining data from the "backbone"

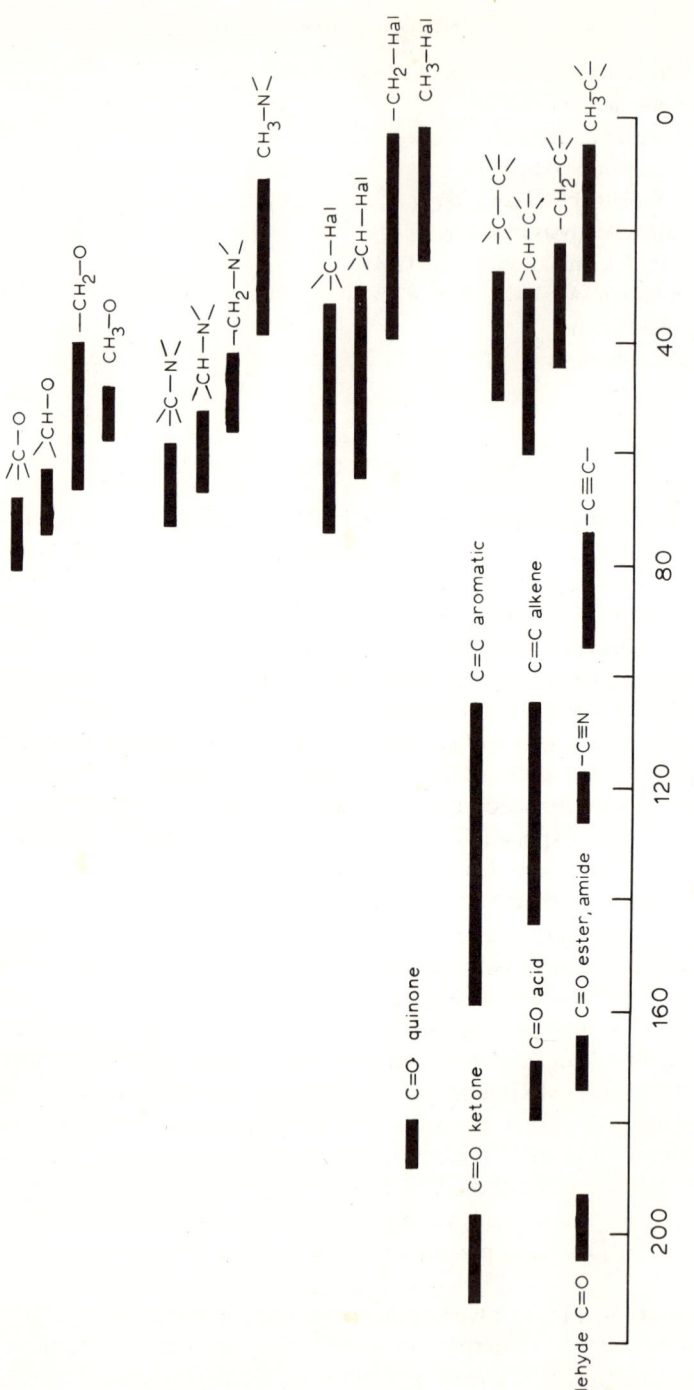

Fig. 5.1 The range of carbon-13 chemical shifts given by carbon in various substitution patterns.

of the molecule rather than from the exterior of the molecule as in proton NMR, there are several additional advantages of CMR. The CMR chemical shift range of the majority of diamagnetic organic compounds is ~200 ppm in comparison with ~10 ppm for proton NMR (Figure 5.1). This means that there is less overlap of peaks in CMR spectra and one has a greater probability of observing individual carbon resonances. For example, it is not uncommon to observe individual carbon resonances for organic compounds with molecular weights in the range 200–400. A second advantage of CMR is that the spectra are not complicated by spin–spin coupling. The probability of having two carbon-13 nuclei adjacent to each other in the same molecule is so low that the possibility of carbon-13–carbon-13 coupling can be ignored. In addition, CMR spectra are usually obtained with proton-decoupling so that only single peaks are observed for each carbon-13 resonance. A third advantage of CMR is that one obtains information about all carbons in the molecule. Thus one can obtain information from carbonyl, nitrile, etc., and functional groups in CMR, whereas similar information is not possible in proton NMR. In addition, information is also obtained from quaternary carbons for which there is no counterpart in proton NMR. Similarly, for inorganic carbonyl and nitrile complexes, the carbon-13 nucleus may be the only nucleus in the molecule conveniently examined by NMR. Finally, since the chemical shift range in CMR is larger than the corresponding range in proton NMR, chemical shifts of nonequivalent carbons are usually more widely separated than corresponding proton shifts. This results in one being able to examine faster rate processes with CMR than with proton NMR and to have a better chance of observing separate peaks for a mixture of isomers.

Several excellent review articles on the application of CMR (2–4) and the instrumentation necessary for CMR have been published. In addition two monographs (5,6) dealing with the general aspects of CMR and its application and a collection of reference spectra (7) have been published. In this section we shall present a general survey of CMR along with examples of recent applications. Those interested in further details of CMR are referred to one of the previously mentioned monographs (5,6).

5.2 EXPERIMENTAL TECHNIQUES

5.2.1 OBSERVATION OF SPECTRA

The methods used to obtain CMR spectra closely parallel the developments in NMR instrumentation. Many of the initial CMR spectra were obtained using the dispersion mode for display of the spectra. High rf power and fast scan rates were used to help overcome problems with sensitivity and saturation. Chemical shifts obtained from these spectra were not as accurate as corresponding proton spectra due to line distortions caused by the fast scan rates employed. With the development of field/frequency stabilized spectrometers, more accurate spectra were obtained using the absorption mode display with low rf power levels and slower scan rates. Multiple-scan, time-averaging techniques were used to improve the sensitivity. In addition, the INDOR technique was used to obtain carbon-13 chemical shifts of simple compounds by monitoring the carbon-13 satellites in the proton spectra while sweeping the second rf field through the carbon-13 resonances.

The first major breakthrough in the observation of CMR spectra was the development of broadband, proton decoupling (8). Unlike the heteronuclear decoupling, discussed in Section 4.11.8, where only a single frequency is irradiated, broadband, proton decoupling of CMR spectra utilizes random-noise modulation to generate a decoupling field B_2 sufficient to irradiate the entire proton region. Due to complete decoupling of carbon-13 from all protons using this technique, the sensitivity of CMR spectra is increased due to the collapse of multiplets due to proton–carbon-13 spin–spin coupling. Thus single peaks for the individual carbon resonances are obtained in the CMR spectra of most organic compounds. If the compound contains other magnetic nuclei such as fluorine or phosphorus, of course, multiplets will be obtained for the carbon-13 nuclei spin coupled to these magnetic nuclei.

An additional increase in sensitivity, other than that resulting from the collapse of multiplets, often results from complete proton decoupling. Early work with proton decoupling showed an increase in the peak area of certain carbons over and above that due to the collapse of multiplets. This phenomenon is referred to as the nuclear Overhauser enhancement (NOE) and is due to a nonequilibrium distribution of the carbon nuclear spin states resulting from the decoupling of neighboring protons. For carbons with directly attached protons where the dipolar relaxation mechanism (Section 3.11.1) is the dominant relaxation mechanism, the NOE can lead to an enhancement in the carbon peak area by a factor of ~ 3.0 (9). This enhancement will be identical for methyl, methylene, and methine carbons provided the dipolar relaxation mechanism is dominant. A smaller value will result if other relaxation mechanisms are important. Since the dipolar relaxation mechanism varies with distance as $1/r^6$, carbons with no attached protons will usually not be enhanced due to the NOE. Thus the NOE may provide the basis for distinguishing between carbons with attached protons and quaternary carbons. For carbons in different chemical environments, the NOE may be different for each carbon providing the basis for assignment.

The latest development in NMR instrument has been that of pulse techniques along with Fourier transform (FT) analysis (Section 3.9). Nowdays, almost all CMR spectra are obtained using this technique along with broadband proton decoupling. By using a combination of these techniques, along with 10–20 mm samples tubes, it is possible to obtain CMR spectra of most organic compounds with molecular weight ~ 300–400 within a few minutes to a few hours, depending on the concentration. If larger sample tubes are used, the recording of CMR spectra of small biopolymers becomes feasible. For example, with a 20-mm sample tube, individual carbon resonances have been observed for a 0.012 M solution of hen egg white lysozyme (molecular weight = 14,314) after only 5 hr of signal averaging (10).

5.2.2 CALIBRATION OF SPECTRA

With modern FT-NMR spectrometers, the chemical shift of the peaks in a spectrum may be conveniently calculated with respect to any preselected peak and obtained as part of the computer printout. The resolution and hence the accuracy of the chemical shifts are determined by the size of the computer memory. With an 8K transform, a precision of better than 0.1 ppm in the chemical shifts is routinely obtained. Coupling constants may be obtained from similar spectra to a precision of better than 1 Hz.

A variety of reference compounds have been used to calibrate CMR spectra. These include CS_2, benzene, chloroform, p-dioxane, cyclohexane, and TMS. The chemical shift of TMS appears upfield from most carbon resonances similar to its position in proton spectra. Since it is quite likely that both proton and CMR spectra will be obtained on the compound, there is general argument that internal TMS is the reference of choice for nonaqueous CMR spectra. The chemical shifts in ppm of the previously mentioned reference materials with respect to TMS are as follows: CS_2 192.8; benzene 128.7; chloroform 77.2; p-dioxane 67.4; and cyclohexane 27.7. Several of these references are both concentration and solvent dependent. The chemical shifts quoted in this chapter have been converted to the TMS scale using the values stated with the convention that a positive value indicates a downfield shift from TMS.

With modern spectrometers employing an internal lock and a means of field/frequency stabilization, any magnetic nucleus could in principle serve as the internal lock. It has been shown that carbon-13 nuclei can serve as the lock signal. However, since it is desirable to eliminate any interference of the locking frequency from either the observation or decoupling frequencies, most recent instruments utilize either fluorine-19 or deuterium as the internal lock for CMR studies. As in proton NMR, the lock compound may be either contained in the same solution as the sample or contained in a small capillary tube inside the sample tube.

The use of deuterium as the internal lock offers additional advantages other than providing a heteronuclear lock. One of the difficulties with FT-NMR is the problem of overflow of the computer A–D converter and memory by a large peak in the spectrum (Section 3.9.1). Thus, for dilute samples, the carbon nuclei in the solvent may present a problem in this respect. However, if a deuterium-labeled solvent is used, the carbon nuclei in the solvent will be split due to spin–spin coupling with deuterium and hence will be spread out in the memory and will not overflow as fast as a nondeuterated solvent. Therefore, a deuterated solvent not only serves as the locking compound but also increases the number of scans possible. For most CMR studies, deuterated solvents similar to those used in proton NMR are employed.

5.3 CHEMICAL SHIFTS

One of the major advantages of CMR spectroscopy over proton NMR is the larger chemical shift range (5,6). For diamagnetic compounds, chemical shifts over a range of ~ 600 ppm have been observed. For most organic compounds, the chemical shift range is ~ 200 ppm downfield from TMS. As in proton NMR, the chemical shift of carbon depends on the hybridization of carbon and its structural environment. The general trend from high to low field is sp^3, sp, sp^2 (Figure 5.1).

The chemical shift range of sp^3 carbons is ~ 10–90 ppm. Although it is clear that substituent effects dominate the chemical shifts of sp^3 hybridized carbons, other factors such as steric interactions and geometrical effects are important in determining the actual shift. Overlapping with the low field region of sp^3 carbons is the characteristic region for sp carbons in acetylene. Acetylinic carbons are found within the relatively narrow range of ~ 65–90 ppm. The range of chemical shifts for sp^2 hybridized carbons is ~ 100–165 ppm. Included in this region are olefinic, aromatic, and heteroaromatic carbons. There is considerable overlap of the chemical shifts of olefinic and aromatic carbons. The chemical shift of the center sp hybridized carbon in allenes appears at appreciably lower fields within the range of ~ 200–210 ppm.

Carbons appearing in functional groups also appear in characteristic chemical shift ranges (Figure 5.1). For example, nitriles (—C≡N) appear in the range 110–130 ppm. Isonitriles (—N$^+$≡C$^-$) appear downfield from nitriles by ~ 50 ppm. Although there is some overlap of the ranges for the chemical shifts of carbonyl functional groups, one can usually distinguish carbonyl groups on the basis of CMR spectra. The carbonyl carbon of carboxylic acid esters and amides absorb in the range 165–175 ppm. Carboxylic acids are slightly further downfield in the range 170–184 ppm.

The aldehyde carbonyl carbon appears in the range 190–202 ppm whereas the ketone carbonyl carbon is downfield further in the range 195–215. These may be distinguished by using techniques discussed in Section 5.9.

The theory of carbon chemical shifts has been examined by a number of investigators (5,6). It seems clear that at least three factors are important in determining carbon chemical shifts. These are the diamagnetic term σ_d, the paramagnetic term σ_p, and the anisotropy of neighboring atoms σ_{other} (Section 2.7). Of these terms, the paramagnetic term makes the major contribution to carbon chemical shifts (11). Within a series of related compounds, the term due to the anisotropy of other atoms should be relatively constant and probably makes only minor contributions to carbon chemical shifts. The diamagnetic term can make an appreciable contribution to carbon chemical shifts in certain cases (12).

Although there has been considerable effort devoted to the theory of carbon chemical shifts (5), with the exception of very small molecules, the theory of carbon chemical shifts has proven to be of little value in assigning chemical shifts for structure determination. The primary emphasis in chemical shift assignment has been devoted to empirical correlations derived from data on known compounds. From the examination of a large body of data, it appears that a substituent will exert a more or less constant influence on the chemical shift of neighboring carbons regardless of the type of compound in which the substituent is found. Furthermore, if two carbon atoms in different molecules have identical environments for four or five carbons removed from the carbons in question, the chemical shift of the two carbons will be almost identical. Therefore, additivity schemes for the effect of substituents have been derived and have proven to be invaluable in assigning CMR spectra.

The effect of substituents on carbon chemical shifts has been discussed in terms of incremental shifts caused by the substituent at neighboring carbon atoms. Replacing hydrogen with a substituent in a compound ($X-C_\alpha-C_\beta-C_\gamma-C_\delta-C_\varepsilon-$) can result in shifts of the carbon resonances up to five or six bonds removed from the substituent. There is general agreement that the effect of a substituent on the α-carbon is primarily an electronic effect. However, in many cases the effect on the β-carbon is similar for a variety of substituents and it is clear that factors other than an electrical effect are contributing to the β-shift. The effect of a substituent on the γ-carbon is perhaps the most interesting. For the majority of substituents, the effect at the γ-carbon is a shielding effect (upfield shift). It appears that the γ-effect is a field effect resulting from steric interactions similar to the familiar gauche interactions in organic chemistry. This is a general trend with sterically hindered carbons always appearing upfield from similar carbons that are not sterically hindered.

5.4 sp³ CARBONS

5.4.1 ALKANES

While the proton NMR spectra of alkanes give little information on the structure of isomeric compounds, this is definitely not the case with CMR spectra (Tables 5.1 and 5.2). In many cases, the number and relative intensities of the peaks allow one to easily distinguish among isomeric compounds. For example, of the five isomeric hexanes (C_6H_{14}), only one isomer, 2,3-dimethylbutane, gives only two peaks in the CMR spectrum. Only one isomer, *n*-hexane, gives only three peaks. The two isomers that give a spectrum consisting of four peaks, 2,2-dimethylbutane and 3-methylpentane, may be distinguished on the basis of the relative intensity of the peaks. Finally, there is only one isomer that has five peaks in its spectrum, 2-methylpentane.

$CH_3CH_2CH_2CH_2CH_2CH_3$ (3 peaks)

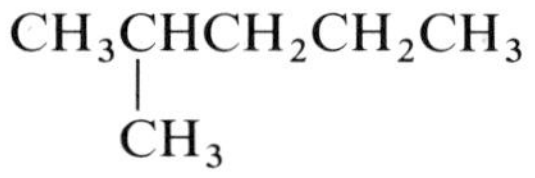

$CH_3CHCH_2CH_2CH_3$
   |
   CH_3 (5 peaks)

$CH_3CH_2CHCH_2CH_3$
   |
   CH_3 (4 peaks)

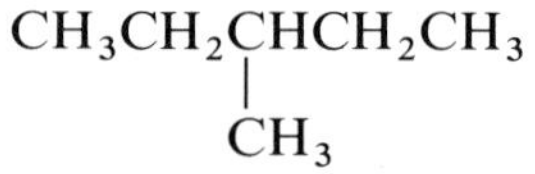

$\quad\quad CH_3\ \ CH_3$
$\quad\quad\ |\quad\ |$
$CH_3CH-CHCH_3$ (2 peaks)

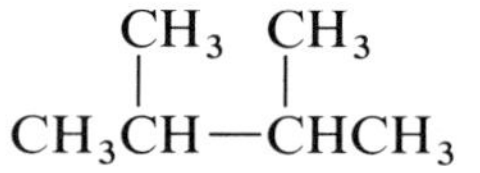

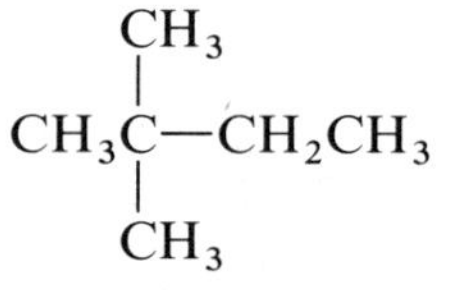

$\quad CH_3$
$\quad\ |$
$CH_3C-CH_2CH_3$ (4 peaks)
$\quad\ |$
$\quad CH_3$

In more complicated cases, the relative positions of the peaks in conjunction with the additivity of substituent effects allow one to distinguish isomeric compounds.

The first additivity scheme for substituent effects was derived for alkanes by Grant and Paul (13) from the data for 17 alkanes. They found that the chemical shifts could be predicted to within ± 1 ppm or better on the basis of the number of α, β, γ, δ, and ε carbon atoms attached to the carbon in question. However, corrective factors were needed for predicting the shifts of branched alkanes.

Lindeman and Adams (14) have extended the additivity scheme of Grant and Paul by employing data for 59 isomers of the alkanes C_5-C_9. Additivity

TABLE 5.1. Carbon-13 Chemical Shifts for Some Linear Alkanes[a]

Compound	C_1	C_2	C_3	C_4	C_5
Methane[b]	2.1				
Ethane[b]	5.9				
Propane[c]	15.6	16.1			
Butane[c]	13.2	25.0			
Pentane	13.5	22.2	34.1		
Hexane	12.7	22.7	32.7		
Heptane	13.7	22.6	32.0	29.0	
Octane	13.6	22.7	32.1	29.4	
Nonane	13.8	22.7	32.0	29.4	29.6

[a] Data taken from Lindeman and Adams (14) unless otherwise indicated.
[b] Data taken from Spiesecke and Schneider (15).
[c] Data taken from Grant and Paul (13).

TABLE 5.2. Carbon-13 Chemical Shifts for Some Branched Alkanes[a]

Compound	C_1	C_2	C_3	C_4	C_5
2-methylpentane	22.7	27.9	41.9	20.8	14.3
3-methylpentane	11.4	29.4	36.8		
			18.7[b]		
2,2-dimethylbutane	28.7	30.3	36.5	8.5	
2,3-dimethylbutane	19.2	34.0			
2,4-dimethylpentane	22.7	25.7	49.0		
2,3-dimethylpentane	20.0	31.9	40.6	26.8	11.6
		17.7	11.6[b]		
3,4-dimethylhexane	11.8	25.8	38.5		
			13.8[b]		
		27.6	39.5	15.8	
3,5-dimethylheptane	10.9	29.5	31.9	44.3	
			18.9[b]		
	11.1	30.5	32.0	44.5	
			19.6[b]		

[a] Data taken from Lindeman and Adams (14).
[b] 3-methyl carbon.

parameters derived from these data permit one to predict the carbon chemical shift of the Kth carbon atom using equation 5.1 to an accuracy of better than 2 ppm for 89% of the shifts.

$$\delta_c(K) = B_s + \sum_{M=2}^{4} D_m A_{sm} + \gamma_s N_{K3} + \Delta_s N_{K4} \quad (5.1)$$

In equation 5.1, B_s, A_{sm}, γ_s, and Δ_s are constants. D_m is the number of carbon atoms bonded to the Kth carbon which in turn have M carbon atoms attached. N_{Ki} is the number of carbon atoms located i bonds away from the Kth carbon. S is the number of carbon atoms attached to the Kth carbon. The B_s represents the base value for each type of carbon atom:

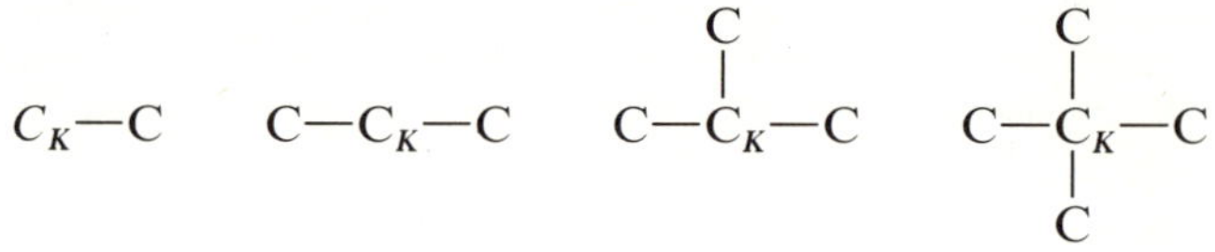

$$\underset{B_1}{C_K-C} \qquad \underset{B_2}{C-C_K-C} \qquad \underset{B_3}{\overset{\overset{C}{|}}{C-C_K-C}} \qquad \underset{B_4}{\overset{\overset{C}{|}}{\underset{\underset{C}{|}}{C-C_K-C}}}$$

The constants are given in Table 5.3. The major difference between this parameter set and that derived by Grant and Paul is that the set in Table 5.1 uses different parameters for each type of carbon atom, B_s.

TABLE 5.3. Carbon-13 Chemical Shift Parameters for Alkanes[a]

Parameters	Shift (ppm)	Parameters	Shift (ppm)
B_1	6.80	B_3	23.46
A_{12}	9.56	A_{32}	6.60
A_{13}	17.83	A_{33}	11.14
A_{14}	25.48	A_{34}	14.70
γ_1	−2.99	γ_3	−2.07
Δ_1	0.49	B_4	27.77
B_2	15.34	A_{42}	2.26
A_{22}	9.75	A_{43}	3.96
A_{23}	16.70	A_{44}	7.35
A_{24}	21.43	γ_4	0.68
γ_2	−2.69		
Δ_2	0.25		

[a] Lindeman and Adams (14).

As an example of the use of Table 5.1, consider the calculation for carbon-3 in 2,4-dimethylpentane:

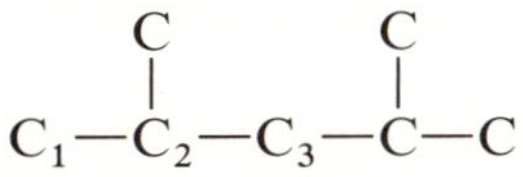

$$C_1-C_2-C_3-\overset{\overset{C}{|}}{C}-C \quad \text{(with C substituent on } C_2\text{)}$$

1. The number of carbon atoms attached to carbon-3 is 2, $S = 2$.
2. The two carbon atoms attached to carbon-3 each have 3 carbon atoms attached to them, $D = 2$, $M = 3$.
3. There are no carbon atoms attached to carbon-3 by 3 or 4 bonds, $N_{33} = 0$ and $N_{34} = 0$.
4. The calculation then follows as

$$\delta_{C(3)} = 15.35 + 2(16.70)$$
$$= 48.74 \text{ ppm (exp} = 49.8 \text{ ppm)}$$

5.4.2 CYCLOALKANES

With the exception of cyclopropane, the chemical shifts of cycloalkanes are not affected appreciably by ring size (Table 5.4). The shift of cyclopropane is shielded with respect to other cycloalkanes and appears in the same region of the spectrum as methane. It has been suggested that a ring current in the three membered ring is responsible for the upfield shift. The chemical shifts of larger cycloalkanes appear over a small range of ~5 ppm and are slightly upfield (~3 ppm) from the chemical shifts of central carbons in linear alkanes. The small variation in the shift with ring size suggest that conformational differences (possibly steric interactions) are influencing the shifts.

TABLE 5.4. Carbon-13 Chemical Shifts for Some Cycloalkanes[a]

Compound	δ_c
Cyclopropane	−2.6
Cyclobutane	23.3
Cyclopentane	26.5
Cyclohexane	27.8
Cycloheptane	29.4
Cyclooctane	27.8
Cyclononane	27.0
Cyclodecane	26.2

[a] Data taken from Burke and Lauterbur (16).

TABLE 5.5. Substituent Parameters for Methyl Cyclohexanes[a]

Substituent	α	β	γ	δ
CH_3 eq	5.6	8.9	0.0	−0.3
CH_3 ax	1.1	5.2	−5.4	−0.1
$(CH_3)_2$ gem	−3.4	−1.2		
$(CH_3)_2$ vic eq–eq	−2.3			
ax–eq	−3.1			

[a] Data taken from Dalling and Grant (17).

The effects of methyl substitution on cyclohexane has been examined by Dalling and Grant (17) and were found to closely parallel the effects on linear alkanes. From the examination of 15 methyl substituted cyclohexanes, they were able to derive the substituent parameters given in Table 5.5. The most interesting aspect of the data is the stereochemical dependence of the substituent effect.

The chemical shift of the carbon to which the methyl group is attached depends on the orientation of the methyl group. A carbon with an axial methyl group attached appears upfield ~4.5 ppm from a carbon with an equatorial methyl substituent. Furthermore, an axial methyl group results in an upfield shift of ~5.5 ppm for the carbons γ to the substituent. This upfield shift results from 1,3 steric interactions between the ring carbon and methyl group. The data also show that the chemical shift of the methyl carbon is indicative of the stereochemistry. Due to steric interactions with hydrogens on carbons 3 and 5, an axial methyl carbon appears upfield from an equatorial methyl carbon. Recent low-temperature CMR spectra of methylcyclohexane (18) and cis-1,4- and cis-1,2-dimethylcyclohexane (19) have confirmed this result and suggest that an axial methyl carbon is ~6.0 ppm upfield from a corresponding equatorial methyl group. Although slightly different in magnitude, similar substituent effects have been found in methyl cyclopentanes (20). These results should prove to be very useful in conformational analysis of five- and six-membered rings.

5.4.3 SUBSTITUTED ALKANES

One of the reasons for the success of CMR spectra for structural determinations has been the consistency of substituent effects on carbon chemical shifts. Empirical correlations of substituent effects have been derived for many common substituents from the examination of known compounds. Regardless of the compound in question, the effect of the substituent on the chemical shifts of carbons up to five bonds removed from the substituent is predictable from the empirical correlations. If one first calculates the chemical shifts of the corresponding alkane using the parameters of Lindeman and Adams (14) and then adds the substituent effects for replacing hydrogen with a substituent, the chemical shifts of the substituted alkane can be predicted to within ±1 ppm in most cases. The deviation in highly branched compounds may be somewhat larger. Nevertheless, the correct trends are reproduced such that chemical shifts assignments may be made.

The effects of some common substituents on carbon chemical shifts of alkanes are summarized in Table 5.6. In some cases, the effect of the substituent was originally derived by considering the substituent to replace a methyl group (CH_3—R → X—R). However, in Table 5.6, these values have been converted such that the substituent is considered to replace hydrogen (H—R → X—R) in the corresponding alkane. Downfield shifts at the α-carbon are observed for substituents more electronegative than carbon whereas upfield shifts are observed for substituents less electronegative than carbon. With the exceptions of deuterium, lithium, and magnesium, upfield shifts are observed at the γ-carbon. The effects of many substituents at the β-carbon are similar indicating that factors other than electronic effects are influencing the β-carbon shift. Small downfield shifts at the δ- and ε-carbons are observed in most cases.

For several of the substituents previously discussed, only a few data are available. For those cases, the substituent parameter sets given in Table 5.6 were derived from alkyl groups containing at least four or more carbon atoms.

5.4.3.1 Deuterium

When hydrogen is replaced by deuterium (21), the α-carbon undergoes an appreciable upfield shift of ~0.5 ppm (Table 5.6). Smaller upfield shifts are observed for the β- and γ-carbons. In addition, the α-carbon resonance appears as a 1:1:1 triplet due to the spin coupling of deuterium (I = 1) with carbon. The resonances of the β- and γ-carbons are broadened somewhat due to unresolved carbon–deuterium coupling.

The introduction of a second or third deuterium on a carbon usually results in a disappearance of the α-carbon signal due to an increase in T_1 and a decrease in the NOE. However, appreciable upfield shifts and line-broadening are observed at the β and γ carbons.

5.4.3.2 Lithium

Only a few CMR spectra of organolithium compounds have been reported (22). It appears that replacing hydrogen with lithium results in an upfield

TABLE 5.6. Substituent Effects on the ^{13}C Chemical Shifts of Alkyl Derivatives with Respect to the Corresponding Alkane

$$X—C_\alpha—C_\beta—C_\gamma—C_\delta—C_\varepsilon$$

X	α	β	γ	δ	ε	Reference
D	−0.44	−0.12	−0.02	—	—	21
D$_2$	—	−0.24	−0.08	—	—	21
D$_3$	—	−0.36	−0.16	—	—	21
Li	−1.9	6.8	6.3	0.6	—	23
Mg	−6.6	6.1	5.5	−0.1	0.3	24
F	60–65					25
Cl	31.2	10.5	−4.6	0.1	0.5	6
Br	20.0	10.6	−3.1	0.1	0.5	6
I	−6.0	11.3	−1.0	0.2	1.0	6
1° OH	48.3	10.2	−5.8	0.3	0.1	28
2° OH	44.5	9.7	−3.3	0.2	0.2	28
3° OH	39.7	7.3	−1.8	0.3	0.3	28
NH$_2$, 1°	28.6	11.5	−4.9	0.3	0.4	29
NH 2°	36.7	7.6	−4.2	0.5	0.3	29
N 3°	40.8	5.2	−4.2	0.5	0.3	29
NH$_3^+$	26.0	7.5	−4.6	—	—	30
O=C—H	31.9	+0.7	−2.3	—	—	31
O=C—CH$_3$	30.9	2.3	−0.9	2.7	1.4	32
CO$_2$H	20.8	2.7	−2.3	1.0	1.2	33
CO$_2^\ominus$	22.5	4.5	−1.7	1.2	—	33
—OCCH$_3$ (O=)	51.1	7.1	−4.8	1.1	0.8	31
CO$_2$Me	20.4	2.3	−1.9	1.2	0.8	31
O=C—Cl	33.7	2.2	−3.3	—	—	34
φ(C$_6$H$_5$)	23	9.5	−2	—	—	6
H$_2$C=CH—	20.7	6.9	−2.2	0.7	—	35
—HC=CH—$_{cis}$	14.2	7.3	−1.5	—	—	35
—HC=CH—$_{trans}$	19.7	7.2	−1.6	—	—	35
H—C≡C—	5.9	7.7	−1.3	—	—	36
NO$_2$	64.5	4.7		—	—	37
CN	3.6	2.0	−3.1	—	—	34
NC	28.0	6.2	−5.4	—	—	34
NCO	29.6	8.5	−5.1	—	—	34
SCN	20.6	6.9	−3.8	0.2	—	34
NCS	31.4	—	—	—	—	6
SH	23	0.7	−3.4	0.3	—	34
OR	58	8.1	−4.7	1.4	—	31

shift for the α-carbon (Table 5.6). However, downfield shifts are observed for the β-, γ-, and ε-carbons (23).

5.4.3.3 Magnesium

The CMR spectra of several Grignard reagents in diethyl ether have been examined by Roberts et al. (24). In ether, the predominant species in solution is the dialkylmagnesium, R_2Mg. An appreciable upfield shift is observed at the α-carbon, (Table 5.6), whereas downfield shifts are observed at the remaining carbons. Apparently, the small upfield shift at the δ-carbon is real, although a downfield shift is observed in some cases. The CMR spectra of allylmagnesium bromide and cyclopentadienylmagnesium bromide indicate that rapid exchange is occurring.

5.4.3.4 Halogens

The effect of halogens on the shielding of α-carbons roughly parallels the electronegativities of the halogens (Table 5.6). However, the effect of iodine (upfield shift) is difficult to explain in terms of electronic properties. Additional iodine substitution results in further shielding [CI_4; $\delta = -292.3$ ppm (26)]. The effect of replacing hydrogen with halogen is almost constant at the β-carbon, regardless of the halogen. The effect at the γ-carbon decreases with increasing size of the halogen atom. It is thought that this is a result of a decrease in the population of the gauche conformation with increasing size of the halogen. The effect of halogen substitution at the δ- and ε-carbons is more or less constant.

The chemical shift additivity scheme for alkyl halides breaks down when applied to vicinal dihalides. Examination of several vicinal dihalides shows that the chemical shifts differ appreciably from those predicted using the additivity parameters (27). Deviations are also observed for branched alkyl halides. However, with the linear 1,3-, 1,4- and 1,5-dihalides, the agreement between "calculated" and observed chemical shifts is reasonably good (6).

5.4.3.5 Alcohols

The shielding effect of the hydroxyl group has been defined in a study of the CMR spectra of several alcohols by Roberts et al (28). For primary alcohols the hydroxyl group results in a deshielding of the α-carbon by 48.3 ppm (Table 5.6). Deshielding is also observed at the β, δ, and ε positions. As expected, shielding is observed at the γ-carbon. Secondary alcohols show similar effects. It is interesting that the α- and β-effects of a secondary hydroxyl group are different for the various types of secondary alcohols, illustrating the dependence of the substituent effect on the location of the substituent along the carbon chain. The data for the tertiary alcohols show trends similar to those observed for secondary alcohols.

The agreement between predicted and observed chemical shifts for glycols is better than that observed for dihalides. For ethylene glycol, the deviation is only 1.0 ppm. Smaller deviations are observed for glycols where the two hydroxyl groups are separated by four or more carbons. For the 2-haloethanols, the deviation is less than 2 ppm, whereas in compounds in which the halogen and hydroxyl are separated by three or more carbons, the deviations are less than 1 ppm (6).

5.4.3.6 Amines

The CMR spectra of 103 amines have been obtained by Eggert and Djerassi (29) in order to derive additivity parameters for the amino substituent. The substituent parameters given in Table 5.6 were derived from the data for n-alkylamines. The increase in the α-effect from primary to tertiary indicates that the alkyl substituent effect is transmitted through nitrogen. The β-effect decreases considerably on going from primary to tertiary, and results from the fact that the α-carbon is also γ to an alkane in the secondary and tertiary amines. The γ-effect is more or less constant.

When secondary or tertiary alkyl groups are attached to nitrogen, the parameter set given in Table 5.6 will lead to significant deviations (29). For example, for 1-methyl alkyl amines, the following parameters set is derived for primary amines: $\alpha = 24.1$; $\beta = 8.3$ (C2), and 10.3 (C2'); $\gamma = -2.8$; and $\delta = 0.2$ ppm. Similar deviations are found for other branched alkanes (29).

TABLE 5.7. Substituent Effects of the Amino Group on Carbon Chemical Shifts[a,b]

	$\delta_C^{Amine} = \delta_C^{Alkane}$		$A + B$
	α	β	γ
Primary amines A	0.846	0.955	0.951
B	23.09	3.14	−1.08
Secondary amines A	0.850	0.958	0.950
B	24.70	1.77	−0.94
Tertiary amines A	0.938	0.946	0.966
B	21.91	1.79	−1.60

[a] The alkane is the one where N is replaced by a CH.
[b] Data taken from Eggert and Djerassi (29).

Eggert and Djerassi (29) suggested that a much better correlation between predicted and observed chemical shifts could be obtained by considering the N to replace CH in the corresponding alkane (Table 5.7). This leads to a significantly better correlation between predicted and observed chemical shifts with a standard deviation of less than 0.5 ppm for most of the data. A consideration of the γ-effect observed for amines suggests that there are no large differences between the size of the NH_2 group as compared to the OH and CH_3 group (29).

Protonation of the amine nitrogen results in an upfield shift of carbons as far as three bonds from the nitrogen, compared to the free amine (Table 5.6) (30). While this trend appears to be general, sufficient data are not available to fully assess the effects of N-protonation.

5.4.3.7 Carbonyl-containing Substituents

The aldehyde functional group $-\overset{\overset{\displaystyle O}{\|}}{C}-H$ results in a deshielding at the α-carbon and a shielding at the γ-carbon as expected (Table 5.6). The deshielding at the β-carbon is smaller than that observed with many other substituents. However, sufficient data are not available to fully evaluate the substituent effects of the aldehyde functional group.

A variety of methyl ketones have been examined (32). The shielding at the α-carbon is much smaller than for the $\overset{\overset{\displaystyle O}{\|}}{C}-H$ group and indicates that the effect of the second alkyl group is transmitted through the carbonyl group (Table 5.6). A small deshielding is observed for the β-carbon which probably results from the fact that the β-carbon is γ to the second alkyl group.

The shielding effects of the carboxylate group have been determined (33) from the examination of the CMR spectra of some n-alkyl carboxylic acids (Table 5.6). The α-effect (deshielding) is smaller than that observed for the above two groups. However, the β-effect is slightly larger. These substituent effects allow one to predict the chemical shifts of dicarboxylic acids with a reasonable degree of accuracy.

Examination of the ammonium salts of the above carboxylic acids yields substituent parameters for the carboxylate anion (33). Formation of the anion leads to a deshielding of all carbons (Table 5.6) with respect to the corresponding carboxylic acid. The deshielding decreases with increasing distance from the carboxylate group. It has been suggested that this deshielding effect is due to polarization of the C—H bonds by the negatively charged CO_2^- group.

Substituent parameter sets for the methyl esters of n-alkyl carboxylic acids and for the n-alkyl acetates may be derived from the CMR data of some n-alkyl aliphatic esters (31). The parameters for the methyl esters are similar to those derived for carboxylic acids. The parameters for the acetates are appeciably different reflecting the difference between oxygen and carbonyl bonded to carbon.

Data from a limited number of acid chlorides yield the parameter set given in Table 5.6 for the acyl group as a substituent (34).

5.4.3.8 Hydrocarbon Substituents

The CMR spectra of several alkylbenzenes, alkenes, and alkynes have been examined. Comparison of these data with that of the corresponding alkanes permit one to derive substituent parameter sets for the phenyl, vinyl, and acetylene groups. The phenyl group causes a deshielding similar to that produced by bromine at the α- and β-carbons (6). Shielding is observed at the γ-carbon.

The effect of the carbon–carbon double bond as a substituent depends on the location of the double bond in the carbon chain and on its stereochemistry (35). The effect of a terminal carbon–carbon double bond is similar to that of a phenyl group with the exception that the γ-effect is smaller. The α-effect for a *cis* double bond is ~6 ppm smaller than for a *trans* double bond. This probably results from 1,4 steric interactions similar to the γ-effect in the *cis* isomer, resulting in an upfield shift compared to the *trans* isomer. This effect should prove useful in distinguishing geometrical isomers.

The α-effect of a terminal carbon–carbon triple bond is considerably less than that of the above two substituents (Table 5.6). In fact, the α-effect is smaller than the β-effect. The effect of the terminal triple bond at other carbon positions is similar to the phenyl group and carbon–carbon double bond. The effect of an internal triple bond is similar to that of the terminal triple bond (36).

5.4.3.9 Miscellaneous Substituents

There are several substituents for which only a few data are available. Although the values given in Table 5.6 are for limited samples, and are probably not as reliable as the previous substituents, the trends are undoubtedly real.

The parameters for ethers given in Table 5.6 were obtained from n-dialkyl ethers (31). The α-effect is larger than that for a primary alcohol and indicates that the effect of the second alkyl group is transmitted through oxygen. The smaller β-effect for an ether linkage compared to hydroxyl supports this view.

The α-effect of the thiol (—SH) group is considerably smaller than that for a hydroxyl group. Furthermore, a small shielding is observed for the β-effect. The γ-effect is only slightly smaller than that for the hydroxyl group (34).

The α-effect of the cyano group is somewhat surprising in view of the electronegativity of —CN. However, the effects of the cyano group at other carbons are similar to those obtained with other substituents. The isonitrile group exhibits normal substituent effects considering its electronegativity. The effects of the thiocyanate and isocyanate groups are approximately what would be expected considering the atom attached to carbon of the alkyl group (34).

Although only a few examples are available, the nitro group exhibits the largest α-effect of any substituent examined (37).

5.4.4 SUBSTITUTED CYCLOALKANES

The CMR spectra of a variety of substituted cyclopropanes have been investigated (38,39). In comparing the effect of substituents in the cyclopropyl series with the n-alkyl series, it is found that there are significant variations in the two series of compounds. Although the general trends are identical, deviations of 11 ppm are found for the α-effect. In most cases, the α-carbon shifts in the cyclopropanes are to higher field than predicted using the substituent constants given in Table 5.6. On the other hand, the β-effect in the cyclopropyl series is slightly larger in most cases than that found for the n-alkyl series.

Several substituted cyclopentanes have been examined by Roberts et al. (20). For the methylcyclopentanes, the effect of methyl substitution on the ring carbon shifts is similar to that found for methylcyclohexanes (Table 5.5). Thus an equatorial methyl group results in shifts of $\alpha = 9.1$, $\beta = 9.2$ and $\gamma = -0.1$ ppm. The corresponding shift for an axial methyl group at the β-carbon is 7.5 ppm. Introduction of a hydroxyl group on the cyclopentane ring causes shifts of $\alpha = 48.0$, $\beta = 9.7$, and $\gamma = -1.9$ ppm. Slightly different shifts were found for the cyclopentyl acetates, namely $\alpha = 51.2$, $\beta = 7.3$, and $\gamma = -1.6$ ppm. The downfield α shift and upfield β shift of the acetates compared to the alcohol should be useful in confirming chemical shift assignments.

A variety of alkylcyclohexanols have been examined. By comparing the shifts with the corresponding alkylcyclohexanes, Roberts et al. (28) have obtained the substituent parameter set given in Table 5.8 for the hydroxyl group. The differences between an equatorial and axial hydroxyl group are approximately the same as those found for methylcyclohexanes (Table 5.5). In using these values for predicting chemical shifts, one should keep in mind

TABLE 5.8. Substituent Effects on the Chemical Shifts of Cyclohexane

Substituent	α	β	γ	δ
—OH eq	43.0	7.9	−1.1	−1.6
—OH ax	39.0	5.5	−6.8	−0.7
C=O	13.0	−0.9	−3.6	

the conformational equilibria of the cyclohexane ring system. Other substituted cyclohexanes such as the diols and dihalides have been examined (40).

The alkylcyclohexanones have been examined by Weigert and Roberts (41). Comparison of the carbon shieldings in cyclohexane and cyclohexanone allow one to derive the substituent parameters given in Table 5.8 for the C=O group as a substituent in the cyclohexane ring. By combining these values with those obtained for the methyl group (Table 5.5), the chemical shifts of the alkylcyclohexanones are predicted to within ~1 ppm.

The cycloalkanols containing from five to eight carbon atoms have been examined by Roberts et al. (28). In general, the shielding effects of the hydroxyl group closely parallel the trends found for cyclohexanol. For the cycloalkanones, the substituent effects of the carbonyl group for the ketones containing rings larger than cyclohexane are similar to that found for cyclohexane. However, for cyclobutanone and cyclopentanone, significant deviations of both the α and β effect are observed.

The most extensive set of data for substituted bicyclic alkanes is that obtained for 2-substituted norbornanes by Grutzner et al. (42). In general, the substituent effects are similar to those found for acyclic alkanes. Furthermore, an exo substituent results in a greater deshielding at the α-position than an endo substituent. In most cases the β and γ effects are similar to those found for acyclic alkanes. Lippmaa et al. have reported data for 50 bicyclic hydrocarbons, alcohols, and ketones (43).

5.4.5 SATURATED HETEROCYCLES

CMR spectra for the saturated heterocycles containing O, S, and N—CH$_3$ have been reported for three through six-membered rings (44,45). Taking the data for the six-membered rings, the substituent parameters in Table 5.9 are derived. For oxygen, the values given in Table 5.9 will reproduce the shieldings in the smaller rings with the exception of the α-effect which varies by ~2 ppm. The parameters given for sulfur apply only to the six-membered ring. Both the α- and β-effects increase with decreasing ring size with the

TABLE 5.9. Substituent Effects for the Carbon-13 Chemical Shifts of Six-Membered Saturated Heterocycles

Substituent	α	β	γ
O	41.9	0.2	−2.7
S	2.2	2.2	2.2
N—H	20.4	0.2	−1.8
N—CH$_3$	29.6	−1.2	−1.2

three-membered ring showing the largest deviation (α = 21.7 ppm). The differences in the parameters given for —N—H and N—CH$_3$ are about what would be expected considering that the effect of the methyl group is transmitted through the nitrogen atom. The α-effect for the N—CH$_3$ group will reproduce the shieldings in the smaller rings to within 2 ppm. The β-effect of N—CH$_3$ is similar for the five-membered ring but a larger value is needed for the four-membered ring (β = −5.6 ppm). Protonation of the ring nitrogen produces an upfield shift of 2–4 ppm at all carbon positions.

A variety of other saturated heterocycles have been examined including various alkyl piperidines (5), dihydropyran (45), dioxanes (46–49), and phosphorus-containing heterocycles (50 53). The CMR spectra of substituted 1,3-dioxanes have been used to investigate the conformational preference of the ring.

5.5 sp^2 HYBRIDIZED CARBONS

5.5.1 ALKENES

The CMR spectra of alkenes have been the subject of extensive investigation. Much of the work has focused attention on the shieldings of the olefinic carbons only, with the result that there is limited data available for the saturated carbons in the alkenes. Nevertheless, useful trends have been observed in the available data which permit one to derive information about an unknown compound. For example, it has been found that the chemical shift difference between the two olefinic carbons is characteristic of the substitution on the double bond. Thus, for 1-alkenes, a shift difference of from 20–28 ppm is observed with carbon-2 being deshielded with respect to the terminal carbon. For 2-alkenes, the shift difference is ∼7 ppm with carbon-3 appearing at lower field. The shift difference for 3-alkenes in ∼3 ppm.

TABLE 5.10. Carbon Chemical Shifts of Some Representative Alkenes[a]

Alkene	C_1	C_2	C_3	C_4	C_5	C_6	C_7	C_8
Ethylene	123.3	123.3						
cis-2-butene	12.1	124.6	124.6	12.1				
trans-2-butene	17.6	126.0	126.0	17.6				
cis-2-pentene	12.3	123.2	132.7	20.5	14.0			
1-octene	124.5	139.4	34.4	29.6	29.5	32.4	23.2	14.2
cis-2-octene	12.7	123.8	131.2	27.3	29.9	32.1	23.1	14.2
trans-2-octene	17.9	124.9	132.1	33.2	30.0	32.1	23.1	14.2
cis-3-octene	14.5	20.9	131.8	129.6	27.3	32.0	22.8	14.1
trans-3-octene	14.1	26.1	132.8	129.8	32.8	32.5	22.7	14.2
cis-4-octene	13.9	23.4	29.8	130.2	130.2	29.8	23.4	13.9
trans-4-octene	13.8	23.3	35.3	130.8	130.8	35.3	23.3	13.8

[a] Data taken from Dorman, Jautelat, and Roberts (35).

The chemical shifts of some selected alkenes are presented in Table 5.10. A more extensive collection of data can be found in reference 5. Although only the olefinic carbon shieldings were considered in much of the early work, it was immediately apparent that alkyl substitution influences the shielding of the olefinic carbons in a manner different from that found for alkanes. Parameter sets for the effect of alkyl substituents on the shielding of olefinic carbons have been developed which permit the chemical shifts to be predicted to within ~ 2 ppm (5).

The most recent approach to calculating olefinic carbon chemical shifts has been that of Dorman et al. (35). In their approach, a distinction is made in the effect of substituents to take into account the transmission of substituent effects through the double bond. An example is given in the following for the calculation of the chemical shift of carbon-5 in 5-decene. Using a least squares analysis of the available data, the

$$C_\delta - C_\gamma - C_\beta - C_\alpha - C_5 = C - C_{\alpha'} - C_{\beta'} - C_{\gamma'} - C_{\delta'}$$

parameter set given in Table 5.11 was derived to calculate the chemical shift of any alkene carbon with respect to ethylene. The correction terms are used to take into account multiple substitution. For example, in a 1,1-disubstituted alkene, the term corr α is included along with α. Similarly, the additional correction terms are used for multiple β-substitution, etc. The cis term is necessary to account for the upfield shift of a cis double bond in comparison to a trans double bond (Table 5.10). The origin of this difference is probably steric in nature. For tetrasubstituted olefins, the cis parameter is added twice. With these parameters, the chemical shifts of 80 olefinic carbons were calculated with a standard deviation of 0.85 ppm (35). Thus,

TABLE 5.11. Substituent Parameters for the Calculation of the Chemical Shifts of Olefinic Carbons[a]

Parameter	Shift (ppm)
α	10.62
β	7.18
γ	1.45
α'	−7.90
β'	−1.84
γ'	1.50
cis	−1.12
corr α	−4.83
corr α'	2.51
corr β	−2.27

[a] Data taken from Dorman, Jautelat, and Roberts (35).

using the parameters in Table 5.11, one can calculate the chemical shifts of olefinic carbons with an accuracy approaching that possible with alkanes (Section 5.4.1).

5.5.2 CYCLIC ALKENES

The chemical shifts obtained for several cyclic alkenes are given in Table 5.12 (35). There is a larger variation in the olefinic carbon shifts with ring size than in the corresponding cycloalkanes. The shift difference between cis- and trans-cyclooctene is approximately the same as that found for acyclic alkenes.

TABLE 5.12. Carbon Chemical Shifts of Some Cycloalkenes[a]

Alkene	C_1	C_3	C_4	C_5	C_6	C_7
Cyclobutene	137.2	31.4				
Cyclopentene	130.8	32.8	23.3			
Cyclohexene	127.4	25.7	23.3			
Cycloheptene	132.7	29.6	28.0	32.7		
cis-cyclooctene	130.4	26.0	27.0	29.8		
trans-cyclooctene	134.0	35.5	[35.2][b]	[29.9]		
cis-cyclododecene	131.9	32.8	26.9	[26.3]	25.6	[25.3]

[a] Data taken from Dorman, Jautelat, and Roberts (35).
[b] Shifts given in brackets may be reversed.

Several alkylcyclopentenes and alkylcyclohexenes have been examined (54). As expected, the largest effects are observed when the alkyl group is attached to the olefinic carbon.

The effect of the double bond on the saturated carbon shielding does not follow the trend observed for linear alkenes (Section 5.5.1). A downfield shift is observed for the α-carbon of cyclopentene, whereas upfield shifts are observed for the other cycloalkenes.

5.5.3 SUBSTITUTED ALKENES

The CMR spectra of a number of substituted ethylene derivatives have been reported (27,55,56). For the vinyl compounds, the α-carbon chemical shift covers a range of ~70 ppm whereas the β-carbon covers a range of ~55 ppm. The only trend that is readily apparent is that when the substituent possesses an unshared pair of electrons such that resonance forms of the type $\overset{+}{X}=CH-\overset{-}{CH_2}$ are important, the β-carbon appears from 25 to 40 ppm upfield from ethylene. The α-carbon chemical shifts do not follow any apparent trends, and it is clear the shifts cannot be rationalized on the basis of the inductive effect alone.

The effect of the geometry of the substituents on the double bond has been examined (27). For some 1,2-disubstituted ethylenes the *trans* isomer absorbs at higher field whereas for others, the *cis* isomer absorption band appears at a higher field. A variety of other substituted ethylenes have been examined including vinyl ethers (57), allyl alcohols (5), α,β-unsaturated aldehydes, ketones, acids, and methyl esters (58,59). Both the α- and β-carbon are shifted downfield when in conjugation with a carbonyl group.

5.5.4 SUBSTITUTED BENZENES

The substituted benzenes were among the first compounds examined by CMR (5,6). A variety of monosubstituted benzenes have been investigated and some representative data are collected in Table 5.13 (60,61). The carbon to which the substituent is attached shows the largest variation with substituent; 96–168 ppm. The *ortho* carbon absorbs in the range 113–139 ppm, whereas the *para* carbon appears in the range 115–136 ppm. The *meta* carbon shows remarkably little variation with substituent (±2 ppm from benzene).

A correlation appears to exist between the *para*-carbon shifts and the Hammett σ *para* constants (60). With respect to benzene, electron-donating substituents shield the *para* carbon whereas electron-withdrawing substituents deshield the *para* carbon. These results suggest that the *para*-carbon

TABLE 5.13. ^{13}C Chemical Shifts For Monosubstituted Benzenes[a]

X	δ_1	δ_2	δ_3	δ_4
H	128.7	128.7	128.7	128.7
OH	155.6	116.1	130.5	120.8
OCH$_3$	158.9	113.2	128.7	119.8
OCOCH$_3$	151.7	122.3	130.0	126.4
NH$_2$	147.9	116.3	130.0	119.2
N(CH$_3$)$_2$	151.3	113.1	129.7	117.2
NHCOCH$_3$	139.8	118.8	128.9	123.1
CH$_3$	137.8	129.0	129.0	125.9
F	163.8	114.6	130.3	124.3
Cl	135.1	128.9	129.7	126.7
Br	123.3	132.0	130.9	127.7
I	96.7	138.9	131.6	129.7
CO$_2$CH$_3$	130.0	128.2	128.2	132.2
CHO	137.7	129.9	129.9	134.7
COCH$_3$	136.6	128.4	128.4	131.6
CN	109.7	130.1	127.2	130.1
NO$_2$	148.3	123.4	129.5	134.7

[a] Data taken from various sources (2,5,6).

shielding is controlled by the electronic effect of the substituents. The correlation previously stated has been used to determine the σ *para* constants for a variety of phosphorus substituents (62). It is clear that additional factors other than electronic are influencing the *ortho*-carbon shifts. In fact, in some cases the *ortho* and *para* shifts are in the opposite direction.

From the results of the early studies on disubstituted benzenes, it appeared that the effect of substituents on carbon chemical shifts was additive, similar to the results for the proton shifts of disubstituted benzenes. A variety of di- and trisubstituted benzenes have now been examined by CMR and, with the exception of *ortho*-disubstituted benzenes, the shifts agree with those predicted assuming additivity to within ±2 ppm (63,64). The deviations shown by the *ortho*-disubstituted benzenes are probably due to steric interactions between the substituents.

The correlation between the *para*-carbon shifts and Hammett σ *para* has prompted investigations of a correlation between carbon shifts and π-electron densities. From the examination of several monocyclic, aromatic ring systems, Spiesecke and Schneider (65) have suggested that carbon shifts depend on π-electron density and derived a correlation of ~160 ppm/electron with respect to benzene. However, work with polycyclic aromatic compounds suggest that both σ- and π-electron densities are important (66).

Several polycyclic aromatic hydrocarbons have been examined (66–71). The results tend to indicate two categories. For alternate hydrocarbons, the chemical shifts appear in the relatively narrow range of ~10 ppm with respect to benzene with quarternary carbons appearing at lowest field. In contrast, the nonalternate hydrocarbons absorb over a range of ~14–20 ppm. It has been suggested that the larger range is due to restricted conjugation in nonalternate systems (72). The chemical shift range has been used to classify acepleiadiene (**1**) and acepleiadylene (**2**) as nonalternate hydrocarbons (73).

5.5.5 AROMATIC HETEROCYCLES

The CMR spectra of the six-membered ring aromatic nitrogen heterocycles have received considerable attention (Table 5.14). Lauterbur has reported the chemical shifts for the parent members of this series along with some methyl derivatives (74). A reasonable correlation was found between the shifts and the π-electron densities for the carbons. However, a better correlation is obtained if both the σ- and π-electron densities are included.

As expected, the introduction of nitrogen into an aromatic ring has a profound effect on the chemical shifts of the ring carbons and results in a larger range of shift values (~120–170 ppm) compared to aromatic hydrocarbons. Comparing pyridine with benzene, nitrogen produces a downfield shift of 21.9 ppm at C-2 and of 7.7 ppm at C-4, and an upfield shift of 4.2 ppm at C-3. The effects of nitrogen are not exactly additive however, as the introduction of a second nitrogen produces a further downfield shift of only ~9 ppm at C-2 (i.e., C-2 of pyrimidine). Protonation effects on the spectra of the six-membered ring heterocycles have been examined (75). Protonation of pyridine produces an upfield shift of 7.8 ppm at C-2 and downfield shifts of 4.4 and 12.7 ppm at C-3 and C-4, respectively.

Friedel and Retcofsky have examined a number of monosubstituted pyridines and compared substituent effects with those found for mono-

sp² HYBRIDIZED CARBONS

TABLE 5.14. Carbon-13 Chemical Shifts for Some Aromatic Heterocycles[a]

[Structures with chemical shift values:

Pyridine: 136.4, 124.5, 150.6
Pyridinium (N-H): 152.6, 127.7
Pyrimidine: 122.6, 157.6, 159.2
Pyrazine: 146.1

Furan: 109.8, 142.8
Thiophene: 127.4, 125.6
Pyrrole (N-H): 108.4, 118.7
Selenophene: 129.2, 130.5

Pyrazole (N-H): 104.7, 133.3
Imidazole (N-H): 121.8, 135.7
1,2,3-Triazole (N-H): 130.4
1,2,4-Triazole (N-H): 147.6

Tetrazole (N-H): 144.0
Purine-type: 144.8, 134.5, 147.9, 160.7
Indole (N-H): 121.3, 128.8, 102.6, 122.3, 125.2, 120.3, 136.1, 111.8

Naphthalene-type with N: 128.5, 128.9, 136.2, 127.0, 121.7, 129.9, 151.1, 130.3, 149.1
Triazine: 166.8]

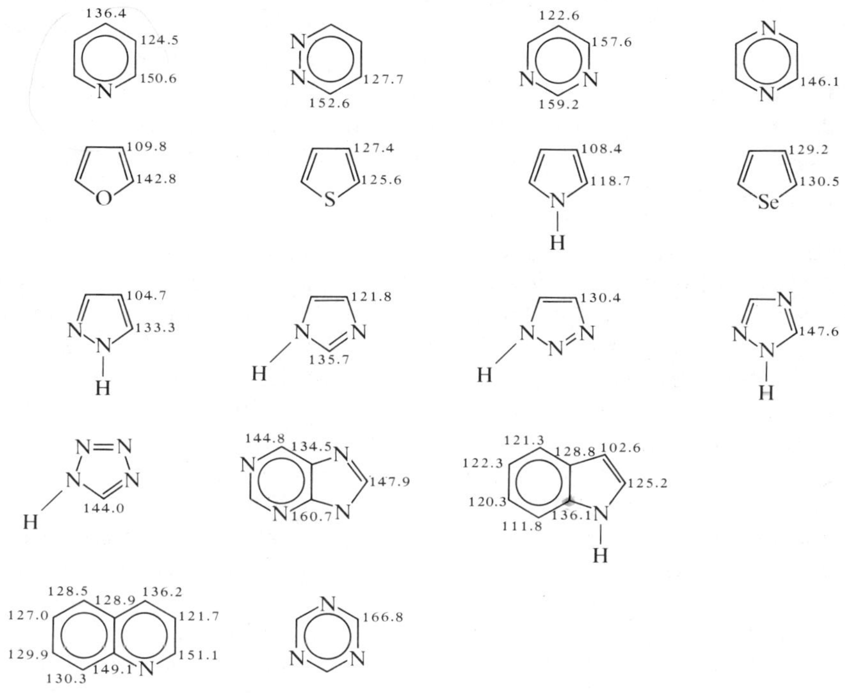

[a] Data taken from various sources.

substituted benzenes (76). Reasonable agreement (±1.5 ppm) was found for the 3-substituted pyridines where the substituent does not interact with the nitrogen. Larger deviations were found for the 4-substituted pyridines although the general trends were the same. Considerable deviations were found for the 2-substituted pyridines. The long range carbon–proton coupling constants in pyridines have also been examined (77).

The CMR spectra of many of the five-membered ring aromatic heterocycles and their methyl derivatives have been reported (78). Both C-2 and C-3 of pyrrole and thiophene are upfield from benzene whereas for furan and selenophene, C-2 is downfield from benzene. The various nitrogen five-membered ring heterocycles have been examined by Weigert and Roberts (79) and it was suggested that the chemical shifts could be predicted using an additivity scheme. Tautomeric exchange of the N—H proton renders C-3,5 equivalent in pyrazole and C-4,5 equivalent in imidazole. Pugmire and Grant (80) have reported the effect of both protonation and deprotonation on the

CMR spectra of pyrrole, pyrazole, and imidazole. Protonation of the anion produces shifts that follow an additivity relationship similar to protonation of six-membered ring heterocycles. However, the magnitude of the shifts are slightly different (Cα to nitrogen shifts upfield by ~ 9 ppm whereas the Cβ to nitrogen shifts downfield by ~ 2 ppm).

A series of 2- and 3-substituted thiophenes have been examined by Takahashi et al. (81). The largest substituent effects are observed at the substituted and α-carbons as expected, with only small effects being observed at the β-carbons. Although the effect of substituents on thiophene chemical shifts is similar to that observed for benzenes, there appears to be no correlation of the substituent effects. Abraham et al. (82) have recently reported the effect of substituents on the carbon shieldings of pyrrole. Additivity of the substituent effects was found. However, the effect of substituents more closely resembles that found for ethylenes than for benzenes. Interactions between substituents were found to be important in determining the shifts.

A number of polycyclic nitrogen heterocycles have been reported by Pugmire et al. (71). An additivity scheme has been developed to predict the effect of nitrogen incorporation into the ring with respect to the corresponding hydrocarbon. Pugmire et al. (71) have also reported the carbon shieldings for pyrrocaline and a number of related azindenes (83). A nitrogen in a bridgehead position produces only minor perturbations on the ^{13}C shieldings and appears to behave more like a pyrrole-type nitrogen than a pyridine nitrogen. The effect of protonation on the shieldings of benzimidazole and purine have been reported (84). Tautomeric averaging of the carbon shift was observed for purine. The protonation shifts for purine are similar to those found for imidazole. Carbon shieldings in various N-methylpurines have been investigated (85). The CMR spectra of four symmetrical naphthyridines have been reported (86). While the carbon shieldings correlate reasonably well with total charge densities, the agreement between observed shielding and those calculated using an additivity scheme is only fair. Parker and Roberts have investigated the carbon shieldings of methyl substituted indoles (87). Using the substituent effects of monomethyl substitution, calculated shieldings of di- and tri-methylindoles were in only fair agreement with observed shieldings.

A number of studies of the effect of shift reagents on the carbon shieldings of nitrogen heterocycles have been reported. These include various pyridines (88), pyridine n-oxides (89), quinoline (90), and isoquinoline (88).

5.6 sp HYBRIDIZED CARBONS

Relatively little data is available on the chemical shifts of sp-hybridized carbons. Dorman et al. (36) have collected the available data on linear

sp HYBRIDIZED CARBONS

TABLE 5.15. Carbon Chemical Shifts of Some Alkynes[a]

Alkyne	C_1	C_2	C_3	C_4	C_5	C_6	C_7	C_8
1-Butyne	68.2	85.9						
1-Pentyne	68.2	83.6	20.1	22.1	13.1			
1-Hexyne	68.6	86.3	18.6	31.1	22.4	14.1		
1-Heptyne	68.6	84.1	18.9	29.3	31.9	23.6	15.2	
1-Octyne	68.9	84.4	18.8	29.2	29.1	32.0	23.1	14.3
2-Butyne		74.8	74.8					
2-Hexyne	2.9	74.9	78.1	20.8	22.8	13.3		
2-Heptyne	3.5	75.4	78.8	18.9	32.4	23.0	14.7	
2-Octyne	3.0	75.2	79.2	19.1	29.5	31.7	22.8	14.1
3-Hexyne	15.6	13.2	81.1					
3-Heptyne	14.9	13.2	81.4	79.5	21.4	23.7	14.3	
3-Octyne	13.7	12.7	81.5	79.3	18.5	31.9	22.4	14.6
4-Octyne	13.5	23.1	20.1	80.1				

[a] Data taken from Dorman, Jautelat, and Roberts (36).

alkynes. Some representative shifts are given in Table 5.15. The shielding of terminal alkynes is relatively constant with C-1 at 68 ppm and C-2 at ~85 ppm. For 2-alkynes, C-2 appears at higher field (~75 ppm) than C-3 (~79 ppm). These differences should prove valuable in structure elucidations.

An additivity scheme for predicting the shieldings of alkyne carbons has been developed. By considering the effect of alkyl substituents similar to that for alkenes, Dorman et al. (36) have arrived at the parameters given in Table 5.16 from

$$C^\gamma - C^\beta - C^\alpha - C^* \equiv C - C^{\alpha'} - C^{\beta'} - C^{\gamma'}$$

TABLE 5.16. Parameters for Calculating the Chemical Shifts of Alkynes[a]

Parameter	Shift (ppm)
α	6.7
β	6.3
γ	−1.6
α'	−5.3
β'	0.6
γ'	0.4

[a] Data taken from Dorman, Jautelat, and Roberts (36).

regression analysis of the available data on alkynes. The shielding in acetylene is predicted to be 66.8 ppm by this analysis and the parameters in Table 5.16 are given with respect to this value. Use of the parameter set yields a standard deviation of 0.3 ppm with the available data. It is pointed out that this parameter set will probably have to be modified to account for branching at the α-carbon.

The carbon shieldings of a number of diynes and cycloalkynes have been reported (9). Ring strain resulting from the incorporation of a triple bond in the ring appears to have only minor effects on the shieldings of the sp-hybridized carbons.

A few results have been reported for substituted acetylenes and 1-substituted-1-hexynes (92–95). Variations in the shieldings over a range of ~100 ppm are observed. The trend appears to be similar to that observed for ethylene derivatives in that substituents with unpaired electrons tend to shield the β-carbon without affecting the α-carbon to any appreciable extent. There appears to be no correlation between the inductive effect of the substituents and the shieldings.

5.7 FUNCTIONAL GROUPS

The fact that information about functional groups may be obtained from CMR spectra represents one of the major advantages of CMR over proton NMR spectroscopy. Of the functional groups studied to date, the carbonyl group in organic compounds has received the most attention (96). The chemical shift of the carbonyl carbon is dependent on the type of functional group such that the functional group may be identified in many cases from only the chemical shift value. Although the chemical shift range of some of the carbonyl functional groups overlaps to a certain extent, additional experiments can serve to identify the functional group.

The carbonyl–carbon chemical shift of some representative examples are given in Table 5.17. In most cases, the effect of the size of the alkyl group attached to the carbonyl carbon is smaller (± 1 ppm) than its effect on alkyl carbon shieldings. Unsaturation in conjugation with the carbonyl group shifts the carbonyl resonance upfield by ~9 ppm from its position in non-conjugated systems. This is presumably due to the importance of resonance structures of the type $C=C-\overset{\overset{\displaystyle O}{\|}}{C}- \leftrightarrow \overset{\overset{\displaystyle O-}{|}}{\overset{+}{C}}-C=C-$ which tend to decrease the polarization of the carbonyl carbon. The carbonyl chemical shift of amides is intermediate between carboxylic acid and esters (acetamide 178.1 ppm) (97). Alkyl substitution on nitrogen shifts the resonance upfield by ~3 ppm.

TABLE 5.17. Carbonyl–Carbon Chemical Shifts in Some Representative Functional Groups[a]

R	R—CHO	RCOMe	RCO_2H	RCO_2Me
Ethyl	201.8	206.3	180.4	173.3
Hexyl	201.8	206.8	180.2	173.4
Cyclohexyl	201.8	209.3	182.1	174.2
Vinyl	192.4	197.2	169.7	165.3
Phenyl	191.0	196.0	174.9	165.9

[a] Data taken from various sources (5,6).

The limited data available on functional groups containing carbon multiply bonded to nitrogen also indicate characteristic chemical shift ranges (Table 5.18).

A variety of data are available on the carbonyl shieldings of inorganic metal carbonyls. The shieldings vary over a range of ~40 ppm (190–230 ppm). The entire area of application of CMR to inorganic compounds has been reviewed recently and those interested in this aspect of CMR are referred to this article (4).

TABLE 5.18. Chemical Shifts of Carbon Multiple Bonded to Nitrogen[a]

R	RCN	RNC	RNCO	RNCS
Methyl	117.2	158.7	121.5	128.7
Ethyl	120.7	156.4	122.6	130.8
Cyclohexyl	121.6	156.7	123.7	132.4

[a] Data taken from various sources.

5.8 COUPLING CONSTANTS

Although most routine CMR spectra are obtained with proton decoupling in order to increase the sensitivity, coupling constants between carbon and hydrogen can often yield useful information concerning the structure of a molecule and aid in the assignment of spectra. In cases involving magnetic nuclei other than hydrogen, the coupling between the other nuclei and carbon is not eliminated and one can make use of these coupling constants.

One-bond couplings to carbon appear to be dominated by the Fermi contact mechanism (Section 2.9). Both valence bond and M.O. theories have been used to treat carbon coupling constants. The most successful method appears to be the M.O. method developed by Pople and Santry (98) and extended by others (99). Within this framework, the Fermi contact contribution for the one-bond coupling can be approximated as

$$J_{AB} = \frac{\gamma_A \gamma_B \alpha_A^2 \alpha_B^2 S_A^2(O) S_B^2(O)}{\Delta E} \tag{5.2}$$

where γ_A and γ_B are the magnetogyric ratios of nuclei A and B, α_A^2 and α_B^2 are the s electron densities of the orbitals used by nuclei A and B in forming the bond, $S_A^2(O)$ and $S_B^2(O)$ represent the magnitude of the valence atomic s orbital at nuclei A and B, and ΔE is the average energy approximation. Further calculations have shown that the terms for orbital electronic currents and dipole–dipole interactions should be included with the Fermi contact term to reproduce trends in the data (100). However, it is clear that the Fermi contact term makes the major contribution.

5.8.1 CARBON–HYDROGEN COUPLINGS

Most of the one-bond carbon hydrogen coupling constant data were obtained from the carbon-13 satellites in proton spectra. It was suggested early that $^1J_{C-H}$ is related to the hybridization of carbon (101). Some representative values in Table 5.19 tend to support this. Substituents also play an important role in determining $^1J_{C-H}$. For example, in substituted methanes, $^1J_{C-H}$ varies from 118 Hz for Si(CH$_3$)$_4$ to 151 Hz for CH$_3$Br. No clear correlation exists between substituent electronegativity and $^1J_{C-H}$. Furthermore, $^1J_{C-H}$ in substituted methanes varies with solvent (102). Ring strain and stereochemistry also appear to play an important role in determining $^1J_{C-H}$ values. This is illustrated by the data in Table 5.20 for some representative cyclic systems (5).

Substituent effects on $^1J_{C-H}$ have been examined for a large variety of structural and hybridization types. These include cyclopropanes (103), saturated heterocycles (104,105), toluenes (106), ethylenes (107), aromatic

TABLE 5.19. Some Representative $^1J_{C-H}$ Values

Compound	$^1J_{C-H}$ (Hz)
Ethane	125.0
Ethylene	156.2
Benzene	157.5
Acetylene	248.7

COUPLING CONSTANTS 227

TABLE 5.20. $^1J_{C-H}$ in Some Cyclic Alkanes

Compound	$^1J_{C-H}$(Hz)	Compound	$^1J_{C-H}$(Hz)
cyclopropane–H	161	bicyclobutane (bridgehead)–H	205
cyclobutane–H	134	norbornane–H	142
cyclohexane–H	123		
bicyclo[1.1.0]butane–H	169	bicyclo[2.1.0] –H	130
bicyclopropyl–H	153		178.5

heterocycles (79), cations (108), and acetylenes (109). In each case, a sizeable range of values due to substituent effects is found for each type of carbon hybridization.

Long-range carbon–hydrogen couplings, $^2J_{C-H}$ and $^3J_{C-H}$ have also received considerable attention. Some typical results are given in Table 5.21. The data indicate that $^2J_{C-H}$ depends on the location of the substituent in the molecule. A study of some substituted cyclopropanes also shows that $^2J_{C-H}$ depends on the orientation of the proton with respect to the substituent (112).

TABLE 5.21. $^2J_{C-H}$ and $^3J_{C-H}$ in Some Aliphatic Systems

Compound	$^2J_{C-H}$	$^3J_{C-H}$	Reference
$CH_3CH_2\underline{C}D_2OH$	4	6.4	110
$(CH_3CH_2)_2\underline{C}DOH$	4.0	5.3	110
$(CH_3CH_2)_3\underline{C}OH$	3.8	4.5	110
$\underline{C}H_3CH_2Br$	4.0		111
$CH_3\underline{C}H_2Br$	2.1		111
$\underline{C}H_3CH_2Cl$	4.2		111
$CH_3\underline{C}H_2Cl$	2.6		111

The effect of substituents on $^3J_{C-H}$ has been examined with *t*-butyl derivatives (113). The results indicate that steric and substituent effects are important in determining $^3J_{C-H}$. Recently, it has been shown that $^3J_{C-H}$ may be used in conformational analysis studies since $^3J_{C-H}$ depends on the dihedral angle. Rotamer populations have been determined for some amino acids and the results agree well with the results obtained from proton NMR (114).

The trends in the long-range couplings of aromatic compounds have also received attention. For benzene, Weigert and Roberts (115) reported the following values: $^2J_{C-H} = 1.0$; $^3J_{C-H} = 7.4$; and $^4J_{C-H} = 1.1$ Hz. The large difference in $^2J_{C-H}$ and $^3J_{C-H}$ has been used as an aid in the assignment of the CMR spectra of aromatic compounds (Section 5.9.5). Long-range couplings in some dihalobenzenes have also been reported (64).

5.8.2 CARBON–CARBON COUPLINGS

Studies of carbon–carbon couplings have generally been performed on labeled compounds due to the low probability of having two carbon-13 nuclei adjacent in the same molecule. Only recently have studies been carried out on natural abundance materials. One-bond carbon–carbon couplings show a dependence on the hybridization of the carbons as expected from equation 5.2. Typical values are given in Table 5.22. The effect of substituents on $^1J_{C-C}$ has been investigated in several series of compounds (117). It appears from the data on ethyl derivatives that $^1J_{C-C}$ is not very sensitive to substituent. However, there are indications that steric interactions may be important.

Long-range carbon–carbon couplings have been determined in a number of cases using labeled compounds (117). These couplings are similar to the long-range carbon–hydrogen couplings in that $^2J_{C-C}$ is much smaller

TABLE 5.22. $^1J_{C-C}$ in Some Representative Compounds

Compound	$^1J_{C-C}$
CH_3CH_3	34.6
$CH_3C_6H_5$	44.2
$CH_3C\equiv CH$	67.4
$CH_2=CH_2$	67.2
$CH_2=C=C(CH_3)_2$	99.5
$HC\equiv CH$	170.6

(~1 Hz) than $^3J_{C-C}$ (~5 Hz). Furthermore, there appears to be a dependence of $^3J_{C-C}$ on dihedral angle with the minimum appearing near 90°.

5.8.3 CARBON–COUPLING WITH OTHER NUCLEI

A variety of data have been reported for carbon–other nuclei coupling constants (4,118). As expected, the majority of data are for phosphorus and fluorine couplings, due to the high natural abundance of these nuclei. Although it is clear that J_{C-F} values depend on structural and electronic characteristics of the particular molecule, the values vary over a large range and do not appear to correlate with any one parameter. Furthermore, carbon–fluorine couplings appear to be an exception in that $^2J_{C-F}$ is usually larger than $^3J_{C-F}$, in contrast to couplings to carbon by other nuclei. Nevertheless, carbon–fluorine couplings can yield useful information concerning structure and assignment of CMR spectra.

Carbon–phosphorus couplings follow trends similar to carbon–hydrogen couplings. The one-bond coupling $^1J_{C-P}$ varies over a range of ~150 Hz, depending on the substitution on phosphorus. Some typical values are given in Table 2.23. For trivalent phosphorus, an interesting dependence of long-range carbon–phosphorus couplings on the orientation of carbon with respect to the lone-pair has been reported (119). It is clear that the magnitude of carbon–phosphorus couplings can provide useful information concerning the structure and stereochemistry of a compound.

TABLE 5.23. Some Typical Carbon–Other Nuclei Coupling Constants

Compound	$^1J_{C-X}$	$^2J_{C-X}$	$^3J_{C-X}$	$^4J_{C-X}$	Reference
n-hexyl F	−166.6	19.9	5.3		118
n-butyl$_3$P	−10.9	11.7	12.5		118
n-butyl$_4$P$^+$Br$^-$	47.6	4.3	15.4		118
(n-butyl O)$_3$PO		5.9	6.5		121
n-butyl$_2$Hg	656	26.3	100		118
n-butyl$_4$Sn	310	25	52		122
n-butyl$_4$Pb	189.2	26.9	74.5		34
phenyl F	−245.3	21.0	7.7	3.3	118
phenyl$_3$P	12.4	19.6	6.7	0	118
phenyl$_3$P$^+$MeI$^-$	88.4	10.9	12.7	2.9	118
(phenyl O)$_3$P		3.0	4.9		121
phenyl$_4$B$^-$	49.4		2.6		118
phenyl$_2$Hg	1186	88	101.6	17.8	118
phenyl Sn Me$_3$	474.4	36.6	47.4	10.8	123
phenyl$_4$Pb	480.9	68.1	80.6	19.5	34

TABLE 5.24. Typical One-Bond Carbon Metal Couplings[a]

Compound	$^1J_{C-X}$ (Hz)
CH_3Li	15
$CH_3{}^{14}NC$	7.6
$CH_3C^{15}N$	17.5
$(CH_3)_4Si$	52
$(CH_3)_2Se$	62
$(CH_3)_3Se^+$	50
$(CH_3)_2Cd$	513
	536
$(CH_3)_4Sn$	340
$(CH_3)_2Te$	162
$(CH_3)_2Hg$	687
$(CH_3)_4Pb$	250

[a] Data taken from various sources (4,5).

A variety of carbon-metal couplings has been reported. Most of the data have been for the one-bond coupling constant (41). Some typical data are given in Table 5.24. An interpretation of these couplings in terms of the Fermi contact coupling mechanism has been given. Further work in this area should provide some interesting correlations of these couplings with structure. In fact, it has recently been shown that carbon–tin couplings depend on stereochemistry similar to some of the more widely studied carbon coupling constants (120).

5.9 ASSIGNMENT OF CMR SPECTRA

While it is clear that CMR spectroscopy offers many advantages over proton NMR for the structural elucidation of organic and inorganic compounds, one must assign the peaks in the spectrum to specific carbons in the compounds in order to realize the full potential of CMR spectroscopy. Since most routine CMR spectra are obtained with broadband noise, proton decoupling, the coupling constant information is usually not available and one is therefore left with only the chemical shifts to work with. The carbons in uncharged organic compounds usually absorb over a range of ~200 ppm downfield from TMS. Within this range, one can usually distinguish skeletal carbons on the basis of hybridization and carbons in functional groups. However, even with this distinction, one is often faced with the problem of assigning the spectra of molecules that contain several, similar-type carbons.

In these cases, one must rely on additional experiments in order to completely assign a CMR spectrum.

A variety of techniques have been developed over the past few years to aid in the assignment of CMR spectra (124). These techniques may be summarized as follows:

1. Chemical shift
2. Off-resonance decoupling
3. Spectral comparison
4. Specific labeling
5. Selective decoupling
6. Undecoupled spectra
7. Spin-lattice relaxation.

Each of these techniques is briefly discussed in the following. The reader is referred to an excellent paper by Roberts and associates (125) where most of these techniques are illustrated in the assignment of the CMR spectra of a variety of steroids. It should be recognized from the beginning that, even with these additional experiments, it may not be possible to completely assign every CMR spectrum.

5.9.1 CHEMICAL SHIFT

Carbon shieldings are determined primarily by the hybridization of carbon. While one can usually distinguish sp^3 from sp from sp^2 carbons on the basis of the chemical shift, the overlap of olefinic and aromatic carbon peaks presents a more difficult assignment problem than the corresponding problem of distinguishing protons bonded to olefinic carbons from protons bonded to aromatic carbons. Furthermore, for compounds containing several saturated carbons, one cannot make the assumption that methyl carbons are more shielded than methylene carbons, which in turn, are more shielded than methine and quarternary carbons.

Substituent effects play an important secondary role in determining the carbon shieldings within a given hybridization type. The examination of a wide variety of compounds has shown that the shielding of a carbon is influenced by substituents up to five bonds away. Empirical correlations of the effect of substituents has permitted additivity parameters to be derived which allow one to calculate carbon chemical shifts for many classes of compounds. Additivity parameters for substituent effects are discussed further in Section 5.4. By starting with some base value, say the chemical shifts of the corresponding hydrocarbon for substituted alkanes, one can add in the substituent parameters and calculate the chemical shifts for the molecule. This procedure will usually give a calculated spectrum which

5.9.2 OFF-RESONANCE AND SELECTIVE DECOUPLING

Since routine CMR spectra consist of single peaks due to complete proton decoupling, one cannot readily distinguish methyl from methylene, methine, or quaternary carbons without additional information. Fortunately, it is possible to determine the number of protons attached to each carbon without losing all the sensitivity enhancement due to decoupling by using the off-resonance decoupling technique (8).

With this technique, the decoupler is adjusted for single frequency irradiation. The frequency of the decoupler is offset a few hundred hertz from the proton resonances such that the carbon–hydrogen couplings are not completely eliminated. The decoupler rf power level is maintained sufficiently high, however, such that long-range carbon–hydrogen couplings are eliminated and the direct, one-bond carbon–hydrogen couplings are only partially decoupled. This results in spectra in which the peaks appear as multiplets or singlets, depending on whether the carbon giving rise to the peak has protons attached to it. Methyl carbons appear as quartets, methylene carbons as triplets, methine carbons as doublets, and quarternary carbons as singlets (Figure 5.2). Under these decoupling conditions, there is still an appreciable NOE such that this aspect of sensitivity enhancement is not lost completely. Usually, from two to four times the number of scans required for the normal spectrum are required for the off-resonance decoupled spectrum.

The magnitude of the splitting of the multiplets in the off-resonance decoupled spectrum depends on the offset frequency, Δv, of the decoupling frequency from the proton resonance and the magnitude of the decoupler rf power level, γB_2. Ernst has given the following expression for the magnitude of the reduced splitting, J_r (8):

$$J_r = \frac{{}^1H_{C-H}\Delta v}{\gamma B_2} \qquad (5.3)$$

This expression is valid provided $|\gamma B_2| \gg |\Delta v|$ and $2\pi |{}^1J_{C-H}|$.

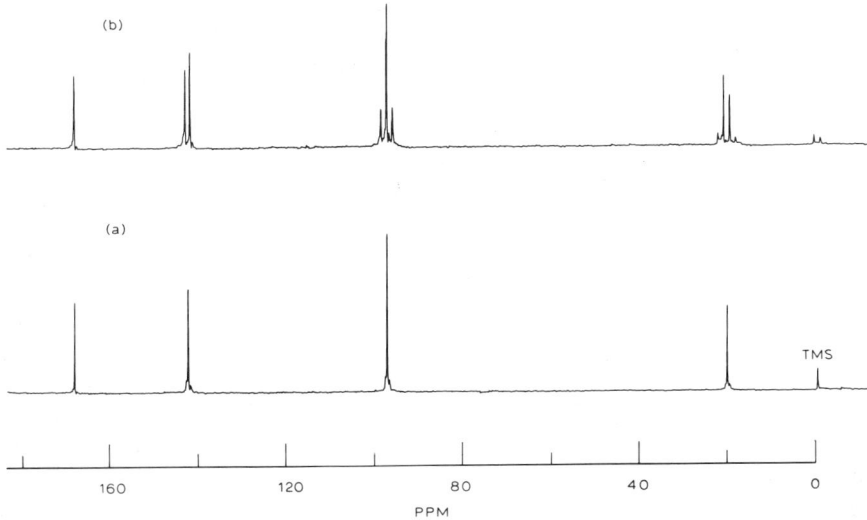

Fig. 5.2 (a) The proton-decoupled, natural abundance CMR spectrum of vinyl acetate. (b) The single-frequency, off-resonance proton-decoupled CMR spectrum of vinyl acetate.

Provided one can assign, or partially assign, the corresponding proton spectrum and can estimate the one-bond carbon–hydrogen coupling constants, it may be possible to make accurate assignments of the CMR spectrum using the off-resonance decoupling technique. The decoupler rf power level and the decoupling frequency are first determined using a known sample (5). The decoupling frequency is then related to the reference TMS. With the sample in question, the decoupling frequency is offset, say 5 ppm upfield from TMS, and the CMR spectrum obtained. Protons absorbing to lower field will be less effectively decoupled and hence, the magnitude of the residual coupling in the CMR spectrum will be larger than for protons absorbing at higher field. With a knowledge of the proton chemical shifts, the frequency offset, and the carbon–hydrogen coupling constants, the residual splittings can be calculated using equation 5.3 and compared with the experimental values. In this manner, specific assignments in the CMR spectrum may be made. This method of assignment works best when the proton spectrum is close to first order. However, even in cases where the proton spectrum is not first order, it may be possible to assign selected peaks.

An alternative approach to calculating residual couplings for special assignment is to run a series of spectra where the frequency offset is increased between successive spectra. If the decoupler frequency is offset upfield from TMS, the residual coupling in the CMR spectrum will increase at a faster rate for carbons whose attached protons absorb at lower field than for carbons whose attached protons appear at higher field.

In cases where the protons being irradiated are part of a tightly coupled spin system ($J/\delta > 0.3$), broadening and additional splitting may appear in the CMR spectrum (126). This normally occurs when a proton bonded to carbon is tightly coupled to a proton on an adjacent carbon and the broadening will appear in the resonances of both carbons. Grutzner (126) has discussed these effects and used them to aid the assignment of the CMR spectrum of 2-methylnorbornan-2-ol.

When the proton spectrum is first order, or when a proton multiplet is separated from other proton peaks, it may be possible to assign certain peaks in the CMR spectrum by obtaining spectra in which only selected protons are decoupled. If a proton multiplet can be assigned, irradiation at this frequency will result in a CMR spectrum in which only the carbon attached to the proton irradiated will appear as a singlet (Figure 5.3). This method of assignment of CMR spectra requires a considerable number of spectra to completely assign the carbon resonances. However, it does provide an unambiguous assignment provided there is no uncertainty in the assignment of the proton spectrum (124).

5.9.3 SPECTRAL COMPARISON

A great deal of information concerning the assignment of a CMR spectrum can often be obtained by comparing the spectrum with the spectra of related compounds. This technique was used most successfully by Roberts and associates (125) in their assignment of the CMR spectra of various steroids. Structurally similar compounds will have identical CMR spectra (± 1 ppm) with the exception of the carbons four or five bonds removed from the point of difference in the structure. If a sufficient number of related compounds are available, complete assignments of the CMR spectra can often be achieved.

The basis for this method of assignment of CMR spectra is the specific effect of substituents on carbon shieldings. For example, carbons in the immediate vicinity of a hydroxyl group can usually be assigned from the changes which occur in the spectrum upon acetylation of the hydroxyl group. Acetylation produces a downfield shift (see Table 5.6) for the α- and γ-carbons and an upfield shift for the β-carbon. Similarly, predictable changes occur in the spectrum upon dehydration of the alcohol. Furthermore, for six-membered rings the effect of substituents is sufficient to establish the stereochemistry in many cases. For other functional groups, derivatization will permit the assignment of carbons in the vicinity of the functional group.

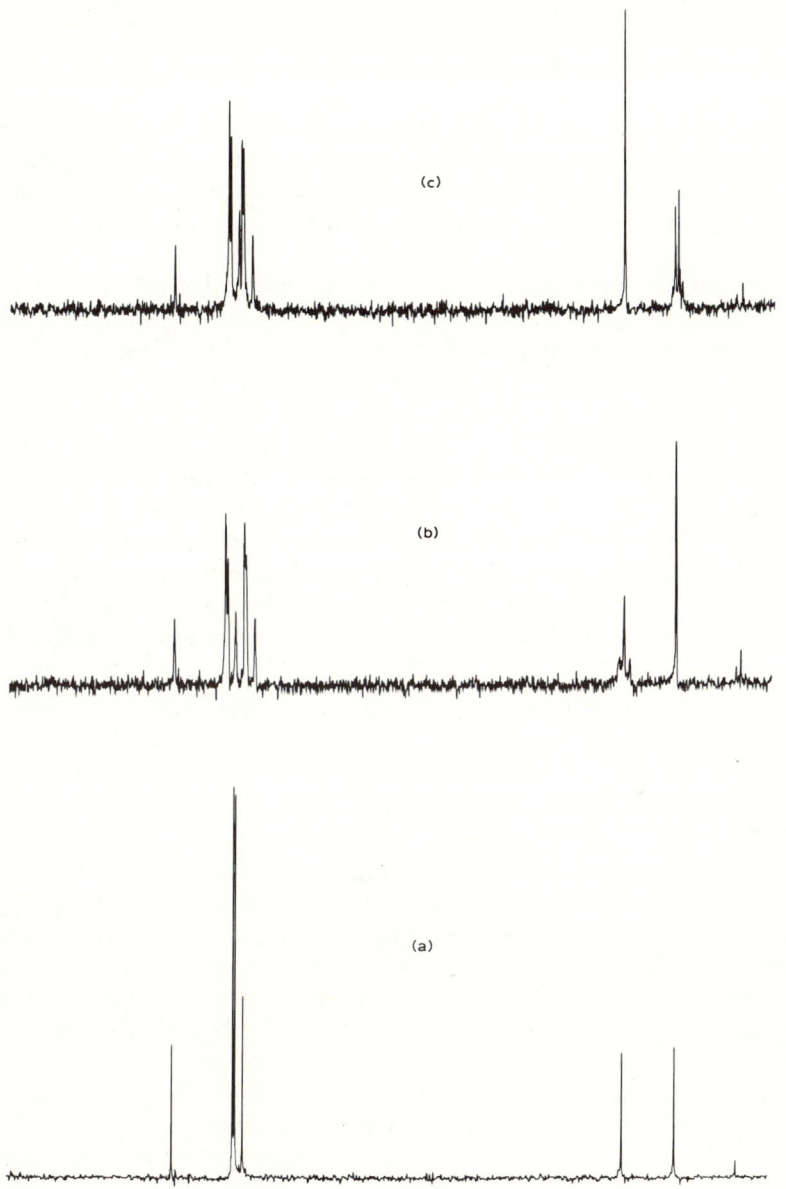

Fig. 5.3 (a) The proton-decoupled, natural abundance CMR spectrum of ethylbenzene. (b) The CMR spectrum as a result of selectively decoupling the methyl protons. (c) The CMR spectrum as a result of selective decoupling the methylene protons.

5.9.4 SPECIFIC LABELING

Unambiguous assignments of selected peaks in a CMR spectrum can be made provided suitable labeled derivatives are available. If one can prepare a carbon-13-labeled sample such that the position of the label is not in question, the peak due to the labeled position can be identified immediately from its CMR spectrum by the increased intensity of that peak. Usually, only $\sim 5\%$ enrichment at a carbon is required to identify the peak due to that carbon.

A synthetically more convenient approach to specific labeling for peak assignment is the substitution of deuterium for hydrogen in a compound. Deuterium substitution may produce a variety of effects, depending on the particular compound. Since deuterium possesses a spin $I = 1$, the carbon attached to deuterium will be split into a 1:1:1 triplet ($J \sim 20$ Hz) due to coupling with deuterium. In addition, carbons β and γ to deuterium will exhibit broadened peaks due to unresolved coupling with deuterium (Section 5.4.3.1) (21). In many compounds, a peak for the carbon attached to deuterium will not be observed, or an extremely broadened peak will be observed due to broadening of the carbon peak by the quadrupole moment of deuterium and the decreased NOE. The exact effects one will observe upon deuterium substitution cannot be predicted in advance. However, regardless of the specific effects, they are such that the peak due to carbon attached to deuterium can be identified.

5.9.5 UNDECOUPLED SPECTRA

CMR spectra without proton decoupling are normally not obtained due to the complexity of the spectra and the decreased sensitivity. However, in certain cases undecoupled spectra may provide a basis for the unambiguous assignment of the spectra. A decoupling method has been developed recently that allows one to obtain the carbon hydrogen couplings and still retain some of the sensitivity enhancement due to the NOE (127,128). This method is referred to as gated decoupling. The broadband-noise proton decoupler is turned on during the time that data is not being collected. When the decoupler is turned off for data collection, the coupling constant information returns immediately, whereas the NOE, being related to relaxation, decays at a slower rate.

In addition to the large one-bond coupling $^1J_{C-H}$, long-range couplings can be observed. For aliphatic compounds, $^2J_{C-H}$ and $^3J_{C-H}$ are of comparable magnitude. However, for aromatic compounds, the *meta* coupling $^3J_{C-H}$ is considerably larger than the *ortho* coupling $^2J_{C-H}$ (~ 8.0 versus

ASSIGNMENT OF CMR SPECTRA

1.0 Hz) (115). One can make use of these differences in long-range couplings to specifically assign CMR spectra.

As an example, consider the spectrum of diphenylether given in Figure 5.4. The four peaks appear at 118.65, 122.99, 129.58, and 157.04 ppm. Relative intensity considerations suggest that the peaks at 122.99 and 157.04 ppm are due to one carbon each. The peak at 122.99 ppm is assigned to C-4 and the peak at 157.04 ppm to C-1 on the basis of the differences in intensity due to the NOE. These assignments are confirmed by the gated decoupled

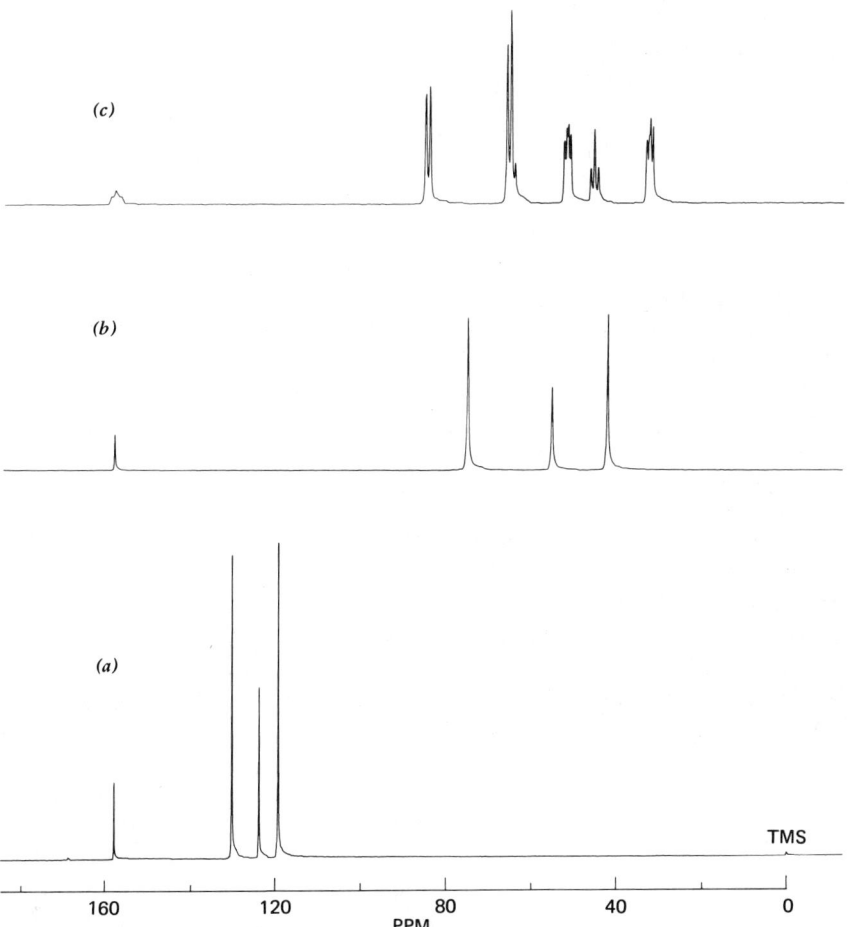

Fig. 5.4 (a) The proton-decoupled, natural abundance CMR spectrum of diphenylether. (b) An expansion of the aromatic region. (c) An expansion of the CMR spectrum of the aromatic region taken under conditions of gated-decoupling.

spectrum which shows the peak at 122.99 ppm as a large doublet split further into triplets due to the long-range couplings $^3J_{C-H}$. The assignment of C-2 and C-3 is not straightforward using the proton decoupled spectrum. However, the gated decoupled spectrum shows differences that allow an unambiguous assignment to be made. The peak at 118.65 ppm appears as a doublet split further into triplets, whereas the peak at 129.58 ppm appears as a doublet split further into doublets (to a first-order approximation). Consideration of the long-range couplings shows that C-2 has two *meta* protons whereas C-3 only has one *meta* proton. Therefore, the peak at 118.65 ppm is assigned to C-2 and the peak at 129.58 ppm to C-3.

This method of assignment of CMR spectra works best when the peaks in the spectra are well separated such that the overlap of peaks in the undecoupled spectrum is at a minimum. However, even in cases of complex spectra, it may be possible to make specific assignments using this technique.

5.9.6 SPIN-LATTICE RELAXATION

The use of spin-lattice relaxation, T_1, for structural assignment has been demonstrated by Allerhand et al. (129). Although T_1 values depend on the motion of the molecule, for cases where the dipolar relaxation mechanism is predominant, the T_1 values depend on the number of hydrogens attached to the carbons. Quaternary carbons usually exhibit the longest relaxation times and can usually be identified by their decreased relative intensity in spectra in which the delay between repetition of the pulses is not sufficient to allow full recovery of the magnetization after a pulse.

The decrease in relative intensity is such that by using the progressive saturation technique, quaternary carbons can be identified with only a few additional spectra. With the progressive saturation technique, a series of spectra are obtained in which the delay between pulses is decreased between successive spectra. As the delay time is decreased, the intensity of the quaternary carbons will decrease at a faster rate due to their long relaxation times, which prevents the reestablishing of the equilibrium magnetization before the next pulse is applied. Controlled experiments of this type also

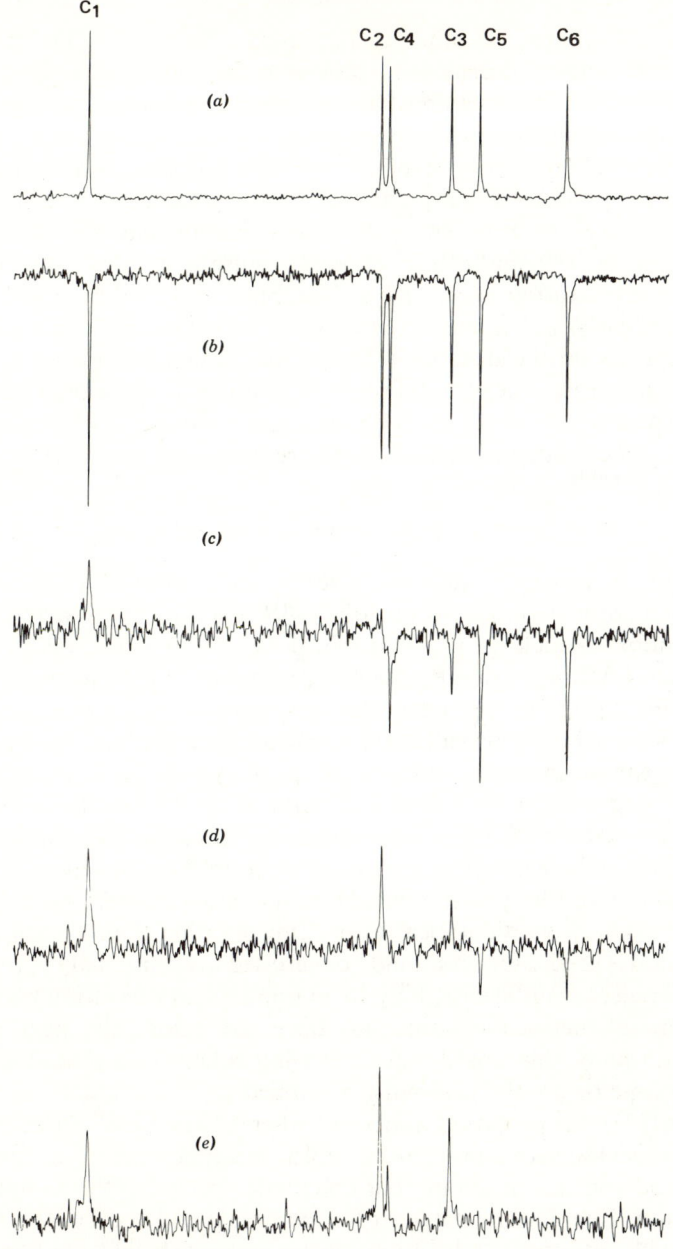

Fig. 5.5 (a) An expansion of the proton-decoupled, natural abundance CMR spectrum of hexanol. (b) The spectrum as a result of applying a $180°$-t-$90°$ pulse sequence with $t = 0.1$ sec. (c) The spectrum under the same conditions as b but with 0.01 M added $Gd(dpm)_3$. (d) The spectrum as in c but with $t = 0.3$ sec. (e) The spectrum as in c but with $t = 0.5$ sec.

permit the assignment of other carbon peaks provided some initial assumptions are made concerning the motion of the molecule (124).

Another method for assigning ^{13}C resonances on the basis of relaxation times has appeared (130). In this procedure, a normal spectrum is obtained. Second, an inversion recovery ($180° - \tau - 90°$) spectrum is obtained where τ is of the order of 0.5 sec. Since many ^{13}C T_1s are > 1 sec, this spectrum shows all the resonances inverted since the magnetization has not had time to recover from the $180°$ pulse before the $90°$ pulse is applied. Addition of a paramagnetic relaxation agent such as Gd(dpm)$_3$ to the solution will shorten the T_1s of the carbons. The effectiveness of the reduction in T_1 for a particular carbon depends on the distance ($1/r^6$) of the carbon from the Lewis base site in the substrate where gadolinium is binding. If additional inversion-recovery spectra are obtained with various τ values, assignment of the spectrum can be made on the basis of the reduction in the T_1s (Figure 5.5).

5.10 QUANTITATIVE ANALYSIS BY CMR

Although, in principle, one could use the area under CMR peaks (integration) in a manner similar to proton NMR (Section 3.5) to determine the relative number of nuclei giving rise to the peaks, the experimental conditions under which CMR spectra are obtained prevent accurate integrations from being obtained. Due to the decreased sensitivity of ^{13}C, CMR spectra are routinely obtained using broadband noise proton decoupling. This technique leads to an enhancement in some ^{13}C peaks due to the NOE effect. The enhancement depends on the relaxation times of the ^{13}C nuclei and may be as large as a factor of 3 (9). This enhancement could be eliminated by obtaining the CMR spectrum without decoupling. However, the time required to obtain the spectrum would be increased further, due to the fact that the resonances would appear as multiplets rather than as single peaks. If this method is used, caution should be exercised to ensure that the repetition time of the pulses is sufficiently long to guarantee that relaxation back to the equilibrium magnetization value has occurred before the next pulse is applied. Normally, this would require waiting at least four times the longest relaxation time before the next pulse is applied.

La Mar (131) has proposed a method whereby the Overhauser enhancement is removed such that proton-decoupled spectra give accurate intensities. This method consists of adding paramagnetic species to the sample under investigation. Interactions between the paramagnetic species and the substrate provide additional relaxation mechanisms such that the Overhauser enhancement is reduced to almost zero (132). Furthermore, the concentration of paramagnetic species required to eliminate the enhancement does not usually produce any additional line broadening (Section 3.3). Experiments with dioxane and acetone (131) show that the addition of ~0.05 M di-*tert* butyl nitroxide reduces the Overhauser enhancement such that accurate intensities can be obtained. One obvious disadvantage with this method is that it cannot be used if the substrate reacts with the added paramagnetic species.

A second method whereby the enhancement is eliminated, yet decoupled spectra are retained, has been proposed (133). This method uses pulse modulation of the decoupler similar to that in gated decoupling; the difference being in that the decoupler is turned on just before the excitation pulse and turned off again after the acquisition of the FID. The build-up of the Overhauser enhancement occurs exponentially with $1/T_1$ whereas the carbon-proton coupling is eliminated as soon as the decoupler is turned on.

Neither of the above methods have been utilized to any appreciable extent for quantitative work with CMR, however. Usually, the assumption is made that similar carbons in isomers, for example, have similar relaxation times and hence, similar Overhauser enhancement. A direct comparison is then made between the peak heights of the peaks due to each isomer. This technique has been shown to provide reasonable results in determining the ratio of *cis-trans* isomers of long-chain fatty acids (134).

5.11 STRUCTURAL DETERMINATIONS USING CMR

Most of the early work with CMR was devoted to establishing the trends in ^{13}C parameters using known compounds. These background data were necessary to establish the structure-parameter correlations discussed in previous sections which are used today for structural studies. The use of CMR in structural studies has increased during the past few years and will continue to increase in the future. No single, spectroscopic technique provides the kinds of information that can be obtained from CMR spectra. For example, from the proton decoupled spectrum and the off-resonance decoupled spectrum, one can, in most cases, immediately determine not only the number of carbons in the compound, but also the types of carbons (i.e., methyl, methylene, methine, and quaternary carbons). From the chemical shifts, one can further classify the carbons as to the type of hybridization and functional groups involved. Within each type of hybridization, one can use the substituent parameter sets (Table 5.6 for saturated carbons) to determine possible molecular fragments and start putting together various pieces of the structure. Additional experiments may be required before a structure can be completely determined.

As an example of the use of CMR for structure determination consider the spectrum (Figure 5.6) of the 4,8,13-duvatriene-1,3-diol (**3**), a diterpene of the

242   CARBON-13 NMR

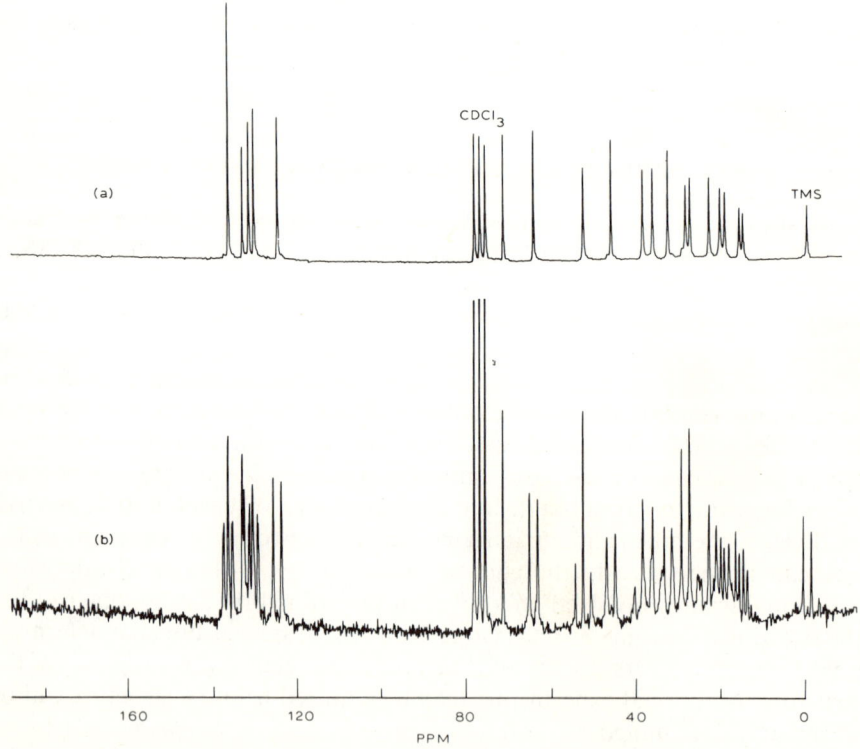

embrene family which is a plant growth inhibitor isolated from immature tobacco leaves (135). The proton–decoupled spectrum exhibits 19 peaks. The off-resonance decoupled-spectrum shows the low-field peak to consist of two overlapping resonances. Furthermore, the off-resonance decoupled spectrum clearly shows the presence of five methyl carbons, five methylene carbons, seven methine carbons, and three quaternary carbons. The position

Fig. 5.6 (a) The proton-decoupled, natural abundance CMR spectrum of 4,8,13-duvatriene-1,3-diol. (b) The corresponding single-frequency, off-resonance decoupled spectrum.

of the resonances establishes that there are fourteen saturated carbons and six sp² hybridized carbons. The absence of peaks further downfield eliminates the possibility of a carbonyl group. Since there are no aromatic protons in the proton spectrum of (**3**), the low-field ^{13}C peaks are established as being olefinic carbons. Although an x-ray structure was required to finally establish the structure of (**3**), the CMR spectrum provided invaluable information.

Other examples of the application of CMR for structural studies rely on the presence or absence of characteristic peaks in the spectrum and on the number of peaks observed in the spectrum. In cases where two or more possible isomers might be expected from a given reaction, it is not uncommon for one to be able to determine which isomer is formed from the CMR spectrum alone. For example, if two possible isomers differ in that one contains a carbonyl group whereas the other does not, the presence or absence of a peak in the low-field region (160–210 ppm) of the spectrum would allow one to distinguish the isomers. An example of this type has been given

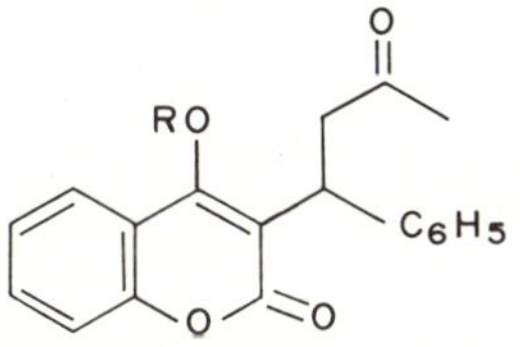

 4 R = H
 5 R = Na

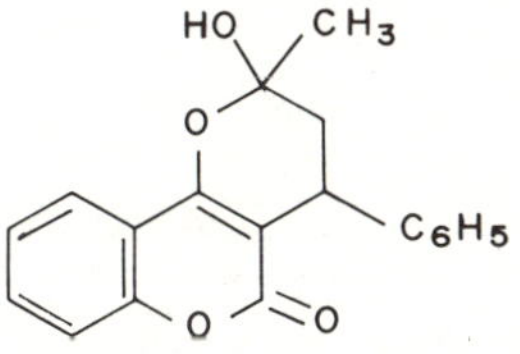

 6

recently. Warfarin (**4**) has been widely used as a rodenticide whereas its sodium salt (**5**) is widely used as an anticoagulant. Chemical and spectroscopic studies have suggested structures (**4**) and (**5**) for warfarin and its sodium salt. However, diffraction data indicate that the cyclic hemiketal structure (**6**) is present in the solid state. CMR spectra of the sodium salt (**5**) show a peak at 216.5 ppm for a ketone carbon which is only consistent with the open chain structure (**5**). However, this peak is absent in the spectrum of warfarin (**4**) and a new peak is present at ~100 ppm. Furthermore, there is a doubling of several of the peaks. This strongly supports the hemiketal structure, (**6**), for warfarin. The doubling of the peaks indicates that it exists as a mixture of diastereoisomers.

If two isomers differ in symmetry such that one would be expected to give a larger number of peaks than the other, the number of peaks in the CMR spectrum would allow one to determine which isomer is formed. An elegant example of an application of this type is given by the *trans-anti-trans-anti-trans* isomer of perhydrotriphenylene (**7**), a compound possessing D_3

symmetry. An intermediate in the attempted synthesis of (**7**) (136) was the hexasubstituted cyclohexane, (**8**), with the all *trans* configuration. The CMR spectrum of the methyl ester of (**8**) shows only four peaks and thus confirms that the cyclohexane carbons possess the proper configuration (136). Cyclization of (**8**) was carried out to give the triketone, (**9**). The CMR spectrum of (**9**) exhibits only three peaks thus confirming the stereochemistry. This is the only isomer out of ten possibilities that would give only three peaks in the CMR spectrum.

       7   **8**   **9**

Another area where CMR provides useful information is in the structure determination of polymers (137). With the aid of model compounds, it is often possible to determine diad, triad, tetrad and pentad units in the polymer and to obtain information on polymerization statistics. Similar information can also be obtained from the spectra of copolymers (138,139).

The examples previously given illustrate the potential of CMR for applications in structure determination. Additional examples may be found in the literature reviews of CMR spectroscopy (138,139). In addition, while we have focused our attention primarily to substituted hydrocarbons, there are several other classes of compounds which have been investigated by CMR. These include amino acids (140), carbohydrates (141,142), nuclei acid derivatives (143–148), alkaloids (146–149), and steroids (125). Applications of CMR in biochemical studies have been reviewed (150). There can be no doubt that this area of CMR will continue to increase at a rapid rate in the future.

5.12 CONFORMATIONAL ANALYSIS

The use of CMR in studies of rate processes has proven to be as valuable as its use in structure elucidations. Although more time is usually required for CMR studies compared to proton studies, due to the low sensitivity of CMR, the additional effort required to obtain the CMR data is usually more than compensated for by the ease with which the data can be treated. The advantages of CMR over proton NMR in conformational analysis studies have been discussed by Dalling et al. (17). First, since single peaks are observed for the carbons with proton decoupling, there is no need for deuterium substitution to reduce the complexity of the spectra. Second, there is usually more than one pair of carbons undergoing exchange such that data may be obtained from more than one site in the molecule. Third, the nonequilibrating carbons in the molecule may be used as a reference for the natural line width. Fourth, the chemical shift difference between equilibrating sites is usually larger than the corresponding difference in the proton resonances. Because of the larger chemical shift difference, the coalescence temperature is usually higher for CMR than for proton NMR and is therefore more accessible to study by NMR.

Anet et al. (18) have observed carbon resonances due to both axial and equitorial conformations of methylcyclohexane at low temperature. The shift differences are in reasonable agreement with those predicted from the substituent effects given previously (Section 5.4.2). The room temperature CMR spectrum of cis-1,2-dimethylcyclohexane consists of four peaks of equal intensity (19). On cooling to $-100°$, the spectrum consists of eight peaks of equal intensity indicating that the exchange between the two conformations of equal energy has been slowed with respect to the NMR time scale. Similar results were obtained for cis-1,4-dimethylcyclohexane. Several additional substituted cyclohexanes have been examined by CMR (151–153). In those cases where proton studies had been carried out, the free-energy differences calculated from both sets of data were in reasonable agreement.

Several additional cycloalkanes have been examined. Low temperature CMR spectra of cyclooctane (154,155) are consistent with the boat-chair conformation being the low energy conformation. Similar studies of 1,5-cyclooctadiene (156) and the C_9—C_{16} cycloalkanones (157) have been reported. Examination of several substituted 1,3-dioxanes have led to substituent parameters which may be used to predict chemical shifts (48). Discrepancies between calculated and observed shifts were taken as evidence for a boat conformation.

Another area of application of CMR to rate processes is the study of restricted rotation. Gansow et al. (158) have examined N,N-dimethyl trichloroacetamide and find that the activation energy for rotation (16.6 kcal/mole^{-1}) is in agreement with that obtained from proton studies.

The barrier to rotation around the phenylcarbonylbond in substituted benzaldehydes has recently been determined by CMR (159). It seems clear that the applications of CMR to studying rate processes will continue to increase in the future.

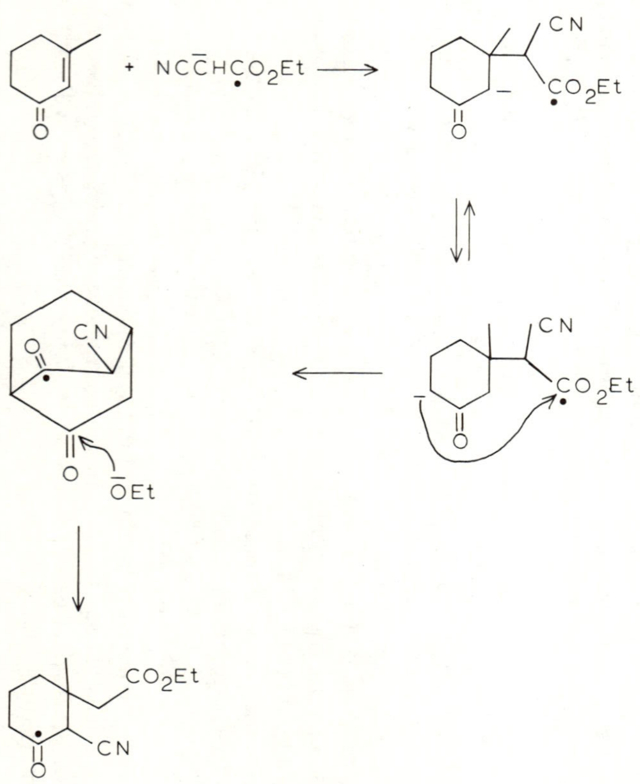

Scheme 1.

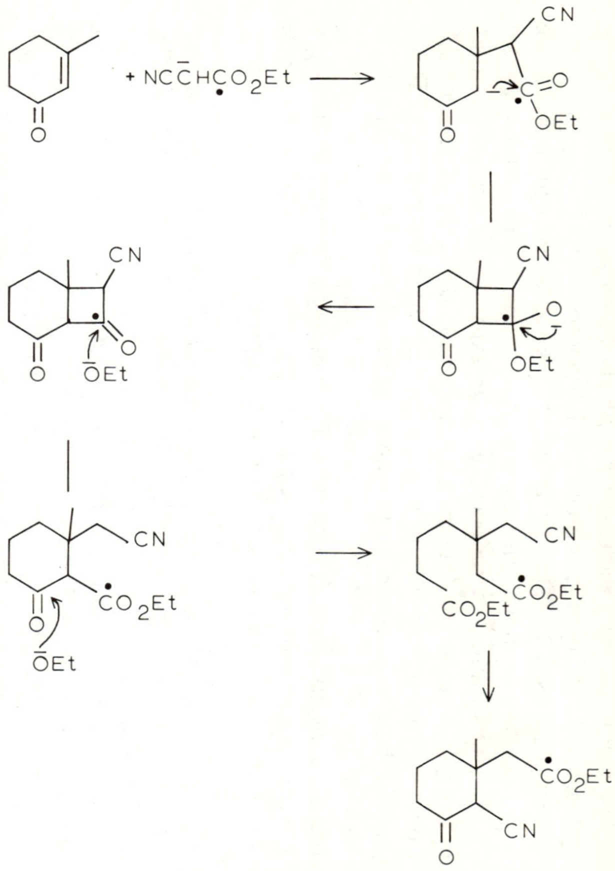

Scheme 2.

5.13 MECHANISTIC STUDIES

CMR is an extremely powerful tool for investigating reaction mechanisms. If one of the starting materials for a given reaction can be prepared such that a specific carbon is enriched with ^{13}C, CMR can be used to determine the position of the label in the product of the reaction. The peak due to this carbon will be more intense than the other peaks in the spectrum. Usually, an enrichment of $\sim 3-5\%$ is needed to unambiguously distinguish the peak from the labelled carbon from the other carbons. Of course, this method requires that the CMR spectrum of the natural abundance material be completely assigned. The advantage of this method over other tracer techniques is that degradation of the compound to determine the position of the label is eliminated.

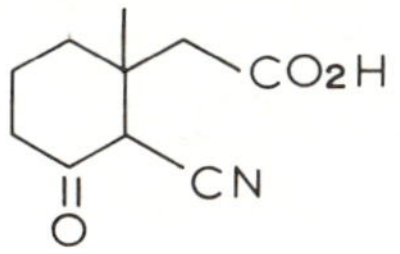

10

An excellent example of the use of CMR for mechanistic studies is provided by a study of the products formed in the Michael reaction between methyl cyclohexenone and ethyl cyanoacetate. An "abnormal" Michael product, the α-cyanoketone (**10**) was obtained in 21% yield. The two most popular mechanisms (160,161) proposed for the formation are shown in Schemes 1 and 2. In an attempt to distinguish between these two mechanisms, Hill and Ledford (162) carried out this reaction using methyl cyanoacetate labelled with ^{13}C at the carbonyl carbon. The dot in Schemes 1 and 2 is used to follow the position of the labeled carbon. The CMR spectrum of the labeled α-cyanoketone clearly shows that the label is present only at the ketone carbonyl. This result is consistent only with the mechanism given in Scheme 1. Mechanistic studies of this type will most likely increase in the future.

5.14 BIOSYNTHETIC STUDIES

The development of FT-NMR techniques has greatly increased the use of carbon-13 tracer techniques for the elucidation of biosynthetic pathways. The major advantages of CMR over radioactive tracer techniques is the elimination of degradative schemes to locate the tracer in most cases. Disadvantages of CMR are the relatively low sensitivity of CMR and the assignment of specific peaks in complex spectra of large molecules. Enrichments of carbon-13 of the order of from 2% to 5% at specific positions are needed to distinguish these carbons from unlabeled positions. In cases where the quantity of material is limited, radioactive tracer techniques may still be the method of choice.

Nevertheless, a variety of biosynthetic pathways have been examined using CMR. The labeled acetates have been the most common starting materials. However, a variety of carbon-13 labeled starting materials are now commercially available and others are likely to become available in the near future. Two examples discussed in the following serve to illustrate the CMR tracer technique.

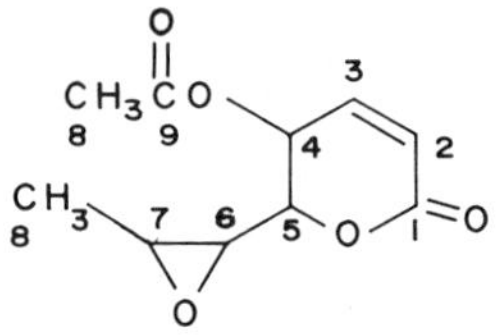

11

The biosynthesis of the antibiotic lactone asperlin (**11**), isolated from growing cultures of *Aspergillus nidulans*, has been examined using a medium containing sodium acetate labeled at carbon-2 (*CH_3CO_2Na, 61%) (163). The CMR spectrum of the labeled material clearly shows that the label has been incorporated at carbons 2, 4, 6, 8, and 10. Comparison of signal intensities indicated on enrichment of ~9% per labeled carbon.

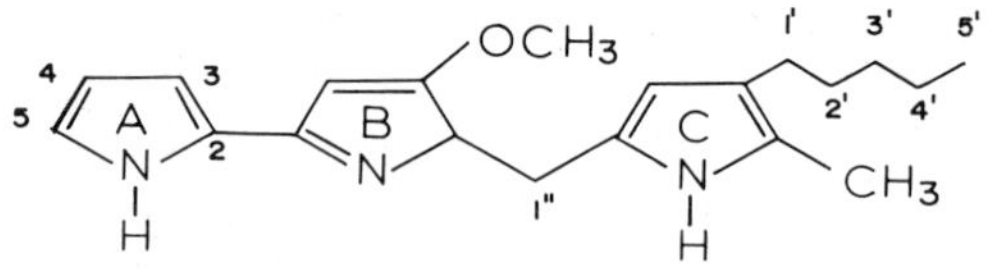

12

An elegant example of the use of CMR for the elucidation of a biosynthetic pathway has been reported by Wasserman et al. (164). By using several labeled precursors, the biosynthetic origin of all 20 carbons of prodigiosin (**12**), a metabolite of the bacterium *Serratia marcescens*, has been determined. Biosynthesis using C-1 labeled acetate shows the label at B-3, C-3, C-5, 2', and 4' while C-2 labeled acetate shows incorporation at B-4, C-4, 3', 1', and

5'. Experiments with [CD3]-D, L-methionine show that the methoxyl on ring B is derived from methionine. Incorporation of carboxyl-labeled proline indicates that B-5 and the A ring are derived from proline. The incorporation of C-3 labeled alanine yields a pattern of labeling identical to that of C-2 labeled acetate plus the methyl group on ring C. Both C-2 labeled glycine and C-3 labeled serine show incorporation of the label at B-2 and 1". These results show that prodigiosin is derived from the condensation of two major components, a methoxybipyrrolecarboxyaldehyde derived from proline, serine, acetate, and the methyl group of methionine, and a methylamylpyrrole derived from alanine and a polyacetate moiety.

Additional examples of the use of CMR in biosynthetic studies may be found in a recent review of applications of CMR in biochemistry (150). It seems clear that activity in this area will continue to increase in the future.

5.15 OTHER NUCLEI

While there is no doubt that ^{1}H and ^{13}C are the most important nuclei in terms of the relative number of compounds containing these nuclei, NMR has by no means been restricted to ^{1}H and ^{13}C. In fact, a large body of data has been accumulated in the past (and will continue in the future) on the NMR of nuclei other than ^{1}H and ^{13}C. The principles involved with these nuclei are identical to those with ^{1}H and ^{13}C. Since a detailed treatment of these nuclei is beyond the scope of this book, we will simply list below (Table 5.25) references to review articles concerning these nuclei. In addition, the literature of nuclei other than ^{1}H and ^{19}F is reviewed each year in the journal, *Magnetic Resonance Reviews* and in the *Chemical Society Specialist Periodical Reports on Nuclear Magnetic Resonance*.

TABLE 5.25. References to Review Articles on the NMR of Nuclei Other than ^{1}H and ^{13}C

Nucleus	Reference
^{2}D, ^{3}T	165
^{11}B	166
^{14}N	167–169
^{15}N	170
^{19}F	171–174
^{29}Si	175
^{31}P	176–179
^{199}Hg, ^{207}Pb, and other metals	180–181

REFERENCES

1. P. C. Lauterbur, *J. Chem. Phys.*, **26**, 217 (1957).
2. P. S. Pregosin and E. W. Randall, in *Determination of Organic Structures By Physical Methods*, F. C. Nachod and J. J. Zuckerman, Eds., Vol. 4, Academic, New York, 1971, p. 263.
3. E. F. Mooney and P. H. Wilson, in *Annual Review of NMR Spectroscopy*, E. F. Mooney, Ed., Vol. 2, Academic, New York, 1969, p. 153.
4. B. E. Mann, in *Advances in Organometallic Chemistry*, F. G. A. Stone and R. West, Eds., Vol. 12, Academic, New York, 1974, p. 135.
5. J. B. Stothers, *Carbon-13 NMR Spectroscopy*, Academic, New York, 1972.
6. G. C. Levy and G. L. Nelson, *Carbon-13 Nuclear Magnetic Resonance for Organic Chemists*, Wiley, New York, 1972.
7. L. F. Johnson and W. C. Jankowski, *Catalog of Carbon-13 NMR Spectra*, Wiley, New York, 1972.
8. R. R. Ernst, *J. Chem. Phys.*, **45**, 3845 (1966).
9. K. F. Kuhlmann and D. M. Grant, *J. Am. Chem. Soc.*, **90**, 7355 (1968).
10. A. Allerhand, R. F. Childers, and E. Oldfield, *J. Mag. Resonance*, **11**, 272 (1973).
11. M. Karplus and J. A. Pople, *J. Chem. Phys.*, **38**, 2803 (1963).
12. J. Mason, *J. Chem. Soc.*, **A**, 1038 (1971).
13. D. M. Grant and E. G. Paul, *J. Am. Chem. Soc.*, **86**, 2984 (1964).
14. L. P. Lindeman and J. Q. Adams, *Anal. Chem.*, **43**, 1245 (1971).
15. H. Spiesecke and W. G. Schneider, *J. Chem. Phys.*, **35**, 722 (1961).
16. J. J. Burke and P. C. Lauterbur, *J. Am. Chem. Soc.*, **86**, 1870 (1964).
17. D. K. Dalling and D. M. Grant, *J. Am. Chem. Soc.*, **89**, 6612 (1967).
18. F. A. L. Anet, C. H. Bradley, and G. W. Buchanan, *J. Am. Chem. Soc.*, **93**, 258 (1971).
19. H. J. Schneider, R. Price, and T. Keller, *Angew. Chem., Internat. Edit.*, **10**, 730 (1971).
20. M. Christl, H. J. Reich, and J. D. Roberts, *J. Am. Chem. Soc.*, **93**, 3463 (1971).
21. A. P. Tulloch and M. Mazurek, *J. Chem. Soc., Chem. Comm.*, 692 (1973).
22. J. P. C. M. van Dongen, H. W. D. van Dijkman, and M. J. A. de Bie, *Rec. Trav. Chim. Pays-Bas*, **93**, 29 (1974).
23. E. L. Eliel, private communication.
24. D. Leibfritz, B. O. Wagner, and J. D. Roberts, *Justus Liebigs Ann. Chem.*, **763**, 173 (1972).
25. W. S. Brey, 26th Southeastern Regional A.C.S. Meeting, S16, Norfolk, Va., 1974.
26. O. W. Howarth and R. J. Lynch, *Mol. Phys.*, **15**, 431 (1968).
27. G. Miyazima and K. Takahashi, *J. Phys. Chem.*, **331**, 3766 (1971).
28. J. D. Roberts, F. J. Weigert, J. I. Kroschwitz, and H. J. Reich, *J. Am. Chem. Soc.*, **92**, 1338 (1970).

REFERENCES

29. H. Eggert and C. Djerassi, *J. Am. Chem. Soc.*, **95**, 3710 (1973).
30. W. Horsley, H. Sternlicht, and J. S. Cohen, *J. Am. Chem. Soc.*, **92**, 680 (1970).
31. E. Lippmaa and T. Pehk, *Eesti NSV Tead. Akad. Toim. Keem. Geol.*, **17**, 210 (1968).
32. L. M. Jackman and D. P. Kelly, *J. Chem. Soc.*, **B**, 101 (1970).
33. R. Hagen and J. D. Roberts, *J. Am. Chem. Soc.*, **91**, 4504 (1969).
34. R. H. Cox and C. W. Pape, unpublished results.
35. D. E. Dorman, M. Jautelat, and J. D. Roberts, *J. Org. Chem.*, **36**, 2757 (1971).
36. D. E. Dorman, M. Jautelat, and J. D. Roberts, *J. Org. Chem.*, **38**, 1026 (1973).
37. P. C. Lauterbur, *Ann. N. Y. Acad. Sci.*, **70**, 841 (1958).
38. P. H. Weiner and E. R. Malinowski, *J. Phys. Chem.*, **71**, 2791 (1967).
39. K. M. Crecely, R. W. Crecely, and J. H. Goldstein, *J. Phys. Chem.*, **74**, 2680 (1970).
40. A. S. Perlin and H. J. Koch, *Can. J. Chem.*, **48**, 2639 (1970).
41. F. J. Weigert and J. D. Roberts, *J. Am. Chem. Soc.*, **92**, 1347 (1970).
42. J. B. Grutzner, M. Jautelat, J. B. Bence, R. A. Smith, and J. D. Roberts, *J. Am. Chem. Soc.*, **92**, 7107 (1970).
43. E. Lippmaa, T. Pehk, J. Paasivirta, N. Belikova, and A. Platé, *Org. Magn. Resonance*, **2**, 581 (1970).
44. G. E. Maciel and G. B. Savitsky, *J. Phys. Chem.*, **69**, 3925 (1965).
45. T. Pehk and E. Lippmaa, *Eesti NSV Tead. Akad. Toim. Keem. Geol.*, **17**, 291 (1968).
46. F. G. Riddell, *J. Chem. Soc.*, **B**, 331 (1970).
47. E. L. Eliel and M. C. Knoeber, *J. Am. Chem. Soc.*, **90**, 3444 (1968).
48. G. M. Kellie and F. G. Riddell, *J. Chem. Soc.*, **B**, 1030 (1971).
49. A. K. Jones, E. L. Eliel, D. M. Grant, M. C. Knoeber, and W. F. Bailey, *J. Am. Chem. Soc.*, **93**, 4772 (1971).
50. S. I. Featherman, S. O. Lee, and L. D. Quin, *J. Org. Chem.*, **39**, 2899 (1974).
51. W. R. Purdum and K. D. Berlin, *J. Org. Chem.*, **39**, 2904 (1974).
52. R. B. Wetzel and G. L. Kenyon, *Chem. Commun.*, 287 (1973).
53. L. D. Quin, S. G. Borleske, and R. C. Stocks, *Org. Magn. Resonance*, **5**, 161 (1973).
54. T. Pehk, S. Rang, O. Eisen, and E. Lippmaa, *Eesti NSV Tead. Akad. Toim. Keem. Geol.*, **17**, 296 (1968).
55. G. E. Maciel, *J. Phys. Chem.*, **69**, 1947 (1965).
56. K. M. Crecely, R. W. Crecely, and J. H. Goldstein, *J. Mol. Spectrosc.*, **37**, 252 (1971).
57. T. Higashimura, S. Okamura, I. Morishima, and T. Yonexawa, *J. Polym. Sci., Part B*, **7**, 23 (1969).
58. D. H. Marr and J. B. Stothers, *Can. J. Chem.*, **43**, 596 (1965).
59. C. Rappe, E. Lippmaa, T. Pehk, and K. Andersson, *Acta. Chem. Scand.*, **23**, 1447 (1969).
60. H. Spiesecke and W. G. Schneider, *J. Chem. Phys.*, **35**, 731 (1961).

61. G. E. Maciel and J. J. Natterstad, *J. Chem. Phys.*, **42**, 2427 (1965).
62. H. L. Retcofsky and C. E. Griffin, *Tetrahedron Letts.*, 1975 (1966).
63. K. S. Dhami and J. B. Stothers, *Can. J. Chem.*, **45**, 233 (1967).
64. A. R. Tarpley, Jr., and J. H. Goldstein, *J. Mol. Spectrosc.*, **39**, 375 (1971); **37**, 432 (1971).
65. H. Spiesecke and W. G. Schneider, *Tetrahedron Letts.*, 468 (1961).
66. A. J. Jones, T. D. Alger, D. M. Grant, and W. M. Litchman, *J. Am. Chem. Soc.*, **92**, 2386 (1970).
67. N. Defay, D. Zimmermann, and R. H. Martin, *Tetrahedron Letts.*, 1871 (1971).
68. A. J. Jones and D. M. Grant, *Chem. Commun.*, 1670 (1968).
69. T. D. Alger, D. M. Grant, and E. G. Paul, *J. Am. Chem. Soc.*, **88**, 5397 (1966).
70. H. L. Retcofsky, J. M. Hoffman, and R. A. Friedel, *J. Chem. Phys.*, **46**, 4545 (1967).
71. R. J. Pugmire, D. M. Grant, M. J. Robins, and R. K. Robins, *J. Am. Chem. Soc.*, **91**, 6381 (1969).
72. P. C. Lauterbur, *J. Am. Chem. Soc.*, **83**, 1838 (1961).
73. A. J. Jones, P. D. Gardner, D. M. Grant, W. M. Litchman, and V. Boekelheide, *J. Am. Chem. Soc.*, **92**, 2395 (1970).
74. P. C. Lauterbur, *J. Chem. Phys.*, **43**, 360 (1965).
75. R. J. Pugmire and D. M. Grant, *J. Am. Chem. Soc.*, **90**, 697 (1968).
76. H. L. Retcofsky and R. A. Friedel, *J. Phys. Chem.*, **71**, 3592 (1967); **72**, 290 (1968); **72**, 2619 (1968).
77. M. Hansen and H. J. Jakobsen, *J. Magn. Resonance*, **10**, 74 (1973).
78. T. F. Page, Jr., T. Alger, and D. M. Grant, *J. Am. Chem. Soc.*, **87**, 5333 (1965).
79. F. J. Weigert and J. D. Roberts, *J. Am. Chem. Soc.*, **90**, 3543 (1968).
80. R. J. Pugmire and D. M. Grant, *J. Am. Chem. Soc.*, **90**, 4232 (1968).
81. K. Takahashi, T. Stone, and K. Fujieda, *J. Phys. Chem.*, **74**, 2765 (1970).
82. R. J. Abraham, R. D. Lapper, K. M. Smith, and J. F. Unsworth, *J. Chem. Soc., Perkin II*, 1004 (1974).
83. R. J. Pugmire, M. J. Robins, D. M. Grant, and R. K. Robins, *J. Am. Chem. Soc.*, **93**, 1887 (1971).
84. R. J. Pugmire and D. M. Grant, *J. Am. Chem. Soc.*, **93**, 1880 (1971).
85. R. J. Pugmire, D. M. Grant, L. B. Townsend, and R. K. Robins, *J. Am. Chem. Soc.*, **95**, 2791 (1973).
86. A. C. Boicelli, R. Danieli, A. Mangini, L. Lunazzi, and G. Placucci, *J. Chem. Soc. Perkin II*, 1024 (1974).
87. R. G. Parker and J. D. Roberts, *J. Org. Chem.*, **35**, 996 (1970).
88. O. A. Gansow, P. A. Loeffler, R. E. Davis, M. R. Willcott, and R. E. Lenkinski, *J. Am. Chem. Soc.*, **95**, 3389 (1973).
89. K. Tori, Y. Yoshimura, M. Kainosho, and K. Tjisaka, *Tetrahedron Letts.*, 1573 (1973).

90. A. A. Chalmers and K. G. R. Pachler, *J. Chem. Soc. Perkin II*, 748 (1974).
91. C. Charrier, D. E. Dorman and J. D. Roberts, *J. Org. Chem.*, **38**, 2644 (1974).
92. D. Rosenberg and W. Drenth, *Tetrahedron*, **27**, 3893 (1971).
93. D. Rosenberg, J. W. de Haan, and W. Drenth, *Rec. Trav. Chim. Pays-Bas*, **87**, 1387 (1968).
94. K. Frei and H. J. Bernstein, *J. Chem. Phys.*, **38**, 1216 (1963).
95. D. D. Traficante and G. E. Maciel, *J. Phys. Chem.*, **69**, 1348 (1965).
96. J. B. Stothers and P. C. Lauterbur, *Can. J. Chem.*, **42**, 1563 (1964).
97. D. E. Dorman and F. A. Bovey, *J. Org. Chem.*, **38**, 1719 (1973).
98. J. A. Pople and D. P. Santry, *Mol. Phys.*, **8**, 1 (1964).
99. G. E. Maciel, J. W. McIver, Jr., N. S. Ostlund, and J. A. Pople, *J. Am. Chem. Soc.*, **92**, 11 (1970).
100. A. C. Blizzard and D. P. Santry, *Chem. Commun.*, 1085 (1970).
101. N. Muller and D. E. Pritchard, *J. Chem. Phys.*, **31**, 768, 1471 (1959).
102. A. W. Douglas, *J. Chem. Phys.*, **45**, 3465 (1966).
103. K. M. Crecely, V. S. Watts, and J. H. Goldstein, *J. Mol. Spectrosc.*, **30**, 184 (1969).
104. F. S. Mortimer, *J. Mol. Spectrosc.*, **5**, 199 (1960).
105. E. Lippert and H. Prigge, *Ber. Bunsenges, Phys. Chem.*, **67**, 415 (1963).
106. C. H. Yoder, R. H. Tuck, and R. E. Hess, *J. Am. Chem. Soc.*, **91**, 539 (1969).
107. R. E. Mayo and J. H. Goldstein, *J. Mol. Spectrosc.*, **14**, 173 (1964).
108. G. A. Olah and A. M. White, *J. Am. Chem. Soc.*, **91**, 5801 (1969).
109. M. P. Simonnin, *Bull. Soc. Chim. France*, 1774 (1966).
110. G. J. Karabatsos, J. D. Graham, and F. M. Vane, *J. Am. Chem. Soc.*, **84**, 37 (1962).
111. G. Miyazima, Y. Utsumi, and K. Takahashi, *J. Phys. Chem.*, **73**, 1370 (1969).
112. K. M. Crecely, R. W. Crecely, and J. H. Goldstein, *J. Phys. Chem.*, **74**, 2680 (1970).
113. G. J. Karabatsos and C. E. Orzech, Jr., *J. Am. Chem. Soc.*, **87**, 560 (1965).
114. J. Feeney, P. E. Hansen, and G. C. K. Roberts, *Chem. Commun.*, 465 (1974).
115. F. J. Weigert and J. D. Roberts, *J. Am. Chem. Soc.*, **89**, 2967 (1967).
116. G. E. Maciel, in *Nuclear Magnetic Resonance Spectroscopy of Nuclei Other Than Protons*, T. Axenrod and G. A. Webb, Eds., Wiley, New York, 1974, p. 187.
117. J. L. Marshall, D. E. Miiller, S. A. Conn, R. Seiwell, and A. M. Ihrig, *Accts. of Chem. Res.*, **7**, 333 (1974).
118. F. J. Weigert and J. D. Roberts, *J. Am. Chem. Soc.*, **91**, 4940 (1969).
119. G. A. Gray and S. E. Cremer, *Chem. Commun.*, 451 (1974).
120. D. Doddrell, I. Burfitt, W. Kitching, M. Bullpitt, C.-H. Lee, R. J. Mynott, J. L. Considine, H. G. Kuivila and R. H. Sarma, *J. Am. Chem. Soc.*, **96**, 1640 (1974).
121. F. J. Weigert and J. D. Roberts, *Inorg. Chem.*, **12**, 313 (1973).
122. T. N. Mitchell, *J. Organometal. Chem.*, **59**, 189 (1973).
123. C. D. Schaeffer, Jr., and J. J. Zuckerman, *J. Organometal. Chem.*, **55**, 97 (1973).

124. F. W. Wehrli, in *Nuclear Magnetic Resonance Spectroscopy of Nuclei Other Than Protons*, T. Axenrod and G. A. Webb, Eds., Wiley, New York, 1974, p. 157.
125. H. J. Reich, M. Jautelat, M. T. Messe, F. J. Weigert, and J. D. Roberts, *J. Am. Chem. Soc.*, **91**, 7445 (1969).
126. J. B. Grutzner, *Chem. Commun.*, 64 (1974).
127. O. A. Gansow and W. Schittenhelm, *J. Am. Chem. Soc.*, **93**, 4294 (1971).
128. R. Freeman and H. D. W. Hill, *J. Magn. Resonance.*, **5**, 278 (1971).
129. A. Allerhand, D. Doddrell, and R. Komoroski, *J. Chem. Phys.*, **55**, 189 (1971).
130. J. W. Faller, M. A. Adams, and G. N. La Mar, *Tetrahedron Letts.*, 699 (1974).
131. G. N. La Mar, *J. Am. Chem. Soc.*, **93**, 1040 (1971).
132. D. F. S. Natusch, *J. Am. Chem. Soc.*, **93**, 2566 (1971).
133. R. Freeman, H. D. W. Hill, and R. Kaptein, *J. Magn. Resonance*, **7**, 327 (1972).
134. F. E. Barton II, D. S. Himmelsbach, and D. Burdick, *J. Magn. Resonance*, **18**, 167 (1975).
135. J. P. Springer, J. Clardy, R. H. Cox, H. G. Cutler, and R. J. Cole, unpublished results.
136. D. D. Giannini, K. K. Chan, and J. D. Roberts, *Proc. Natl. Acad. Sci.* USA, **71**, 4221 (1974).
136a. D. W. Ladner, Ph.D., Dissertation, University of Georgia, 1974.
137. A. R. Katrizky and D. E. Weiss, *Chem. Commun.*, 401 (1974).
138. I. D. Robb and G. J. T. Tiddy, in *Nuclear Magnetic Resonance*, R. K. Harris, Ed., Vol. 3, Specialist Periodical Reports, The Chemical Society, London, 1974, p. 279.
139. R. H. Cox, *Mag. Res. Rev.*, **1**, 271 (1972); **3**, 207 (1974).
140. W. Voelter, G. Jung, E. Breitmaier, and E. Bayer, *Z. Naturforsch, B*, **26**, 213 (1971).
141. D. Doddrell and A. Allerhand, *J. Am. Chem. Soc.*, **93**, 2779 (1971).
142. D. E. Dorman and J. D. Roberts, *J. Am. Chem. Soc.*, **93**, 4463 (1971).
143. A. J. Jones, D. M. Grant, M. W. Winkley, and R. K. Robins, *J. Am. Chem. Soc.*, **92**, 4079 (1970).
144. A. J. Jones, D. M. Grant, M. W. Winkley, and R. K. Robins, *J. Phys. Chem.*, **74**, 2684 (1970).
145. D. E. Dorman and J. D. Roberts, *Proc. Natl. Acad. Sci., USA*, **65**, 19 (1970).
146. A. J. Jones and M. H. Benn, *Can. J. Chem.*, **51**, 486 (1973).
147. W. O. Crain, Jr., W. C. Wildman, and J. D. Roberts, *J. Am. Chem. Soc.*, **93**, 990 (1971).
148. P. W. Sprague, D. Doddrell, and J. D. Roberts, *Tetrahedron*, **27**, 4857 (1971).
149. E. Wenkert and B. L. Buckwalter, *J. Am. Chem. Soc.*, **94**, 4367 (1972).
150. G. A. Gray, *Crit. Rev. Biochem.*, **1**, 247 (1973).
151. G. W. Buchanan, D. A. Ross, and J. B. Stothers, *J. Am. Chem. Soc.*, **88**, 4301 (1966).
152. G. W. Buchanan and J. B. Stothers, *Can. J. Chem.*, **47**, 3605 (1969).

153. J. D. Roberts, F. J. Weigert, J. I. Kroschwitz, and H. J. Reich, *J. Am. Chem. Soc.*, **92**, 1338 (1970).
154. F. A. L. Anet and U. J. Basus, *J. Am. Chem. Soc.*, **95**, 4424 (1973).
155. H. J. Schneider, T. Keller, and R. Price, *Org. Magn. Resonance*, **4**, 907 (1972).
156. F. A. L. Anet and L. Kozerski, *J. Am. Chem. Soc.*, **95**, 3407 (1973).
157. F. A. L. Anet, A. K. Cheng, and J. Krane, *J. Am. Chem. Soc.*, **95**, 7877 (1973).
158. O. A. Gansow, J. Killough, and A. R. Burke, *J. Am. Chem. Soc.*, **93**, 4297 (1971).
159. T. Drakenberg, R. Jost, and J. Sommer, *Chem. Commun.*, 1011 (1974).
160. E. H. Farmer and J. Ross, *J. Chem. Soc.*, 3233 (1926).
161. P. R. Shafer, W. E. Loeb, and W. S. Johnson, *J. Am. Chem. Soc.*, **75**, 5963 (1953).
162. R. K. Hill and N. D. Ledford, *J. Am. Chem. Soc.*, **97**, 666 (1975).
163. M. Tauabe, T. Hamasaki, D. Thomas, and L. Johnson, *J. Am. Chem. Soc.*, **93**, 273 (1971).
164. H. H. Wasserman, R. J. Sykes, P. Peverada, C. K. Shaw, R. J. Cushley, and S. R. Lipsky, *J. Am. Chem. Soc.*, **95**, 6874 (1974).
165. P. Diehl, in *Nuclear Magnetic Resonance Spectroscopy of Nuclei Other Than Protons*, T. Axenrod and G. A. Webb, Eds., Wiley, New York, 1974, p. 275.
166. W. G. Henderson and E. F. Mooney, in *Annual Review of NMR Spectroscopy*, E. F. Mooney, Ed., Academic, New York, 1969, p. 219.
167. E. F. Mooney and P. H. Winson, in reference 166, p. 125.
168. E. W. Randall and D. G. Gillies, in *Progress in Nuclear Magnetic Resonance Spectroscopy*, J. W. Emsley, J. Feeney, and L. H. Sutcliffe, Eds., Vol. 6, Pergamon, New York, 1970, p. 119.
169. M. Witanowski and G. A. Webb, in reference 166, Vol. 5A, 1972, p. 395.
170. R. L. Lichter, in *Determination of Organic Structures by Physical Methods*, F. C. Nachod and J. J. Zuckerman, Eds., Vol. 4, Academic, New York, 1971, p. 195.
171. E. F. Mooney and P. H. Winson, in reference 166, Vol. 1, 1968, p. 244.
172. K. Jones and E. F. Mooney, in reference 166, Vol. 3, 1970, p. 261.
173. K. Jones and E. F. Mooney, in reference 166, Vol. 4, 1971, p. 391.
174. R. Fields, in reference 166, Vol. 5A, 1972, p. 99.
175. G. C. Levey and J. D. Cargioli, in reference 165, p. 251.
176. G. Mavel, in reference 168, Vol. 1, 1966, p. 251.
177. G. Mavel, in reference 166, Vol. 5B, 1973, p. 1.
178. M. M. Crutchfield, C. H. Dungan, J. H. Letcher, V. Mark, and J. R. van Wazer, in *Topics in Phosphorus Chemistry*, M. Grayson and E. J. Griffith, Eds., Vol. 5, Interscience, New York, 1967.
179. J. R. van Wazer, in reference 170, p. 323.
180. P. R. Wells, in reference 170, p. 233.
181. G. E. Maciel, in reference 165, p. 347.

CHAPTER

6

METHODS OF QUANTITATIVE MEASUREMENTS

6.1	Chemical Shift	257
6.2	Peak Height	258
6.3	Peak Width	259
6.4	Peak Area	260
	6.4.1 Instrumental Parameters	260
	6.4.2 Sample Preparation and Procedure for Peak Area Measurements	263
	6.4.3 Accuracy and Precision	270
6.5	Special Methods and Techniques	271
	6.5.1 Signal Averaging	271
	6.5.2 Pulsed NMR	275
	6.5.3 Process Control	276
6.6	Equilibrium Studies	277
6.7	Chemical Kinetics	279
	6.7.1 Slow Reactions	279
	6.7.2 Fast Reactions	281
	6.7.2.1 Two-Site Exchange	283
	6.7.2.2 Complete Line-Shape Analysis for Multisite Exchange	289
	6.7.3 Transient Methods for Rate Studies	291
	6.7.4 Experimental Procedures for Rate Studies	293
6.8	Conclusion	294

Nuclear magnetic resonance is a powerful tool for quantitative analysis. Because proton and ^{19}F magnetic resonance has dominated the applications of NMR, and ^{13}C NMR is rapidly growing with the advent of Fourier transform (FT) spectrometers, most examples of quantitative NMR deal with organic compounds in solution. These are usually determined by measurement of the peak area of the NMR resonance. However, there are many potential quantitative measurements that may be performed using NMR methods on nuclei other than hydrogen, carbon, and fluorine, and using techniques other than peak integration. The use of pulse techniques for the determination of liquids in solids is a further example of additional expansion of the potential quantitation using NMR.

Although the most common and usually the most practical method of quantitation by high-resolution NMR is peak area measurement, there are several potentially useful methods of quantitation. Chemical shift, peak width, and peak heights may also be of use under special circumstances. Each of these techniques will be considered in turn. A few examples of applications will be given in this section to illustrate the principles of the techniques. A more extensive discussion of applications will be given in Chapter 7.

6.1 CHEMICAL SHIFT

Use of chemical shift has limited applications for quantitative measurements in NMR. Although chemical shift measurements can be made with high precision, the origin of the chemical shift is often a result of many factors (Section 2.7). Chemical shift is useful as a quantitation method when the analyst is confident that the observed changes in the chemical shift arise only from the parameter to be measured. The most frequent application of chemical shift measurements are found for binary mixtures that are participating in fast chemical exchange. Under these conditions a single peak is observed which has a mean chemical shift determined by the mole fraction weighted chemical shifts of the pure materials. For example, an acid HA in water will undergo rapid proton exchange according to equation 6.1, where H is used to symbolically label a particular proton.

$$HA + \bar{H}OH \longrightarrow \bar{H}A + HOH \qquad (6.1)$$

If the chemical shifts of the HA and HOH protons in pure acid and water are represented as δ_{HA} and δ_{HOH}, respectively, the average chemical shift observed for these exchanging protons is given as

$$\delta_{obs} = X_{HA}\delta_{HA} + X_{HOH}\delta_{HOH} \qquad (6.2)$$

where X_{HA} and X_{HOH} are the mole fraction of protons that originate from the acid and water, respectively (1). In most cases it is more practical to prepare a calibration curve than to attempt direct calculations. In the best cases, such as the analysis of acetic acid–water mixtures (2), a linear curve is obtained for a plot of mole fraction acid versus δ_{obs}. A relative precision of a few percent can usually be obtained. In principle, the technique can be applied to any binary mixture that undergoes rapid chemical exchange. However, there are many factors that limit the practical utility of this method. A linear relationship between the observed chemical shift and concentration will be obtained only if there are no significant changes in the degree of dissociation of either component with concentration. Dissociation equilibria of strong acid–water mixtures such as concentrated sulfuric acid solutions (1,3) may also be investigated.

Quantitative determinations may also be made on aprotic media using chemical shift. For example, Mengenhausen has determined the aromatic components in petroleum from the chemical shifts of methylene chloride or nitromethane that serve as acceptors for molecular complexes (4). The aromatic content of petroleum products may be determined with a precision of 0.2–0.5% if the chemical shift is measured with a precision of 0.4 Hz. Considerable precaution must be taken to avoid secondary effects that may contribute to chemical shift change such as temperature. Acid dissociation equilibria can be significantly affected by temperature changes which may cause at least a loss of precision unless the temperature of the sample is controlled.

An interesting variation on the use of chemical shift in order to determine the number of repetitive functional groups on a molecule is the use of the number of peaks at different chemical shifts. A nice example has been given by Sudmeier in which the number of amino acid units in a peptide was determined by dissolving the peptide in a "super acid" (5). The carbonyl groups are protonated and the exchange rate is sufficiently slow that separate peaks are observed for these protons. By simply counting the peaks, the number of amino acid units is determined.

Quantitative determinations using chemical shift measurements may also be made in systems not undergoing chemical exchange. A few cases have been reported in which the chemical shift of a nonlabile proton has been found proportional to the number of substitutents in the molecule (6,7). Although this technique would be of great value in process and synthesis monitoring, it will likely find limited application because of the many factors that can influence chemical shift.

6.2 PEAK HEIGHT

Although it is the area of an NMR peak which is proportional to the concentration of nuclei giving rise to the peak, the peak height may occasionally be useful for quantitation. Peak height measurements have been used in other instrumental methods such as gas–liquid chromatography before the availability of inexpensive electronic integrators. The assumption made when utilizing peak heights is that the peak is approximated as a triangle and the area is then proportional to the height. This assumption is reasonably valid if all the measured peaks have the same width and shape. In NMR, the observed peak widths may be the result of the value of T_2 or the inhomogeneity of the field. When the peak width is determined by the field inhomogeneity, peak height measurements can be used with success. Because of the ease and accuracy of peak area measurements by integration on modern instruments, peak height is now rarely used. Only in cases where peak overlap limits the accuracy of direct integration will an alternative be desired. In such cases, it is probably best to resort to a separate device which can synthesize peak shapes and resolve the relative areas (8). These devices are usually analog computers with hardware-programmable functions such as gaussian or lorentzian functions, or a linear combination of the two. In this way peak shapes identical to those obtained experimentally may be simulated; NMR peaks are usually a combination of the lorentzian function predicted by resonance theory and the gaussian function generated by random field inhomogeneities. Of the three curve parameters needed for each peak, the position and width may be estimated with some accuracy from the original spectrum to be integrated. The third parameter, the area, may be adjusted for all simulated peaks until an "acceptable" fit for the superposition of the simulated and experimental spectrum is obtained. Digital computers may also be used for this purpose. As the digital computer is programmed by software, any combination of functions to simulate the curve may be obtained easily using these devices. However, because the parameter adjustments must be made digitally by entering numerical weighting factors, and human visual response is analog, the analog curve resolvers are usually faster and easier to use. A more complicated digital computer program can iterate the parameters to a "best fit" as determined by a set of criteria entered in the program. Either method is suitable for curve resolving and the choice will usually be determined by the equipment available.

6.3 PEAK WIDTH

Peak width is a potential quantitative tool in high resolution NMR which has received relatively little attention which is somewhat justified. With modern instrumentation, the linewidth of proton magnetic resonance peaks is very nearly determined by the T_2 of the sample. Depending on the tuning and condition of the instrument, field inhomogeneity may play an important part in determining the linewidth. However, certain chemical and physical processes can dramatically affect the observed line width. For example, NMR spectroscopists have long been aware that the presence of paramagnetic substances in the sample will cause line broadening. Protons bonded to atoms which have nuclear quadrupole moments, such as nitrogen, will frequently show broadened lines.

Day and Reilley showed that the width of the proton magnetic resonance line of water increased in a linear relationship to the concentration of paramagnetic metal ions over a limited concentration range (9). The broadening of the solvent resonance lines is a result of rapid exchange between the water molecules in the coordination sphere of the cation and the bulk solvent. The magnitude of the broadening for a given concentration of a particular cation is determined by one or more of several possible conditions (10). The obvious limitation of this technique is that it is nonselective; that is a mixture of cations will lead to a net line broadening. However, addition of a complexing agent such as ethylenediaminetetraacetic acid (EDTA) will cause an extensive decrease in the line width if paramagnetic metal ions are present. This procedure can be used to confirm the presence of as little as 10^{-5} M paramagnetic metal ion in the sample and actual titration of the metal ion(s) with EDTA may be performed directly in the NMR sample tube. As little as 0.5 ml of solution has been titrated in this way (11). In addition to metal cations, the presence of organic free radicals may also be detected by line broadening if rapid electron exchange is occurring between the radical and the parent compound. Methods related to peak width measurements using pulsed techniques to determine T_1 and T_2 values for quantitative purposes will be discussed in Section 6.5.2.

6.4 PEAK AREA

6.4.1 INSTRUMENTAL PARAMETERS

Integration of the NMR resonance line to determine the area beneath the peak is by far the most common and the most practical method of quantitation by NMR. Using modern electronic integrator circuits, integration can be performed with a precision of a few tenths of one percent. However, in most cases the reproducibility of the recorder, less-than-perfect tuning of the instrument, partial saturation of the signal and transient noise results in an overall precision of 1–2%. The precision may be improved by the use of a digital voltmeter with decade scaling on the output of the integrator. It is important for the reader to remember here the fundamental difference between precision and accuracy, which are occasionally used incorrectly as interchangable terms. Precision is only a measure of the reproducibility of the observation; in this case the signal integration. There is no relationship between precision and accuracy as the latter is a measure of the difference between an observed result and a "true" value. Improper standardization or calibration, an error in measurement of the integral value or some other determinate error will affect accuracy, whereas the precision of the same measurement may be excellent.

There are several precautions that must be observed in order to obtain both precise and accurate quantitation by peak integration of the NMR signal. First, the instrument must be in good condition. Poor connections of cables, noisy electronic components, and microphonic vibrations will contribute transient signals. The very nature of the integration process will tend to average to zero any noise which is random and has a period much less than the time required for the integration. However, d.c. transient pulses or low frequency drift will not be averaged and will cause an error in the integration directly proportional to their magnitude, duration, and frequency of occurrence. Proper precaution must be taken to balance the d.c. offset of the integrator circuit using a control that is normally accessible on the instrument console. To do this, the spectrum amplitude may be set at its minimum to reduce input to the integrator. The recorder should be set off any resonance to further reduce input. Some spectrometers have an integrator balance "switch" setting to ground the integrator input during the balancing adjustment. The recorder zero may be used to set the pen near the center of the recorder. With the integrator on, adjust the balance control until the integrator drift seen as pen drift is eliminated or diminished as much as possible. The integrator output gain should be increased as the adjustment proceeds to increase the sensitivity of the balancing operation. The integrator reset may be operated frequently to keep the pen near the center of the page. This adjustment should be checked before any integration is performed as a d.c. balance is difficult to stabilize for long time periods. If a persistent leakage of the integrator appears as indicated by a downward tailing of the integrator output, the amplifier or integrating capacitor should be suspected as failing.

A second important parameter for quality integration is the adjustment of the phase detector (Figure 6.1). Provided the amplifier balance is properly adjusted, the integral step should show a flat initial baseline as well as a flat plateau after the peak has been passed. It is easy to recognize the effect of incorrect phase detector adjustment as the baseline and plateau of the integral will drift in opposite directions. This is a result of the presence of a component of the dispersion mode signal. In contrast, improper integrator amplifier balance will result in baseline and plateau drift in the same direction. As both may be present simultaneously, it is advised to balance the amplifier as a first step as this procedure is independent of other spectrometer parameters. When an instrument that has a field/frequency sweep option is used, integration must be performed in the field sweep mode. In the frequency sweep mode, the detector phase is frequency dependent and proper phase adjustment cannot be maintained during the sweep.

When integration is performed, it is best to set the integrator amplitude (gain) at its maximum and adjust the spectrum amplitude such that the integral step(s) over the peaks to be integrated extends nearly full scale. This is done for the same reason that a time averaging device is used to enhance the spectrum. The integrator will tend to average the high frequency noise. However, if transient d.c. noise is present at the detector, the integral will

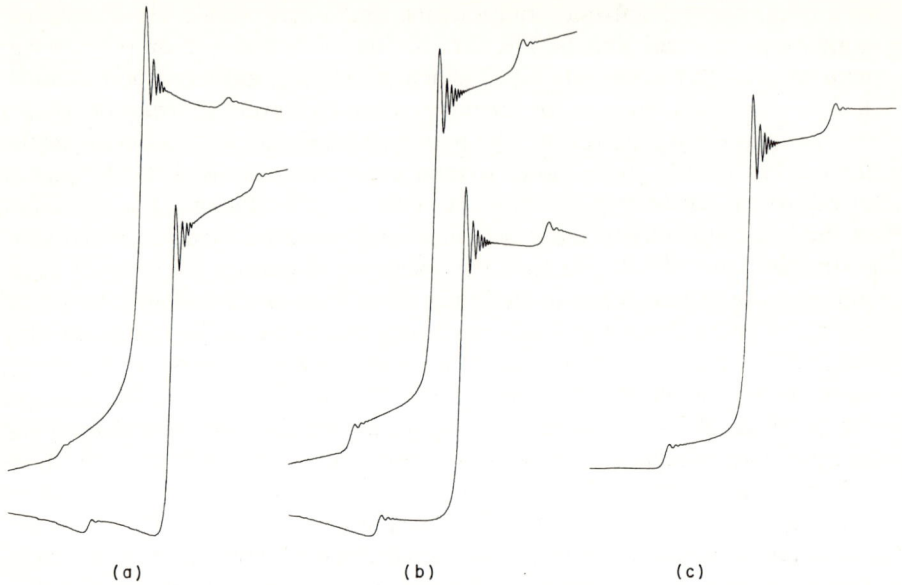

Fig. 6.1 (a) Improper settings of the detector phase. (b) Improper settings of the amplifier balance. (c) Proper integration showing steps resulting from a sharp singlet and the associated spinning side-bands.

be seriously affected; therefore the output of the spectrum amplitude gain is kept at a minimum.

A parameter which may seriously affect the accuracy (and precision) of integration is the ratio of rf power (B_1) to sweep rate. There are many opinions concerning the adjustments of these parameters. The major concern when selecting the value of B_1 and sweep rate is the trade-off between good signal to noise ratio obtained at the higher B_1 setting and slower sweep rates, and the saturation which will likely occur under these conditions. Several theoretical treatments permit calculations of the values needed to prevent saturation (12–14). However, most of these treatments require a knowledge of the value of T_1 and T_2 (or at least their ratio) and a reasonably accurate knowledge of the value of B_1. These parameters can be measured, but to do so would certainly overly complicate the determinations to be performed. The vast literature reporting NMR integrations certainly documents that these tedious procedures are not necessary to obtain good results. However, some empirical scheme is advised to assure accurate results (15). If one only attempts to minimize saturation of the signal, the best selection of B_1 and sweep time is the lowest possible for each. However, poor signals will invariably result. There are at least three practical schemes to select the settings. First the sweep rate selected is normally one of the fastest available on commercial spectrometers (~ 10 Hz/s.). The fast sweep reduces the product of B_1 and sweep time which helps to reduce saturation and decreases the time the integrator must hold its value. The ringing which occurs at the plateau of the integral at these sweep rates does not affect the mean value of the integral. However, one report indicates that integrals with strong ringing tend to be evaluated higher than their true value (16). The value of B_1 may be increased in increments from a low value with an integration taken at each setting. If a single peak is integrated, the value of the integration step height will increase with increasing B_1, reach a maximum, then begin to slowly decrease as saturation begins to occur (15). The value of B_1 just less than that used for the maximum integration is best; however, the value at the maximum may give more precise results as it is less dependent on B_1 fluctuations. If several resonance peaks in a spectrum are to be integrated, the value of B_1 at which saturation begins may be different for nuclei in different groups of the molecule. In this case, the lowest value of B_1 that will saturate the resonance which has the longest $T_1 T_2$ product can be detected by a change in the ratio of the integrated values. For the most cautious spectroscopist, especially when long T_1 values (such as aromatic protons in degassed samples) are encountered, the area ratios of integrated peaks is most accurately determined using several B_1 settings and extrapolation of the area to zero B_1. In general a sweep rate of approximately 10 Hz/s. and a B_1 of 0.1–0.2 mG is satisfactory for most purposes.

The precautions taken in considering the previously stated instrumental parameters are necessary, but not sufficient, for maximum quality in quantitation by NMR peak integration. Several other factors are of equal importance. As already mentioned, the quality, condition and environment of the instrumentation are of importance. The factors which affect resolution and sensitivity of the instrument, discussed in Chapter 3, will also be of importance in the use of the instrument for quantitation. There is no set of rules or procedures which will take the place of experience. Although it is of great value to know the principles, an experienced user of NMR spectrometry can perform excellent measurements without an in-depth knowledge of the theory. The actual practices of sample preparation and calibration are of equal importance to the quality of the final result of the determination.

6.4.2 SAMPLE PREPARATION AND PROCEDURE FOR PEAK AREA MEASUREMENTS

In general the same procedures and precautions discussed in Section 3.3 for the preparation of an NMR sample are used for quantitative measurements. Some general considerations are worthy of further comment. For purposes of high resolution spectra required for integration of closely spaced lines, careful selection of the NMR sample tube is required. To obtain the necessary resolution expected of modern instruments, the sample is spun at a rate of about 30 Hz to average the inhomogeneity in the field. In cases in which peak separation is not a problem, the integration (not the spectrum) may be as accurately obtained without spinning the sample. There is little evidence that sample spinning is of any direct advantage for integration and a report has appeared that it has no effect (17). If the sample is spun, spinning side-bands resulting from modulations generated by the spinning sample in an inhomogeneous field may be observed in the spectrum unless the spectrometer is very well tuned. These "spinning side-bands" should be included in the integral of the peak with which they are associated. Every effort should be made to minimize spinning side-band intensities and therefore the correction made on the area. These side bands will be symmetric about the main peak and removed from it by a frequency shift equal to multiples of the spinning rate of the sample. Their area is derived by modulation of the main peak. Their position may be varied by changing the spinning rate. The sample spinner should be in good condition such that it, or an unbalanced sample tube, does not cause wobbling which can deteriorate the signal and possibly damage the probe. The tube should fit snugly in the spinner such that it will not slip down while in the probe. A gauge should be used to insert the sample tube reproducibly in the spinner. The insertion distance is determined by trial such that the best signal-to-noise (S/N) ratio is obtained. Adequate sample must be placed in the tube to assure filling of the probe coil volume.

It is important that the sample be properly prepared. Trace quantities of paramagnetic materials such as transition metal ions may broaden the spectrum to the point where only very broad bands are observed. The effect may be even worse if small particles of ferromagnetic iron or steel are allowed to enter the sample. The use of glass wool or micropore filters to remove the solid particles is highly recommended. Although stainless-steel hypodermic needles are commonly used by some spectroscopists to fill sample tubes, their use is not recommended. If used, they should not come in contact with acids or strong oxidizing agents such that metal ions are formed which may leach into sample solutions. The presence of dissolved oxygen as a paramagnetic line broadening agent has long been recognized by NMR spectroscopists. When the best resolution is desired, the oxygen should be removed by one of many techniques such as alternate freezing and evacuating the thawing sample. After several freeze-evacuation cycles, the sample is sealed. For samples which are to be used for only a short time, a well fitted rubber serum cap will suffice. The pumping may be done through a small hypodermic syringe inserted in the cap. When ready, the syringe may be removed and a partial vacuum will persist for several hours. For samples which are needed for longer time periods, the sample tube should be sealed by fusion of the glass. Care should be taken to draw the seal to a tip along the axis of the sample tube so that wobbling is not caused by spinning the sample in the probe.

Once the instrumental parameters have been properly adjusted and the sample prepared, the spectrum is recorded and the integration performed. There are several details of the technique of recording the integration which may be considered. As mentioned earlier, the use of a digital voltmeter can improve the precision of the integration mainly by elimination of recorder errors. However, a permanent record should always be made which can be examined later if an error is suspected in recording the value of the digital readout. Some spectroscopists prefer to record the integral directly on the spectrum. This is best done by adjusting the sweep offset before integration such that the rising part of integration step will not fall on the absorption peak. When a complex spectrum is integrated this procedure is not always practical, but when it can be done the peak positions can be measured easily even in the presence of the recorded integration. Others prefer to place an unlined sheet of paper over the absorption spectrum and record the integration separately. Either technique is satisfactory. In any event, it is poor practice to measure the integral step heights by counting squares on the lined paper as the probability of error is high. The best procedure is to extrapolate the base line and plateau of the integral step and draw a line through the inflection point of the integration curve perpendicular to the bottom edge of the paper. In this way the noise may be somewhat visually averaged, drift in the integrator is partially compensated and a reproducible technique is used.

A very common use of quantitative NMR peak area measurements is to perform proton counting to determine the relative number of protons in different sets of resonance lines from protons in a single molecule. The characteristic chemical shift, the spin–spin splitting pattern and the number of protons in each magnetically equivalent group can then be used in conjunction in order to determine the structure of the molecule. Relative area determinations may be performed easily by integration because of the fortunate result that the area of proton resonance lines depend only upon the number of protons giving rise to the signal and not upon their chemical nature. However, it is important to remember that nuclei in different chemical environments may have significantly different T_1 or T_2 values. Therefore a B_1 value which is sufficiently low as to avoid saturation in one set of resonance lines may not be sufficiently low for another set (15). Provided the precautions discussed above are taken, relative area measurements may be obtained directly from the integrals. This same or a similar procedure

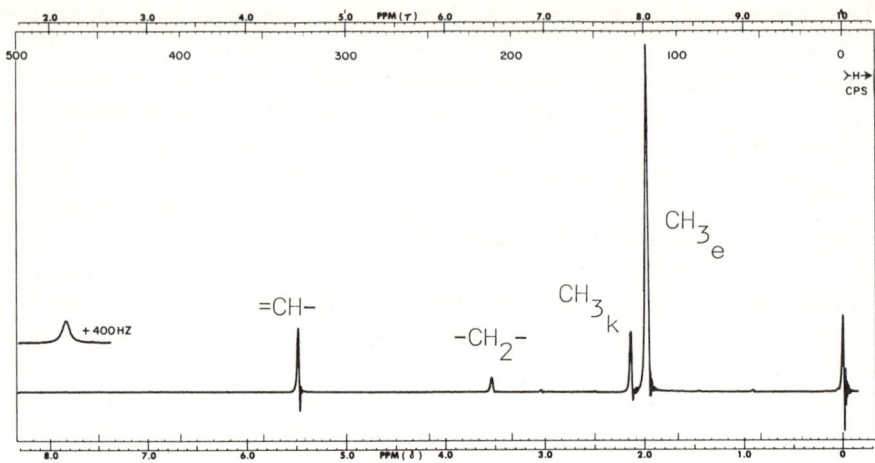

Fig. 6.3 NMR spectra of aspirin, phenacetin, caffeine, and mixture: (a) Aspirin, (b) Phenacetin, (c) Caffeine, and (d) APC mixture. Reprinted with permission from D. P. Hollis, *Anal. Chem.*, **35**, 1682 (1963). Copyright by the American Chemical Society.

may be employed when peak area measurements are used to investigate equilibria or slow chemical reaction rates. For example, the keto-enol tautomerization equilibria of β-diketones may be easily investigated as a function of concentration, solvent and temperature using the relative areas of the $-CH_2-$ and $=CH-$ protons in the two tautomeric forms, as illustrated in Figure 6.2 (18).

In situations in which absolute quantitative measurements are desired some form of calibration of the instrumental system is required. All of the instrumental parameters should be selected prior to calibration of the instrument and none changed until the analysis is completed. There are three principle methods of calibration. First, a calibration curve technique may be used provided a sample of the pure material, which is to be determined in unknown solutions, is available. In this technique, solutions of the compound(s) to be determined are prepared which encompass the concentration range expected in the unknowns. The area of one or more resonance lines is plotted versus the concentration in each standard. The plot should be linear and "least squares" methods may be applied to the line if desired. The concentration of the unknown sample may be read directly from the curve once the area has been obtained. The primary advantage of this technique is an improvement in the statistics of the measurement. If many samples are to be determined at one time, it is an excellent method. However, because the calibration curve is time consuming to prepare, the procedure may not be suitable for a single or infrequent determination.

A second method which may be used if a pure sample of the unknown is available is the method of standard additions. For this method two samples are prepared. One is the unknown as received. The second is a mixture of a known volume of the unknown and a known volume of a standard solution prepared from the pure material. The integrated resonance of the unknown (A_x) and mixture (A_m) is given as

$$A_x = kC_x \tag{6.3}$$

$$A_m = k\frac{V_x C_x + V_s C_s}{V_x + V_s} \tag{6.4}$$

where k is a proportionality constant, C_x and C_s are the concentrations of the unknown and standard, respectively, and V_x and V_s are their respective volumes in the mixture. Solving for C_x:

$$C_x = \frac{A_x V_s}{A_m(V_x + V_s) - A_x V_x} C_s \tag{6.5}$$

This method has the advantage that fewer samples have to be prepared than for the calibration curve method and it also has a tendency to compensate for any unsuspected chemical effects. However, an even less complicated method is the one most frequently used and is described next.

Because all nuclei of a given type (viz. protons) give resonance areas proportional to their concentration and independent of their chemical environment, it is, in fact, not required that the reference standard be the same compound as the unknown. The analyst should be alert to possible chemical effects however. For example, if a single resonance line is used to determine a compound that may exist as tautomers, an error may result if the resonance line chosen for integration does not exist for all tautomeric forms. In the absence of such chemical effects, any suitable compound may be used as a standard. The method of internal standard is a simple and direct method which is ideally suited for quantitative NMR using peak area measurements. This method is based on the addition of a known amount of an inert compound to the sample. The relative area of the integrated resonance of the unknown and the internal standard are directly compared to a single run. Obviously, to improve the reliability of the determination, more than one sample should be prepared. The selection of the compound to be used as the standard is based upon practical considerations such as solubility in the solvent used. It should be chemically inert to both the solvent and the sample. It should give no resonance lines which overlap those of the sample. Preferably, it should give a single line which has similar saturation characteristics as the line(s) used for the sample. It should have a moderately high molecular weight so that small mole quantities may be weighed with accuracy. It should be easily removed from the sample if needed. Many of

268   METHODS OF QUANTITATIVE MEASUREMENTS

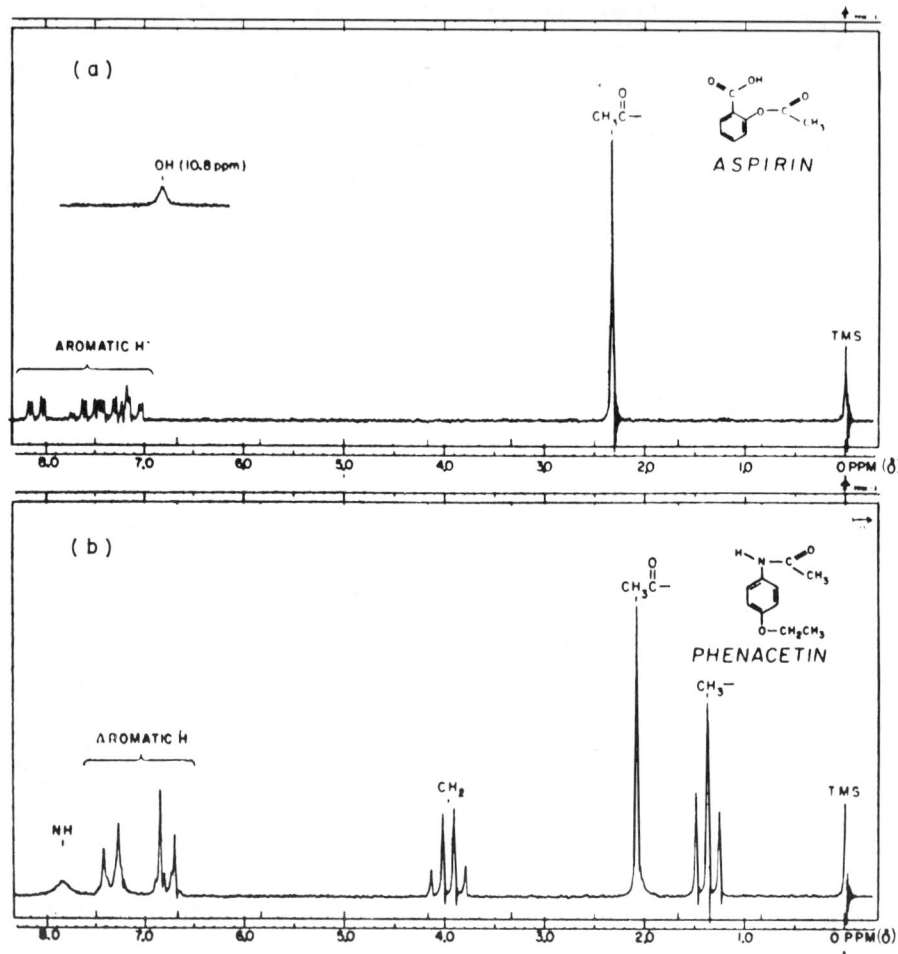

Fig. 6.3 NMR spectra of aspirin, phenacetin, caffeine, and mixture: (a) Aspirin, (b) Phenacetin, (c) Caffeine, and (d) APC mixture. Reprinted with permission from D. P. Hollis, *Anal. Chem.*, **35**, 1682 (1963). Copyright by the American Chemical Society.

these properties are the same as those desired for chemical shift reference compounds. The concentration of the unknown may be calculated as

$$C_x = C_s \frac{N_s A_x}{N_x A_s} \qquad (6.6)$$

where C_x and C_s are the concentrations (mole units) of the unknown and standard, respectively. N_x and N_s are the number of protons giving rise to the respective integrated areas A_x and A_s.

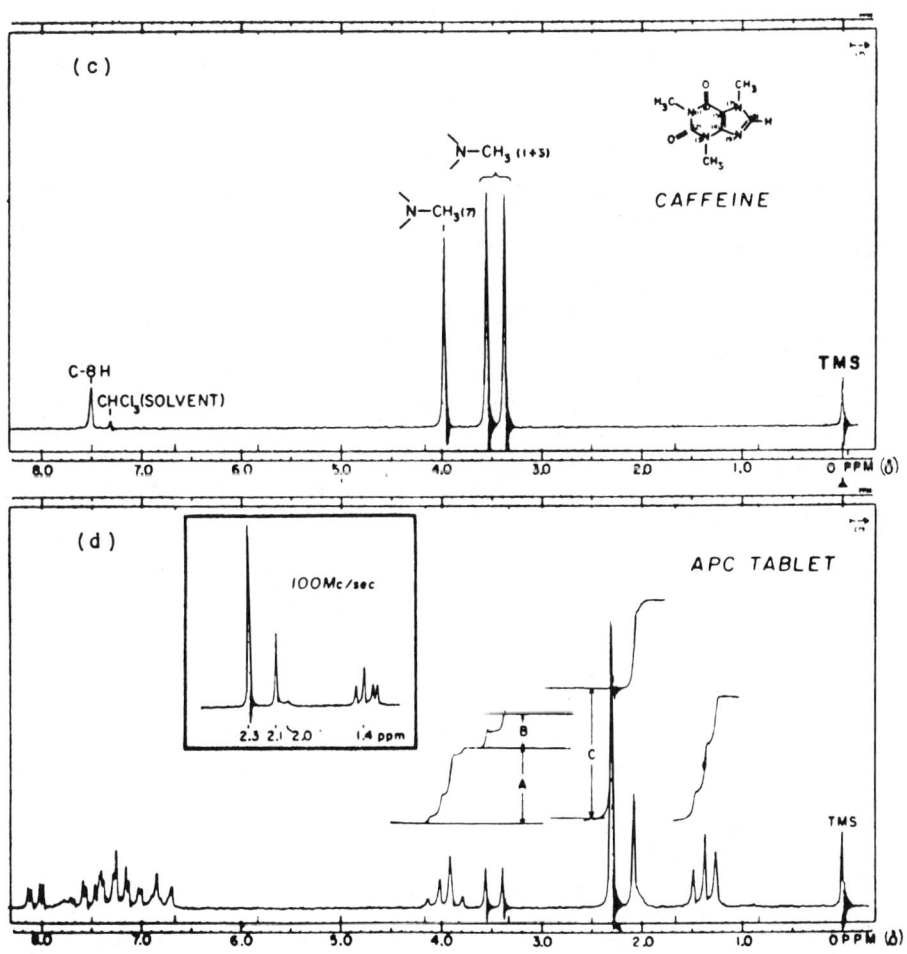

In the proton magnetic resonance spectra of organic compounds as neat samples and concentrated solutions, satelites resulting from the spin–spin interaction between the proton and ^{13}C in the molecule are frequently observed. The natural abundance of ^{13}C is 1.1%. Therefore corrections must be made for the area of these satelites in two ways. First, their area must be added to the central peak from which they originated; second, their area must be subtracted from that of any lines which they overlap. An excellent illustration of these corrections as well as a general example of quantitative NMR is seen in the analysis of APC tablets for the aspirin, phenacetin, and caffeine content (19).

Figure 6.3 shows the NMR spectra of each component and an APC mixture in deuterochloroform. A sharp single peak at 2.3 ppm (versus tetramethylsilane) which arises from the ester methyl protons of aspirin is shown in spectrum A. This peak is free of major overlap with others in the mixture and was chosen as an analytical peak for aspirin. Phenacetin gives a quartet at about 4.0 ppm, a triplet at about 1.3 ppm, and a sharp peak at about 2.1 ppm, as shown in spectrum B. The latter peak consists of two closely spaced peaks as shown resolved in the insert in spectrum D. The triplet at 1.3 ppm was not used in the analyses as an impurity in commercial preparations interfered with the high-field side, as may be seen in spectrum D, and the insert in spectrum D. The quartet at about 4.0 ppm is suitable as an analytical peak for phenacetin even though it overlaps with a caffeine peak. Corrections for this overlap can be made by subtraction of the area of one of the remaining caffeine peaks at 3.4 or 3.6 ppm.

The number of milligrams of caffeine per milligram of sample is calculated as follows:

$$\frac{\text{mg. caf.}}{\text{mg. sample}} = \frac{\frac{1}{2}(13 - 0.0055C)}{\text{area}_{\text{caf.}}} \times \left[\frac{\text{mg. caf.}}{\text{ml.}}\right]_{\text{STD}} \times \left[\frac{\text{ml.}}{\text{mg.}}\right]_{\text{sample}} \quad (6.7)$$

where $\text{area}_{\text{caf.}}$ is the area of one methyl group of the caffeine standard. The factor $0.0055C$ arises from a correction for the overlap of a ^{13}C satelite of the aspirin peak that overlaps the methyl resonance of caffeine at 3.4 ppm. This correction is required because of the large amount of aspirin in APC preparations. Corrections for the ^{13}C satelite of caffeine on the area on the aspirin resonance is not as critical. The aspirin and phenacetin content may be calculated in a manner similar to caffeine.

$$\frac{\text{mg. aspirin}}{\text{mg. sample}} = \frac{c}{\text{area}_{\text{caf.}}} \times \frac{\text{M. WT. asp.}}{\text{M. WT. caf.}} \times \left[\frac{\text{mg. caf.}}{\text{ml.}}\right]_{\text{STD}} \times \left[\frac{\text{ml.}}{\text{mg.}}\right]_{\text{sample}} \quad (6.8)$$

$$\frac{\text{mg. phen.}}{\text{mg. sample}} = \frac{A - \frac{1}{2}(B - 0.0055C)}{\text{area}_{\text{caf.}}}$$

$$\times \frac{3}{2} \times \frac{\text{M. WT. phen.}}{\text{M. WT. caf.}} \times \left[\frac{\text{mg. caf.}}{\text{ml.}}\right]_{\text{STD}} \times \left[\frac{\text{ml.}}{\text{mg.}}\right]_{\text{sample}} \quad (6.9)$$

These results may be obtained using a single caffeine standard, or a standard of some other compound. A careful examination of this example will illustrate many of the factors to consider when devising a quantitative scheme using NMR.

6.4.3 ACCURACY AND PRECISION

Although the integration recorder system is capable of a precision of 1–2%, there are no generalized statements on the accuracy or precision of quantitative results which can be given. These will vary between specific instrument quality and condition as well as operator techniques. Reports of studies on a single instrument (17) and on interlaboratory studies (20) generally indicate that a standard deviation of the mean of 0.5% is obtainable for several repetitive runs, and 1–2% should be routine. Occasionally, authors make the statement that repeated integrations improve the precision of the result. This, of course, is not true. What is true is that the mean of several measurements is a better estimate of the "true" result than a single measurement, and that the confidence interval decreases with the square root of the number of measurements for a given confidence level. That is to say, the precision is estimated more accurately with an increasing number of measurements. The precision is determined by random variations in the total measurement system, including sample preparation, and is a measure of the reproducibility of that system. Also, a highly precise (i.e., reproducible) measurement will not be accurate if, for example, the standard were not correctly prepared. In one study the percent of total hydrogen in chemically different sites of over twenty compounds were determined with an absolute error between theoretical and found values no greater than 0.5% (21). The relative error is much less. This example illustrates the accuracy that may be obtained via NMR peak area measurements.

6.5 SPECIAL METHODS AND TECHNIQUES

6.5.1 SIGNAL AVERAGING

One of the limiting aspects of continuous-wave high-resolution NMR is the inherent lack of good lower limits of detection. The minimum detectable quantity of sample which gives sufficiently good S/N ratio such that accurate quantitative measurements can be made will depend on the molecular weight of the compound, the number of hydrogen atoms giving rise to a clean, easily integrated signal, the degree of spin–spin splitting and, of great importance, the instrument used. No one value of detection limit can be given with any certainty, but for a point of discussion the latest commercial proton magnetic resonance spectrometer should be able to provide quantitative data on about 1–10 μg of a compound of a molecular weight of 200. Very often, however, the analyst is frustrated by possessing insufficient sample. There are some current aids to this problem and some hope for the future. If routine detection limits of NMR methods can be lowered by improvements in instrumentation, there would likely be a stampede of biomedical and clinical applications.

There are two fundamental approaches to improving detection limits of any instrumental method: increase the signal or lower the noise. The only variable parameter which may be used to increase a magnetic resonance signal is to increase the field strength B_0 such that the Boltzmann distribution of spin states results in a larger population difference. At higher magnetic fields the fraction of spins in the lower energy state increases and the rate of absorption of the rf energy (signal) increases. The use of currently available high field superconducting magnets offer theoretical improvement of about a factor of 10 over conventional electromagnets. However, their main objective is to increase the chemical shift between magnetically nonequivalent nuclei. These instruments are overly complicated and expensive for what one hopes to be a routine analytical tool. Therefore, in practice it is a decrease in the noise which is the practical approach.

Assuming that the instrument manufacturer has done all that is possible to reduce electronic noise in the instrument and proper installation has been performed to avoid ground loops and line noise, reduction of the remaining noise may be approached using statistics. If a response consisting of a d.c. signal such as an NMR resonance line is present in conjunction with an a.c. noise, the S/N ratio may be improved by signal-averaging. The signal accumulates in direct proportion to the number of scans recorded, whereas the noise accumulates as the square root of the number of scans. Therefore, n repeated scans improves the S/N by a factor of $\sqrt{n}$. A signal may be averaged by an analog integration over a period of time. With this regard the NMR integrator is a signal averaging device. High frequency noise may be reduced by using a large RC time constant in the integrator. This method is practical for spectroscopic methods involving stationary states with short relaxation times. However, the real danger of saturation of the NMR signal requires a transient signal rather than lengthy averaging of the peak. Although some crude analog signal averaging devices were constructed using the integrator (22), digital averaging is much more practical.

Hardwired digital data acquisition equipment such as a computer of average transients (CAT) has become very common as signal enhancement devices for NMR. These devices contain some form of core memory, perhaps in association with a recording device such as magnetic tape or cassettes. The devices perform the function of multichannel (viz. 1024 channels) digital data storage. The NMR spectrum is automatically recorded a preset number of times. Each time, the spectrum amplitude is converted into digital form at equal frequency increments and stored in the corresponding memory address in the CAT. The operator may view the spectrum between scans by displaying the readout on a cathode ray tube. The spectrum may be recorded directly on the NMR spectrometer recorder when desired. The spectrum may be integrated using the spectrometer integrator while recording the spectrum from the CAT. As there are many commercial vendors for these devices, each having its own set of operational procedures, no attempt will be made here to give detailed procedures. One set of procedures has been given for the operation of the Varian C-1024 CAT with the Varian A-60 spectrometer (23).

The advantages of using a computer of average transients to improve the S/N ratio is readily seen because their performance is very close to the theoretical limit. For example, the accumulated spectrum of 100 scans will have very close to ten times the signal to noise ratio of a single scan. The few percent deviation from the limit is usually a result of instrumental drift. A decision on the number of scans that will be recorded must be made. The improvement in S/N beyond 400 scans diminishes rapidly and is usually not worthwhile. When the highest resolution is not necessary, it is usually profitable to sweep at a higher rate than is normally used (~ 25 Hz/s.) and employ a relatively high B_1 setting. In this way the signal to noise is improved over a shorter period of total accumulation time than using slower sweep rates and lower B_1 settings. This procedure exercises the frequently encountered trade between resolution and S/N.

The signal averaging process using a CAT is a very time consuming process that occupies a spectrometer and a rather expensive hardware item for long periods of time. The interruption of a scan procedure by one of the many events that plague instruments can be annoying. The availability of digital computers (to almost all laboratories) has evolved other methods of "data smoothing." These procedures may be applied to a single scan, or to spectra which has been averaged using a CAT. The spectrum must be converted to a digital form and entered into a digital computer (in a way similar to the CAT). This is best done by interfacing an A–D converter to the spectrometer and recording spectrum amplitude values at increments of frequency on magnetic tape, paper tape, disk, or directly into computer core memory. Programs may then be written in a high level language (FORTRAN, BASIC, etc.) to smooth the data. One form of computer averaging is known as boxcar averaging (24). This technique replaces the set of two or more data points with the mean of those points. For example, five adjacent amplitudes are averaged and the mean replaces the point originally occupied by the third point in the set. The next five points are treated in the same way, and so on. Obviously this leads to serious loss of resolution. A simple refinement of the boxcar method is the "moving window" method (24). This technique again replaces a set of data points with the mean. However, instead of advancing an entire set, the next set is formed by dropping the first point of the previous set and adding the point following the previous set. The window is thus moved through the entire data set. This technique does not suffer the severe loss of resolution that the boxcar method does.

These various methods of S/N improvement assume the noise is random in phase and amplitude with respect to the signal. The smoothing techniques

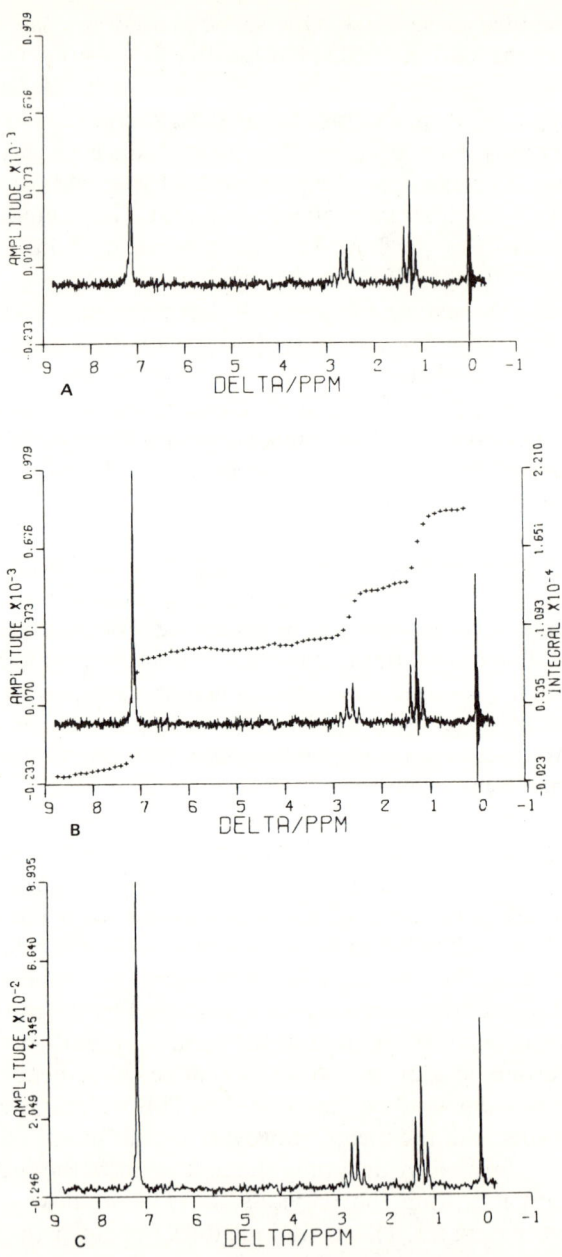

Fig. 6.4 Computer reproduced nmr spectra of ethylbenzene: (A) raw data, (B) with integral trace (+ + + +), and (C) after one smoothing. Reprinted from *American Laboratory*, volume 5, number 9, page 53 (1973). Copyright 1973 by International Scientific Communications, Inc.

can be of great usefulness when a computer data acquisition system is available. However, they must be used with discretion so that real data is not overly removed and artifacts generated. The programmable computer is much more powerful than the CAT because it may be used not only to acquire the data (25,26), but also to treat the data, store several spectra as a function of elapsed time, perform digital integration and control the instrument including maintaining tuning (27). Figure 6.4 shows a simple but illustrative example of the acquisition, smoothing, and digital integration using an on line computer (28).

6.5.2 PULSED NMR

Pulsed NMR techniques may be employed in two ways for quantitative analysis. First, the amplitude of the free induction decay (FID) signal at zero time interval following a 90° pulse (Section 3.9) is proportional to the total magnetization of the spin, which is in turn proportional to the number of nuclei in the probe coil volume (29). Therefore the amplitude of a FID curve at various times following the pulse sequence may be extrapolated to zero time and related to the concentration of nuclei leading to the signal. By using a calibration curve obtained from known samples, quantitative determinations can be performed. This method is useful in the determination of magnetically active nuclei in solids where the NMR linewidths are so broad as to obscure the high-resolution spectrum, and to other wide-line resonance signals. This method measures the total spin magnetization of the nuclei observed by a given spectrometer (viz. protons) and therefore does not permit any qualitative distinction between chemical differences. The technique is useful for the determination of water in solids for example. One difficulty with the technique is the problem of reproducibly packing solids in the sample tube.

A modification of the technique that permits distinction between physical states of compounds in solids or viscous media is based on differences in T_1 and/or T_2 values which may be observed using pulse or "spin-echo" methods. In cases in which solids contain water in both a "bound" and "free" state, or water and an oil, or amorphous and crystalline mixtures, there may be significant differences in the T_1 and/or T_2 values. The FID signal will have an initial amplitude proportional to the total proton concentration in the solid. However, the decay curve will be a composite of two curves with different decay time constants. Figure 6.5 shows an example of the determination of the water content in margarine. The T_1 of the solid is very short and its contribution to the FID signal is nil after about 1.2 sec. The remaining FID signal is due to the water which has a much longer T_1. Extrapolation of the latter portion to zero time yields the contribution of the water to the initial

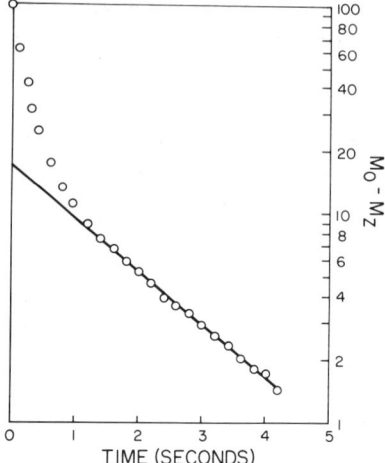

Fig. 6.5 Amplitude of the free induction decay signal versus time for the determination of water in margarine.

FID amplitude. Assuming the only components in the sample are organic solids and water, the percentage water may be estimated by plotting the ratio of the water-to-total initial amplitudes versus the percentage water in known samples.

Several pulse NMR spectrometers are available for this type of quantitation and are relatively inexpensive. One currently available (30) is a desk top, readily portable unit especially suitable for routine process monitoring. The unit has a two channel "averager" which enables a digital display of the ratio of the amplitude of the FID signal at two selected time intervals following the pulse. In this way routine quality control of such items as moisture content in creams, solids, greases, and similar materials may be rapidly monitored by technicians. Several pulse sequences may be selected, including rapid T_1 measurements which would be valuable in monitoring phase changes. A second instrument (31) is designed to serve both as a routine quantitative instrument and as a moderate research instrument. Both instruments are easily interfaced to a computer for process control and data recording and reduction.

6.5.3 PROCESS CONTROL

Process control in the chemical, pharmaceutical, petroleum, and other industries are important functions. Modern industry relies on either rapid manual analytical methods or totally automated systems for quality control. The use of NMR can play a significant role in process monitoring. Anderson has reported an evaluation of the Varian PAT-20 as a process control monitor (32). Reinhart has discussed the use of NMR as a monitor for water in coal (33). Rollwitz and Persyn have discussed on-line water monitoring in starch production processes (34). A technique for monitoring the rehydration of dried foods has been presented (35).

An application of NMR which has received relatively little attention is use of flowing samples. The effect of flowing samples was investigated early by Bloom and Shoolery (36) and Suryan (37). Suryan found that for nuclei having relaxation times between 0.1 and 0.05 s., flowing samples at a rate that flushed the coil volume about each 0.01 s. resulted in a higher signal than static samples with B_1 near that of saturation. NMR flow meters have been based on this general effect (38). McIvor has reported quantitative determinations in flowing systems (39). More work has been done on flowing systems in the USSR and a book has been published there on the subject (40). Further developments in the area of flowing samples in NMR should lead to increased applications in process monitoring.

6.6 EQUILIBRIUM STUDIES

Investigation of chemical equilibria using NMR can be performed in two ways. First, equilibria which, although chemically dynamic, involve slow (Section 6.7.1) chemical exchange processes may be studied using peak area measurement. In this approach, the relative area of resonance lines are determined, corrected for any differences in the number of nuclei (equivalent or otherwise) giving rise to the peaks, and the equilibrium constant calculated directly from the corrected peak area ratios. An excellent illustration of the application of this approach is given by the determination of keto–enol tautomerism in β-diketones. Figure 6.2 shows the spectrum of 2,4-pentanedione. The —CH$_2$— and =CH— proton resonance lines are readily identified for the keto and enol tautomers respectively.

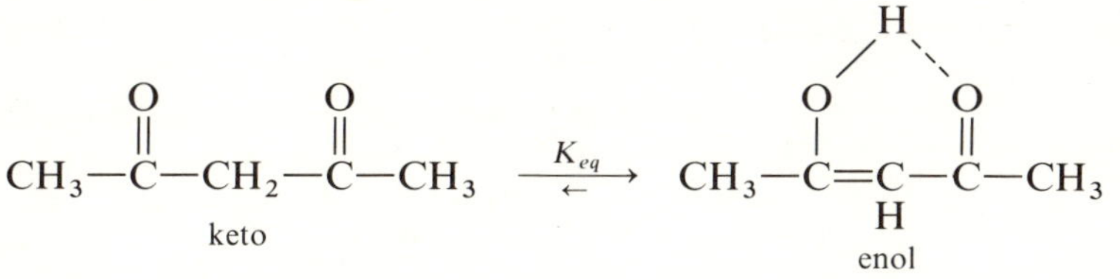

$$\underset{\text{keto}}{CH_3-\overset{\overset{O}{\|}}{C}-CH_2-\overset{\overset{O}{\|}}{C}-CH_3} \underset{\leftarrow}{\overset{K_{eq}}{\rightarrow}} \underset{\text{enol}}{CH_3-\overset{\overset{O-H}{|}}{\underset{H}{C}}=\overset{\overset{}{|}}{C}-\overset{\overset{O}{\|}}{C}-CH_3} \qquad (6.10)$$

The rate of the keto–enol interchange is insufficiently slow that sharp resonance lines are observed for each proton. The equilibrium constant is then calculated as

$$K_{eq} = \frac{A_E}{0.5\,A_K} \qquad (6.11)$$

where A_K and A_E are the areas of the —CH_2— and =CH— resonance lines, respectively (18). This equilibrium may be investigated as a function of temperature, solvent, or concentration without the need for extraneous standardization. Amine complexes with 2,4-pentanedione have been investigated in this way (41).

A second approach to equilibrium studies may be used in the case of fast chemical exchange between the chemical species involved. This approach is an extension of the example given in Section 6.1 for quantitative measurements of binary mixtures undergoing rapid chemical exchange. If two or more species are in equilibrium and rapidly interchanging their chemical form or exchanging labile nuclei such as an acidic proton, the observed chemical shift will be the population-weighted average of the chemical shifts of the individual species. The observed phenomena may be divided into two subsets. In one case, the chemical shift of a resonance line of nonlabile protons in the molecule participating in the equilibrium is observed. An early example of this approach was given by Grünwald, Loewenstein, and Meiboom (42). The chemical shift of the methyl-proton line of aqueous solution of methyl amine was observed as a function of pH. Figure 6.6 shows that a plot of chemical shift versus pH gives the familiar sigmoid "titration curve." A plot of chemical shift versus fraction $CH_3NH_3^+$ is linear. Obviously, the acid dissociation constant, K_a, for $CH_3NH_3^+$ may be calculated from these data. The observed chemical shift of the $CH_3NH_3^+$ and CH_3NH_2 mixture is given as:

$$\delta_{obs} = x_{BH^+} \delta_{BH^+} + x_B \delta_B \qquad (6.12)$$

where x_{BH^+} and x_B are the mole fractions of $CH_3NH_3^+$ and CH_3NH_2 in a solution at a given pH, and δ_{BH^+} and δ_B are the chemical shifts of the methyl protons of $CH_3NH_3^+$ and CH_3NH_2, respectively. The latter quantities are obtained by extrapolation of the chemical shifts at low and high pH values, respectively. In these simple cases, it is useful to arbitrarily set one chemical shift, for example, δ_B, equal to zero giving:

$$\delta_{obs} = x_{BH^+} \delta_{BH^+} \qquad (6.13)$$

$$K_a = \frac{(1 - x_{BH})(H^+)}{x_{BH^+}} = \frac{\left(1 - \dfrac{\delta_{obs}}{\delta_{BH^+}}\right)[H^+]}{\delta_{obs}/\delta_{BH^+}} \qquad (6.14)$$

This particular type of equilibrium is better studied with more appropriate techniques such as potentiometry or spectrophotometry. However, with refinements, the NMR method can provide information difficult to obtain in other ways such as microscopic protonation schemes of polyfunctional bases (43). Examples will be given in Chapter 7. One of the limitations of

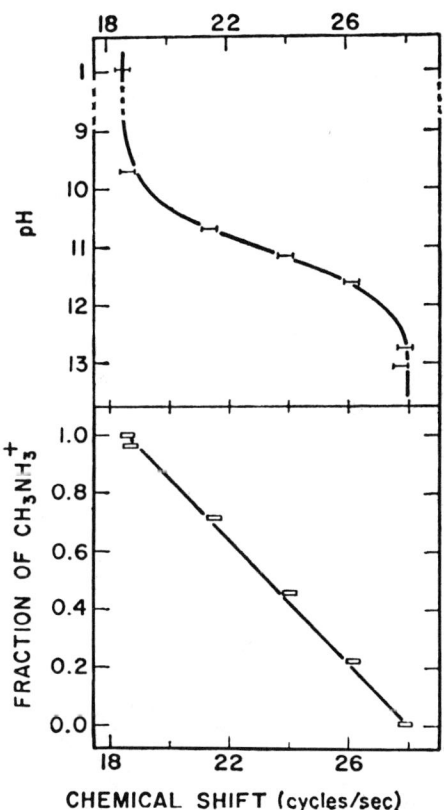

Fig. 6.6 Chemical shift of methyl protons (cycles per second at 7,430 G) relative to $(CH_3)_4N^+$ in $CH_3NH_2\text{-}CH_3NH_3^+$ equilibrium mixture as a function of pH (upper part) and concentration ratio of acid to base plus acid. Reprinted from E. Grünwald, A. Loewenstein, and S. Meiboom, *J. Chem. Phys.*, **27**, 641 (1957), with permission of the American Institute of Physics and Grünwald, Loewenstein, and Meiboom.

NMR for this type of study has been the need for solutions more concentrated than desired for determination of thermodynamic equilibrium constants. This problem has been solved to a great degree with modern signal enhancement techniques. For example, FT-NMR methods may be used to obtain chemical shift data on dilute solutions such that ionic strength of electrolyte solutions fall within the limits of applicability of the Debye-Hückel expression for activity coefficients.

6.7 CHEMICAL KINETICS

6.7.1 SLOW REACTIONS

NMR techniques may be applied to a wide range of studies of chemical kinetics. Similar to equilibrium studies, two subgroups of techniques may be employed depending upon the rate of the chemically dynamic process. First, the NMR spectrometer may be used as a quantitative instrument to investigate dynamic processes, such as chemical reactions including isotopic exchange reactions, with half-lives on the order of minutes. In these cases, resonance peaks may be repetitively integrated at known time intervals following initiation of the reaction. The absolute concentration at any time of each chemical species (viz. reactant and product) may be calculated from the relative peak areas and mass balance of the nuclei leading to the resonance. The data may then be treated according to conventional methods of rate-data analysis. There are obvious applications of this procedure such as the study of isotopic exchange reactions (^{2}H for ^{1}H), hydrolysis reactions, and so on. In these cases, NMR offers convenience more than any other advantage. Using a CAT or computer to acquire rapidly-changing spectra permits slightly faster reactions to be investigated (44). If fast sweep rates (~ 25 Hz/s.) can be tolerated without detrimental loss of resolution, an on line-computer can store repetitive scans of a peak approximately each 5 s. The peak may then be integrated digitally by the computer. For nuclei with short relaxation times, FT spectra may be collected each few seconds (45). These latter methods have the advantage of presenting changes in the entire spectrum during the time of the reaction. In that way changes at all sites containing resonant nuclei may be monitored.

A variation useful to the study of chemical exchange involving half-lives of a few seconds has been demonstrated (46,47). The technique is based on the sudden application of a saturating rf field (B_2) at the resonance frequency of one of two reversibly exchanging nuclei, and the observation of the time dependence of the intensity of the second resonance line. B_2 may be applied using double resonance techniques. The perturbation of B_2 on the spin distribution (i.e., the extent of saturation) of the nuclei irradiated will be transferred to the second magnetic environment by chemical exchange. Forsén and Hoffman presented an excellent example of the application of this technique (47). The base-catalyzed exchange of —OH and —CH= protons in the enol tautomer of 2,4-pentanedione (Figure 6.2) was known and proposed to occur though the keto tautomer as an intermediate (41). An example of the experimental results is shown in Figure 6.7. The —OH and —CH$_2$— signals were suddenly saturated and the decay of the —CH= signal to a new equilibrium value was followed with time. Using several combinations of saturation and observation, the authors were able to conclude that the —OH and —CH= proton exchange does occur through the keto intermediate. It should be pointed out that the range of rates which may be studied by this technique is relatively small. Therefore limited numbers of applications are expected. However, it is a simple method when applicable.

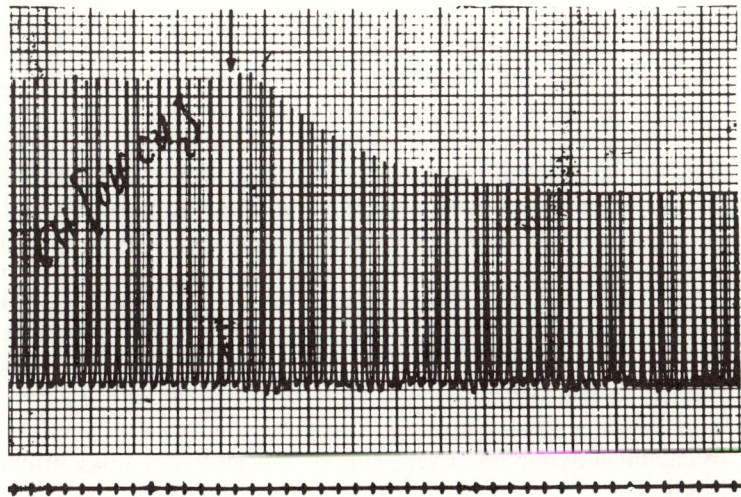

Fig. 6.7 Experimental recording of the decay of signal A (—CH=) to a new equilibrium value after the saturation of signals B (—OH), and C (—CH$_2$—) from a sample of acetylacetone containing a small amount of triethylamine. The markers are spaced at 1-sec intervals and the small arrow indicates the beginning of saturation. Reprinted from S. Forsén and R. A. Hoffman, *J. Chem. Phys.*, **40**, 1189 (1964), with permission of the American Institute of Physics and Forsén and Hoffman.

6.7.2 FAST REACTIONS

In the previous sections, the application of NMR as a quantitative tool to follow concentration changes during the course of chemical reactions with half-lives on the order of minutes was discussed. However, the real significance of the application of NMR to investigations of chemical kinetics lies in the ability to provide information on very fast chemical reactions, even those of a virtual type where reactants and products are chemically identical. Of perhaps even greater importance is the fact that the measurements may be taken while the system is at chemical equilibrium, thus many fast chemical exchange processes may be investigated using conventional CW NMR spectrometers. There have been many reports of the theory of the effects of chemical exchange on NMR spectra. It is beyond the scope of this work to attempt to develop the theoretical treatment. Several reviews and the references therein will provide an excellent introduction to the theory (48-50). Johnson has pointed out the two major factors which make possible the application of NMR to studies of chemical dynamics (48). First, if chemical dynamic processes lead to magnetic field fluctuations at a nucleus under observation, the precessional frequency of the nucleus will fluctuate accordingly. If the *difference* in the precessional frequency of the two or more environments is comparable in magnitude to the frequency of the fluctuations, there may be an observable effect on the NMR spectrum. Second, the width of NMR resonance lines in the absence of chemically dynamic processes (or other factors such as paramagnetic impurities) is very small so these effects are readily observed. Even with instrumental contributions, linewidths are typically only a few tenths of a hertz. It is important to point out here that within this section which is labeled "Fast Reactions," relative reaction rates will be termed fast and slow. Within this section, a fast reaction will be one which occurs at a rate larger than the differences in the precessional frequencies (chemical shift) between the nuclei under observation. A slow reaction refers to the converse.

Before presenting a brief introduction of the theory of the application of NMR to chemical dynamics, a very simple conceptual analogy drawn from the Heisenberg uncertainty principle may be of value. Heisenberg's principle states that the product of the uncertainty in the values of the conjugate variables which describe the energy and position of a particle is approximately equal to Planck's constant. The uncertainty of the energy of a nucleus precessing in a magnetic field may be described by the uncertainty in the pressional frequency, and the uncertainty in its position described by the uncertainty in its residence time in a given magnetic environment.

$$\Delta E \cdot \Delta \tau = \hbar = \frac{h}{2\pi} \tag{6.15}$$

$$h \cdot \Delta v \cdot \Delta \tau = \frac{h}{2\pi} \tag{6.16}$$

$$\Delta v = \frac{1}{2\pi \Delta \tau} \tag{6.17}$$

The value of Δv is an approximation of the linewidth, and $\Delta \tau$ may be expressed as the mean lifetime of the nucleus in a given magnetic environment, τ. Thus the linewidth due to chemical exchange is approximated as:

$$\Delta v = \frac{1}{2\pi \tau} \tag{6.18}$$

From this simple approximation, one can see that for a mean time lifetime of 10^{-2} s. a linewidth of approximately 16 Hz would be observed, which is much greater than linewidths obtained in the absence of chemical exchange. Thus, if the nucleus in this case were exchanging between two equally populated sites, an exchange rate $(1/\tau)$ of approximately 10^2 sec^{-1} could easily be observed. This simple approach greatly underestimates the applicability of NMR to rate measurements and in diamagnetic systems rates in the range of $10-10^5$ exchanges/sec may be investigated. It would be easy to develop a large section of this book devoted to the vast number of NMR techniques and applications to rate studies. However, to maintain a balance in content, a description of the techniques with a few illustrative examples will be given in this section. In Chapter 7 further examples of applications will be given. Rather than attempt to present a historical and complete development of the theory, an example of an early theory of a two-site exchange case will be given followed by an example of a multisite problem. The purpose of this approach is to show that for simple systems rate-data may be obtained with very little mathematical or experimental effort, and that almost amazing insight may be gleaned from complex chemical processes using NMR.

6.7.2.1 Two-Site Exchange

The early theoretical treatment of the effect of chemical exchange on NMR spectra was given by Gutowsky, McCall, and Slichter and has become known as the GMS theory (51). These authors approached the description of chemical dynamics by considering the effect on the Bloch equations given in Section 2.5. The Bloch equations in their original form described the equations of motion of a macroscopic moment of nuclei in the presence of a magnetic field B_0. If the exchange between two sites of different magnetic (and perhaps chemical) environment are to be considered, a Bloch equation for each magnetically nonequivalent site must be considered. A complex moment defined as

$$G = u + iv \tag{6.19}$$

where u and v are the respective transverse components of the magnetic moments along and perpendicular to a rotating field B_1. The absorption intensity is proportional to the component v. If there is no exchange between the two magnetically nonequivalent sites A and B, separate Bloch equations may be written for each site.

$$\frac{dG_A}{dt} + \alpha_A G_A = -i\gamma B_1 M_{0A} \tag{6.20}$$

$$\frac{dG_A}{dt} + \alpha_B G_B = -i\gamma B_1 M_{0B} \tag{6.21}$$

where

$$\alpha_A = \frac{1}{T_{2A}} - i(\omega_A - \omega) \tag{6.22}$$

$$\alpha_B = \frac{1}{T_{2B}} - i(\omega_B - \omega) \tag{6.23}$$

and T_{2A} and T_{2B} are the spin–spin relaxation times of nuclei at the respective sites. The Bloch equations must be modified to account for exchange between the two sites. Some assumptions are made. It is assumed that nuclei remain at one site until a sudden jump is made to a nonequivalent site and the nuclear moment during the exchange is neglected. Exchange between equivalent sites will have no effect. The term "interchange" is sometimes used to emphasize that only exchange between magnetically nonequivalent sites are observable. It will be assumed that the mean residence time of a given nucleus on a particular site is constant and expressed as τ_A and τ_B for sites A and B, respectively. A statistical correction may be necessary if the two sites are not equally populated;

$$\frac{P_A}{P_B} = \frac{\tau_A}{\tau_B} \tag{6.24}$$

where P_A and $P_B (= 1 - P_A)$ are the fractional populations of sites A and B. McConnell (52) presented a simplified form of the GMS theory giving a modified form of the Bloch equation:

$$\frac{dG_A}{dt} + \alpha_A G_A = -i\gamma B_1 M_{0A} + \frac{G_B}{\tau_B} - \frac{G_A}{\tau_A} \tag{6.25}$$

$$\frac{dG_B}{dt} + \alpha_B G_A = -i\gamma B_1 M_{0B} + \frac{G_A}{\tau_A} - \frac{G_B}{\tau_B} \tag{6.26}$$

The last two terms on the right-hand side represent the modification. Qualitatively, G_B/τ_B in equation 6.25 represents the increase in magnetization (G_A) resulting from the transfer of spin (nuclei) from site B to site A at a mean rate of τ_B^{-1} s^{-1}. G_A/τ_A represents the decrease in magnetization at site $A(G_A)$ resulting from the transfer of spins from site A to site B. The reverse is the case for equation 6.26.

Before proceeding with equations, a pause to examine the chemical significance of these terms may be helpful. The Bloch equations describe the time dependence of the magnetic moment of a nucleus precessing in a magnetic field as a function of several parameters. One parameter which will be seen to be important is $(\omega_0 - \omega)$, the frequency difference between the resonance frequency (ω_0) and the frequency at which the spectrometer is scanning at any instant (ω). Remember, the complex moment G contains a component v which is proportional to the absorption intensity at any frequency of observation (ω). The effect of nuclei exchanging between two nonequivalent sites is to alter the value of v for that nucleus. The way this exchange will affect the spectrum in this two-site case can be seen by solving for v for values of τ_A, τ_B, and ω; that is to calculate a theoretical resonance line. As a specific example, consider the exchange of protons between an alcohol and water represented by

$$\text{RO}\bar{\text{H}} + \text{HOH} \rightleftharpoons \text{ROH} + \bar{\text{H}}\text{OH} \tag{6.27}$$

where $\bar{\text{H}}$ labels a particular proton. (Obviously there are several possible exchange mechanisms (53)). If we call the alcohol proton site A and the water proton site B, each interchange between these sites will affect the absorption intensity of the alcohol and water protons in a way described by the solution of equations 6.25 and 6.26 for the v component.

Two very significant points should be made. First, the exchange process is assumed to be at steady state. This permits the spectrum to be recorded on a solution at equilibrium. Second, if a small rf field (B_1) is used, there will be no change in the total magnetization during a slow sweep of the spectrum (slow passage conditions) and

$$\frac{dG_A}{dt} = \frac{dG_B}{dt} = 0 \tag{6.28}$$

Solving equations 6.25 and 6.26 for G_A and G_B, the total complex moment G is

$$G = -i\gamma B_1 M_0 \frac{\tau_A + \tau_B + \tau_A \tau_B (\alpha_A P_A + \alpha_B P_B)}{(1 + \alpha_A \tau_A)(1 + \alpha_B \tau_A) - 1} \tag{6.29}$$

as was originally obtained by Gutowsky, McCall, and Slichter (51). Of more practical interest to chemists using high-resolution NMR spectrometers in the absorption mode is the "imaginary part" of G, that is the v or absorption mode component of G. Gutowsky and Holm (54) found this to be

$$v = \frac{\omega M_0 [(1 + \tau T_2^{-1})P + QR]}{P^2 + R^2} \tag{6.30}$$

where

$$P = \tau\{T_2^{-2} - [\tfrac{1}{2}(W_A + W_B) - W]^2 + \tfrac{1}{4}(W_A - W_B)^2\} + T_2^{-1} \tag{6.31}$$

$$Q = \tau[\tfrac{1}{2}(W_A + W_B) - W - \tfrac{1}{2}(P_A - P_B)(W_A - W_B)] \tag{6.32}$$

$$R = [\tfrac{1}{2}(W_A + W_B) - W](1 + 2\tau T_2^{-1}) + \tfrac{1}{2}(P_A - P_B)(W_A - W_B) \tag{6.33}$$

Equation 6.30 may be used directly to calculate the intensity (v) versus frequency (ω) for various values of τ, a simple process using a digital computer. However, the rate of exchange may be considered in three rather arbitrary groups of slow, fast, and intermediate. The first two cases permit some simplifying assumptions that are of value. Before considering these, the reader is reminded that equation 6.30 was specifically derived for an exchange between two nonequivalent sites and is not general. The case of intermediate exchange rates will be considered first and simplifying approximations shown for the limits of fast and slow exchange.

The intermediate exchange rate condition will be considered first as it is the most general in that no approximations may be made in equation 6.30. Consideration of this condition first will lead to the other categories. Several authors (50,55) have presented solutions to equation 6.30 using the assumptions

$$P_A = P_B = \tfrac{1}{2} \quad \text{(equal population of sites)} \tag{6.34}$$

so that

$$\tau_A = \tau_B = 2\tau \quad \text{(Note: } \tau \text{ is half the mean lifetime of each site)} \tag{6.35}$$

and

$$\frac{1}{T_{2A}} = \frac{1}{T_{2B}} \cong 0 \tag{6.36}$$

The last assumption is that the transverse relaxation times are long; that is, the linewidth in the absence of exchange is small compared to the separation between the lines. Pople, Schneider, and Bernstein show calculated line shapes for this case for illustration (55). However, this assumption would not be made in most real applications. Figure 6.8 shows some line-shapes calculated from equation 6.30 for various parameters, as shown on the figure. By comparing experimental spectra of systems undergoing simple two-site exchange reactions, the value of τ may be determined. This method has become known as "line shape analysis" although further development to the N-site exchange case to be discussed and called "complete line shape analysis" has further important connotations.

As Figure 6.8 shows, there are extremes of τ values that lie on either end of a central value. First, consider the central case in which $2\pi\tau(v_A - v_B) = \sqrt{2}$. At this point the two peaks representing the resonance lines have coalesced into a broad single peak. In practice, this region is the most accurate in determining rate-data because the line shape is most sensitive to changes in τ. The extreme values of rate are defined with respect to $2\pi(v_A - v_B)$. When $\tau > 10/2\pi(v_A - v_B)$ the spectrum shows two fully resolved lines and the exchange is "slow." When $\tau < 0.1/2\pi(v_A - v_B)$ the spectrum shows a single sharp resonance at a chemical shift equal to $P_A v_A + P_B v_B$ and the exchange is "fast." The range of τ values which may be determined is then seen to be from approximately $0.1/2\pi(v_A - v_B)$ to $10/2\pi(v_A - v_B)$ or roughly a factor of 10^2, with the central value of τ at $\sqrt{2}/2\pi(v_A - v_B)$. This rule of thumb is very helpful in a prior estimate of whether NMR may be applied to a problem as well as in the initial adjustments of parameters such as pH or temperature once some information concerning the system has been obtained. The range of rates available is sometimes called the "NMR kinetic window."

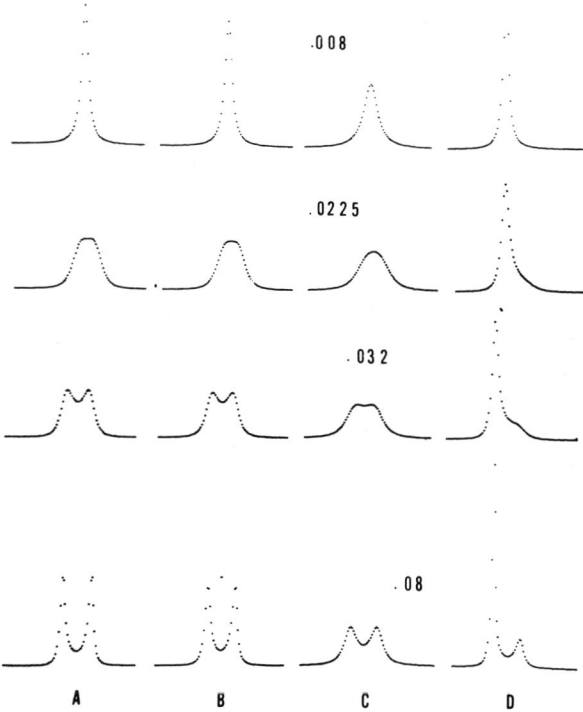

Fig. 6.8 Calculated NMR line-shapes for a two-site chemical exchange. The chemial shift difference between the two sites is 10 Hz and the values of τ (seconds) used in the calculations are shown. (A) $T_2 = 15$ sec; $P_a = 0.5$. (B) $T_2 = 1$ sec; $P_a = 0.5$. (C) $T_2 = 0.1$ sec; $P_a = 0.5$. (D) $T_2 = 1$ sec; $P_a = 0.75$.

In the limit of slow exchange, the linewidth at half maximum (assuming a lorentzian line) is given by (56)

$$(\Delta v)_{\frac{1}{2}} = \frac{1}{\pi T_2} + \frac{(1 - P_A)}{\tau} \qquad (6.37)$$

which holds for either line by appropriately substituting either P_A or P_B. One can view the term $(1 - P_A)/\tau$ as a quantitative representation of the contribution that random spin exchanges make to the dephasing of nuclear spins. The term $1/T_2$ is a "blank" correction for all other processes leading to line-broadening including magnetic field inhomogeneity and here are assumed equal for both sites and independent of the exchange processes. This is obviously an exceedingly simple method of obtaining rate data; however, caution should be used in extending the method beyond the limits in which the approximations are valid.

A second technique of obtaining τ in the slow exchange limit was introduced by Gutowsky and Holm (54). By setting the derivative of v with respect to $\frac{1}{2}(W_A + W_B) - W$ in equation 6.30 equal to zero, the expression

$$\frac{1}{\tau} = \frac{1}{\sqrt{2}}(\delta_\omega - \delta_{\omega_e})^{\frac{1}{2}} \qquad (6.38)$$

was obtained where $\delta_\omega = 2\pi(v_a - v_B)$ is the separation in the limit of no exchange and δ_{ω_e} is the corresponding value in the presence of exchange. The apparent peak separation, δ_{ω_e}, decreases as τ decreases as a result of each peak overlapping the other as they become broader. A plot of computer calculated ratios of $\delta_{\omega_e}/\delta_\omega$ versus τ for various values of T_2 can serve as a working curve. This method may be useful in some applications but is of limited general value.

Work done before the availability of computers to almost all laboratories stimulated other techniques of obtaining τ-values in the slow exchange limit. The intensity ratio method is a further example (56). From equation 6.30, it can be found that the ratio (R) of the peak intensities to the intensity midway between the peaks of a symmetrical two-site exchange $(P_a = P_B)$ case is related to τ by

$$\frac{1}{\tau} = \pm \frac{\delta_\omega}{\sqrt{2}}[(R \pm (R^2 - R)^{\frac{1}{2}}]^{-\frac{1}{2}} \qquad (6.39)$$

Again, as in the peak separation method, working curves must be computed. This method suffers the same limitations but gives more accurate values of τ than the peak separation method; probably because the intensity ratios are more precisely measured than peak separations for almost coalesced lines.

As the value of τ decreases (the exchange rate increases) the NMR spectrum of a two-site exchange coalesces into a single line and becomes increasingly narrow. In this region (57)

$$\frac{1}{\tau} = \frac{(\delta_\omega)^2}{4}\left[\frac{1}{T_2'} - \frac{1}{T_2}\right]^{-1} \qquad (6.40)$$

where $\delta_\omega = 2\pi(v_a - v_B)$ and T_2' is the contribution of the exchange to the transverse relaxation. This equation assumes $P_A = P_B$ and $T_{2A} = T_{2B}$. In principle equation 6.40 should permit a larger range of rates to be measured than stated previously. However, as T_2' approaches T_2, the linewidth of a slow passage line cannot be used to determine T_2'. The use of pulse or rapid passage methods of T_2 measurement can extend this range.

The previous example of the two-site exchange theories were presented primarily to provide a conceptual view of the applications of high resolution (slow passage) NMR techniques and spectrometers to the study of chemical kinetics. The same type of formulation may be used for exchange processes which lead to the coalescence of weakly coupled spin–spin doublets (58); quantum corrections may be required in some cases (59). Although similar closed solutions have been given for three-site (60–62), four-site (63), triplets (64), quartets (65), AB patterns (66,67) and a proton–deuterium coupled (68) system, this approach has been shown in recent years to be of limited value. The primary reasons for this is that new theoretical approaches to line-shape calculations have developed and the availability of computer hardware and programs permit great flexibility in designing varied solutions to exchange problems. The next section represents a transition of almost twenty years in the theory of NMR line-shape calculations.

6.7.2.2 Complete Line-Shape Analysis for Multisite Exchange

The most general treatment of multisite chemical exchange theories is based on writing the modified Bloch equations in a "density matrix" which is a format for expressing the relative probabilities of a set of possible exchange paths (48,49,66,67). An introduction of this concept, including specific examples has been given by Johnson and Moreland (69). The use of this approach was developed independently by Kubo (70) and Sack (71). The equation used may be represented as (69):

$$I(v) \propto Re(\mathbf{P} \cdot \mathbf{A}^{-1} \cdot \mathbf{1}) \tag{6.41}$$

where $I(v)$ is the intensity as a function of frequency, $\mathbf{P}$ is a row vector of the N fractional populations of the nuclear sites, $\mathbf{A}$ is an $N \times N$ complex matrix and $\mathbf{1}$ is a column unit vector. Re represents the "real part" of the result. For an N site exchange, there would be N equations of the form of equation 6.41 which may be represented as

$$\frac{dG_i}{dt} + \alpha_i G_i = -i\gamma B_1 M_{0i} - \frac{G_i}{\tau_i} + \sum_{j \neq i} \frac{P_{ij} G_j}{\tau_j} \tag{6.42}$$

where P_{ij} is the probability of an exchange occurring from site j to site i. The relative equilibrium magnetization M_{0i} is proportional to the fractional population of that site $P_i(\sum_{i=1}^{N} P_i = 1)$. Retaining the steady-state assumption ($dG_i/dt = 0$), equation 6.42 written in the matrix representation would be

$$i\gamma B_1 M_0 \begin{pmatrix} P_a \\ P_B \end{pmatrix} = \begin{pmatrix} -\left(\alpha_A + \frac{1}{\tau_A}\right) & +\frac{1}{\tau_B} \\ +\frac{1}{\tau_A} & -\left(\alpha_B + \frac{1}{\tau_B}\right) \end{pmatrix} \begin{pmatrix} G_A \\ G_B \end{pmatrix} \tag{6.43}$$

after taking the transpose of both sides:

$$i\gamma B_1 M_0 (P_A P_B) = (G_A G_B) \begin{pmatrix} -\left(\alpha_A + \dfrac{1}{\tau_A}\right) & +\dfrac{1}{\tau_A} \\ +\dfrac{1}{\tau_B} & -\left(\alpha_B + \dfrac{1}{\tau_B}\right) \end{pmatrix} \quad (6.44)$$

Because $I(v) = Im(\mathbf{G} \cdot \mathbf{1})$, equation (6.44) is multiplied from the right by $\mathbf{A}^{-1}$ followed by $\mathbf{1}$ to obtain

$$I(v) = \gamma B_1 M_0 \, Re(P_A P_B) \begin{pmatrix} -\left(\alpha_A + \dfrac{1}{\tau_A}\right) & +\dfrac{1}{\tau_A} \\ +\dfrac{1}{\tau_B} & -\left(\alpha_A + \dfrac{1}{\tau_A}\right) \end{pmatrix} \begin{pmatrix} 1 \\ 1 \end{pmatrix} \quad (6.45)$$

This method can be readily generalized (72) and a transformation matrix which diagonalizes the $\mathbf{A}$ matrix can be generated (73) which provides for efficient computer simulation of the spectrum. To perform this generalization, $\mathbf{A}$ is defined as

$$\mathbf{A} = \mathbf{\Omega} + \mathbf{D} \quad (6.46)$$

when

$$\Omega_{ij} = -\delta_{ij}\alpha_i \quad (6.47)$$

$$D_{ij} = \frac{P_{ij}}{\tau_i} \quad (j \neq 1) \quad (6.48)$$

$$D_{ii} = -\frac{1}{\tau_i} \quad (6.49)$$

and δ_{ij} is the Kronecker delta. The $\mathbf{A}$ matrix may then be constructed as

$$\mathbf{A} = \begin{pmatrix} -a_1 & 0 & \cdots & 0 \\ 0 & -a_2 & & \vdots \\ \vdots & & & \\ 0 & \cdots & & -a_N \end{pmatrix} + \begin{pmatrix} P_{11} & P_{12} & \cdots & P_{1N} \\ P_{21} & P_{22} & & \vdots \\ \vdots & & & \\ P_{N1} & \cdots & & P_{NN} \end{pmatrix} \times \frac{1}{\tau} \quad (6.50)$$

where $1/\tau$ represents the total rate of nuclear exchange of all sites. The D matrix is the probability distribution of exchanges between particular sites. Using the general solution devised by Gordon and McGinnis (73), equation (6.50) may be efficiently solved by a computer as a function of v, τ, and P_{ij}.

The first problem is to construct the **D** matrix by obtaining the probability elements P_{ij} for a particular reaction scheme. Once this matrix is constructed, various values of τ are chosen and the spectrum simulated by calculation of $I(v)$ over the range of v values which encompasses the experimental spectrum. Johnson and Moreland have written a set of rules for obtaining the **D** matrix and give examples (69). On comparison of the simulated spectra with the experimental spectra over a range of exchange rates (brought about by changes in temperature, concentration, etc.), one decides if they agree. If they do not, a different reaction scheme is chosen and tried. This approach provides a tremendously powerful tool for elucidating the reaction schemes of very fast and complex reactions such as ring rearrangements, hydride shifts, and multischeme exchanges such as proton exchange reactions. Further specific examples will be given in Chapter 7 and are given by Johnson (48).

6.7.3 TRANSIENT METHODS OF RATE STUDIES

The techniques of application of NMR previously described were for the most part slow passage methods. An exception is the transient observation of signal decay in the double resonance method of Forsén and Hoffman, discussed in Section 6.7.1 (46,47). The use of pulse techniques were limited in early applications primarily because of instrumental limitations. Commercial instruments were available, but difficult to justify as separate instruments for pulse NMR studies. Several modifications of high-resolution instruments were made (74,75). However, with the advent of commercial spectrometers capable of cw high resolution and FT work, various pulsed experiments may be performed more readily. A second consideration in the use of pulsed NMR methods is the limited chemical systems to which it may be applied. This limitation is a direct result of the nature of pulsed experiments.

The principle of the technique (Section 3.9) is that when a strong rf field B_1, with a frequency equal to the resonance frequency of the nuclear spins is applied, the nuclear magnetization is described by the time-dependent Bloch equations. The amplitude and duration of the pulse determines the degree to which the magnetic moment vector is rotated about the x axis. For this reason, the pulses are quantitatively referred to as π or $\pi/2$ pulses. For example, after a $\pi/2$ pulse, the magnetization lies along the y axis in the rotating frame, and decays with a decay time-constant determined by processes that effect T_2 including nuclear exchange. By analysis of the free induction decay of the signal after the pulse, the net T_2' may be determined. As it happens, some clever tricks involving a sequence of pulses can be used to minimize field inhomogeneity and diffusion effects. These sequences are those described by Carr and Purcell (76) and modified by Meiboom and Gill (77). Application of these pulse sequences are followed by a series of magnetization bounces or echos that diminish with time and hence the term "spin-echo" is applied to those methods.

The limitation of spin-echo methods lies in the fact that the strong rf field used for the pulses will have a wide range of frequency components and hence interact with nonexchanging nuclei (of a given isotope) as well as exchanging nuclei. As a result, complicated decay functions are obtained when magnetically nonequivalent nuclei are present. In the extreme, this means that only one type of nucleus (proton for example) may be present in the molecule under study, a serious limitation when organic compounds are investigated. At least one example has been worked out for a case of two nonequivalent protons (78). There are two possibilities to expand the applications. One is to use isotopic substitution for all but the chemical site of interest, for example deuterium for protons. An alternative is to label the exchanging site with deuterium and use an rf pulse at the ^{2}H resonance frequency. The second, and the one which has lead to the most frequent use of spin-echo methods, is to observe one type of proton which is present in a large excess over all others. For example, protolysis exchange reactions of amines have been studied extensively by observing the spin-echo of the exchanging solvent water protons in solutions so dilute that the nonexchanging amine protons do not contribute significantly to the spin-echo signal (79). A second example is the study of paramagnetic metal binding by macromolecules in water, again in dilute solutions (80,81). With the availability of the Fourier transform instruments, more applications of the pulse methods are expected as T_1 and T_2 may be determined for each peak.

The principle advantage of spin-echo techniques should be specifically pointed out. In general, faster reactions may be investigated using these methods. Reaction rates at which "exchange narrowing" of high resolution lines leads to linewidths too narrow to be measured accurately can be studied in many cases using spin-echo. Using the techniques previously described for investigation of protolysis reactions offers a secondary advantage. To observe high-resolution spectra of the amine directly so that line-shape analysis may be performed requires a concentration sufficiently high as to cause viscosity effects on T_2 and diffusion controlled rate constants, and salt effects on the rate of exchange. These effects are minimized using dilute solutions and observing the spin-echo signal of the solvent (provided the solvent participates in the exchange either as a molecule or atomic exchange).

A comparison of spin-echo and slow passage techniques for a study of the internal rotation of N,N-dimethyltrichloroacetamide has been reported (82). The values of the activation energy E_a and the preexponential factor A were found to differ greater than predicted from the standard deviation of the respective methods. Furthermore a systematic difference of rate was observed. The cause of these differences has yet to be completely explained.

6.7.4 EXPERIMENTAL PROCEDURES FOR RATE STUDIES

To obtain reliable rate data for chemical kinetic investigations can be a challenge regardless of technique used. Many parameters which may effect the results have to be considered. This is equally true when using the various NMR methods. It is not possible to describe all the various procedures in detail here. Specific examples from the literature will be given in Chapter 7 and experimental details can be found in the literature references given there. However, comments will be given for those workers attempting these studies for the first time. The comments will be limited to the use of high resolution methods as they are the most frequently used to date. However, the general principles apply to the other methods as well.

The first step is to become knowledgeable of the literature so that the importance of as many parameters as possible can be understood. Some of these will be apparent, others more subtle. Table 6.1 shows some parameters to consider and the effects they may have on results; the table is by no means complete. Temperature can have many effects. It is important to take special precautions to measure and control the sample temperature. In variable temperature experiments, each temperature setting must be given adequate time to stabilize; this may require more than an hour when large changes are made. Temperature will of course effect the rate of the chemical exchange process, but it may also effect the results indirectly by changing the viscosity, diffusion coefficients, equilibrium constants, instrumental

TABLE 6.1. Parameters and Their Effects in Rate Studies Using High Resolution (Slow Passage) NMR

Parameter	Potential Effects
Temperature	η, T_2, D, rate, equilibrium constants, instrument tuning
Concentration	η, T_2, D, J, v, salt effects, μ
rf intensity (B_1)	Saturation
Impurities	Exchange catalysis, participation in exchange

η = viscosity.
D = diffusion coefficients.
J = spin–spin coupling constant.
v = chemical shift.
μ = ionic strength.

tuning and T_2 (viz. T_2 may change with viscosity). Similarly, changes in concentration may effect such things as viscosity, T_2, spin–spin coupling constants, chemical shift, ionic strength, and salt effects in ionic reactions. The value of B_1 used to obtain spectra should be as low as possible to avoid line broadening, which may be present at B_1 values much lower than that required to show significant changes in the amplitude of the signal. Of course, the presence of impurities is an important consideration. It is best not to add any material to the reaction mixture which is not pertinent to the reaction as extraneous reagents may have an unknown effect. The experimental spectra should be taken with a moderate sweep rate (< 1 Hz sec^{-1}) and the useful portion of the spectrum expanded on the frequency and amplitude axis as much as practical. When complete line shape methods are used, it is helpful to take the time to prepare the computer program such that the plot of simulated spectra are on the same scale as the experimental spectra. In this way visual comparisons are much easier to make. Most line shape simulation programs may be written or modified to iterate τ values. If the experimental spectrum is encoded into the computer, τ values may be iterated to some predetermined condition. For example, all experimental and simulated intensities may be normalized relative to the maximum intensity in each spectrum. One arbitrary iteration is to vary τ such that a minimum sum of the squares of the difference between experimental and simulated intensities is obtained as a residual. The value of this residual is a measure of the "fit." Considerable caution must be used however to select the intensity points carefully to cover spectral regions sensitive to τ. The method should never be used without the aid of plotted simulated spectra for visual comparison as a "fit" for τ may be obtained using incorrect spectral parameters.

6.8 CONCLUSION

In this chapter we have attempted to explain some of the fundamental principles on which many of the varied chemical applications of NMR spectroscopy are based. Many volumes have been written on the subject, and as yet no single work has encompassed all applications of NMR techniques. It would be foolish to attempt to do so. Many of the techniques outlined in this chapter are now standard practice in many chemical laboratories. With the availability of inexpensive, powerful, on-line laboratory computers and FT techniques, NMR is likely to develop as rapidly in the next 10 years as it has in the past 10 years. Less abundant isotopes such as ^{13}C and ^{14}N are expected to play a major role in the NMR spectroscopy of the future. Fortunately, the principles remain essentially the same so that a well founded basis in proton magnetic resonance techniques may be extrapolated into these new methods.

REFERENCES

1. H. S. Gutowsky and H. Saika, *J. Chem. Phys.*, **21**, 1688 (1953).
2. G. D. Brabson, *J. Chem. Ed.*, **46**, 754 (1969).
3. G. C. Hood and C. A. Reilly, *J. Chem. Phys.*, **27**, 1126 (1951).
4. J. V. Mengenhauser, U.S. Clearinghouse Fed. Sci. Tech. Inform., AD-711892 (1970).
5. J. L. Sudmeier, K. E. Schwartz, and A. J. Senzel, *Inorg. Chem.*, **8**, 2815 (1969).
6. D. J. Martin and R. H. Pearce, *Anal. Chem.*, **38**, 1604 (1966).
7. L. K. Keefer, L. Wallcone, J. Loo, and R. S. Peterson, *Anal. Chem.*, **43**, 1411 (1971).
8. Dupont Model 310 Curve Resolver, Dupont Instrument Company.
9. R. J. Day and C. N. Reilley, *Anal. Chem.*, **38**, 1323 (1966).
10. R. G. Pearson, J. Palmer, M. M. Anderson, and A. L. Alred, *Z. Elektrochem.*, **64**, 110 (1960).
11. D. E. Leyden and J. F. Whidby, *Anal. Lett.*, **1**, 417 (1968).
12. C. A. Reilley, *Anal. Chem.*, **30**, 839 (1958).
13. R. B. Williams, *Ann. N. Y. Acad. Sci.*, **70**, 890 (1958).
14. R. R. Ernst, in *Advances in Magnetic Resonance*, J. S. Waugh, Ed., Vol. 2, Chap. 1, Academic, New York, 1966.
15. P. J. Paulsen and W. D. Cooke, *Anal. Chem.*, **36**, 1713 (1964).
16. T. Sato and Y. Mikami, *Kog. Kag. Zasshi*, **68**, 1401 (1965).
17. T. G. Alexander and S. A. Koch, *App. Spectrosc.*, **21**, 181 (1967).
18. L. W. Reeves, *Can. J. Chem.*, **35**, 1351 (1957).
19. D. P. Hollis, *Anal. Chem.*, **35**, 1682 (1963).
20. T. G. Alexander and S. A. Koch, *J. Assoc. Offic. Agr. Chem.*, **50**, 676 (1967).
21. J. L. Jungnickel and J. W. Forbes, *Anal. Chem.*, **35**, 938 (1963).
22. R. E. Mayo and J. H. Goldstein, *Rev. Sci. Instr.*, **35**, 1231 (1965).
23. F. Kasler, *Quantitative Analysis by NMR Spectroscopy*, Academic, New York, 1973.
24. G. Dulaney, *Anal. Chem.*, **47** (1), 25A (1975).
25. R. D. Waymack and D. R. Bowman, *Radio. Electron. Eng.*, **41**, 421 (1971).
26. R. G. Jones, P. Partington, B. W. Ready, and T. Trill, *J. Phys. E.*, **5**, 44 (1972).
27. R. R. Ernst, *Rev. Sci. Instr.*, **39**, 998 (1968).
28. G. Beech, *Amer. Lab.*, **5**, 53 (1973).
29. A. Carrington and A. D. McLachlan, *Introduction to Magnetic Resonance*, Harper & Row, New York, 1967, Chap. 11.
30. Praxis Corporation, San Antonio, Texas, Model PR-103.
31. Bruker-Physik AG, Karlsruhe-Forcheim, Am Silberstreifen, West Germany, Mini Spec, P-20.
32. L. O. Anderson, *J. Am. Oil Chem. Soc.*, **48**, 47 (1971).

33. H. Reinhardt, *Bergbautechnik*, **20**, 29 (1970); **20**, 81 (1970).
34. W. L. Rollwitz and G. A. Persyn, *J. Am. Oil Chem. Soc.*, **48**, 59 (1971).
35. G. E. Hall, J. G. Lawrence, and R. J. Simpson, *Nature*, **216**, 474 (1967).
36. A. L. Bloom and J. N. Shoolery, *Phys. Rev.*, **90**, 358(A), 1953.
37. G. Suryan, *Proc. Indian Acad. Sci.*, **33A**, 107 (1951).
38. Badger Meter Manuf., Milwaukee, Wis., Brit. Pat. 1,240,594, July, 1971.
39. M. C. McIvor, *J. Sci. Instrum.*, Sec. 2, **2**, 292 (1969).
40. A. I. Zhernovoi and G. D. Latyshev, *NMR in a Flowing Liquid*, Consultants Bureau, N.Y., 1965.
41. L. W. Reeves and W. G. Schneider, *Can. J. Chem.*, **36**, 793 (1958).
42. E. Grünwald, A. Loewenstein, and S. Meiboom, *J. Chem. Phys.*, **27**, 641 (1957).
43. N. E. Rigler, S. P. Bag, D. E. Leyden, J. L. Sudmeier, and C. N. Reilley, *Anal. Chem.*, **37**, 872 (1965).
44. E. W. Firth, D. J. E. Ingram, *J. Sci. Instrum.*, **44**, 821 (1967).
45. T. C. Farrar and E. D. Becker, *Pulse and Fourier Transform NMR*, Academic, New York, 1971, p. 82.
46. S. Forsén and R. A. Hoffman, *Acta. Chem. Scand.*, **17**, 1787 (1963).
47. S. Forsén and R. A. Hoffman, *J. Chem. Phys.*, **40**, 1189 (1964).
48. C. S. Johnson, Jr., in *Advances in Magnetic Resonance*, J. S. Waugh, Ed., Vol. 1, Academic, New York, 1965, pp. 33–102.
49. R. M. Lynden-Bell, in *Progress in Nuclear Magnetic Resonance Spectroscopy*, J. W. Emsley, J. Feeney, and L. H. Sutcliffe, Eds., Vol. 2, Pergamon, New York, pp. 163, 1967.
50. J. W. Emsley, J. Feeney, and L. H. Sutcliffe, *High Resolution Nuclear Magnetic Resonance Spectroscopy*, Vol. 1, Pergamon, New York, 1965, Chap. 9.
51. H. S. Gutowsky, D. W. McCall, and C. P. Slichter, *J. Chem. Phys.*, **21**, 279 (1953).
52. H. M. McConnell, *J. Chem. Phys.*, **28**, 430 (1958).
53. Z. Luz, D. Gill, and S. Meiboom, *J. Chem. Phys.*, **30**, 1540 (1959).
54. H. S. Gutowsky and C. H. Holm, *J. Chem. Phys.*, **25**, 1228 (1956).
55. J. A. Pople, W. G. Schneider, and H. J. Bernstein, *High-Resolution Nuclear Magnetic Resonance*, McGraw-Hill, New York, 1959, p. 222.
56. M. T. Rogers and J. C. Woodbrey, *J. Phys. Chem.*, **66**, 540 (1962).
57. M. Anbar, A. Loewenstein, and S. Meiboom, *J. Am. Chem. Soc.*, **80**, 2630 (1958).
58. H. S. Gutowsky and A. Saika, *J. Chem. Phys.*, **21**, 1688 (1953).
59. E. Grünwald, C. F. Jumper, and S. Meiboom, *J. Am. Chem. Soc.*, **84**, 4664 (1962).
60. H. M. McConnell and S. B. Berger, *J. Chem. Phys.*, **27**, 230 (1957).
61. A. Patterson and R. Ettinger, *Z. Elecktrochem.*, **64**, 98 (1960).
62. T. J. Swift and R. E. Connick, *J. Chem. Phys.*, **37**, 307 (1962).
63. A. Carrington, *Mol. Phys.*, **5**, 425 (1962).
64. J. T. Arnold, *Phys. Rev.*, **102**, 136 (1956).

65. E. Grünwald, A. Loewenstein, and S. Meiboom, *J. Chem. Phys.*, **27**, 630 (1957).
66. J. Kaplan, *J. Chem. Phys.*, **28**, 278 (1958).
67. S. Alexander, *J. Chem. Phys.*, **37**, 966 (1962).
68. R. J. Day and C. N. Reilley, *J. Phys. Chem.*, **71**, 1588 (1967).
69. C. S. Johnson, Jr. and C. G. Moreland, *J. Chem. Ed.*, **50**, 477 (1973).
70. R. Kubo, *Nuovo Cimento, Suppl.*, **6**, 1063 (1957).
71. R. A. Sack, *Mol. Phys.*, **1**, 163 (1958).
72. C. S. Johnson, Jr., *Am. J. Phys.*, **35**, 929 (1967).
73. R. G. Gordon and R. P. McGinnis, *J. Chem. Phys.*, **49**, 2455 (1968).
74. Z. Luz and S. Meiboom, *J. Chem. Phys.*, **39**, 366 (1963).
75. A. Ginsburg, A. Lipman, and G. Navon, *J. Phys. E.*, **3**, 699 (1970).
76. A. Y. Carr and E. M. Purcell, *Phys. Rev.*, **94**, 630 (1954).
77. S. Meiboom and D. Gill, *Rev. Sci. Instr.*, **29**, 688 (1958).
78. S. Alexander, *Rev. Sci. Instr.*, **32**, 1966 (1961).
79. E. K. Ralph and E. Grunwald, *J. Am. Chem. Soc.*, **89**, 2963 (1967).
80. M. Cohn and J. S. Leigh, *Nature*, **193**, 1037 (1962).
81. M. Cohn, *Biochemistry*, **2**, 623 (1963).
82. A. Allerhand and H. S. Gutowsky, *J. Chem. Phys.*, **42**, 1587 (1965).

CHAPTER

7

EXAMPLES OF ANALYTICAL APPLICATIONS

7.1	Introduction	299
	7.1.1 Literature Sources	300
7.2	Functional Group Analysis	300
	7.2.1 Determination of Hydroxyl Groups	301
	7.2.2 Determination of Carbonyl Groups	302
	7.2.3 Determination of Carboxylic Acids and Related Compounds	303
	7.2.4 Determination of Ethers, Epoxides, and Peroxides	305
	7.2.5 Determination of Olefins	305
	7.2.6 Determination of Acetylenic Hydrogen	312
	7.2.7 Determination of Amines	313
	7.2.8 Summary of Functional Group Analysis Using NMR	314
7.3	Industry Related Applications	316
	7.3.1 Synthetic Polymers	317
	7.3.1.1 Quantitative Determinations	317
	7.3.1.2 Structure Determinations	318
	7.3.1.3 Nuclei Other than ^{1}H in Polymer Studies	324
	7.3.1.4 Shift Reagents in Polymer Studies	325
	7.3.2 Coatings	326
	7.3.3 Foods	327
	7.3.3.1 Wide Line Methods	327
	7.3.3.2 Pulsed Methods	329
	7.3.3.3 High-Resolution ^{1}H and ^{13}C Methods	330
	7.3.4 Petroleum	331
	7.3.4.1 Crude Oil	333
	7.3.4.2 Hydrocarbons	333
	7.3.4.3 Summary	344
	7.3.5 Pesticides	344
	7.3.5.1 Organophosphorus Pesticides	346
	7.3.5.2 DDT-type Pesticides	352
	7.3.5.3 Applications of NMR to Pesticides Found in the Environment	353
	7.3.5.4 Future Applications of NMR to Pesticides	354
	7.3.6 Pharmaceuticals	354
	7.3.6.1 Quantitative Applications	355

7.3.6.2	Structure Elucidation	357
7.3.6.3	Molecular Interactions	359
7.3.6.4	Summary	362

7.4 Applications to Fundamental Investigations 362
    7.4.1 Applications to Biochemistry 362
        7.4.1.1 Binding of Small Molecules to Macromolecules 363
        7.4.1.2 Binding Studies Using Quadrupolar Nuclei 367
        7.4.1.3 Studies Using Paramagnetic Ions 371
        7.4.1.4 Applications of ^{13}C-NMR 375
        7.4.1.5 Biomolecular Conformation 377
    7.4.2 Applications to Studies of Structure, Equilibria, and Kinetics of Metal Complexes 381
        7.4.2.1 Protonation of Free Ligands 381
        7.4.2.2 Protonation of Metal Complexes 390
        7.4.2.3 Protolysis Kinetics of Free Ligands 395
        7.4.2.4 Protolysis Kinetics of Metal Complexes 396
        7.4.2.5 Structure and Bond Lability of Metal Complexes 399
        7.4.2.6 Exchange Reactions Involving Metal Complexes 407
        7.4.2.7 Conformational Analysis of Diamagnetic Metal Chelates 411
        7.4.2.8 Conformational Analysis of Paramagnetic Metal Chelates 414

7.1 INTRODUCTION

To say that nuclear magnetic resonance (NMR) is one of the most powerful tools the analytical chemist has available would probably receive little challenge. Few if any other techniques can claim the range of applicability of NMR, both in sample type as well as the kinds of information that may be obtained. The relative simplicity of the theory of NMR spectra permits chemists, untrained in quantum mechanics, to predict with considerable accuracy the NMR spectrum of a known structure, and to interpret with little ambiguity the spectrum of unknown compounds. NMR has been applied to the study of chemical equilibria and kinetics, electron exchange reactions, self-diffusion in liquids, correlation times in molecular motion, electron density distribution in molecules, molecules absorbed on surfaces, determination of liquids in solids, and the list could go on.

Yet as one thinks about the parameters of the technique, Nature almost seemed to be waiting for its discovery. Consider the probable limited use of ^{1}H-NMR if the ^{12}C isotope had a nonzero nuclear spin or ^{13}C had been the more abundant isotope. If nuclear relaxation times were a factor of ten larger the early experiments may have failed as indeed some using solids did. Fortunately these and other natural phenomena are all ideal for the nuclear magnetic resonance experiment.

The previous chapter attempted to present some of the principles of various quantitative measurements using NMR. However, as the previous paragraphs imply, it is an impossible task to completely describe even the areas of application of NMR, not to mention specific examples. This chapter then will attempt to select representative examples of applications that are considered to be "analytical." As discussed in the first chapter, our definition of analytical encompasses a wide variety of studies.

7.1.1 LITERATURE SOURCES

Because this chapter cannot contain an exhaustive review of the applications of NMR, a guide to other sources of information is worthwhile. The annual reviews published by the journal *Analytical Chemistry* are the most comprehensive periodic reviews. On alternate years fundamental and applications reviews are published in April. The operational difference between these in the past has been that the fundamental reviews are organized by technique, and the applications reviews are organized by fields of study or source of sample. For example, if one is interested in the applications of NMR to pharmaceuticals, he may look under that heading in the applications review issue, or under NMR in the fundamental review issue. These reviews tend to be exhaustive and not critical. However, in 1974 over 2,200 papers were screened from 10,000 abstracts concerning NMR.

There are several other reviews or periodic publications as well. Some are highly specialized chapters or papers (1,2), whereas others tend to cover a wider scope of NMR research (3–5). Several reviews have appeared on areas of application of NMR. Papers or books have appeared on the applications of NMR to pharmaceuticals (6–11), petrochemicals (12–14), pesticides (15,16), detergents (17), and one monograph on quantitative analysis by NMR has been published (18). An extensive list of texts, reviews, monographs, and symposia on both general and specific aspects of NMR appears in a quarterly review journal, *Magnetic Resonance Review* (19). Texts and reviews on special topics will be pointed out during the course of this chapter.

7.2 FUNCTIONAL GROUP ANALYSIS

One of the most common problems presented to the practicing analytical chemist is functional group determinations. Therefore, examples of the applications of nuclear magnetic resonance to these problems will be the first considered.

7.2.1 DETERMINATION OF HYDROXYL GROUPS

The simplest and most direct determination of hydroxyl groups is based on replacement of the hydroxyl hydrogen with deuterium (20). The exchanged hydrogen then appears in the HDO line and may be determined from the area of this peak. In the event of "slow" exchange in which case the active hydrogen and the HDO show separate resonance lines, a 50-fold excess of deuterium is added to drive the exchange to completion. A qualitative use of this procedure is a common method of identification of active hydrogen resonance lines. If on the addition of D_2O, a resonance line is seen to decrease in area or disappear, an active hydrogen such as a hydroxyl, amine, carboxylic acid, etc. is suspected. Less labile active hydrogen such as those in esters, methyl ketones, amides, and alkynes may require a catalyst such as pyridine or LiOD. Relative errors using this technique range from near 0% to 5%. However, the method is subject to many interferences and can be used only when pure compounds are to be determined. Because of the strong intermolecular hydrogen bonding of hydroxyl groups, the chemical shifts of their resonance lines are highly temperature and concentration dependent. Therefore, changes in chemical shift, with temperature can confirm their presence. Amine hydrogen will behave in a similar way but usually shows broad lines. The chemical shift as a function of concentration may be used to estimate hydroxyl concentration. However, this method is also highly subject to interference from other compounds.

To increase the specificity of hydroxyl group determination, some form of derivatization is commonly used. This approach frequently has the fringe benefit of permitting the determination of whether the alcohols are primary, secondary, or tertiary. Manatt presented a method of utilizing the trifluoroacetic esters of alcohols (21). The derivatization is performed by the addition of trifluoroacetic anhydride to the neat alcohols (or in an inert solvent) using an excess of the anhydride. The reaction can be monitored by scanning the NMR spectrum and observing the disappearance of the hydroxyl resonance lines. The ^{19}F resonance is then taken. Relative to trichlorofluoromethane, the ester ^{19}F resonance will appear at lowest field for primary alcohols, with those for secondary and tertiary alcohols at progressively higher fields. Amines, phenols, and thiols will interfere. The requirement of ^{19}F NMR is somewhat of a nuisance. The use of dichloroacetic anhydride serves as a similar technique and has the advantage that ^1H-NMR may be used to observe the derivative (22).

An exceedingly simple method is to obtain the spectrum of alcohols in dimethylsulfoxide (23). In this solvent the exchange rate of the hydroxyl proton is sufficiently slow that spin–spin splitting of the hydroxyl proton resonance is observed. The presence of acids or bases may catalyze the exchange and coalesce the splitting pattern. Otherwise, primary alcohols show triplets (except for a quartet for methanol), secondary alcohols show doublets, and tertiary alcohols show singlets for the hydroxyl resonance. In each case the chemical shift is between 5.5 and 4.0 ppm relative to TMS. This method is particularly useful in identification of the source of hydroxyl resonance lines in alkaloids (as quaternary salts), steroids, and terpenes. Polyhydroxyl compounds show separate peaks for each hydroxyl group. The chemical shift may be sensitive to isomerism. For example, the doublets of *trans* and *cis*-4-*t*-butylcyclohexyl alcohol are separated by 0.34 ppm. A cautionary note has been issued when using this technique with alcohols that have strong electron withdrawing groups adjacent to the hydroxyl group (24).

Shift reagents are useful in the investigation of alcohols, particularly in the elucidation of structure. Sanders and Williams showed that complex spectra of alcohols such as benzyl alcohol and *n*-hexanol were reduced to simple first-order spectra by the addition of *tris*(dipivalomethanate)europium (25). Quantitative applications of this technique have been described (26).

The phenolic-hydroxyl group is more easily identified as the chemical shift of the hydroxyl hydrogen appears in the range 8–13 ppm relative to TMS in hexamethylphosphoramide (27). Several precautions must be taken and possible interferences are aldehydes (peak overlap) and acids.

7.2.2 DETERMINATION OF CARBONYL GROUPS

Aldehydes are normally easily determined using ^{1}H-NMR. The chemical shift of the formyl hydrogen resonance is in the range 9.6–10.4 ppm relative to TMS. This region of the spectrum is frequently free of other resonance lines and direct integration of the formyl hydrogen resonance may be performed.

Ketones are less easily determined by NMR because there is no direct proton signal. The use of ^{13}C-NMR with proton decoupling would appear to be an obvious choice. However, the chemical shift of the ^{13}C resonance of the carbonyl carbon varies widely and must be identified prior to integration. Cautions of using ^{13}C-NMR for quantitation have been given in Section 5.10.

Recognizing that infrared spectra may surpass the importance of NMR in the identification and determination of carbonyl groups, there are ^{1}H-NMR methods that may be useful. The most common method is the derivatization of the functional group, which has been discussed in detail (28). The obvious derivatives are hydrazones and many such derivatization reagents may be used. Methyl ketones are usually easily determined by integration of the methyl proton singlet.

An interesting possibility for ketone mixtures is the technique used by Sudmeier for "counting" peptide moieties (29). By dissolving the carbonyl compound in a super strong acid, protonation of the carbonyl oxygen is accomplished. If the proton exchange rate is sufficiently slow, a single resonance line is observed for each protonated carbonyl. Quantitation should be possible because the acidity required to sufficiently reduce the exchange rate is more than adequate to guarantee complete protonation of the carbonyl sites.

7.2.3 DETERMINATION OF CARBOXYLIC ACIDS AND RELATED COMPOUNDS

There are several approaches to the determination of carboxylic acids and the choice will be in part determined by the solvent and the presence of other compounds. An obvious and widely used method is the exchange of the active hydrogen for deuterium, as already discussed for alcohols in Section 7.2.1 (20). However, the solvent in which the sample is obtained plays an important role in this choice. In protic solvents, the ^{1}H-NMR resonance lines of the acidic proton may be coalesced with the solvent line and a mixture of acids may show only one merged resonance. In this case, the deuterium substitution technique will be of limited value. In aprotic solvents however, this method is of significant use.

The use of resonance lines originating from protons on the carbon in the alpha position is important in the determination of carboxylic acids. These proton resonance lines may be directly integrated. However, because there is considerable overlap between the chemical shifts of alpha-CH_2 and alpha-CH protons in primary and secondary acids considerable caution must be exercised in using this procedure. All of the procedures previously described become difficult in mixtures more complicated than binary.

There are some possibilities that can be explored when the sample form permits. For example, detailed studies have been conducted on "protonation shifts" of amines, carboxylic acids, and aminocarboxylic acids (30–32). The protonation shift is the chemical shift change observed for nonlabile (e.g., α-CH_2—) protons on protonation of a basic functional group with the addition of acid. The pH (aqueous solution) at which the protonation occurs depends on the pK_a of the functional group. Therefore, the control of pH can effect the chemical shift of the nonlabile protons and provide some control over overlapping peaks in mixtures of acids or bases with different pK_a values. The magnitude of the shift on protonation is an aid in assigning the resonance line. A second and perhaps more useful approach to effect chemical shift changes is the esterification of the acids and use of shift reagents to separate overlapping lines (33).

The use of the chemical shift of the merged solvent-acidic proton resonance lines in acetic acid–water mixtures has already been described in Section 6.1. This method is applicable to other binary acid–water mixtures if the samples are known to be that simple. Formic acid is unique and may be determined in small amounts by integration or peak height measurements of the formyl hydrogen, which is free from most spectral interferences (34).

Salts of carboxylic acids present a problem when NMR determinations are desired, as the methods based on active hydrogen replacement cannot be used unless the sample is first converted to the acid by ion-exchange or some other procedure. The use of an adjacent nonlabile hydrogen is usually the method of choice.

The determination of esters by NMR spectroscopy is normally accomplished by integration or peak height measurement. The complication with esters, however, is that frequently there is spectral overlap between the protons in the acid residue and those from the alcohol residue. Kan has investigated the effect of the polar substituent constant σ^* of the alcohol group on the chemical shift of the methyl protons (35). The methyl proton resonance of the acetic acid moiety showed shifts of about -1 Hz for higher n-alkyl esters to 14 Hz for aromatic esters relative to methyl acetate. It is easily seen that for mixtures of esters, peak overlap can be a significant difficulty. The use of shift reagents for the determination of mixtures of esters should prove valuable.

Amides of carboxylic acids have received a great deal of investigation using NMR spectroscopy. However, much of the attention has been given to the study of the hindered rotation about the carbon–nitrogen bond. Amide-NH proton resonance lines are observed between 5 and 9 ppm relative to TMS. These peaks are somewhat broad as a result of the effect of the nitrogen quadrupole moment on the relaxation time of the —NH proton and ^{14}N—^1H coupling. However, these protons are easily distinguished from amine protons by chemical shift (0–2 ppm for aliphatic amines and 2–5 ppm for aromatic amines) and the persistence of spin coupling between the —NH proton and protons on N-alkyl groups in monosubstituted amides. Hydrogen bonding of amide hydrogens is strong, therefore solvent, concentration, and temperature may affect the —NH proton chemical shift.

A useful method of quantitation was reported by Leader (36). This method is based on the formation of an adduct between hexafluoroacetone (HFA) and compounds that have active hydrogen atoms. In addition to amides, the technique may be applied to alcohols, mercaptans, amines, oximes, and other functional groups. Leader tabulated the ^{19}F chemical shifts (Δ) of the adducts measured relative to the HFA-water adduct. There is some correlation of these chemical shift values with the functional group forming the adduct. Some of these are shown in Table 7.1. Once identified, the ^{19}F

resonance line for the adduct of interest can be integrated. A fringe benefit is the six fluorine atoms per functional group that amplifies the sensitivity.

7.2.4 DETERMINATION OF ETHERS, EPOXIDES, AND PEROXIDES

This unusual classification of these functional groups is made for a simple reason. One characteristic of a good analytical chemist is that he does not force his problem to be solved by a given technique. In this case the facts are that functional groups such as the ones in this section are probably better determined by other analytical methods. This is not to say that NMR is useless when identification and/or determination of these functional groups is required. Obviously, the direct application of an NMR spectrum will aid in the elucidation of the structure of compounds bearing these groups. Some direct applications are possible.

For example, vinyl ethers are easily identified in the presence of other vinyl compounds because the ethylenic protons appear at a higher field in the ether than in other vinyl derivatives (37).

Organic peroxides have been determined in mixtures using proton resonance lines of protons either alpha or beta relative to the functional group (38). This method was applied to mixtures of peroxides, hydroperoxides, and alcohols.

7.2.5 DETERMINATION OF OLEFINS

The NMR spectra of olefinic compounds have been studied extensively. The technique is most applicable to the identification and determination of olefins when there is one or more protons on a carbon containing the double bond. Olefinic hydrogen atoms give NMR resonance lines in the range 4–7 ppm from TMS. Only protons on carbon atoms containing highly electronegative substituents or aromatic compounds interfere. When 1,2-disubstituted compounds are present, the *cis/trans* components can be easily determined from the difference in the splitting patterns of the vicinal protons. This is because J_{cis} lies in the range 6–12 Hz whereas J_{trans} shows values between 12 and 19 Hz (Section 4.12.2). The group electronegativity of substituents affect the value of these coupling constants. However, because of the minor overlap in the values of J_{cis} and J_{trans}, NMR is one of the fastest methods of elucidating *cis/trans* isomers in 1,2-disubstituted olefins.

Frequently, olefinic compounds give complicated NMR spectra because there is great opportunity for magnetic nonequivalence on substitution about the double bond. The spectrum must be analyzed by the methods given in Chapter 4 to obtain accurate chemical shift and coupling constant data.

TABLE 7.1A. Chemical Shifts Relative to HFA(H$_2$O) for Alcohol Adducts of HFA in Ethyl Acetate

		Δ (ppm)
1.	Primary alcohols	
	Methanol	2.68
	Ethanol	2.45
	1-Propanol	2.57
	1-Butanol	2.58
	2-Methylpropanol	2.62
	1-Pentanol	2.57
	2-Mercaptoethanol (OH)	2.50
	2-Mercaptopropanol (OH)	2.56
	2-Methyl-2-nitropropanol	2.67
	2-Phenylethanol	2.55
	Allyl alcohol	2.54
	2-Propyene-1-ol	2.47
	Benzyl alcohol	2.80
	Glycolic acid	2.27
	Trifluoroethanol	2.12
2.	Glycols and related compounds	
	Ethylene glycol	2.36 (1.36)[a]
	Ethylene glycol monomethyl ether	1.51
	Tetrahydrofurfuryl alcohol	1.49
	1,3-Propanediol	2.46
	1,4-Butanediol	2.60
	1,4-Butenediol	2.53
	2-Methyl-2-nitro-1,3-propanediol	2.54
	1,2-Propanediol	
	Pri. (OH)	2.52
	Sec. (OH)	1.70[b]
	Glycerol	
	Pri. (OH)	2.46 (1.97)
	Sec. (OH)	1.71
	Ethanolamine (OH)	2.38
	Diethanolamine (OH)	2.74 (1.95) [2.84][c]
	3-Amino-1-propanol (OH)	2.55
	2-Methylaminoethanol (OH)	2.50 [2.91]
	Dimethylaminoethanol	0.57 [3.1]
3.	Secondary alcohols	
	2-Propanol	1.80
	2-Butanol	1.94
	1,3-Dichloro-2-propanol	1.88
	Cyclohexanol	1.78
	Cyclopentanol	2.00
	1,4-Cyclohexanediol	1.87, 1.72
	Hexafluoro-2-propanol	1.99
	2,2,4,4-Tetramethyl-1,3-cyclobutanediol	2.49, 2.59
4.	Tertiary alcohols	
	tert-Butyl alcohol	1.20

TABLE 7.1A. (*Continued*)

		Δ (ppm)
	tert-Amyl alcohol	1.41
	Diacetone alcohol	1.35
	Citric acid	3.08
5.	Phenols	
	Phenol	2.73
	Hydroquinone	2.70
	Resorcinol	2.74 (2.61)
	o-Cresol	3.07
	o-Chlorophenol	3.04
	p-Cresol	2.71
	p-Chlorophenol	2.66
	4-Chloro-3-methylphenol	2.65

[a] Δ values in parentheses are for an additional HFA adduct line which appears when compound equivalent concentration exceeds that of HFA.
[b] Center of complex pattern.
[c] Δ values in brackets are for HFA adduct line position when TFA sufficient to neutralize the basic group is present.
Reprinted with permission from G. R. Leader, *Anal. Chem.*, **42**, 16 (1970).

TABLE 7.1B. Chemical Shifts Relative to HFA for Mercaptan Adducts of HFA in Ethyl Acetate

		Δ (ppm)
1.	Primary mercaptans	
	Ethanethiol	0.95
	1-Propanethiol	0.92
	1-Butanethiol	0.93
	2-Methyl propanethiol	0.87
	2-Mercaptoethanol (SH)	1.18
	3-Mercaptopropanol (SH)	1.03
	Ethyl thioglycolate	1.46
	Benzyl mercaptan	0.68
2.	Secondary mercaptans	
	2-Propanethiol	1.00
	2-Butanethiol	0.96
3.	Tertiary mercaptans	
	tert-Butyl mercaptan	1.51
4.	Aromatic mercaptan	
	Benzenethiol	0.04

Reprinted with permission from G. R. Leader, *Anal. Chem.*, **42**, 16 (1970).

TABLE 7.1C. Chemical Shifts for HFA Adducts of Amines in Ethyl Acetate

	Δ (ppm)
1. Aliphatic primary amines	
Methylamine	4.11
N-Propylamine	3.96
Isopropylamine	3.48
Benzylamine	3.93
α-Methylbenzylamine	3.52
Ethanolamine (NH_2)	3.60 (2.95)[a]
3-Amino-1-propanol (NH_2)	3.85
Monoisopropanolamine (NH_2)	3.86
β-Ethoxyethylamine	2.70
3-Methoxypropylamine	3.63
Ethylenediamine	
(As EDA·2HFA)	3.74
(As EDA·HFA)	2.55 [3.5][b]
Glycine	3.38
2. Aromatic primary amines	
Aniline	3.74
o-Chloroaniline	3.03
p-Chloroaniline	3.61
p-Aminophenol (NH_2)	3.74
p-Aminobenzoic acid	3.51
o-Aminobenzoic acid methyl ester	3.28
Sulfanilamide (NH_2)	3.50
2-Aminopyridine	[1.05]
3-Aminopyridine	[3.20]
4-Aminopyridine	[3.42]
3. Secondary amines	
Dimethylamine	−1.85
Diethylamine	−0.67
Di-N-propylamine	−0.86
Diethanolamine	+0.16
N-Methylaniline	−0.60
2-Methylaminoethanol	−1.22
Piperidine	−2.06
Morpholine	−1.70
Piperazine	−1.77

[a] Δ values in parentheses are for an additional HFA adduct line which appears when compound equivalent concentration exceeds that of HFA.
[b] Δ values in brackets are for HFA adduct line position when TFA sufficient to neutralize the basic group is present.
Reprinted with permission from G. R. Leader, *Anal. Chem.*, **42**, 16 (1970).

Table 7.1D. Chemical Shifts of HFA Adduct Lines for Miscellaneous Compounds in Ethyl Acetate

Compound	Δ (ppm)
Hydrogen sulfide	4.84
Ammonia (or NH_4^+)	0.86
Hydrogen peroxide	
As $HFA \cdot H_2O_2$	5.08
As $(HFA)_2 \cdot HO_2$	5.55
Hydrazine	
As $HFA \cdot N_2H_4$	2.90
As $(HFA)_2 \cdot N_2H_4$	4.55
Methylhydrazine	5.26
1,1-Dimethylhydrazine	2.41
2,4-Dinitrophenylhydrazine	4.65
Hydroxylamine	
OH	4.11
NH_2	6.29
Dimethylglyoxime	4.01
Acetophenone oxime	3.93
Methyl ethyl ketoxime	3.70, 3.80
Diethylhydroxylamine	2.74
Acetamide	0.62
Acrylamide	0.85
Chloroacetamide	1.27
Benzamide	1.43
Sulfanilamide (SO_2NH_2)	1.74
p-Toluene sulfonamide	1.76
Methane sulfonamide	1.60
Methyl thiourea	1.61
Phenyl urea	0.36
Methyl urea	−0.05

Reprinted with permission from G. R. Leader, *Anal. Chem.*, **42**, 16 (1970).

However, many experienced NMR spectroscopists can extract useful and often sufficient information by inspection of these spectra.

Several studies on the determination of specific olefins and the extent of unsaturation of a mixture have been conducted. Stehling and Bartz have published an excellent paper in which chemical shifts, spin–spin coupling constants, and spectral patterns were correlated to the structure of 60 mono-olefins (39). These authors used the results to demonstrate that the structure of oligomers of mono-olefins, polymers, diolefins, and monomer sequence distribution in isobutylene–isoprene copolymers could be determined. This paper is a good starting point for those becoming involved in the use of NMR

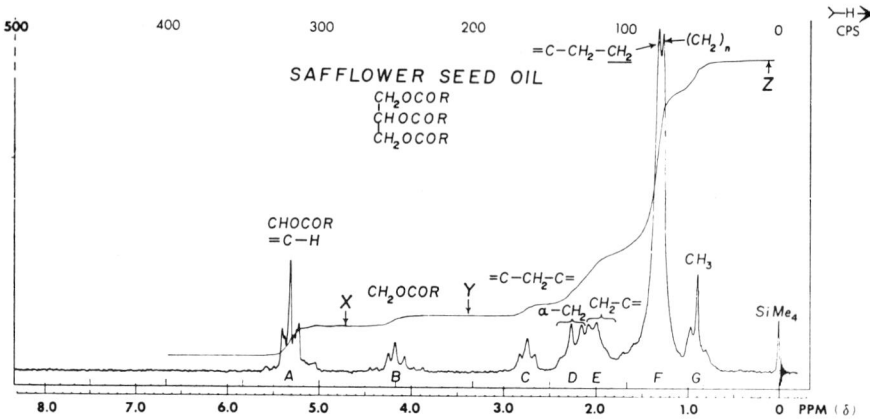

Fig. 7.1 60-Mc spectrum of safflower seed oil (Reprinted with permission from L. F. Johnson and J. N. Shoolery, *Anal. Chem.*, **34**, 1136 (1962). Copyright by the American Chemical Society.)

to investigate olefinic compounds. By careful examination of the spectra as described by Stehling and Bartz, structure of olefinic compounds may be determined. However, using the correlations found by these authors, worthwhile results may be obtained by looking at the NMR data of the substituent groups in less detail than total spectrum analysis.

An early paper which is one that effectively demonstrates how NMR can replace an old, tedious wet chemical procedure was published by Johnson and Shoolery (40). This paper reports a method for the determination of the average molecular weight of fatty acids and triglycerides as well as the unsaturation of these compounds. Figure 7.1 shows an example in the 40 MHz spectrum of safflower seed oil. The integrated area X is proportional to the olefinic proton content, the area given by Y is the area X plus the area resulting from the two methylene groups (4 protons) in each glyceryl moiety. The area at Z is the total area of proton resonance lines in the spectrum. Because it is known that the area given by $Y - X$ results from the four methylenic protons, the area per proton is

$$\text{area per proton} = \frac{Y - X}{4} \tag{7.1}$$

The number of olefinic protons is

$$V = \frac{X - (Y - X)/4}{(Y - X)/4} \tag{7.2}$$

and the total number of protons is

$$T = \frac{Z}{(Y - X)/4} \qquad (7.3)$$

These relationships prove to be approximate because of overlap of ^{13}C satellite lines and the authors give corrections for this overlap. The authors show that these data may be used to calculate the average molecular weight with a final equation

$$\text{Molecular Weight} = 120.0 + 7.013\,T + 6.006\,V \qquad (7.4)$$

The degree of unsaturation may be expressed as an iodine number using the relationship

$$\text{Iodine Number} = \frac{12691\,V}{\text{Molecular Weight}} \qquad (7.5)$$

Table 7.2 gives some comparison of the NMR values and WIJS method results. The technique has been developed as a routine process monitor by interfacing a minicomputer to an A-60D or T-60 spectrometer (41). The computer maintains instrument adjustment, acquires the spectrum and integral and performs the calculations. Any analyst who has performed a WIJS iodine number will agree that this is progress. The method must be used with discretion when triglycerides that have conjugated double bonds are determined, as illustrated by tung oil in Table 7.2. The NMR method is probably more accurate; however the olefinic protons associated with the

TABLE 7.2. Iodine No. Values of Various Fats

Oil	NMR No.	WIJS No.
Coconut	10.5 ± 1.3	8.0–8.7
Olive	80.8 ± 0.9	83.0–85.3
Peanut	94.5 ± 0.6	95.0–97.2
Soybean	127.1 ± 1.6	125.0–126.1
Sunflower seed	135.0 ± 0.9	136.0–137.7
Safflower seed	141.2 ± 1.0	140.0–143.5
Whale	150.2 ± 1.0	149.0–151.6
Linseed	176.2 ± 1.2	179.0–181.0
Tung	225.2 ± 1.2	146.0–163.5

(L. F. Johnson and J. N. Shoolery, *Anal. Chem.*, **34**, 1136 (1962) with permission.)

TABLE 7.3. Average Molecular Weights of Fats from Saponification Values and NMR

Oil	Sap. Value	Mol. Wt.	NMR Mol. Wt.
Olive	189.3	887.1	873.7 ± 5.3
Peanut	188.8	891.5	882.3 ± 7.4
Safflower seed	191.5	879.0	874.9 ± 9.3

(L. F. Johnson and J. N. Schoolery, *Anal. Chem.*, **34**, 1136 (1962), with permission.)

conjugated double bond are shifted 0.4–1 ppm downfield from those olefinic protons not in a conjugated system (Table 7.3).

On occasion, it is desirable to combine chemical reactions with NMR spectroscopy to gain information concerning olefins. For example, the addition of sulfenyl chlorides across double bonds can be performed and monitored in situ (42). Substituents determine whether the addition is Markovnikov or anti-Markovnikov and is useful in structure elucidation. The reaction between iodium nitrate and olefins has been shown as a useful tool to shift resonances downfield because of strong deshielding in the adducts (43). Adducts between mercuric acetate and fatty acid olefinic sites have been shown to permit direct determination of the *cis/trans* ratio in these compounds (44). The olefinic proton resonance of the two isomers are separated in the adduct (~0.05 ppm), whereas they overlap in the parent compounds. The choice of solvent is important and peak height serves as the method of quantitation.

7.2.6 DETERMINATION OF ACETYLENIC HYDROGEN

Acetylenic hydrogen nuclei show strong diamagnetic anisotropy effects resulting in a chemical shift of 2.4–2.7 ppm from TMS. This chemical shift range encompasses the values for methylene and methine protons as well as highly deshielded methyl groups. As a result, the direct identification of acetylenic protons by NMR faces limitations. In addition, the identification of substituted acetylenes is limited by a small acetylenic proton chemical shift range of ~0.3 ppm for a wide variety of substituted acetylenes. There are, however, some useful aids to the identification of acetylenic protons. For example, addition of pyridine to a dilute solution of monosubstituted acetylene in CCl_4 results in a downfield shift of ~1 ppm for the acetylenic proton (45). Acetylenes exhibit long-range coupling (viz.

$J_{CH_2-C\equiv C-H} \sim 2.9$ Hz), which is an aid to the identification of acetylenic proton resonance lines. When sufficient concentration or neat samples are available, the ^{13}C satellite may be useful in identifying acetylenic protons. The $J_{^{13}C-H}$ between ^{13}C nuclei bearing the acetylenic proton is ~ 250 Hz, much larger than the corresponding value for ethylene, benzene, or cyclopropane (~ 160 Hz).

Once identified, quantitation of the acetylenic proton may be performed by direct integration. The choice of solvent or addition of base such as pyridine may effect a shift to avoid overlap with other lines. The technique of deuterium substitution for active hydrogen using LiOD as a catalyst has been applied to phenylacetylene by Paulson and Cooke (20).

7.2.7 DETERMINATION OF AMINES

The use of NMR for the determination of amines would at first appear suspect because the chemical shift of the NH proton resonance lines may appear over a wide range of values depending upon solvent, concentration, presence of acids or bases, temperature, and other factors. In addition, the line shape will vary with similar factors as they determine the rate of chemical exchange of the NH protons. There are, fortunately, simple solutions to these problems. The chemical shift may be reproducibly controlled by the addition of a strong acid which protonates the amine forming a salt and reduces the rate of proton exchange to the point that spin–spin coupling between the NH protons and protons on the carbon alpha to the nitrogen is observed. Trifluoroacetic acid is commonly used as it is a sufficiently strong acid, does not contribute interfering peaks to the proton resonance signal, and is soluble in many organic solvents. The coupling constant between NH protons and the nitrogen is usually approximately 5 Hz. Because of some splitting from the nitrogen nucleus the N—H resonance lines may not show clear first-order spectra. However, these lines may be integrated for quantitation; remember the protonation contributes a proton to the area. Observation of the resonance from the proton on the alpha carbon is frequently as useful. In trifluoroacetic acid, a primary amine shows a quartet, a secondary amine a triplet, and a tertiary amine a doublet for these lines. This is of particular use for methyl amines.

The use of the hexafluoroacetone (HFA) adducts described by Leader is of importance for amines (36). As with alcohols, thiols, and amides, the adducts between amines and HFA show characteristic fluorine resonance shifts from the free HFA. Many amines may be identified from this shift value. However, the technique suffers, but only slightly with modern instrumentation, from the need for fluorine resonance equipment.

Under the heading of amines it is important to note that this functional group has been widely studied by NMR in terms of the kinetics of proton exchange. Both pulse and continuous-wave techniques have been employed. Grünwald has presented a brief but effective review of this topic (46). Caughman has prepared an extensive collection of the results of all N—H proton exchange studies to 1970 (47). Also, considerable attention has been devoted to the study of the inversion at nitrogen centers in amines using NMR methods (48–51).

Amino acids represent an important class of amines. Many properties of these compounds may be studied by NMR techniques. The quantitative determination of these compounds by NMR can be performed in a straightforward manner provided the sample is not too complex of a mixture and the compounds are adequately soluble. Again, the use of trifluoroacetic acid as a solvent as suggested by Bovey and Tiers provides the advantage of reproducible chemical shift values (52). Care must be taken, however, as some amino acids such as glycine and cysteine have limited solubility in this solvent. The use of NMR as a routine tool for the determination of amino acids may be questioned as there are many other techniques that are suitable, some of which are automated.

7.2.8 SUMMARY OF FUCTIONAL GROUP ANALYSIS USING NMR

This section has attempted to present a representation of the variety of methods that may be employed for functional group determinations using NMR. The emphasis has been placed on the use of proton magnetic resonance for the determination of functional groups in monomeric organic compounds. This was done for two reasons. First, these represent the most common applications of NMR. Second, during the course of the remainder of this chapter, further examples will be given for nuclei other than ^{1}H, and in samples from a variety of fields of research and industrial productions.

The use of proton NMR has had tremendous impact on organic functional group determinations. Continuous wave NMR is sufficiently simple in concept and practice that those with limited experience may use it effectively. However, it is significant that many of the procedures reported to date employ chemical derivatization or judicious choice of solvent to enhance selectivity, implying that input of chemistry by chemists is still of great significance even when using such a powerful tool as NMR.

Yet even the most current techniques are being explored to extend the range of applications of NMR. For example, Brunner, et al. have proposed a novel pattern recognition approach to the identification of functional

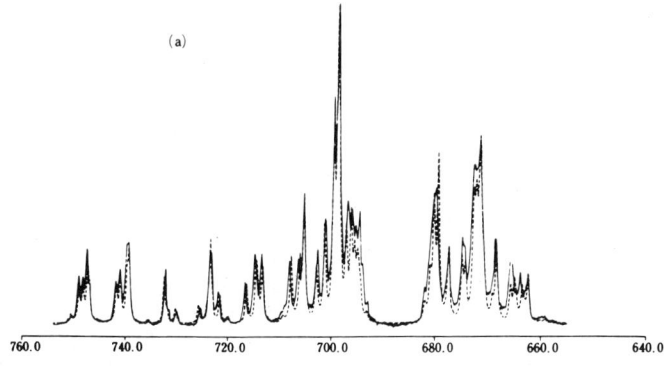
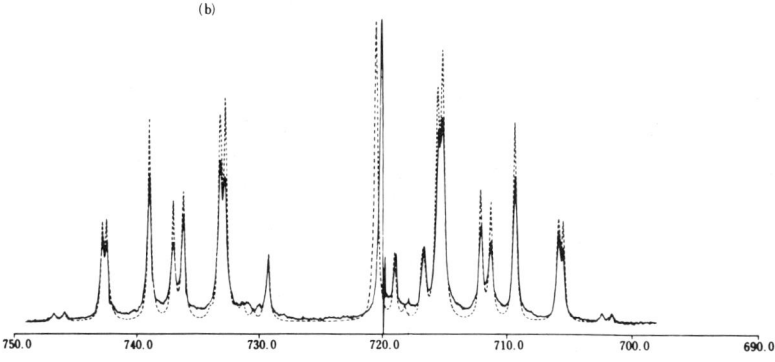
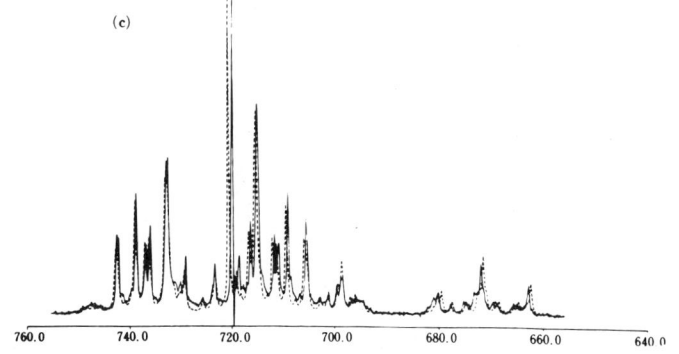

Fig. 7.2 Observed (solid line) and calculated (dashed line) spectra obtained from QUANTNMR. (a) Bromoanisole mixture (Run No. 13). (b) dichlorobenzene mixture (Run No. 6), and (c) bromoanisole and dichlorobenzene mixture (Run No. 12). (Reprinted with permission from O. Yamamoto and M. Yanagisawa, *Anal. Chem.*, **47**, 697 (1975). Copyright by the American Chemical Society.)

groups from ^{13}C-free induction decay data (53). Although this idea was first suggested by Kowalski and Reilly for ^{1}H, it was impractical primarily because of the high degree of spin–spin interaction and its effect on the ^{1}H-FID waveform (54). The ^{13}C-FID data that are usually lacking in multiplicity resulting from spin coupling are more suitable. Early results by Brunner et al. indicate potentially high accuracy in the use of pattern recognition for the identification of functional groups from ^{13}C-FID data.

The computer has been effectively used to assist in the direct quantitation of known mixtures showing severe peak overlap. This has always been a serious problem with any instrumental method. However, NMR spectra are easily simulated if a pure sample of each component in a mixture is available so that the spectrum may be analyzed. Yamamoto and Yanagisawa have taken advantage of this for the quantitative determinations of mixtures (55). The procedure is simple in principle and similar to other "simultaneous" spectrometric determinations. Provided the NMR spectral parameters for each component in the mixture are known, the spectrum for each may be simulated, the spectra for all components are summed with intensities weighed for relative concentrations. The resulting computed spectrum is compared with the experimental one until a minimum sum of the squares of residuals is obtained. The weighting factors then represent the relative concentrations. An example is shown in Figure 7.2.

7.3 INDUSTRY RELATED APPLICATIONS

Along with chromatography and mass spectrometry, NMR is one of the most powerful tools that may be employed in a wide variety of industry related problems. Areas to which the technique may be applied are almost too extensive to be listed, much less to survey, in a single volume. An attempt to list some of those areas is given in Table 7.4, which is by no means exhaustive.

TABLE 7.4. Applications of Nuclear Magnetic Resonance to Industrial Problems[a]

Polymers	Pharmaceuticals
Paints and coatings	Foods
Petroleum	Detergents
Pesticides	Perfumes
Explosives	Flavors

[a] References to most of these topics may be found in the Appendix.

In this section we have presented selected examples of the applications of NMR to analytical problems in a few industrial fields. The organization is arbitrary; however, it provides an opportunity to review some applications from a variety of industrial examples. Further examples are certain to develop as new techniques and imaginative chemistry are combined. Some attempt has been made to select the examples from recent literature at this writing as these will provide for references to earlier work.

7.3.1 SYNTHETIC POLYMERS

The desire to apply NMR techniques to the study of polymers began as soon as the technique was made available. Early reports of successful applications began quickly leading to a review of the subject in 1958 (56). Growth in the field was at first slow. However, as magnets of higher field strength were made available, shift reagents were introduced, sensitivity improved, and the advantages of pulsed techniques recognized, rapid growth in the field was observed. In this section we have attempted to provide the reader with examples of the information that may be obtained about synthetic polymers using magnetic resonance methods. We have elected to do this by first reviewing some various types of studies, followed by specific results on several types of polymers.

The general applications of NMR to the study of polymers has been discussed in monographs by Slonim and Lyubimov (57) and Bovey (58). The book by Bovey also discusses bio-macromolecules. Several reviews of the applications of NMR to polymer studies have also been published (59–62). Although most applications of NMR to polymer studies have been ^{1}H-NMR, the use of ^{13}C and other nuclei is growing rapidly. The emphasis of the use of high resolution NMR has been on characterization of polymers. Identification of polymers, structure determination, chain configurations (tacticity studies), and quantitative determinations have received most attention. To a lesser extent has been the use of NMR to investigate molecular motions in polymers.

7.3.1.1 Quantitative Determinations

One of the first precautions to observe in quantitative NMR spectroscopy of polymers is to ascertain that the sample is completely in solution. In a report by Wilkes on a study of the tacticity of poly(vinylchloride), samples prepared by different dissolution techniques gave different percentages of polymer resulting in a high-resolution NMR spectrum (63). The highly crystalline samples showed zero solubility in o-dichlorobenzene, and greater than 90% solubility was observed for other samples. Resonance line intensity measurements were made versus an internal standard and studies were conducted over a temperature range of 413–443°K. Wilkes observed that solutions that gave high resolution spectra from greater than 80% of the polymer in the sample were clear. However, there may still be sufficient crystalline material present such that as much as 20% of the polymer does not show high resolution spectra. Lack of attention to these details may result in unreliable results.

As the precautionary note previously given immediately illustrates, quantitative measurements of polymers by NMR requires attention to detail. However, the principles of these measurements are the same as with monomers. Peak height or peak areas may be employed. Several types of samples may be encountered. In the simplest case, the concentration of a homogeneous solution of a single polymer may be quantitatively measured. These determinations yield higher accuracy if done from a calibration curve prepared from samples of known composition.

A second type of sample is a two-phase system containing two different magnetically active nuclei, one in each phase; for example, ^{1}H and ^{19}F. In this case, each phase may be determined in the one case by ^{1}H-NMR and in the other using ^{19}F-NMR. Borodin and Skripov have reported an ingenious instrumental adaptation to determine hydrogen impurity in fluoro-polymers using wide line NMR (64). Two rf coils were constructed in a probe in series with the circuit to the rf transmitter. One coil contains a sample X and the other a standard S_1. The peak heights of the resonance signal from each (A_x, A_{s_1}) is obtained. Then the sample is replaced by a second standard of concentration C_{s_2} and the experiment repeated. The concentration of the nuclei in the sample is given as

$$C_x = C_{s_2} \left(\frac{A_x}{A_{s_1}}\right)\left(\frac{A_{s_1}}{A_{s_2}}\right) \tag{7.6}$$

This technique is well suited for the determination of impurities of one phase in another. For example as little as 0.01% hydrogen in compounds of the type $C_4F_6Cl_2$ was determined within about 10% relative error.

If a two-phase system contains only one type of nucleus, then the determination is more complicated unless there is sufficiently high resolution to obtain a resonance line characteristic of each component. If the polymer is sufficiently soluble, the problem may be readily solved.

7.3.1.2 Structure Determinations

Of equal importance, and likely of greater significance than quantitative determinations of polymers, is the applications of NMR to the determination of structural features of polymers. A major effect has been made in this direction. As an introduction some brief comments will be given on the structural features of polymers. The seriously interested reader should consult a text such as *The Stereochemistry of Macromolecules* (65). The events which may occur during polymerization of a vinyl monomer will be used as an example. A similar but more detailed discussion is given in Chapter 11 in a book by Bovey (58). Other monomers may yield similar or additional structural features.

Considering a vinyl monomer of the type

$$\begin{array}{c} A \\ \diagdown \\ C=C \\ \diagup \\ B \end{array} \begin{array}{c} H \\ \diagup \\ \\ \diagdown \\ H \end{array} \qquad \text{II}$$

the propagation of the polymer may occur by:

(1) "Head-to-Tail" addition leading to

$$\cdots -\underset{B}{\overset{A}{\underset{|}{C}}}-\underset{H}{\overset{H}{\underset{|}{C}}}-\underset{B}{\overset{A}{\underset{|}{C}}}-\underset{H}{\overset{H}{\underset{|}{C}}}-\underset{B}{\overset{A}{\underset{|}{C}}}-\underset{H}{\overset{H}{\underset{|}{C}}}- \cdots$$

(2) "Head-To-Head Tail-To-Tail" addition leading to

$$\cdots -\underset{B}{\overset{A}{\underset{|}{C}}}-\underset{H}{\overset{H}{\underset{|}{C}}}-\underset{H}{\overset{H}{\underset{|}{C}}}-\underset{B}{\overset{A}{\underset{|}{C}}}-\underset{B}{\overset{A}{\underset{|}{C}}}-\underset{H}{\overset{H}{\underset{|}{C}}}-\underset{H}{\overset{H}{\underset{|}{C}}}-\underset{B}{\overset{A}{\underset{|}{C}}}- \cdots$$

with the possibility of one or more monomic units entering out of order leading to

$$\cdots -\underset{B}{\overset{A}{\underset{|}{C}}}-\underset{H}{\overset{H}{\underset{|}{C}}}-\underset{H}{\overset{H}{\underset{|}{C}}}-\underset{B}{\overset{A}{\underset{|}{C}}}-\underset{H}{\overset{H}{\underset{|}{C}}}-\underset{B}{\overset{A}{\underset{|}{C}}}-$$

↗ propagation in a new sense

↘ revert back to original propagation sense

(3) Chain branching of monomers leading to

```
        A—C—B                          A—C—B
        |                              |
        H—C—H                          H—C—H
        A     A  H                 A       A  H
        |     |  |                 |       |  |
       —C—C—C—C...              —C—C—C—C---
        |     |  |                 |       |  |
        B  H  B  H                 B       B  H
                                   H—C—H
                                   |
                                   A—C—B

   Trifunctional               Tetrafunctional
   Branch Point                 Branch Point
```

```
        A  H  A  H
        |  |  |  |
       —C—C—C—C—...
        |
        B
        A—C—B
        |
        H—C—H
        A     A  H
        |     |  |
  ...—C—C—C—C—...
        |     |  |
        B  H  B  H
```

Crosslinking

(4) Tacticity or the relative configuration of a sequence of monomers leading to

```
     A  H  B  H  A  H  B  H
     |  |  |  |  |  |  |  |
...—C—C—C—C—C—C—C—C—...     "Syndiotactic"
     |  |  |  |  |  |  |  |
     B  H  A  H  B  H  A  H
```

```
  A   H   A   H   A   H   A   H
  |   |   |   |   |   |   |   |
…—C — C — C — C — C — C — C — C—…      "Isotactic"
  |   |   |   |   |   |   |   |
  B   H   B   H   B   H   B   H


  A   H   A   H   B   H   B   H
  |   |   |   |   |   |   |   |
…—C — C — C — C — C — C — C — C—…      "Heterotactic"
  |   |   |   |   |   |   |   |
  B   H   B   H   A   H   A   H
```

These examples (58, with permission) represent the simplest cases. Obviously many combinations may occur in real polymers including variations within a single chain. Fortunately, NMR can provide significant results in distinguishing between these isomeric forms and information about the length of each tacticity class in a polymer. As yet, however, little information concerning branching sites has been obtained. In addition to the simple examples previously given, polymerization of dienes lead to a maze of geometrical (1,4 or 1,2 enchainment) and configurational (*cis* or *trans*) isomerism. Clearly such possibilities as *cis*-isotactic, *trans*-syndiotactic and other combinations exist. These examples only show the tremendously complicated possibilities which exist for the structural and tacticity problems in polymers. It is well beyond the scope of this volume to enter into an in-depth discussion that will lead to a clear understanding of the impact of NMR on polymer chemistry. However a few reports will be cited in a qualitative way.

One of the first NMR tools that is brought to bear on structural studies of polymers is chemical shift. Equivalence or nonequivalence of chemical shift may result from incidental screening effects or molecular symmetry. Tiers and Bovey showed that the central CF_2 fluorine nuclei in meso-$CF_2ClCFClCF_2CFClCF_2$ have different chemical shifts and couplings, whereas those in the racemic isomer are identical (58). The spectra of racemic and meso-2,4-diphenylpentane shown in Figure 7.3 illustrate a further example and represents a model of polystyrene. In these spectra the methylene group resonances differ by 0.21 ppm in the meso isomer. They are, however, equivalent in the racemic isomer as is qualitatively evident from the complexity of the methylene resonance at about 1.8 ppm. The effect of the sequencing of triad, tetrad, etc. units on chemical shift has been observed for a variety of polymers. The position of the chemical shifts of lines in the ^{1}H-NMR spectra may then be used to determine sequences present in a polymer and the intensity of each line related to the relative population of a given sequence. These techniques are discussed in detail in the book by Bovey (58).

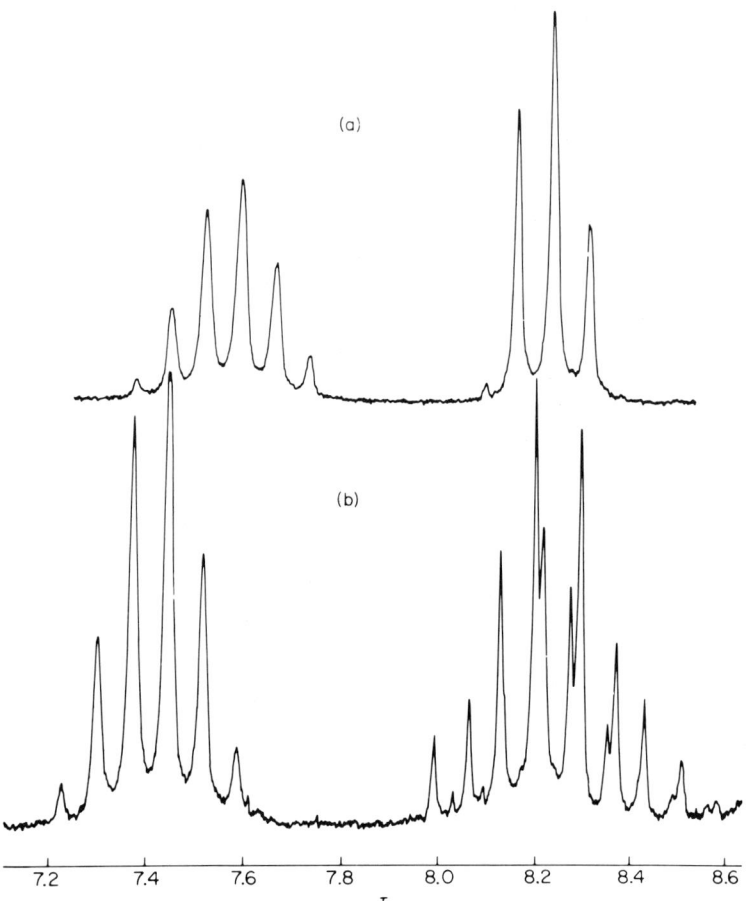

Fig. 7.3 The 100-MHz spectra of (a) *racemic* and (b) *meso*-2,4-diphenylpentane, 10% (v/v) in chlorobenzene at 35°. (F. A. Bovey, *High Resolution NMR of Macromolecules*, Academic, New York, 1972, with permission.)

The NMR spectra of polymers prepared after specific deuteration of the monomers is a very significant aid in determining the conformation of reactants during polymerization. A detailed study of the polymerization of vinyl methyl ether was so conducted. In one study the —CH_2— spectra of deuterated polymers was assigned in terms of tetrad sequences (66). This was made possible by the simplified spectrum of the deuterated polymer. Later, a study was conducted using various polymerization conditions (67). From careful comparisons of sequence ratios it was concluded that both isotactic-like and syndotactic-like monomer–polymer cation reactions occur with slight preference to the isotactic-like presentation.

The effect of tacticity (essentially microtacticity) on reactions of polymers may be investigated using ^{1}H-NMR. A study of the conversion of poly(isopropyl[α,β-^{2}H$_2$]acrylate) into poly(methyl[α,β-^{2}H$_2$]acrylate) is an example (68). The tacticity of poly(isopropylacrylate) was determined for polymers prepared at different temperatures and the ΔH and ΔS differences between isotactic and syndiotactic addition thereby determined. The rate of hydrolysis of poly(isopropylacrylate) was found to be independent of macrotacticity and dependent on molecular weight. However, the NMR study indicated that the rate of racemization of the isotactic sequences in the polymer was slower than that for the syndiotactic.

In addition to the many examples of specific deuteration found in the literature as a method to simplify polymer NMR spectra, this technique used in conjunction with proton NMR as 220 MHz has proved especially useful. The sequencing of poly(propylenebutadiene) was studied in this way (69).

A report which may prove to be of some importance involves the spinning of solid samples at an angle of 54.7°, frequently referred to as the "magic angle." If there is some degree of molecular motion and the spinning rate is large enough (greater than a few KHz), substantial narrowing of the NMR lines is achieved by removal of dipolar coupling. In reports by Schneider and associates a theory was presented that related line shape to correlation time for molecular motion, spinning rate, and the second moment of the resonance line (70,71). The theory was tested and it was demonstrated that the relative amounts of two types of amorphous polymer and a crystalline polymer in a mixture could be determined. The second paper demonstrated that the resonance linewidths of acetonitrile and benzene in the presence of cross-linked polystyrene spheres were reduced from 120 and 58 Hz for nonspinning to 4.1 and 3.3 Hz respectively, for magic angle spinning. In the case of methanol in a styrene-divinylbenzene resin in the acid form, the methyl resonance linewidth was reduced to high resolution, but the OH line remained broad, possibly a result of proton exchange. The use of magic angle spinning should prove valuable not only in the direct examination of polymer mixtures, but in the study of solvents in contact with macromolecules, for example chromatographic column materials.

The previously described examples are only a superficial indication of the types of information that can be obtained by high resolution ^{1}H-NMR spectroscopy of polymers. In summary, stereochemistry, sequeching, microtacticity, polymerization kinetics (rate and mechanism), molecular motion, thermodynamic parameters of sequencing, and many other properties may be investigated as well as quantitative determinations and molecular weight measurements. It would be impossible to survey even one application to all polymers. In the following we have cited a few references, recent at this writing, to indicate further examples. Review articles are cited where known to us.

Acrylics

NMR has been used to determine the tacticity of poly(alkyl methacrylated) and poly(methacrylic acids) prepared under varied conditions (72). The precision and accuracy of determining the stereoregularity of poly(methyl methacrylate) using 60 and 100 MHz NMR spectrometers have been compared (73). A complete assignment of the 20 possible triads in methacrylic-methyl methacrylate has been reported (74).

Cellulose

A review of NMR spectroscopy applied to the investigation of the structure of cellulose has been given (75).

Acrylonitriles (AN)

Tacticities of acrylonitrile-methyl methacrylate copolymers were determined using 60-MHz ^{1}H-NMR (76). AN-butadiene and butadiene-butadiene sequences were established (77).

Styrene-butadiene (SBR)

Sytrene sequences have been measured in SBR (78) and an analog computer was employed to resolve the spectra to facilitate the differentiation of short sequences (< 3 units) from longer ones (79).

7.3.1.3 Nuclei Other Than ^{1}H in Polymer Studies

^2H

With the growing availability of spectrometers capable of rapid conversion from one operating frequency to another, nuclei other than ^{1}H will play an increasingly important role in the study of polymers by NMR. The use of ^{2}H to simplify the spectra of synthetic polymers has already been mentioned. The ratio of $J_{^1H-^1H}$ to $J_{^1H-^2H}$ representing a ^{2}H substitution is approximately 0.16 ($\gamma_{^2H}/\gamma_{^1H}$). As vicinal couplings in polymers are approximately 6 Hz, the substitution of ^{2}H almost eliminates these couplings. Residual line broadening may be removed by double resonance at the ^{2}H frequency. It is rare, to date, to directly observe the ^{2}H resonance.

^{19}F

The use of ^{19}F-NMR spectra is now common-place. This is due in part to the similar resonance frequencies between ^{1}H and ^{19}F at a fixed magnetic field. Therefore, instruments are easily switched from one nucleus to the other. Two major applications of ^{19}F-NMR are found in synthetic polymers. First, poly(tetrafluoroethylene) (PTFE) and poly(vinylfluoride) (PVF) are obviously targets for this technique. Structure studies have been conducted on PVF (80) and chlorinated-PVF (81) using ^{19}F-NMR. PTFE has been studied extensively by ^{19}F-NMR and reports are referenced in a paper by McBierty and associates who conducted pulsed ^{19}F-NMR measurements on oriented PFTE fibers (82). These authors reported that relaxation measurements showed that the fibers contained randomly oriented amorphous regions and crystalline regions.

Naylor and Lasoshi have investigated a number of polymers containing fluorine (83). The large chemical shifts seen in ^{19}F spectra assist in assignment.

^{13}C

The use of ^{13}C offers great advantage to the investigation of polymer structure. Many instrumental problems have been overcome with the development of almost routine ^{13}C-FT spectrometers with proton decoupling. The resulting spectra are simplified as coupling is essentially eliminated, natural abundance ^{13}C spectra are acquired in a few hours, and like ^{19}F the large chemical shift values of ^{13}C facilitate interpretation of the spectra. One remaining problem is the NOE (Section 5.10) enhancement of the ^{13}C resonance intensities. As discussed earlier, progress is being made in this area. The enhancement factors have been calculated to be 2.988 and 1.153 for ^{13}C, H dipolar relaxation in the extremes of very fast and very slow molecular motions. Attempts to calculate these effects for known molecular correlation times have been made (84). In principle, ^{13}C-NMR may be used to obtain structural information similar to that given by ^{1}H-NMR. Such reports have already appeared for poly(vinylchloride) (85), polypropylene (86), poly(methyl methacrylate) (87,88), poly(acrylonitrile) (89).

7.3.1.4 Shift Reagents in Polymer Studies

The use of paramagnetic shift reagents in NMR spectroscopy has been discussed in Section 4.11.9. Here we will simply say tht these reagents have already been applied to polymer studies (90–92) and will find continued use in those polymers that have functional groups capable of interacting with the paramagnetic ion. This technique is well suited to those polymers that have suitable terminal functional groups as the neighboring proton resonances can be shifted under controlled conditions to a clean spectral region. This will permit easy integration and thereby quantitation and number-average molecular weight determinations.

7.3.2 COATINGS

The applications of NMR spectroscopy in the coatings industry can be classified arbitrarily into the three different areas of solvents, resins and polymers, and fatty acids. Characterizations, identifications, and quantitative determinations have been performed in these areas using NMR. Pusey has reviewed the applications of NMR spectroscopy for the analysis (qualitative and quantitative) citing such examples as the identification of unknown expoxy resins, the determination of the relative amounts of monomer types in copolymers, saponification values for glycerides and iodine numbers (93). Obviously, many of the examples that might be mentioned are very similar to those given in the polymer industry, food industry (fatty acids), and functional group sections of this chapter. However, a few studies will be cited here as they relate to the coatings industry.

The use of NMR spectroscopy is an ideal choice for the identification of components in paint and coating solvents and thinners. These solvents are frequently no more complicated than ternary mixtures and quantitative analysis can usually be performed by integration of the resonance signals. An example is given by Afremow for the quantitative analysis of a toluene/methyl ethyl ketone solvent (94). The identification and determination of toxic components in solvents is of considerable importance and Gruenfield has reported an application of NMR spectroscopy to this problem (95). When mixtures are moderately complex, double resonance techniques are very useful to assign spectral patterns arising from a given molecule.

The area of fatty acid investigations as related to coatings is very similar to those mentioned in Section 7.2.5. As one example, Wineburg and Sivern have used shift reagents to confirm the structures of methyl oleate, methyl petroselinate, methyl ricinoleate, and methyl 12-hydroxy-sterate by NMR spectroscopy (96,97). An interesting result of this work is that the selection of shift reagent was important. The shift reagent $Eu(fod)_3$ permitted distinction of only the resonance patterns of the first four methylene groups in the alkyl chain of methyl oleate in the ^{1}H-NMR spectrum at 100 MHz. This shift reagent usually causes shifts to lower field. However, if the shift reagent $Pr(thd)_3$ is used instead, shifts to higher fields are observed, and all methylene groups are observed (98).

The compounding of raw materials in the manufacturing of coating resins requires careful quality control for which NMR spectroscopy is well suited. Because one is usually dealing with known mixtures, spectral analyses can be performed in advance, proper quantitative procedures developed and checked, and rapid routine determinations with accuracies within a few percent relative error made without separation or other pretreatment than solvent removal. Each specific resin will require a developed technique as to which resonances will serve for quantitation. For example in the preparation of polyester resins from polyols, acidic and hydroxyl proton resonance lines disappear and —OCH— and —OCH$_2$— protons show a shift to lower field (99,100). The relative or absolute amounts of each component (polyol, glycol, alkyl carboxylic acid and isophthalic acid) may be determined in the initial mixture. The polymerization may be monitored in situ if desired. The rate of formation of an acrylic resin formed with methyl-methacrylate (MMA), maleic anhydride (MA), and vinyl isobutyl ether (VIBE) has been effectively investigated in this manner (101). The characteristic vinyl monomer proton resonances were monitored as their area decreased during polymerization in order to determine the overall rate. The polymer that formed consisted of alternating (MMA)$_x$ and (MA-VIBE)$_y$ regions that were linked. From these studies the authors concluded that:

1. Methyl-methacrylate adds to methyl-methacrylate monomers and maleic anhydride but not to vinyl isobutyl ether. The ratio of rates of reaction of methyl-methacrylate radical with methyl methacrylate and maleic anhydride is three to one.

2. Maleic anhydride radical adds to methyl-methacrylate monomer and reacts readily with vinyl isobutyl ether monomer, but not to maleic anhydride.

3. Vinyl isobutyl ether radical adds to maleic anhydride monomer, but not to vinyl isobutyl ether monomer.

7.3.3 FOODS

Applications of NMR have not been as extensive in the food industry as in some other areas; however, several reports are available and rapid growth is expected as sensitivity of NMR spectrometers increases and methodology for smaller sample size is developed.

7.3.3.1 Wide Line Methods

To date, the principal applications of NMR to foodstuffs have been based on either wide line (low resolution) continuous-wave spectra or pulse techniques. In the main, samples have been substances that contain high molecular weight oils or fats and water bound to solid or semisolid materials. Many of the applications take advantage of the short-spin relaxation time (T_1) of water bound to solids compared with "unbound" water. Others make use of sudden changes in relaxation time with phase changes. In many cases,

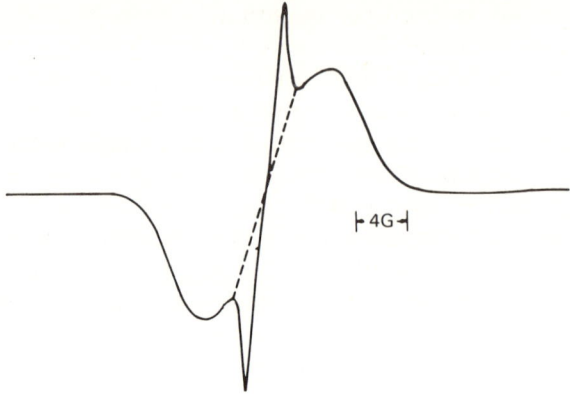

Fig. 7.4 Typical wide-line NMR spectrum of moist cellulose. (R. A. Pittman and V. W. Tripp, *Applied Spec.*, **25**, 235 (1971), with permission.)

the rf power may be adjusted such that the resonance of liquid or "unbound" water (long T_1) may be saturated with no significant saturation of the bound water. At times, rather arbitrarily but useful definitions of bound and liquid states are used.

An example that illustrates the type of information that may be obtained is a study of water–cellulose systems by Pittman and Tripp (102). Figure 7.4 shows a derivative mode wide line NMR spectrum of moist cellulose. As readily seen, this spectrum has two components of different linewidth. Pittman and Tripp separated the wider components into three groups, the total intensity of a wet sample (cw), the intensity obtained on a dry sample (c), and the difference between the two (i). They described elsewhere the use of a curve synthesizer for this purpose (103). The component cw represents all protons in the wet cellulose, and the fraction of noninterchanging protons is given by the area of the c component divided by the area of the cw component (A_c/A_{cw}). The fraction of protons interchanging between the water and cellulose is given by A_i/A_{cw}. The narrow component of the spectrum represents the noninterchanging water in the wet cellulose. From these data the authors calculated the fraction of four different types of protons in wet cellulose: noninterchanging water-associated protons, noninterchanging cellulose-associated protons, interchanging water protons, and interchanging cellulose protons. There is no other technique that can provide this type of information with such ease.

Wide line techniques are now routinely applied as illustrated by several reports. Wiggall et al. have applied the technique to the determination of fat in chocolate (104). Shanbhag et al., used wide line NMR to determine oil in aqueous emulsions (105). NMR has been used to estimate liquid fats in cream and butter (106). A review of the use of chromatography and magnetic resonance methods for the determination of fats and fatty acids in oil-bearing foodstuff was given by Harris (107). Several applications of low resolution NMR to the determination of oil in seeds have been published using wide line techniques (108–110). As with other NMR methodologies, the non-destructive nature of these determinations is of value. The oil content of seeds may be determined, and those same seeds then used for agricultural experimentation. The sources of error in the use of NMR for the determination of solid fat content has been examined (111).

7.3.3.2 Pulsed Methods

Pulsed NMR techniques are of even more value than wide line methods. The wide-line technique is convenient if only gross oil or fat content is sought. However, pulsed techniques using simple instrumentation provides for resolution of different T_1 or T_2 values in the sample and thereby detection of different phases. Figure 7.5 shows an example. The FID signal of a soybean sample is shown versus time following a 90° rf pulse. The proton-bearing components such as solid carbohydrates and proteins and water bound to these solids have short T_2 values and are represented by the steep portion of the curve below 50 μsec. The more mobile oils show longer relaxation times.

Extrapolation of the two curves to the time origin shows that roughly 50% of the total proton content is present in the oil in the seed. The pulsed NMR technique offers great advantage in these types of determinations when paramagnetic materials such as iron are present. By using a suitable two pulse sequence where "spin-echo" decays are seen, the effect of the paramagnetic material can be greatly reduced. Continuous-wave NMR would be

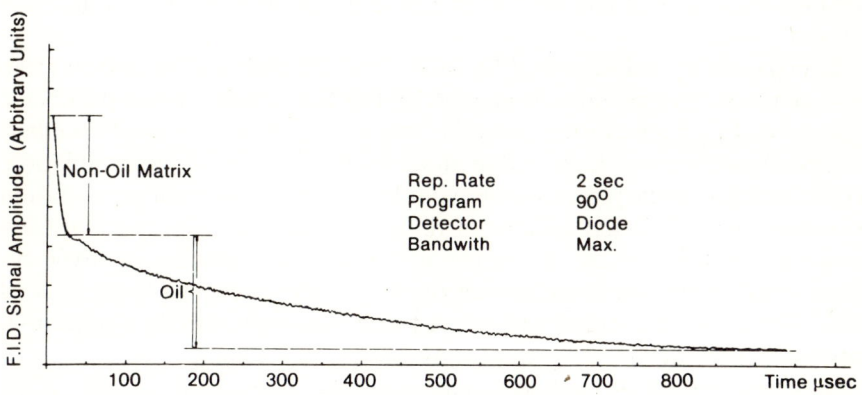

Fig. 7.5 FID signal of soybean. (Courtesy of Bruker Scientific Instruments, with permission.)

TABLE 7.5. Applications of Pulsed NMR to Food Products

Solid content of fats, creams, and oils
Oil content of seeds
Oil and water in butter, margarine, and cream
Hydrogenation of unsaturated oils
Polymorphism of fats
Moisture content in dairy products
Rehydration rate of dehydrated foods
Biodegradation of food
Moisture content of meat products
Solidification of gelatins

very suspect under these conditions. Table 7.5 gives a partial list of applications of pulsed NMR to determinations in the food industry. Special instruments are available for these applications and were discussed in Section 6.5.3.

7.3.3.3 High-Resolution ^{1}H and ^{13}C Methods

As discussed earlier, a *cw* high-resolution NMR spectrometer interfaced to a minicomputer has been used to determine the average molecular weight, degree of unsaturation (iodine number) and the average number of vinyl groups in vegetable oils (41). This configuration provides for research capability as well as a quality control instrument. All calculations are performed and reports prepared by the computer. The disadvantage of this approach is the lack of sensitivity. This problem is considerably reduced by the use of FT-NMR. Limited use of ^{1}H-FT-NMR has been made at this writing, but that will change. Of immense potential is natural abundance ^{13}C-FT-NMR.

Two papers by Schaefer and Stejskal (112,113) and one by Barton et al. (114) demonstrate the application of ^{13}C-FT-NMR. Using a 10-mm diameter receiver coil and quadrature detection, the first authors obtained excellent ^{13}C-NMR spectra on single viable intact soybeans in 20 min accumulation. These authors employed computer simulation and analog curve resolution to isolate the ^{13}C-NMR spectra of the vinyl region of oleic, linoleic, and linolenic acids. These are the major unsaturated triglycerides in soybeans. Assignments were confirmed by comparison of the spectra with the isolated acids. The quantitative results compared with values obtained by chromatography are given in Table 7.6 and the agreement is excellent.

Of course, the NMR method is nondestructive. A brief comparison of the ^{13}C-FTNMR of soybean and radish seeds was also given, quickly confirming

TABLE 7.6. Comparison of ^{13}C-FT-NMR and Chromatographic Results of Relative Amounts of Oleic, Linoleic, and Linolenic Acids in Soybeans

Acid	^{13}C-FT-NMR	Chromatograph
Oleic	0.30	0.33
Linoleic	0.60	0.57
Linolenic	0.10	0.10

the higher fraction of linoleic acid in the radish seeds. The authors comment on the pulse repetition rate during the acquisition of the FID data. This is an important factor in quantitative ^{13}C-FT-NMR because of the nuclear Overhauser effect (NOE) which is mentioned several times in this book. Before using ^{13}C-FT-NMR as a quantitative method, spectroscopists should become very familiar with NOE factors that affect line intensities.

Barton et al. (114) showed that ^{13}C-FT-NMR of methyl oleate and methyl eladiate may be applied to the direct quantitative determination of *cis-trans* isomers in these compounds without the use of shift reagents as used by Walters and Horvat (115). Although the samples used were not taken from food products, in principle they could be. Figure 7.6 shows a ^{13}C-NMR spectrum of an 80:20 *cis:trans* mixture. The determinations were (in some cases) made on as little as 8 mg of sample using a microcell. The overall standard deviation of the determination was ± 0.4 mole %. These authors gave special attention to NOE factors in establishing pulse repetition rate.

In summary, pulse and wide-line NMR techniques are routine tools for the determination of "phase components" in foods such as oils, fats, emulsions, etc. FT-NMR will, without a doubt, increase the applications of high-resolution NMR as current detection limits approach a few hundred nanograms for ^{1}H resonance and a few hundred micrograms for natural abundance ^{13}C resonance.

7.3.4 PETROLEUM

The applications of nuclear magnetic resonance to problems in the petroleum and fuel industry began to appear in the late 1950s and have been numerous. Of importance to the knowledge in these fields are the chemical composition and structure of the compounds in petroleum crude and refined products. NMR has provided much information of this type. In one of the earliest applications it was shown that NMR could identify two types of hydrogen in alkyl concentrates of gas oil (methyl and acyclic methylene), and three types (aromatic, alkyl alpha to an aromatic ring and other alkyl)

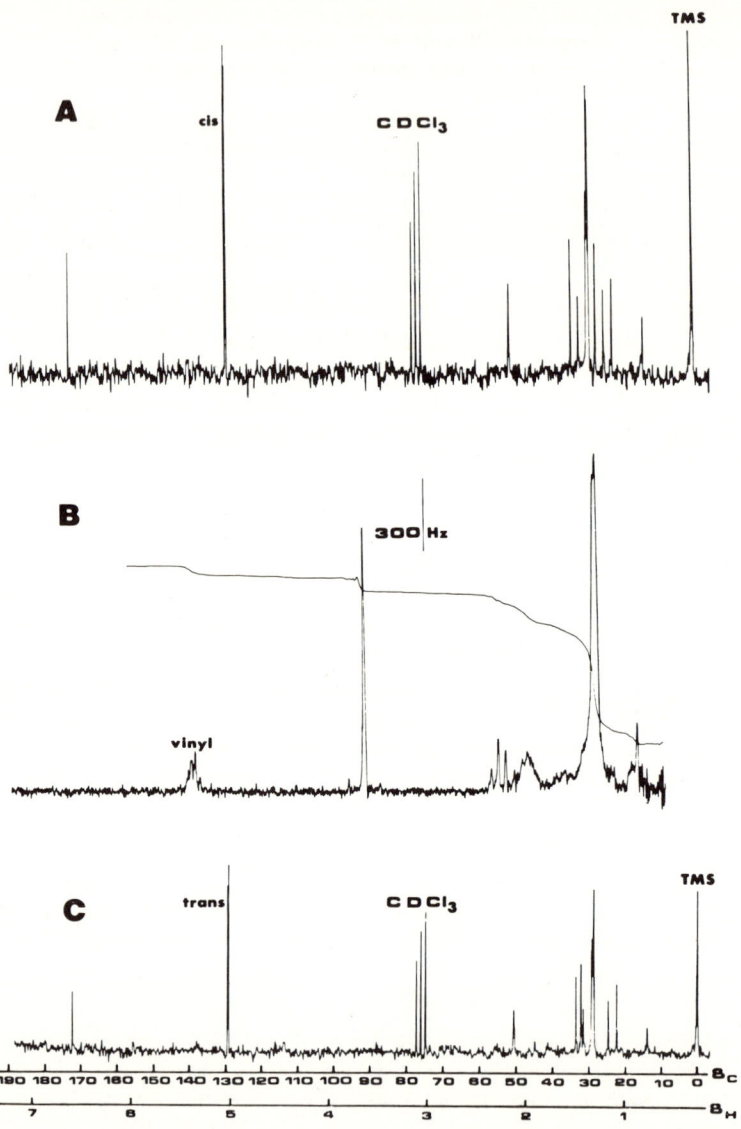

Fig. 7.6 (A) 5000 Hz ^{13}C spectra of methyl oleate *cis* C-18:1. (B) 1000 Hz ^{1}H spectra of a 50:50, ratio of methyl oleate and elaidate C-18:1. (C) 5000 Hz ^{13}C spectra of methyl elaidate *trans* C-18:1. (F. E. Barton II, D. S. Himmelsbach, and D. Burdick, *J. Magn. Res.*, **18**, 167 (1975), with permission.)

in catalytic cracker cycle stock and asphalt (116). These results were found to compare well with those obtained from mass spectrometry data. Further applications were to crude oil fractions containing asphaltenes in which the relative abundance of methylene, aliphatic methyl, benzylic and aromatic hydrogens were found to decrease in that order (117,118).

7.3.4.1 Crude Oil

The characterization of crude oil usually requires many physical and chemical considerations. In recent years NMR spectroscopy has played an increasingly important role in these characterizations, often in conjunction with other techniques. For example, Zimina et al., used gas chromatography, mass spectrometry, NMR, and distillation to characterize crude oils (119). The study resulted in establishing relationships between structures and the physical and chemical properties of the crude oils. A relationship between hydrocarbon structure and the depth of the source oil deposit was also found. Such results may lead to more rapid characterization of crude oils.

NMR has been applied to the urgent problem of the need for better and more rapid methods for the characterization of shale oils. To date, most reports of the applications of NMR to shale oils deal with samples extracted from the shale, and which in many cases, have been fractionated. Examples of these studies are given by Lille et al. (120,121). A rapid wide-line NMR technique has been devised which may be used to estimate the oil yield from shale samples (122). This technique also gives information on total organic carbon in the sample. Thirty samples per hour may be evaluated and the NMR signal has been correlated to the Fisher assay oil yield value. Using a relationship developed by Cook (123), the Fisher assay oil yield may be used to estimate the organic carbon content. Thus low resolution NMR can play an important part in the characterization of crude oils as well as in the quantitative estimation of yields. The usually simplified interpretation of proton decoupled ^{13}C-NMR over ^{1}H-NMR will likely find the former technique growing in popularity in the investigation of crude oil and shale fractions. The fact that ^{13}C-NMR chemical shift values are much more sensitive to subtle structural variations than ^{1}H-NMR chemical shift values is of immense value for such complex mixtures as crude oils.

7.3.4.2 Hydrocarbons

There have been many applications of NMR spectroscopy to the investigation of the hydrocarbons derived from petroleum. An in-depth discussion of the use of NMR spectroscopy for the analysis of organic compounds, including many petroleum-related materials, has been given by Trogolo (124). The review by Trogolo includes discussions of: the determination of structure of hydrocarbon components and fractions from petroleum; the analysis of aromatics and polycyclics, investigations of alkyl side chains in cyclic hydrocarbons; the analysis of fractions containing oxygen, nitrogen, and sulfur; structure of asphalt components; and quantitative analysis of mixtures. This review is an excellent starting point to survey the applications of NMR in the petroleum industry.

Several specific determinations have been reported. For example, the use of double resonance to simplify spectra of complex mixtures for quantitative measurements, was applied to olefin mixtures of which were monosubstituted, 1,1-disubstituted, 2,2-disubstituted, and in certain mixtures 1,1,2-trisubstituted ethylenes (125). As mentioned in Section 6.1, the aromatic content of fuels was estimated from the proton chemical shift of the ^{1}H-NMR resonance from acceptor molecules in donor–acceptor complexes (126). Suitable acceptors are CH_2Cl_2 and CH_3NO_2. The chemical shift of the acceptor is related to the volume percent aromatic content. A novel application of wide-line NMR has been reported by Herington and Lawrensen (127) and Lawrensen (128). Based on the long relaxation times in a solid, the spectrometer can be adjusted to saturate the solid component spin distribution and thereby only the liquid form is detected. Impurities that are liquids are detected in solids by taking measurements at a temperature just below the melting point of the major component in the sample. Perhaps a more precise technique would be the use of one of the various pulse mode operations to distinguish solids and liquids as well as liquids "bound" to solids in mixtures. Potentially, these techniques could be applied to viscosity measurements as well.

The application of ^{13}C-NMR to the analysis of petroleum hydrocarbon samples will likely prevail over ^{1}H-NMR in a few years. Qualitative applications are being reported with increasing frequency at the time of this writing, and in view of current progress in overcoming the nuclear Overhauser effect problems, quantitative applications are near. At this point in time, ^{13}C-NMR chemical shift correlations with structure are being obtained similar to the history of ^{1}H-CMR. Caser has published a rewiew that includes ^{13}C-NMR chemical shift correlations and the structure of linear and branched paraffins, cycloparaffins, olefins, acetylenic and aromatic hydrocarbons (129).

A report that may prove to be useful in the determination of the structure of hydrocarbons using ^{13}C-NMR was published by Lindeman and Adams (130). These authors illustrated the increased potential of ^{13}C-NMR over ^{1}H-NMR with the spectra of pristane shown in Figure 7.7. Compared with ^{1}H-NMR spectrum which does not resolve CH and CH_2 proton types, the ^{13}C-NMR shows a resolved resonance for each of the nine different carbon atoms in the molecule. The ^{13}C-NMR chemical shift for all isomers of C_5—C_8 and some C_9 paraffin hydrocarbons are given in Table 7.7. These data alone

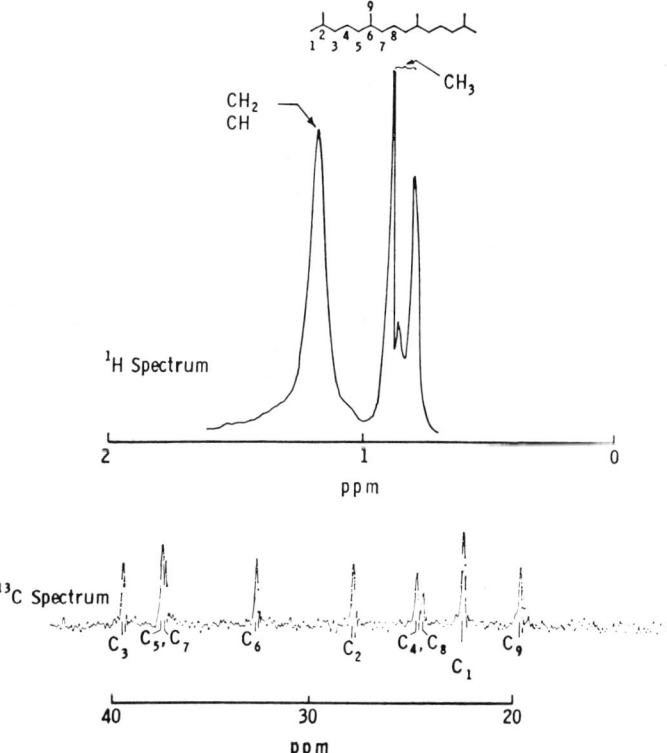

Fig. 7.7 ^{1}H and ^{13}C NMR spectra of pristane (Reprinted with permission from L. P. Lindeman and J. Q. Adams, *Anal. Chem.*, **43**, 1245 (1971). Copyright by the American Chemical Society.)

will be of use in identifying structural subgroups. Using these data, the ^{13}C-NMR chemical shift parameters, proposed by Grant and Paul (131), were modified for the linear and branched paraffin hydrocarbons. Grant and Paul had proposed the expression (Section 5.4.1)

$$\delta_c(k) = B + \alpha N_{k_1} + \beta N_{k_2} + \gamma N_{k_3} + \Delta N_{k_4} + \varepsilon N_{k_5} + \sum_{M=2}^{4} D_m A_{sm} \quad (7.7)$$

where $\delta_c(k)$ is the ^{13}C-NMR chemical shift of the kth carbon nucleus; B, α, β, γ, Δ, ε and A_{sm} are constants, N_{k_p} is the number of carbon atoms that are p bonds removed from the kth carbon atom; D_m is the number of carbon atoms which have m carbon atoms attached and are bonded to the kth carbon atom; and s is the number of carbon atoms bonded to the kth carbon atom. The terms A_{sm} and D_m were called "corrective terms" by Grant and Paul.

TABLE 7.7. ^{13}C Chemical Shifts for C_5 to C_9 Paraffins
TMS = 0, $DS_2 = +192.8$

	Identification	Chemical shift								
		C_1	C_2	C_3	C_4	C_5	C_6	C_7	C_8	C_9
C_5H_{12}	$C_1-C_2-C_3-C$	13.5	22.2	34.1						
	$C_1-C_2-C_2-C_4$ C	21.9	29.9	31.6	11.5					
	C_1-C_2-C C	31.6	28.0							
C_6H_{14}	$C_1-C_2-C_3-C-C$	13.7	22.7	31.7						
	$C_1-C_2-C_3-C_4-C_5$ C	22.7	27.9	41.9	20.8	14.3				
	$C_1-C_2-C_3-C$ C_4	11.4	29.4	36.8	18.7					
	$C_1-C_2-C_3-C_4$ C	28.7	30.3	36.5	8.5					
	C_1-C_2-C-C CC	19.2	34.0							

C_7H_{16}						
$C_1-C_2-C_3-C_4-C-C-C$	13.7	22.6	32.0	29.0		
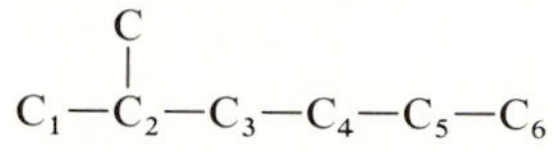 $C_1-C_2-C_3-C_4-C_5-C_6$, C on C_2	22.4	28.1	38.9	29.7	23.0	13.6
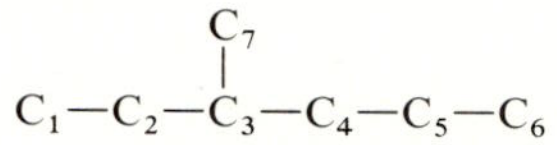 $C_1-C_2-C_3-C_4-C_5-C_6$, C_7 on C_3	10.9	29.5	34.3	39.0	20.2	13.9
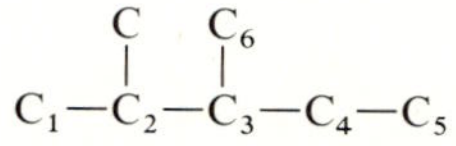 $C_1-C_2-C_3-C_4-C_5$, C_6 on C_2	20.0 / 17.7	31.9	40.6	26.8	11.6	14.5
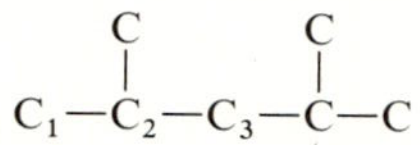 $C_1-C_2-C_3-C_4-C_5$, C on C_2, C on C_3	22.7	25.7	49.0			
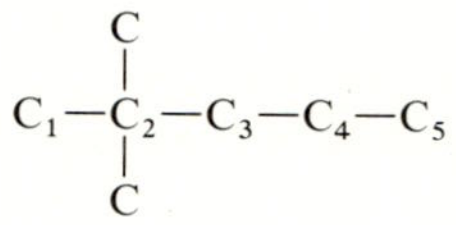 $C_1-C_2-C_3-C_4-C_5$, two C on C_2	29.5	30.6	47.3	18.1	15.1	
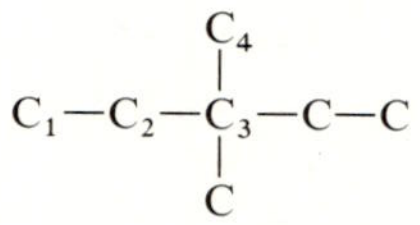 $C_1-C_2(C_4)-C_3-C$, C on C_2	7.7	33.4	32.3	25.6		
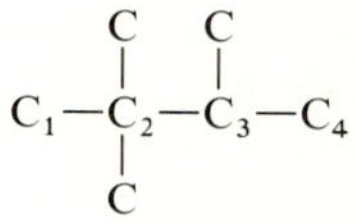 $C_1-C_2-C_3-C_4$, C on C_2, C on C_3	27.0	32.7	37.9	17.7		
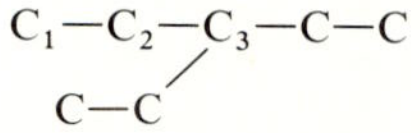 $C_1-C_2-C_3-C$, C–C branch	10.5	25.2	42.4	18.8		

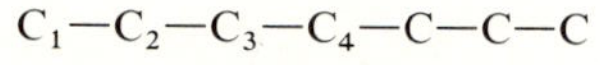

337

TABLE 7.7. (*Continued*)

	Identification	Chemical shift								
		C_1	C_2	C_3	C_4	C_5	C_6	C_7	C_8	C_9
C_8H_{18}	$C_1-C_2-C_3-C_4-C-C-C-C$	13.6	22.7	32.1	29.4					
	$C_1-C_2-C_3-C_4-C_5-C_6-C_7$ with C on C_2	22.4	28.1	39.3	27.2	32.4	22.8	13.8		
	$C_1-C_2-C_3-C_4-C_5-C_6-C_7$ with C_8 on C_2	11.3	29.7	34.7	36.5	29.7	23.3	14.1	19.3	
	$C_1-C_2-C_3-C_4-C-C-C$ with C_5 on C_3	14.1	20.2	39.5	32.3	19.3				
	$C_1-C_2-C_3-C_4-C_5-C_6$ with C, C_7 on C_3	17.8 / 20.0	32.8	38.5	36.7	20.7	14.0	15.1		
	$C_1-C_2-C_3-C_4-C_5-C_6$ with C, C_7 on C_4	22.2 / 23.2	25.4	46.6	32.1	29.9	11.0	19.0		
	$C_1-C_2-C_3-C-C$ with C, C on C_3	22.4	28.4	36.9						
	$C_1-C_2-C_3-C-C$ with C_4, C on C_3		25.8	38.5	13.8					
	$C_1-C_2-C_3-C-C$ with C_4, C on C_2	11.8	27.6	39.5	15.8					

Structure	C1	C2	C3	C4	C5	C6	C7
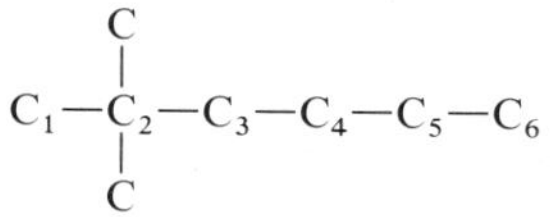 C₁–C₂(C)(C)–C₃–C₄–C₅–C₆	29.2	30.1	44.1	27.0	23.7	13.9	
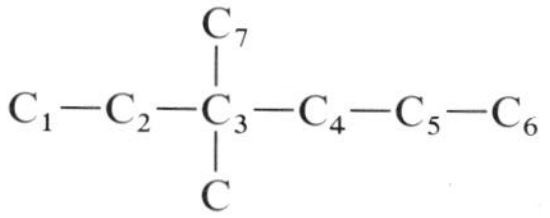 C₁–C₂(C₇)–C₃(C)(C)–C₄–C₅–C₆	8.1	34.3	32.8	44.3	17.3	14.8	26.5
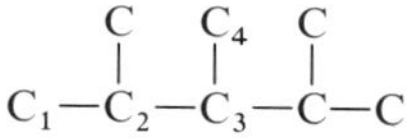 C₁–C₂(C₄)(C)–C₃–C	21.4 / 18.1	29.8	45.3	10.4			
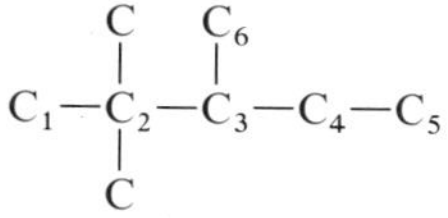 C₁–C₂(C₆)–C₃(C)–C₄–C₅	27.1	33.0	45.4	24.4	13.0	13.3	
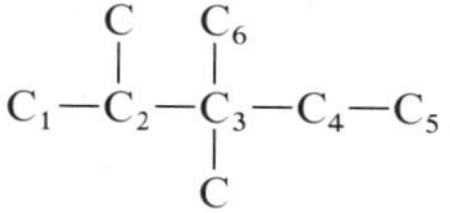 C–C₁–C₂(C₆)–C₃–C	17.1	35.1	34.9	32.6	7.9	23.3	
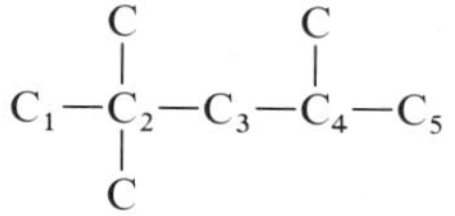 C–C₁–C₂–C₃(C)–C₄–C₅	29.9	30.9	53.3	25.3	24.7		
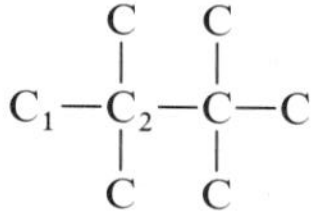 C–C₁–C₂(C)–C(C)–C	25.6	35.0					

339

TABLE 7.7. (Continued)

Identification				Chemical shift					
	C_1	C_2	C_3	C_4	C_5	C_6	C_7	C_8	C_9
$C-C$ $\vert$ $C_1-C_2-C_3-C_4-C_5-C_6$	10.6	25.6	40.6	35.4	20.0	14.1			
$C-C$ $\vert$ $C_1-C_2-C_3-C_4-C_5$	19.0	29.1	47.6	22.6	11.8				
C_4 $\vert$ $C_1-C_2-C_3-C$ $\vert$ C	7.5	30.6	34.8	23.2					
C_9H_{20}									
$C_1-C_2-C_3-C_4-C_5-C-C-C-C$	13.8	22.7	32.0	29.4	29.6				
C $\vert$ $C_1-C_2-C_3-C_4-C_5-C_6-C_7-C_8$	22.3	28.0	39.2	27.4	29.7	32.0	22.7	13.6	
$C_1-C_2-C_3-C_4-C_5-C_6-C_7-C_8$ $\vert$ C_9	11.1	29.7	34.6	36.7	26.9	32.4	22.7	13.8	19.0
$C_1-C_2-C_3-C_4-C_5-C_6-C_7-C_8$ $\vert$ C_9	14.0	19.4	39.6	32.6	36.8	29.3	23.0	13.7	20.2
C C_8 $\vert$ $\vert$ $C_1-C_2-C_3-C_4-C_5-C_6-C_7$	20.1 17.9	32.3	38.8	34.0	30.0	23.1	13.8	15.2	

340

Structure	C1	C2	C3	C4	C5	C6	C7	C8	C9
$C_1-C_2-C_3(C_8)(C)-C_4-C_5-C_6-C_7$	22.1 / 23.1	25.3	47.0	30.2	39.9	19.9	14.0	19.4	
$C_1-C_2(C)-C_3(C_8)-C_4-C_5-C_6-C_7$	22.3 / 22.5	28.4	36.5	34.4	34.8	29.5	11.0	19.0	
$C_1-C_2-C_3(C)-C_4-C$	22.4	28.1	39.5	25.2					
$C_1-C_2(C_8)-C_3(C_9)-C_4-C_5-C_6-C_7$	11.9	25.8 / 27.6	38.9 / 39.8	36.3 / 37.2	35.5 / 37.5	20.8	14.2	13.9 / 15.8	16.3 / 14.2
$C_1-C_2-C_3(C_5)(C)-C_4$	10.9 / 11.1	29.5 / 30.5	31.9 / 32.0	44.3 / 44.5	18.9 / 19.6				
$C_1-C_2-C_3(C_8)(C)-C_4-C_5-C_6-C_7$	29.2	30.2	44.4	24.4	33.0	22.8	13.8		
$C_1-C_2(C_8)(C)-C_3-C_4-C_5-C_6-C_7$	8.0	34.2	32.5	41.3	26.4	23.7	13.7	26.4	
$C_1-C_2-C_3-C_4(C_5)(C)-C$	14.9	17.3	44.8	32.8	27.0				
$C_1-C_2(C_7)-C_3(C)-C_4-C_5-C_6$	17.8 / 20.0	32.4	36.2	43.9	25.7	21.9 / 23.5	15.3		

TABLE 7.7. (*Continued*)

Identification	Chemical shift								
	C_1	C_2	C_3	C_4	C_5	C_6	C_7	C_8	C_9
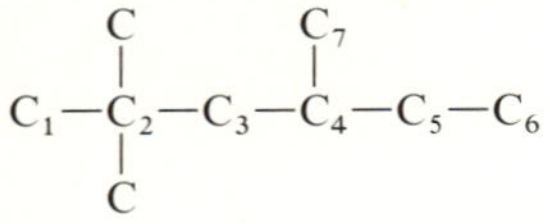 $C_1-C_2-C_3-C_4-C_5-C_6$ with C_7 and C branches on C_2	29.9	31.0	51.0	31.9	31.0	11.2	21.9		
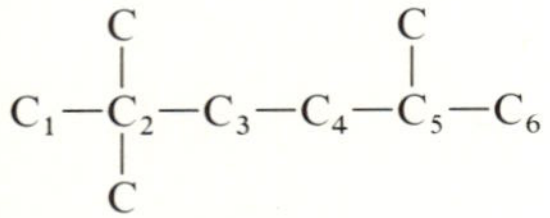 $C_1-C_2-C_3-C_4-C_5-C_6$ with C branches on C_3	29.3	30.1	42.0	33.9	28.9	22.5			
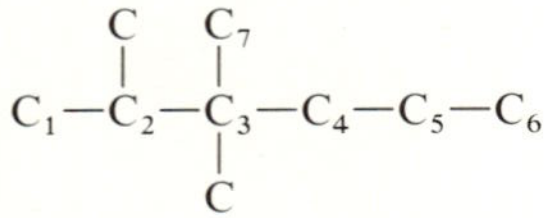 $C_1-C_2-C_3-C_4-C_5-C_6$ with C_7 and C branches on C_3	17.1	35.0	35.5	43.1	17.0	14.8	23.8		
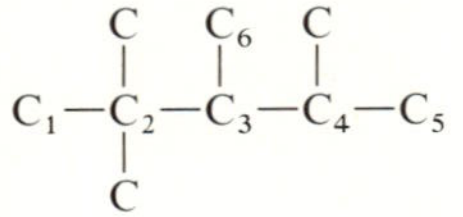 $C_1-C_2-C_3-C_4-C_5$ with C_6, C branches	28.2	34.0	47.9	27.4	17.3 / 24.5	11.6			
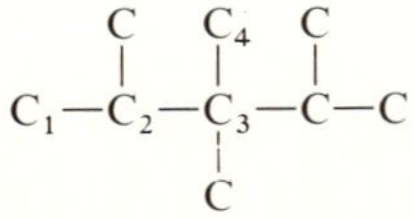 $C_1-C_2-C_3-C-C$ with C_4, C branches	17.2	37.1	33.6	18.9					

342

Structure							
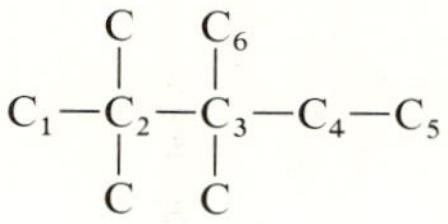 C₁—C₂(C)(C)—C₃(C₆)(C)—C₄—C₅	25.6	36.0	37.3	28.8	9.0	20.6	
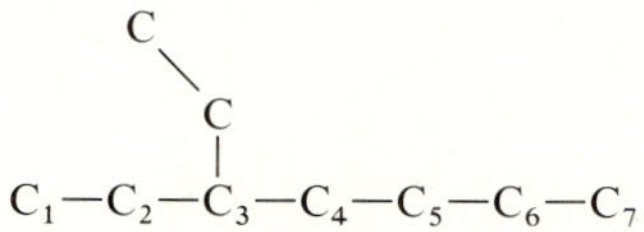 C₁—C₂—C₃(C)—C₄—C₅—C₆—C₇	10.6	25.6	40.7	32.7	29.2	23.1	13.7
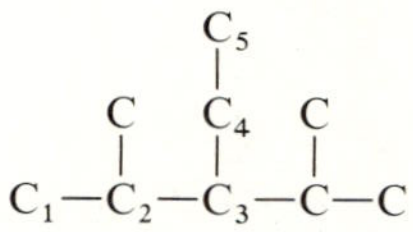 C₁—C₂—C₃(C₄)(C₅)—C	19.0 / 20.0	29.0	56.8	21.1	14.5		
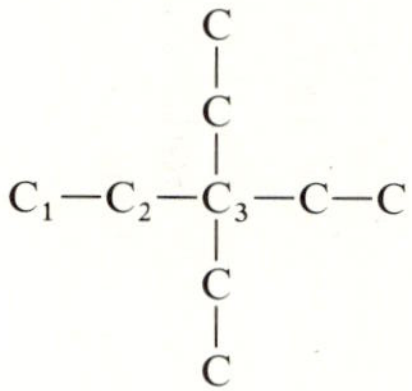 C—C₂—C₃(C)(C)—C	7.1	27.1	37.1				
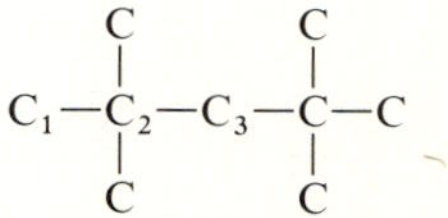 C₁—C₂(C)(C)—C₃(C)(C)—C	31.8	32.4	56.5				

(L. P. Lindeman and J. Q. Adams, *Anal. Chem.*, **43**, 1245 (1971), with permission.)

Lindeman and Adams tested equation 7.7 using the Grant and Paul parameters by comparing calculated versus experimental chemical shift values for all ^{13}C-NMR resonance lines for 59 paraffins. The overall standard deviation of the error was 2 ppm in chemical shift with very poor predictions of tertiary and quaternary carbon chemical shift values. A revised expression was developed which has the form

$$\delta_c(k) = \beta_s + \sum_{M=2}^{4} D_m A_{sm} + \gamma_3 N_{k_3} + \Delta_s N_{k_4} \qquad (7.8)$$

Each of the primary, secondary, tertiary, and quaternary carbon types are represented by $s = 1, 2, 3$, and 4, respectively, and a set of parameters is used for each carbon type. The A_{sm} term is not treated in equation 7.8 as a corrective term, but is a parameter that is used to describe the steric configuration of the next nearest neighbor carbon atoms and replaces the β term in equation 7.8. Table 5.3 gives the parameter set used in the final test that predicted ^{13}C-NMR chemical shifts with a standard deviation of the error of 0.79 ppm. In highly branched hydrocarbons, the predictions show a larger error, possibly a result of steric interactions.

7.3.4.3 Summary

Obviously NMR (^{1}H and ^{13}C) have and will play a significant part in the identification, characterization, and quantitation of compounds in the petroleum industry. The applications mentioned here only touch on the potential. As the limits of detection of FT-NMR are pursued to the hundred nanogram range for protons, and milligram range for natural abundance carbon-13, the application of combined preconcentration or enrichment techniques and NMR will likely play an increasingly important role in the identification of petroleum-based pollutants. As an example, Novotny et al., have used enrichment techniques, chromatographic and solvent partition fractionation in conjunction with ^{1}H-NMR to identify polynuclear aromatic hydrocarbons in air (132). A 1.88-mg fraction showed ^{1}H-NMR spectra that indicated the presence of methylpyrenes. Only about 0.1 mg of the methylpyrenes were thought to be present in the fraction. The samples were drawn from Indianapolis and Gary, Ind., city air. Although NMR may never become a routine tool for trace determinations, the initial identification of compounds in environmental pollution is an appropriate application, especially when enrichment techniques are employed.

7.3.5 PESTICIDES

With growing world population, the demands on agricultural production have increased constantly. To partially offset uncontrollable natural disasters and to increase profitable yields, pesticides have been employed world wide in agricultural endeavors. The realization that many of these compounds are stable toward biodegradation and create lingering effects on the ecological system has caused considerable concern. The concentrating effect of large animals eating smaller animals which have consumed pesticides or herbicides, and man eating flesh taken from animals fed from grain treated with these materials has been discussed many times. The detailed study of pesticides was then required in order to determine which compounds showed long residual effects as opposed to those which were readily biodegradable or underwent photochemical degradation. Furthermore, the question as to whether the degradation or metabolic products were compound types which were known to be harmful required answers. Obviously, NMR could be easily used in the determination of the structure of newly prepared compounds. Plenty of sample was available and the standard parameters of chemical shifts and coupling constants were easily obtained.

However, in the mid 1960s several groups began to use ^{1}H-NMR to investigate pesticide residues and degradation products. Many of these experiments were performed under controlled conditions in a laboratory, and again plenty of sample was available. The question could have easily been raised at that time as to the practicality of using NMR for studies of samples taken from the natural environment. However, the relatively widespread availability of FT-NMR has proved the early work valuable. Using FT-NMR, the spectrum of as little as 1 μg of these compounds may be obtained. This is approaching quantities which may be isolated from the environment. Furthermore, a variety of NMR techniques has been used to study the chemical and physical properties of pesticides, such as their ability to form molecular complexes, the stereoselective chemical and biological transformations which they undergo, and the nature of pesticide molecules and their molecular motions on membrane surfaces. Each of these properties likely plays a role in determining the toxicity of the compound.

A rather substantial amount of literature has been published on the application of NMR to pesticides. Half of the book edited by Hague and Biros is devoted to this topic and presents a representation of a wide range of applications (133). Several reviews have been published. Two are general reviews which use specific examples to illustrate how NMR may be used to confirm structural and stereochemical features of several pesticide types (134,135). Another illustrates how NMR may be used to confirm pesticide presence in environmental water samples (136). Three others deal in detail with organophosphorus pesticides (137), DDT-type pesticides (138) and carbamate pesticides (139). These reviews include the trade and chemical names, structures and NMR parameters for large numbers of each type of pesticide. An additional publication catalogs the NMR spectra of over one hundred pesticides (140). Most of the use of NMR toward pesticide chemistry is covered in these reviews. The specific research papers usually involve highly detailed arguments of structure proof from chemical shifts and coupling constants which cannot be re-argued here except for a few illustrative examples.

7.3.5.1 Organophosphorus Pesticides

The organophosphorus pesticides consist mainly of phosphate (or thiophosphate) esters with the general formula

$$R_1O-\underset{\underset{OR_2}{|}}{\overset{\overset{X}{\|}}{P}}-Y-R_3$$

where R_1 and R_2 are most often methyl or ethyl groups whereas R_3 is a large organic group. In these compounds X is oxygen or sulfur and Y is oxygen, sulfur, or nitrogen. The ^{1}H-NMR spectra of the phosphomethoxy class of compounds is characterized by methyl resonances at $\delta = 3.7$ to 3.9 ppm, ^{31}P—^{1}H spin coupling (J_{P-CH_3}) of 11–16 Hz. In the case of the phosphoethoxy class, the methyl resonance is near $\delta = 1.4$ ppm and methylene resonance $\delta = 4.2$ to 4.4 ppm. The J_{P-CH_2} and J_{P-CH_3} values are near 10 Hz and 0.8 Hz, respectively. With various substituents at R_3, interpretation of the spectra may require considerable effort. One example will be given here as an illustration. Figure 7.8 shows the NMR spectra of *o,o*-dimethyl-5-(1,2-dicarbethoxyethyl)phosphorodithionate which is sold under the commercial name Malathion. The methyl quartets a and a' are readily seen as two overlapping triplets. The methylene protons at d and d' are confirmed as being two closely spaced triplets centered at about 4.2 ppm by proton decoupling at the methyl resonance frequency. The resulting singlets for the d and d' protons are shown in the insert in Figure 7.8. The methylene protons at b and b' are nonequivalent from symmetry considerations. In addition, these protons are spin-coupled to the methine proton at e and therefore the resonance of the b and b' protons appear as the AB part of an ABX pattern. As noted in the insert of Figure 7.8, the protons at e give resonance lines that lie beneath the pattern of the d and d' protons. Figure 7.9 shows the spectrum of Malathion in deuterobenzene in which the e proton signal is shifted to lower field and the eight-line pattern resulting from coupling to the b and b' protons as well as the phosphorus is clearly identified and confirmed by double decoupling from the b and b' protons. The NMR parameters for many

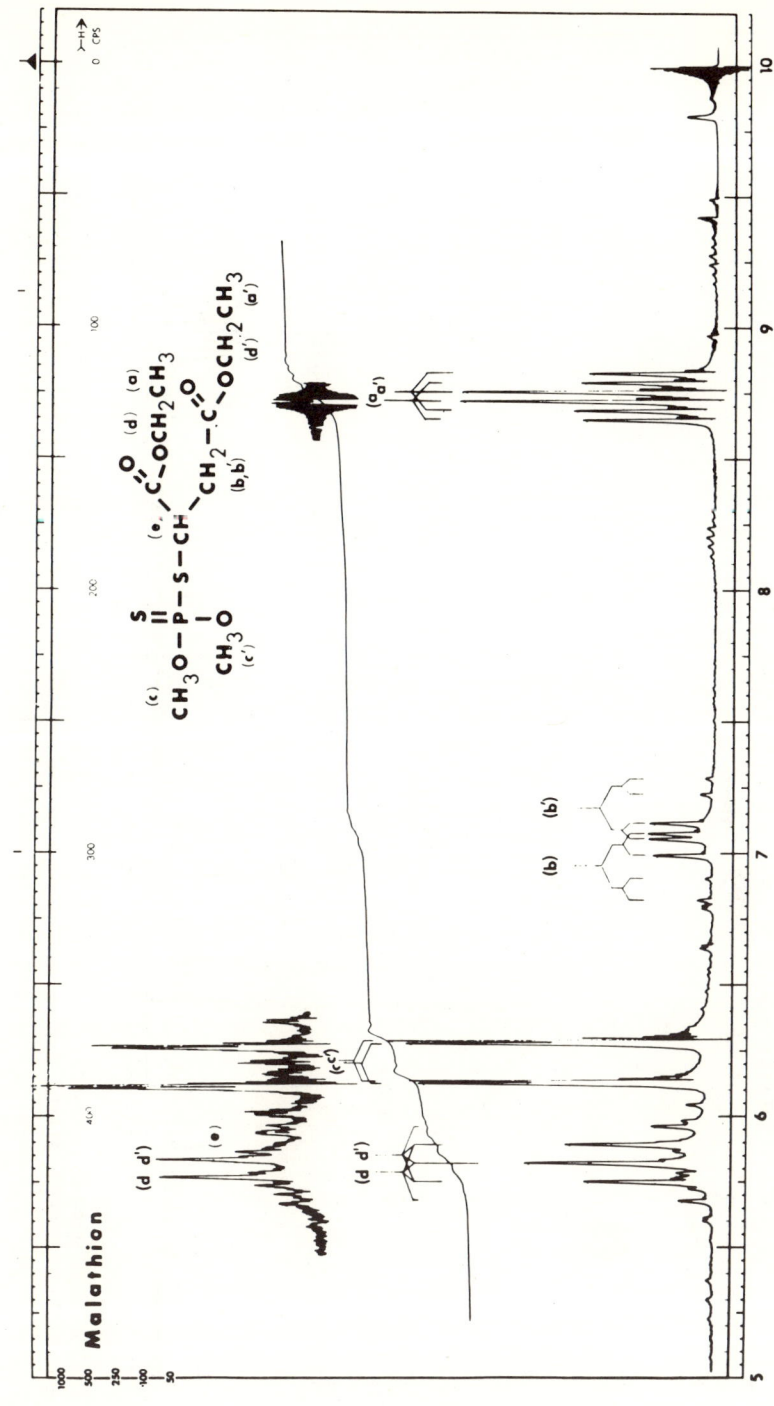

Fig. 7.8 NMR spectrum of malathion in CDCl$_3$. (L. H. Keith, A. W. Garrison, and A. L. Alford, *J. Assoc. of Offic. Anal. Chem.*, **51**, 1063 (1968). Reprinted with permission of AOAC.)

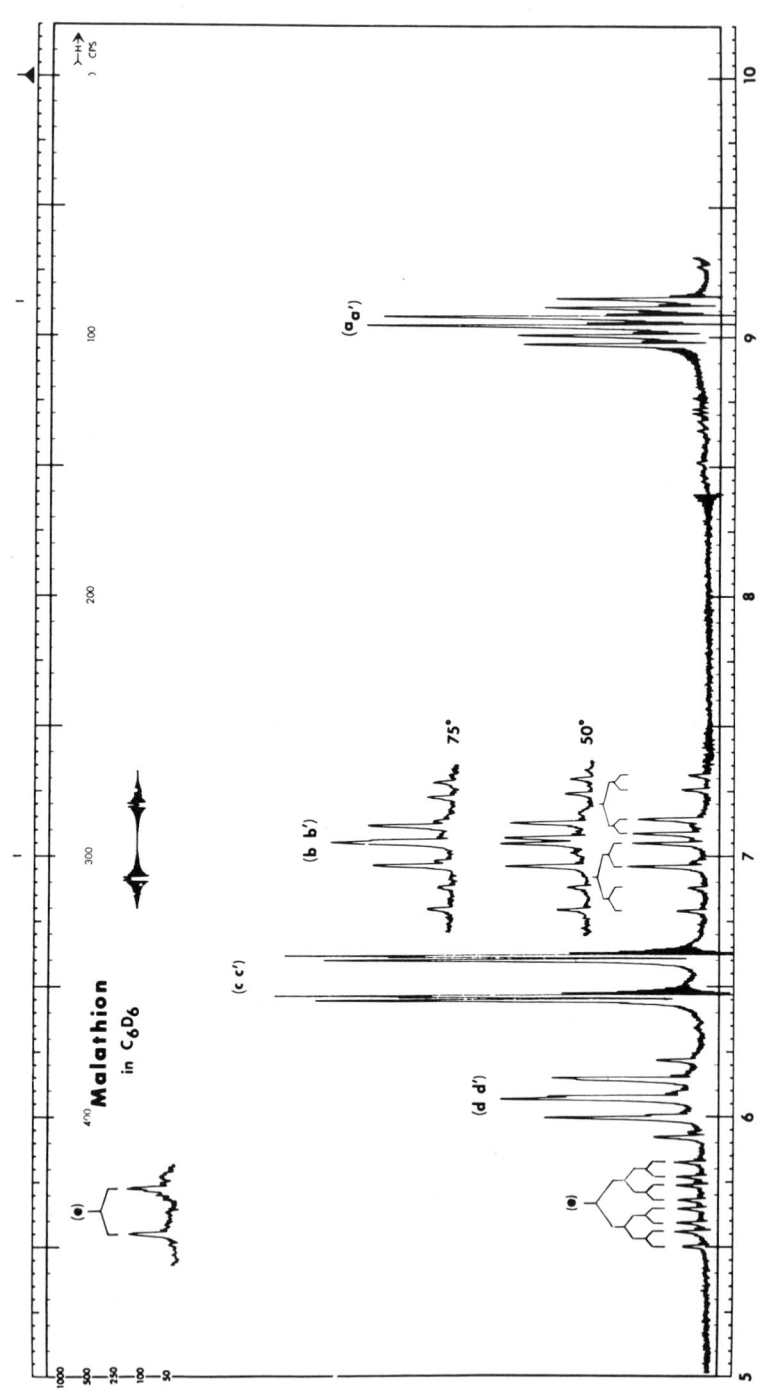

Fig. 7.9 NMR spectrum of malathion in C_6D_6. (L. H. Keith, A. W. Garrison, and A. L. Alford, *J. Assoc. of Offic. Anal. Chem.*, **51**, 1063 (1968). Reprinted with permission of AOAC.

types of organophosphorus pesticides have been tabulated (135,137) and some of these are given in Table 7.8.

NMR has been employed to investigate equilibrium and thermal effects in organophosphorus pesticides. In separate studies 12 years apart Miller and Goldenson (141) and Babad and Taylor (142) used NMR to investigate the composition of pesticides in solution. In the first study it was found that a solution of the pesticide Dematon consists of an equilibrium mixture of the tautomers (7.I) and (7.II):

$$\begin{array}{cc} C_2H_5O\diagdown\quad\diagup S \\ P \\ C_2H_5O\diagup\quad\diagdown OC_2H_4SC_2H_5 \end{array} \quad\rightleftharpoons\quad \begin{array}{cc} C_2H_5O\diagdown\quad\diagup SC_2H_4SC_2H_5 \\ P \\ C_2H_5O\diagup\quad\diagdown O \end{array}$$

7.I   7.II

The thiono tautomer (7.I) shows a ^{31}P-NMR resonance at -67.7 ppm whereas the thiol (7.II) form was at -25.9 ppm relative to phosphoric acid. The equilibrium ratio was determined from the area beneath the respective peaks.

Babad and Taylor found that commercial preparations of Dematon contained the *o,o,o,o*-tetraethyldithiopyrophosphate (142).

$$\begin{array}{c} C_2H_5O\diagdown\quad\diagup S \quad\quad S\diagdown\quad\diagup OC_2H_5 \\ P\text{---}O\text{---}P \\ C_2H_5O\diagup\quad\quad\quad\quad\quad\diagdown OC_2H_5 \end{array}$$

7.III

These authors devised a set of equations to determine the three compounds (7.I, 7.II, and 7.III) in Dematon by integration of the ^{1}H-NMR resonance curves of the methyl protons (1.1–1.5 ppm), the thio ether protons (2.3–3.1 ppm) and the oxy-ether protons (3.8–4.3 ppm). The mole fraction of each component is given by

$$X_{7.I} = \tfrac{3}{8}A_{OH} + \tfrac{1}{8}A_{SH} - \tfrac{1}{4}A_{CH} \tag{7.9}$$

$$X_{7.II} = \tfrac{1}{6}A_{CH} - \tfrac{1}{4}A_{OH} \tag{7.10}$$

$$X_{7.III} = \tfrac{5}{48}A_{CH} - \tfrac{3}{32}A_{SH} - \tfrac{3}{32}A_{OH} \tag{7.11}$$

where A_{CH}, A_{OH}, and A_{SH} are the respective areas of resonances due to protons in aliphatic groups, near ether linkages and near thioether linkages.

TABLE 7.8. Trends in NMR Parameters as Related to the Structure of Organophosphorus Pesticides

Compounds with $CH_3O-\overset{\overset{O}{\|}}{\underset{|}{P}}-$

Compound	$\tau\ CH_3$	$J_{P-CH_3 (Hz)}$
Ciodrin	6.20	11
Neguvon	6.09, 6.10	11
Bidrin	6.17	11
Dibrom	6.08, 6.09	12
Phosphamidon	6.12, 6.20	12
Phosdrin	6.25	11
Malaoxon (oxygen analog of malathion)	6.17, 6.18	13
DDVP	6.14	11
Meta-Systox-R	6.18	13
Dimethoxon	6.17	12.5
Guthion oxygen analog	6.19	13
Ruelene	6.19	11

Compounds with $CH_3O-\overset{\overset{S}{\|}}{\underset{|}{P}}-$

Compound	$\tau\ CH_3$	$J_{P-CH_3 (Hz)}$
Tiguvon	6.19	14
Ronnel	6.10	14
Dicapthon	6.08	14
Methyl Trithion	6.28	15
Guthion	6.24	15
Malathion	6.19, 6.20	15.5
Cygon	6.20	15
Imidan	6.23	15
Methyl parathion	6.13	14
Zytron	6.20	14

Compounds with $-\overset{\overset{O}{\|}}{\underset{|}{P}}-O-\overset{\overset{R}{|}}{C}H-$

Compound	$\tau\ CH$	$J_{P-CH (Hz)}$
Dibrom (R = Br)	3.30	9
DDVP (R = =CCl$_2$)	3.00	5.5

TABLE 7.8 (*Continued*)

Compounds with $-\overset{\overset{O}{\|}}{\underset{|}{P}}-S-\overset{\overset{R}{|}}{CH}-$

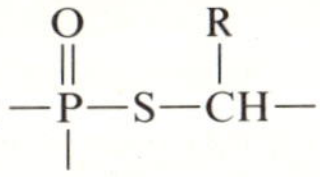

Compound	τ CH	$J_{P-CH(Hz)}$
Malaoxon $\left(R = C\diagdown\overset{\overset{O}{\|}}{\diagup}_{O-}\right)$	5.9	—
(oxygen analog of malathion)		
Dimethoxon (R = H)	6.49	17
(oxygen analog of Cygon)		
Meta-Systox-R (R = H)	6.8	—
Guthion oxygen analog (R = H)	4.20	15
DEF (R = H)	7.01	15

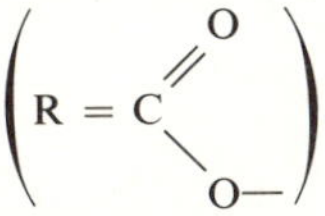

Compounds with $-\overset{\overset{S}{\|}}{\underset{|}{P}}-S-\overset{\overset{R}{|}}{CH}$

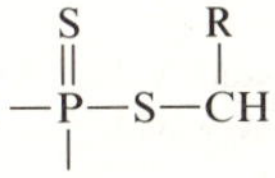

Compound	τ CH	$J_{P-CH(Hz)}$
Guthion (R = H)	4.23	16
Delnav (R = —O—)	4.40	16
Ethion (R = H)	5.74	17
Thimet (R = H)	5.97	13
Malathion 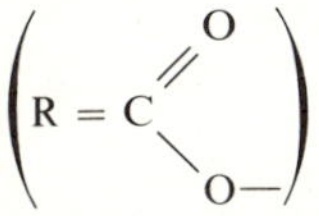 $\left(R = C\diagdown\overset{\overset{O}{\|}}{\diagup}_{O-}\right)$	5.9 (in $CDCl_3$) 5.67 (in C_6D_6)	17.5 (C_6D_6)
Cygon (R = H)	6.44	18
Imidan (R = H)	4.98	15
Phencapton (R = H)	5.66	14
Trithion (R = H)	5.72	13.5
Methyl Trithion (R = H)	5.74	14
Di-Syston (R = H)	6.5–7.5	—

Compounds with $CH_3CH_2O-\overset{\overset{O}{\|}}{\underset{|}{P}}-$

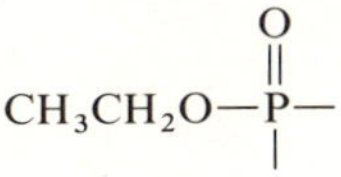

Compound	τ CH_2	$J_{P-CH_2(Hz)}$	$J_{H-H(Hz)}$	τ CH_3	$J_{P-CH_2(Hz)}$
Paraoxon	5.74	8.5	7	8.61	0.9
(oxygen analog of parathion)					
Diazoxon	5.65	8	7	8.61	1.2
(oxygen analog of Diazinon)					

TABLE 7.8. (*Continued*)

Compounds with $CH_3CH_2O-\overset{\overset{S}{\|}}{\underset{|}{P}}-$

Compound	$\tau\,CH_2$	$J_{P-CH_2(Hz)}$	$J_{H-H(Hz)}$	$\tau\,CH_3$	$J_{P-CH_3(Hz)}$
Co-Ral	5.75	10	7	8.60	0.8
Zinophos	5.65	9	7	8.59	0.8
Ethyl parathion	5.75	10	7	8.63	0.8
Diazinon	5.66	9.5	7	8.62	0.8
Di-Syston	5.83, 5.85	10	7	8.65	0.7
Trithion	5.88, 5.91	10	7	8.67	0.8
Phencapton	5.82, 5.84	10	7	8.64	0.8
Thimet	5.815, 5.83	10	7	8.64	0.8
Ethion	5.80, 5.82	10	7	8.64	0.8
Delnav	5.83	10	7	8.64	—
EPN	5.73	10	7	8.64	0.5

(L. H. Keith, A. W. Garrison and A. L. Alford, *J. Assoc. of Offic. Anal. Chem.*, **51**, 1063 (1968). Reprinted with permission of AOAC.)

7.3.5.2 DDT-type Pesticides

The DDT (dichlorophenyl trichloroethane) family of pesticides has the skeletal structure of substituted diphenylmethane as illustrated specifically

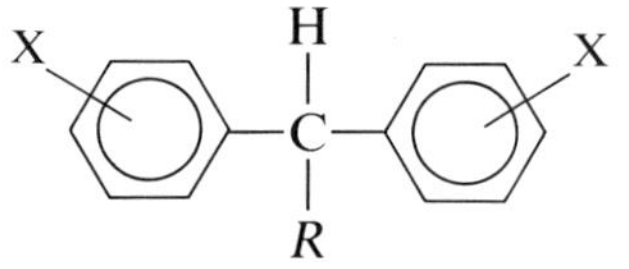

7.IV

by *p,p'*-DDT [1,1,1-trichloro-2,2-bis(*p*-chlorophenyl)ethane] (7.V). The

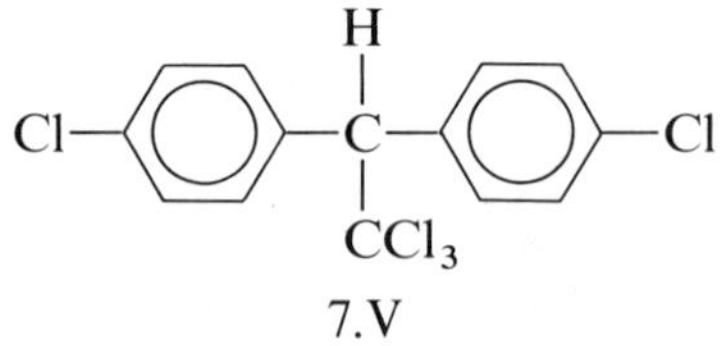

7.V

majority of this class of pesticide contains *para*-substituted benzene rings leading to an $AA'BB'$ pattern for the aromatic proton ^{1}H-NMR spectrum. These protons show slight variations in their chemical shift and coupling constants with various substituents, R. However, the portion of the spectrum due to the substituents is more useful in identification of the specific compound. Of course, careful analysis of the complex patterns of the aromatic protons must be conducted in order to determine the substitution on the benzene rings.

7.3.5.3 Applications of NMR to Pesticides Found in the Environment

NMR offers two important advantages for the identification of pesticides and their degradation products in the environment. First, from the background information that has been collected, positive identification of pesticides is less ambiguous by NMR than by some other techniques. Second, the method is nondestructive so other tests may be applied to limited samples after NMR data are collected. The relatively large sample required even with FT-NMR techniques is a limitation that will restrict NMR to a "confirmation" tool rather than a primary one for the identification of pesticides in the environment. Garrison, Keith, and Alford have reported several such examples (136).

Following extensive fish kills in the waters of Charleston Harbor, S.C., an organic thiophosphate was extracted from the water. In this case it was known that the pesticide Merphos (7.VI) was likely to be discharged into

$$(CH_3CH_2CH_2CH_2S)_3P$$
7.VI

these waters. The mass spectrum of the compound extracted showed a parent ion peak at mass 314, which is 16 mass units greater than Merphos. The ^{1}H-NMR spectrum of the extract was similar to that of Merphos, but the signal produced by the six methylene protons on the carbons bonded to the sulfur atoms was shifted downfield from the standard. Based upon the mass spectral data and the direction of this shift, it was reasonable that oxidation of the compound to (7.VII) had occurred. Comparison of the

$$(CH_3CH_2CH_2CH_2S)_3P\!=\!O$$
7.VII

NMR spectra of a known sample of 7.VII with that of the extract showed they were identical (143).

Degradation of Phygon (7.VIII) was observed in benzene solution under

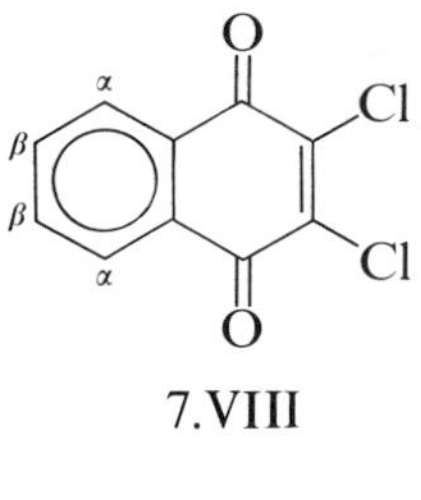

7.VIII 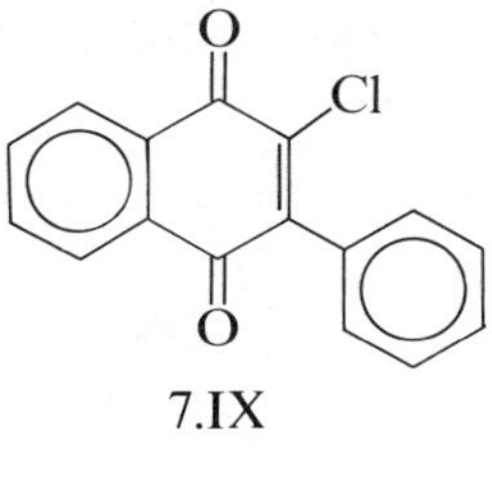   7.IX

laboratory conditions (144). The most abundant degradation product was isolated by gas chromatography. The NMR spectrum of this component showed three types of aromatic protons in the ratio 2:2:5. Phygon shows only two types of aromatic proton resonances, from the α- and β-protons. Mass spectral data show that only one chlorine atom remains in the degradation product. This evidence suggested the degradation involved substitution of an aromatic ring for one chlorine atom. The compound 7.IX was synthesized and comparison of NMR, IR and mass spectral data showed it was identical to the degradation product (144).

7.3.5.4 Future Applications of NMR to Pesticides

The enhanced sensitivity achieved by FT-NMR and its growing availability will surely lead to further applications of NMR to the study of pesticides. However, new directions will be taken to determine fine structural features of the pesticides and to investigate the mechanism of their biological activities. For example, shift reagents have been effectively employed to determine the stereochemical features of pesticides and to observe their changes during degradation (145). These determinations are experimentally simple, whereas they would be exceedingly tedious, if possible at all by other techniques.

The nature of molecules adsorbed on various surfaces may be a principal factor in determining the rate and pathways of their degradation. Although this research has not yet yielded much information on pesticides to date, the principles and techniques, as well as goals for the research, have been considered (146). The type of information obtained deals with the correlation time (τ) for molecular motion using NMR relaxation methods. The type of molecular motion and the significance of the correlation time depends upon the nature of the surface interaction and the symmetry of the absorbed molecule. However, indications as to whether preferential orientation of the molecule on the surface exist and the energy of these orientations may be obtained.

A more recent technique developed by Waugh and associates (147) which negates the large nuclear dipolar broadening of NMR lines in solids and enhances the signal of dilute nuclear spins (viz. ^{13}C) may prove very valuable in these studies. This technique of "cross-polarization" NMR spectroscopy can provide values of the chemical shift tensors; thus more subtle molecular electronic perturbations of adsorbed molecules may be observed.

7.3.6 PHARMACEUTICALS

The applications of NMR spectroscopy to the pharmaceutical industry covers a broad scope of investigations. Several reviews have been published (148–150). The most common applications of NMR in the pharmaceutical industry can be grouped into the three categories of structure elucidation, molecular interaction studies, and quantitative determinations.

7.3.6.1 Quantitative Applications

The determination of aspirin, phenacetin, and caffine in APC tablets discussed as an example in Section 6.4.2 was one of the earliest drug analyses performed using NMR (151). This determination has been used frequently as an undergraduate laboratory experiment as it illustrates well several of the considerations that must be given to the use of NMR for multicomponent determinations on real samples. Most of the quantitative applications of NMR in drugs and pharmaceuticals are straightforward applications of peak height or peak area measurements. Kasler (152) has given details of reported procedures for several drug analyses. Some of the reported applications include corticosteroids (153), estrogens (154), barbiturates (155), hypnotics (156), cryptopine in thebaine (157), mebromate (158), and chlorophenosin-2-carbamate as an impurity in Maolate (chlorophenosin-1-carbamate), a muscle relaxing agent (159). Typically, direct integration of the spectrum or a portion of the spectrum is employed with relative accuracies of 1–3% and standard deviations about the mean of near 0.5%. Occasionally, some chemical or other spectroscopic aid complements the use of NMR in the determination. For example, in the determination of the impurities in Maolate (159), the spectrum of both carbamate isomers was simplified by exchanging the active hydrogens with deuterium. The exchange rate was promoted by the addition of HCl gas to a d_6-acetone solution. Figure 7.10

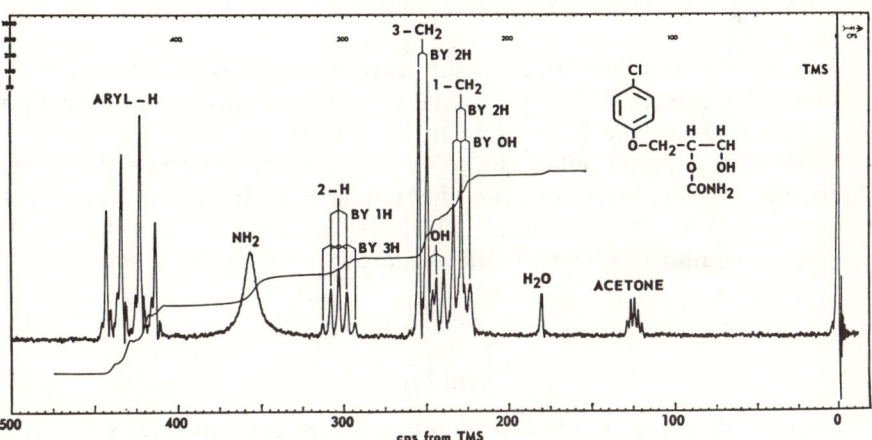

Fig. 7.10 NMR spectrum of secondary isomer in d_6-acetone. (Reprinted with permission from G. Slomp, R. H. Baker, and F. A. Mackeller, *Anal. Chem.*, **36**, 375 (1964). Copyright by the American Chemical Society.)

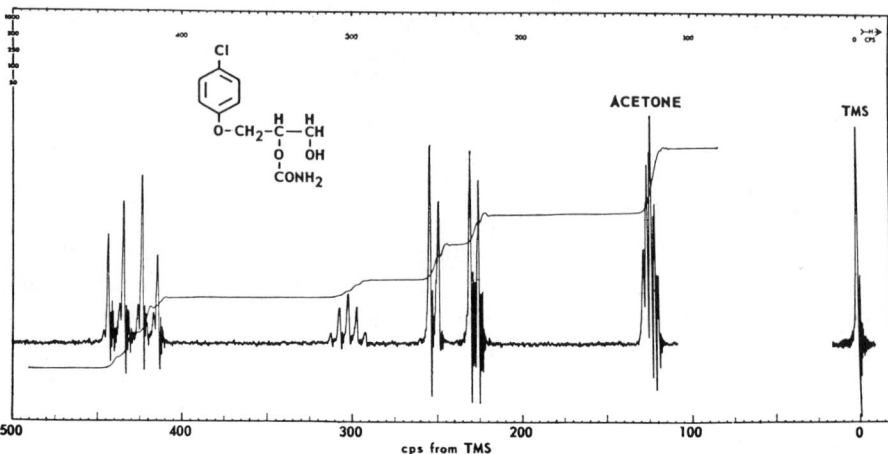

Fig. 7.11 NMR spectrum of secondary isomer in acidified d_6-acetone. (Reprinted with permission from G. Slomp, R. H. Baker, and F. A. Mackeller, *Anal. Chem.*, **36**, 375 (1964). Copyright by the American Chemical Society.)

shows the ^{1}H-NMR spectrum of the impurity, chlorophenosin-1-carbamate, in d_6-acetone, whereas Figure 7.11 shows the same compound in acidified d_6-acetone. The NH$_2$ resonance is completely removed as is the OH resonance in the solvent acidified with DCl. The splitting of the 1-CH$_2$ proton resonance by the hydroxyl protons is also eliminated. Thus the quantitation of a mixture of the two isomers may be conducted by measuring the peak height at 4.06 ppm for Maolate and 3.75 ppm for the chlorophenosin-2-carbamate.

For the determination of small quantities of barbiturates, Rücker employed an unusual approach (155). An external standard was prepared of 5 mg/ml in *d*-chloroform of the barbiturate to be determined. A 0.2-ml microcell was used that required only 1 mg of the standard. A computer of average transients was employed to scan the standard *n* times. An instrument constant, *k*, was then obtained which is the integral step height in mm/mg/ml · number of scans · number of protons giving the signal

$$k = \frac{\text{mm}}{\left(\dfrac{\text{mg}}{\text{ml}}\right)_{\text{std}} \cdot n \cdot v} \tag{7.12}$$

The concentration in an unknown solution is then calculated as:

$$\frac{\text{mg}}{\text{ml}} = \frac{mm}{k \cdot n \cdot v} \tag{7.13}$$

where mm is the integral step height after n scans of a peak resulting from v protons. The authors suggest that the product, $n \cdot v$, should exceed 600 and a precision of 5–10% was obtained. Obviously, the standard should be checked frequently and the instrument must be very stable toward drift. FT-NMR techniques will greatly simplify this type of application, and only a few micrograms may be determined in the same time period as 1 mg with repeated cw-scans.

The quantitative applications of NMR spectroscopy to the drug and pharmaceutical industry will likely increase with the spread of FT-NMR instrumentation. Even so, routine methods using cw-NMR are being developed as simple solutions to sometimes old analytical problems. For example Warren et al. (160) have applied the method to the rapid determination of chloroform in cough syrup formulations. Chloroform appears in these products in the range of 0.2–0.4%. Using peak height measurements and phenylacetic acid as an internal standard, as little as 0.1% chloroform was determined with accuracy and precision of about 5%.

7.3.6.2 Structure Elucidation

Many applications of NMR to the elucidation of the structure of new synthetic and newly discovered natural compounds of pharmaceutical interest are made daily. Techniques such as those described in Chapter 4 are used routinely in pharmaceutical research and most probably go unreported. It is obvious that in the scope of this monograph only a few illustrative examples can be discussed.

Occasionally important structural information may be determined from a single cw-NMR spectrum. For example, when benzaldehyde is condensed with nitroimidazoles, the resulting styrene derivative may be *cis* or *trans* with respect to the olefinic double bond. These products are confirmed as *trans* from the observation that the value of $J_{AB} \sim 16$ Hz (161).

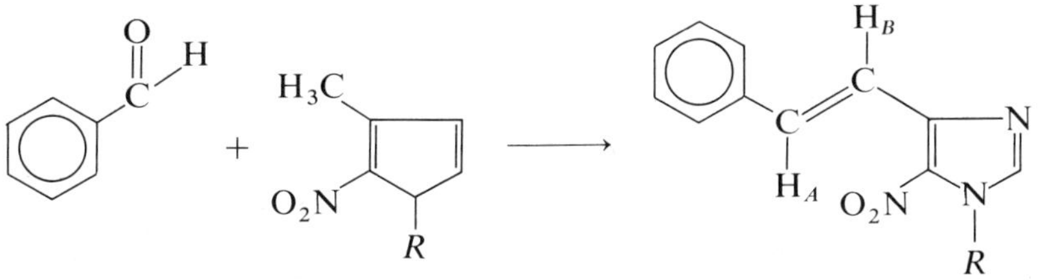

At times the conclusion that appears the most obvious may in fact not be the correct answer. The reaction of 2,3-dihydropyran with 3,6-dicarbomethoxy-s-tetrazine may yield one or both of the isomers represented by

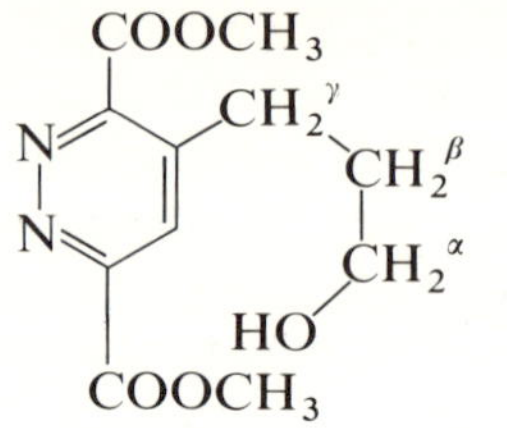

7.X

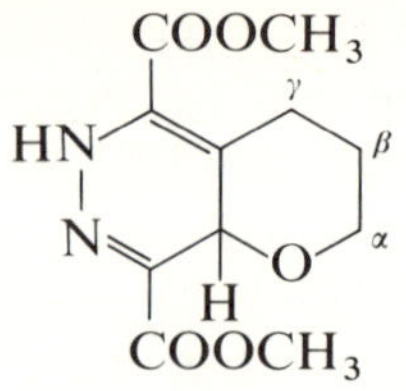

7.XI

structures 7.X and 7.XI. The NMR spectrum of the product in deuterochloroform can be attributed to either isomer. Simple shaking of the deutuochloroform solution with D_2O causes a singlet to disappear. The position of the peak and the lability of the proton giving rise to it would suggest an NH proton. However, the spectrum of the product in deuterodimethylsulfoxide shows that the band that disappears is a triplet in this solvent strongly indicating a hydroxyl proton (Section 7.2.1). The α-methylene proton resonance lines show additional splitting in deuterodimethylsulfoxide as expected, and which confirms structure 7.X (162).

When biologically active compounds are isolated from natural sources a wide variety of structures is often consistent with the available data. An isoleucine antagonist, later to be named Furanonycin, was isolated from culture filtrates of Streptomyces L803 (163). Based on evidence primarily other than nuclear magnetic resonance spectra, three possible structures were proposed. The ^{1}H-NMR spectrum of the isolated compound contains an AB pattern centered at 5.95 ppm with $\Delta\delta_{AB} = 0.43$ ppm and $J_{AB} = 6.3$ Hz.

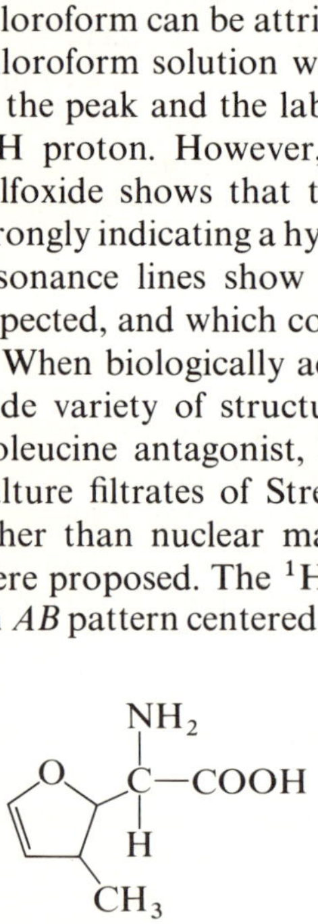

7.XII

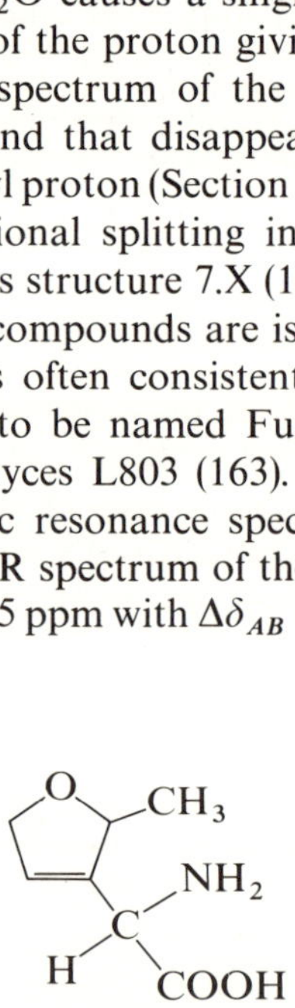

7.XIII

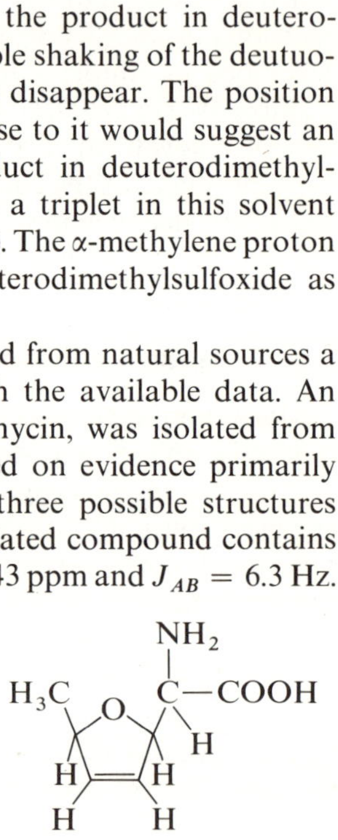

7.XIV

Further splitting of the AB protons is observed. This resonance pattern was assigned to the two olefinic protons. A three-proton doublet at 1.23 ppm ($J = 6.4$ Hz) was assigned as a methyl group attached to a tertiary carbon atom. The conclusion that structure 7.XII can be ruled out is drawn from the presence of a one proton doublet at 3.82 ppm ($J = 2.6$ Hz) which indicates that the amino acid group must be an α-amino acid. Otherwise, other proton resonances would be observed for hydrogen atoms in the α-position. Single proton multiplets at 5.19 and 5.42 ppm were assigned to tertiary ring protons.

One of these multiplets contains a coupling constant equal to that seen for the methyl proton resonance doublet and the other multiplet contains a coupling constant equal to that for the doublet at 3.82 ppm. The low-field position of both these protons indicates strong deshielding by the oxygen linkage and the double bond. Therefore only structure 7.XIV fits the spectral data.

7.3.6.3 Molecular Interactions

A major application of NMR in studies related to pharmaceutical chemistry are those concerning molecular interactions. There are various types of interactions which are significant in the mechanisms of drug action and NMR is a useful tool to use in their investigations. Both high resolution NMR spectroscopy and relaxation methods may be employed.

Hydrogen bonding is one form of inter- or intramolecular interaction which is easily studied by high resolution NMR. A plot of chemical shift of the proton involved versus concentration can provide a measure of the strength of the hydrogen bond formed (164–167) and a careful examination of the curve can discriminate between intra- and intermolecular hydrogen bonding (168). In the case of enantiomeric compounds, the selective binding between the solute and optically active solvents can resolve the NMR spectra of the enantiomers. Compounds of pharmaceutical interest such as cocaine have been investigated in this way using (+) and (−) phenyl methyl carbinol as the solvent (169).

The molecular interaction studies, of the type previously mentioned, apply to small molecules whose high resolution NMR spectra are observable in detail. Another case of interest is one in which one of the species interacting is a macromolecule such as a protein. Fisher and Jardetzky showed that much information concerning the binding of penicillin to serum albumin was obtained from the high-resolution spectrum of penicillin G (170). The theory of the techniques employed have been summarized by Fisher (149). In brief, when a relatively small molecule such as penicillin interacts with a macromolecule, significant reductions in the nuclear relaxation time (T_2) may occur; the principal mechanism being dipolar–dipolar nuclear interaction. The result is a broadening of the linewidth of all nuclei in the smaller molecule influenced by the interaction. It is important to be aware, however, that many factors such as paramagnetic impurities, chemical exchange and viscosity can also lead to line-broadening and detailed studies are required to confirm that the observed line width are a result of only molecular interactions.

Figure 7.12 shows as an example the ^{1}H-NMR spectrum of 0.5M penicillin G in D_2O and in D_2O with 10% albumin (170). The spectral assignments are shown on the spectra. Notice that in the presence of the albumin there are

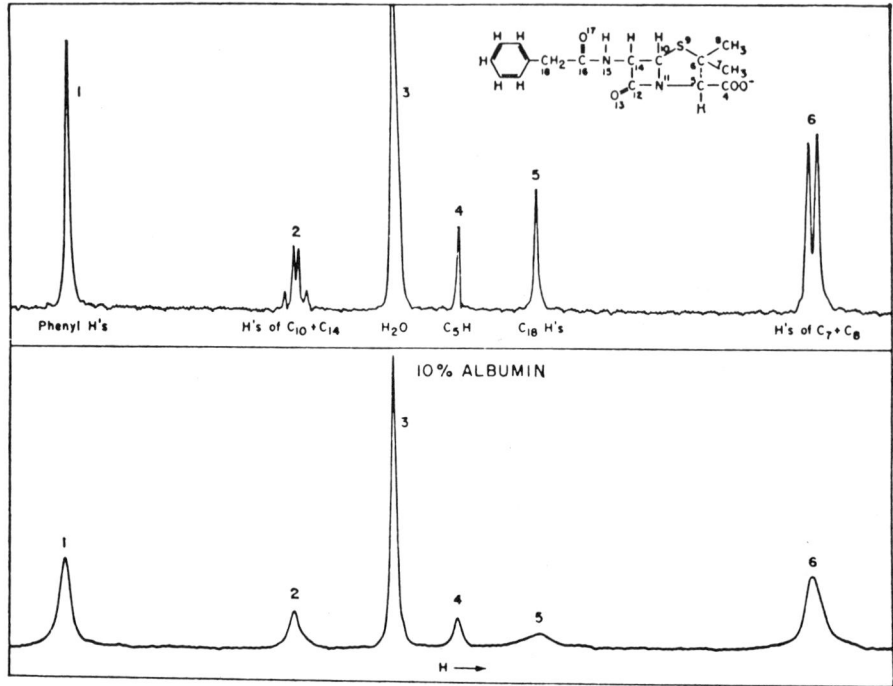

Fig. 7.12 (Reprinted with permission from J. J. Fischer and O. Jardetzky, *J. Am. Chem. Soc.*, **87**, 3237 (1965). Copyright by the American Chemical Society.)

no significant changes in the chemical shifts of the penicillin peaks and that the spectrum of the albumin is not seen. Also notice that the peaks are broadened to varying degrees which aids in ruling out chemical exchange between the "complex" and free penicillin as a major contributor to the broadening. Fisher and Jardetzky studied the line-broadening for several derivatives and over a range of 0.01–0.5M penicillin and 0–10% albumin concentrations. Figure 7.13 shows a plot of the line widths (as $1/T_2$) for three peaks labeled in Figure 7.12. These three peaks represent protons from the penicillin nucleus of the molecule (no. 4), the alkyl side chain (no. 5), and the aromatic ring (no. 1). The results show that the side chain and the aromatic ring are most influenced by the penicillin–albumin interaction. As a result of the complete study, the authors concluded that the interaction was dependent upon the nature of the side chain. Only when the side chain contained aromatic groups did the binding occur to any significant degree.

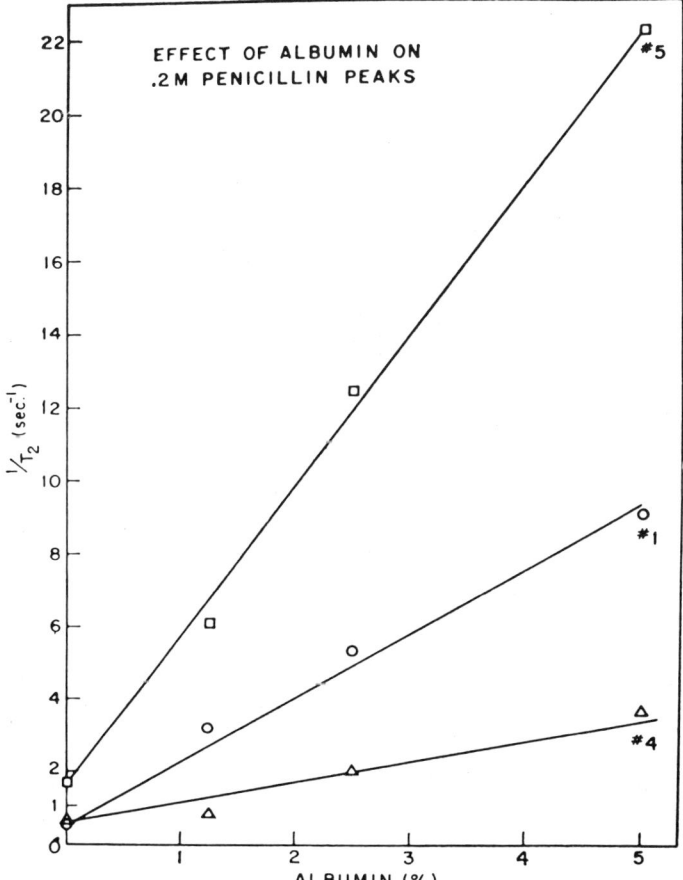

Fig. 7.13 (Reprinted with permission from J. J. Fischer and O. Jardetzky, *J. Am. Chem. Soc.*, **87**, 3237 (1965). Copyright by the American Chemical Society.)

For example, penicillin V (phenoxy derivative) partially displaced penicillin G indicating a specific interaction at sites competed for by the two forms. Studies of this type will play in increasingly important role in the attempt to understand the nature of drug interactions with molecules present in physiological systems.

The techniques of relaxation studies by pulsed NMR and the use of various chemical probe techniques are of importance in drug interaction investigations. These techniques are of equal value in the application of NMR in biochemistry, and will be discussed in Section 7.4.1.

7.3.6.4 Summary

The applications of NMR to the quantitative analysis of drugs and other biologically active chemical compounds has become routine. The use of high-resolution NMR for the elucidation of structure in synthetic and natural drug related compounds is common in the industry. In the investigation of molecular interactions of these types of compounds, NMR spectroscopy is one of the few techniques of value. Both high-resolution and pulsed relaxation methods are useful. Many of the techniques important to drug studies are those of value in biochemical research.

Without doubt, FT ^{13}C-NMR and ^{1}H-NMR will play an increasingly important role. For example, excellent ^{1}H-NMR spectra have been obtained for 10 μg of 3-ethyl-vanillin in 15 min by FT-NMR (171).

7.4 APPLICATIONS TO FUNDAMENTAL INVESTIGATIONS

In the previous sections of this chapter we have presented examples of applications of nuclear magnetic resonance spectroscopy to the determination of organic functional groups and to several industrial topics. In earlier sections, the techniques of structure elucidation were presented. However, nuclear magnetic resonance is a technique of exceptionally broad scope. It may be used in perhaps more different kinds of studies and fields of endeavor than any other instrumental method. Therefore, in this section we will present a few selected types of investigations to illustrate how nuclear magnetic resonance techniques may be used in a variety of chemical studies. Also, some uncommon techniques will be presented which have special application.

7.4.1 APPLICATIONS TO BIOCHEMISTRY

Early attempts to apply nuclear magnetic resonance to research in biochemistry must have been frustrating for the researchers. The organic chemists were determining old structures with greater accuracy and structures of new compounds were rapidly elucidated. With the exception of amino acids and simple peptides, the biochemist could obtain little information from nuclear magnetic resonance spectra. The sensitivity was poor and many of the compounds of interest were not sufficiently soluble, or the quantity needed was not available. When spectra were obtained on macromolecules of biological interest, they were extremely difficult to interpret to the degree that useful information could be obtained. As the instrumentation improved, some of these difficulties were significantly diminished. As will be discussed later, the high-resolution NMR spectra (^{1}H-NMR or ^{13}C-NMR) of macromolecules can now be useful. However during the development of more sensitive instruments with higher magnetic fields, the biochemist began using some more subtle approaches to the problem.

It is our intention here to present a brief survey of NMR methods which are applicable to biochemical studies. The treatment of the theory of these methods will be limited in scope. An excellent treatment in detail has been given in books by Dwek and James devoted to the applications of NMR to biochemistry (172,173).

7.4.1.1 Binding of Small Molecules to Macromolecules

In many cases, the biochemist is interested in the binding of small molecular ligands to biochemical macromolecules. If the ligand is bound to the macromolecule, one expects the same problems as encountered in attempts to observe the macromolecule directly because of the low concentration of the bound ligand. Also, the shorter relaxation times of nuclei bound to large molecules create large linewidths which further decrease sensitivity and degrade the resolution. As it happens, many such studies involve binding of ligands that undergo rapid chemical exchange with free ligand in solution. If we assume for the moment that there is only one binding site on the macromolecule and one form of free ligand in solution, then this becomes an example of the two-site chemical exchange case discussed in Section 6.7.2.1. If the chemical exchange rate is sufficiently large (rate $> 2\pi\Delta v$), then a single resonance will be observed for nuclei in the ligand (neglecting spin–spin coupling). The chemical shift of this resonance will be the mole fraction weighted average of the chemical shifts of the bound and free ligand. The linewidth will be a function of the exchange rate. If the various chemical parameters are favorable, a large excess of ligand over macromolecule may be used and a strong signal obtained. In this way the effects of parameters such as pH, temperature and ligand structure and concentration can be investigated. Observation of the linewidth (better yet, the measurement of T_1 relaxation times) of several nuclei in the ligand can provide more details as to the nature of the binding. Obviously, this approach is not as desirable as direct observation of the high-resolution NMR spectra of the macromolecules so that details of the actual binding site are made available. However, the information obtained is actually acquired with relative ease. Both the chemical shift and the linewidth of rapidly-exchanging bound and free ligands may be used to evaluate binding constants. An example of each will be given.

A mutarotated solution of N-acetylglucosamine (NAG) shows two sets of resonance lines for the anomeric protons of the α- and β-forms of NAG. The N-methyl group hydrogens show only one resonance for both forms.

On addition of lysozyme to the NAG solution, two resonance lines for the methyl groups are observed at higher field than the original resonance indicating that the two anomers of NAG have different binding constants and/or that the bound ligands are at magnetically nonequivalent sites. Adding α-NAG in which the N-methyl protons are replaced by deuterium shows that both methyl resonances decrease as a result of competition. This shows that α-NAG binds at the same site as β-NAG. The difference in the chemical shifts must then be a result of different binding constants. This conclusion is drawn from the considerations of an equilibrium between free ligand, L, and a macromolecule, E, represented by:

$$EL \rightleftharpoons E + L \qquad (7.14)$$

and resulting in an equilibrium constant, K, given by

$$K_L = \frac{[E][L]}{[EL]} \quad \text{liters/mole} \qquad (7.15)$$

If the chemical shift of the bound ligand measured with reference to the free ligand is $\Delta\omega_B$, then the observed chemical shift, $\Delta\omega$, in a rapidly exchanging system is the average of the fraction of ligand which is bound.

$$\Delta\omega = P_B \Delta\omega_B \qquad (7.16)$$

In order to observe the NMR spectrum of the ligand, the conditions are $[L]_t \gg [E]_t$ where the concentrations are the total concentration of the ligand and enzyme, respectively. Under these conditions

$$P_B \cong \frac{[EL]}{[L]_t} \qquad (7.17)$$

and

$$[L]_t \cong [L] \gg [EL] \qquad (7.18)$$

where $[L]_t$ is the total ligand concentration in solution. Combining equations 7.15 and 7.16 and using the previous approximations, the simple relationship

$$[L]_t = \frac{\Delta\omega B[E]_t}{\Delta\omega} - K_L \qquad (7.19)$$

is obtained where $[E]_t$ is the total concentration of the macromolecule. A plot of $[L]_t$ versus $1/\Delta\omega$ is linear if the previous approximations obtain, and the line has a slope of $\Delta\omega_B$ and intercept of K_L. Figure 7.14 shows such a plot for the binding of α- and β-anomers of methyl N-acetyl-β-D-glucosamine to hen white lysozyme (174). Table 7.9 shows some results obtained by this method. Equation 7.19 may be expanded to take into account the

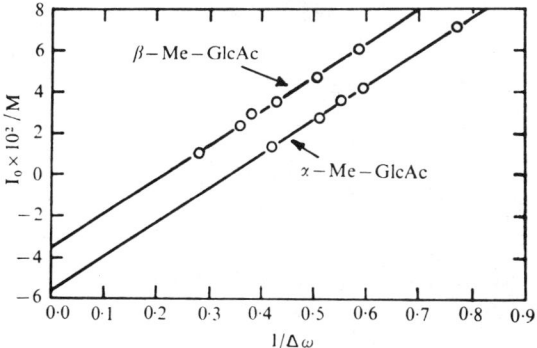

Fig. 7.14 The reciprocal of the observed chemical shift change of the acetamido methyl protons ($1/\Delta\omega$), as a function of total inhibitor concentrations (I_0) on addition of various concentrations of methyl N-acetyl-α-D-glucosamine or methyl N-acetyl-β-D-glucosamine to a constant concentration (0.003 M) of hen egg white lysozyme. (M. A. Raftery, F. W. Dahlquist, S. I. Chan, and S. M. Parsons, *J. Biol. Chem.*, **243**, 4175 (1968), with permission.)

simultaneous competition for the same binding site by two ligands yielding

$$[L]_t = \frac{\Delta\omega_M [E]_t}{\Delta\omega} - K_L - \frac{K_L}{K_2}[L_2]_t \quad (7.20)$$

where K_2 and $[L_2]_t$ are the binding constants and total concentration, respectively, for the second ligand. In cases in which the two binding constants are similar or very small, other techniques must be used.

The use of linewidth measurements is very similar in principle to the use of chemical shift in evaluating ligand binding constants to macromolecules. If a large excess of ligand is present and it is assumed that there is only one

TABLE 7.9. Data for Binding of the Anomers of NAG and Me-NAG to Lysozyme

		$\Delta\omega_M$ (ppm)	
Inhibitor	K_1 (mM)	(a) NH·COMe	(b) OMe
β-NAG	33 ± 2	0.51 ± 0.03	
α-NAG	16 ± 1	0.68 ± 0.03	
β-Me-NAG	33 ± 5	0.54 ± 0.04	0.17 ± 0.03
α-Me-NAG	52 ± 4	0.55 ± 0.02	0

(M. A. Raftery, F. W. Dahlquist, S. I. Chan and S. M. Parsons, *J. Biol. Chem.*, **243**, 4175 (1968).

binding site for the ligand, and that fast chemical exchange is in effect, then:

$$T_{2\text{obs}} = \frac{(T_{2,B} + \tau_B)}{[E]_L}(K_L + [L]_t) \tag{7.21}$$

where $T_{2\text{obs}}$ is the relaxation time observed for the ligand in the presence of the enzyme, $T_{2,B}$ is the relaxation time of the bound ligand, and τ_B is the mean lifetime of the bound ligand before exchange with a free ligand. The remaining terms have the same meaning as before. The value of $T_{2\text{obs}}$ may be taken as $1/\pi\Delta v_{\text{obs}}$ where Δv_{obs} is the full width at half-height of a ligand resonance line expressed in hertz. Substituting $1/\pi\Delta v_{\text{obs}}$ for $T_{2\text{obs}}$ is equation 7.22 and rearranging gives

$$[L]_t = \frac{[E]_t}{(T_{2,B} + \tau_B)} \frac{1}{\pi\Delta v_{\text{obs}}} - K_L \tag{7.22}$$

which is a very similar form as equation 7.19. K_L may be determined from a plot of $[L]_t$ versus $1/\Delta v_{\text{obs}}$. However, in many cases the binding by the ligand is very strong resulting in the need for low concentration of the ligand. This in turn makes the NMR spectra difficult to obtain. In these cases, equation 7.22 may be expanded to account for a competing ligand similar to that described above for chemical shift measurements. The resulting expression is

$$\frac{1}{\pi\Delta v_{\text{obs}}} = \frac{(T_{2,B} + \tau_B)}{[E]_t}\left([L]_t + K_L + \frac{K_L}{K_2}\frac{[L_2]_t}{[L]_t}\right) \tag{7.23}$$

where all terms have their previously defined meaning.

Although there is likely less potential for the use of spin–spin coupling constant measurements as probes for binding studies, in principle they may be used in a fashion similar to linewidth and chemical shift measurements. This is because the coupling constant of vicinal protons is dependent on the dihedral angle as expressed by the Karplus equation (Section 4.12.2.2). In most cases, the bound ligand will have an average conformation different from that of the free ligand as a result of interactions with the binding site. If the bound and free ligand undergo rapid chemical exchange, the observed coupling constant will be the mole fraction weighted average of the coupling constant for the free (J_A) and bound (J_B) ligand.

$$J_{\text{obs}} = P_A J_A + P_B J_B \tag{7.24}$$

This technique is not as useful for the determination of binding constants as chemical shift or linewidth measurements. However, it is of significant value in providing insight as to the conformation of the bound ligand. For example, the value of J_{H_1,H_2} is 7.0 ± 0.1 Hz for NAG-Glu-$C_6H_4 \cdot NO_2$ free in solution

p-Nitrophenyl-4-*O*-(2-deoxy-2-acetamido-β-D-glucopyranosyl)-β-D-glucopyranoside
(NAG-Glu-$C_6H_4 \cdot NO_2$.)

(175). This value corresponds to a dihedral angle ϕ of near 180°. From approximate calculations using the Karplus equation, a change in the dihedral angle to 120° should give a value of $J_{H_1, H_2} = 1.75$ Hz, which is a significant change. However, Rand–Meir and associates found no decrease in the coupling constant when NAG-Glu-$C_6H_4 \cdot NO_2$ is bound to lysozyme. The obvious conclusion is that the ligand does not undergo conformational changes on binding to the enzyme. However, these studies must be done in conjunction with other binding constant measurements so that the fraction of ligand bound may be calculated. If the complex is weak, then there may be a small fraction of bound ligand leading to an imperceptible change in the observed coupling constant.

7.4.1.2 Binding Studies Using Quadrupolar Nuclei

The previous examples have illustrated that straightforward ^{1}H-NMR measurements may be used to obtain important information concerning the binding of ligands such as inhibitors to macromolecules (viz. enzymes). A technique which has been of significant value is the use of the resonance of nuclei other than ^{1}H which have nuclear spin quantum numbers greater than $\frac{1}{2}$. These nuclei possess nuclear quadrupole moments that provide additional relaxation mechanisms. A nuclear quadrupole may interact with electrical field gradients. Such field gradients can be very strong in chemical bonds. Unless the field is highly symmetrical about the nuclear quadrupole, the relaxation time of the nucleus may be shortened to such a degree that the magnetic resonance lines become very broad and perhaps difficult to detect. Therefore, wide-line (low resolution) NMR techniques are generally employed. There are many such nuclei and some of these are listed along with some of their properties in Table 7.10.

The concentration of the enzyme that may interact with an ion of one of the nuclei shown in Table 7.10 will invariably be too low to allow direct

TABLE 7.10. Nuclear Properties of Some Naturally-Occurring NMR-Active Nuclei with Quadrupole Moments Compared with those of Hydrogen

Isotope	Natural Abundance (%)	Resonance Frequency MHz in Field of 10^4 G	Sensitivity at Constant Field Relative to H = 1.000	Nuclear Spin in Units of $\hbar$	Nuclear Electric Quadrupole Moment (Barns)
^{1}H	99.985	42.5759	1.000	$\frac{1}{2}$	0
^{2}D	0.015	6.5357	9.65×10^{-3}	1	2.77×10^{-3}
^{17}O	0.037	5.772	2.91×10^{-2}	$\frac{5}{2}$	-4.0×10^{-3}
^{14}N	99.635	3.076	1.01×10^{-3}	1	2×10^{-2}
^{23}Na	100	11.262	9.27×10^{-2}	$\frac{3}{2}$	0.1
^{25}Mg	10.05	2.606	2.68×10^{-3}	$\frac{5}{2}$	0.22
^{39}K	93.08	1.987	5.08×10^{-4}	$\frac{3}{2}$	0.07
^{41}K	6.91	1.092	8.39×19^{-5}	$\frac{3}{2}$	—
^{43}Ca	0.13	2.865	6.39×10^{-2}	$\frac{7}{2}$	—
^{85}Rb	72.8	4.111	1.05×10^{-2}	$\frac{5}{2}$	0.31
^{87}Rb	27.2	13.932	1.75×10^{-2}	$\frac{3}{2}$	0.15
^{133}Cs	100	5.585	4.74×10^{-2}	$\frac{7}{2}$	-0.3×10^{-3}
^{35}Cl	75.53	4.1717	4.70×10^{-3}	$\frac{3}{2}$	-7.89×10^{-2}
^{37}Cl	24.47	3.472	2.71×10^{-3}	$\frac{3}{2}$	-6.21×10^{-2}
^{79}Br	50.54	10.667	7.86×10^{-2}	$\frac{3}{2}$	0.33
^{81}Br	49.46	11.498	9.85×10^{-2}	$\frac{3}{2}$	0.28
^{127}I	100	8.5183	9.34×10^{-2}	$\frac{5}{2}$	-0.69

From: *Nuclear Magnetic Resonance in Biochemistry* by R. A. Dwek published by the Oxford University Press.

observation of the bound ion. Therefore, as before with the ^{1}H-NMR techniques, one must resort to observing the effect of rapid exchange between bound and free ions on the NMR spectra of a solution of ions of the nucleus to be observed. Some examples will be given here. However, several precautions must be taken to avoid serious errors in the interpretation of results obtained using this technique. The observed linewidth of the nuclear magnetic resonance must be obtained under conditions of fast chemical exchange. Because of the great sensitivity of the nuclear quadrupole to molecular motion, the molecular rotational correlation time, τ_R, may make a significant contribution to the linewidth. For nuclei with I > 1, the relaxation theory becomes complex and the relaxation times are no longer expressed as simple exponential functions. In these cases one will usually rely on changes in the observed linewidth as simply a probe to determine the effect of some parameter which is varied.

Halide ions are among the most commonly used in studies based upon rapid relaxation and chemical exchange of quadrupolar nuclei. The sensitivity of line-broadening increases with the square of the quadrupole moment if all other parameters are equal. Thus the order of sensitivity to relaxation effects is:

$$^{127}I > {}^{79}Br > {}^{81}Br > {}^{35}Cl > {}^{37}Cl$$

The chloride ion is of greater significance in biological systems and is therefore used most frequently in binding studies even though it is the least sensitive. It is important to be aware that many factors may affect the observed linewidth of the magnetic resonance signal of a quadrupolar nucleus. Even under slow exchange conditions, the relaxation rates of bulk halide ions in solution may change in the presence of protein as a result of changes in solvent structure. Ion–ion interactions may alter the linewidths; viscosity effects are also important. A carefully designed experiment with detailed analysis of the results is required to assure accurate conclusions. These ions may be used in binding studies in two major categories: when metal ions are absent in the macromolecule and when metal ions are present.

The direct binding of a halide ion to a macromolecule may be detected. For example, Zeppezauer et al. (176) showed that bromide ion was bound to lysozyme by observing the ^{81}Br-NMR line in aqueous solution. A solution of 0.35 M lysozyme in 0.5 M KBr at pH = 7.5 caused the ^{81}Br line to be three times as wide as in the absence of the enzyme. To prove this was not a viscosity effect, ^{85}Rb Br was used. There was no difference in the ^{85}Rb-NMR linewidth in the presence or absence of lysozyme. If the effect had been a result of increased viscosity, both the ^{85}Rb and ^{81}Br line would have been altered. It is also necessary to prove that the observed halide ion is undergoing fast exchange; that is, that the free and bound halide ion resonance is coalesced into a single line. In this case the ^{81}Br linewidth decreases with increasing temperature. This last observation is strong, but not conclusive, evidence that fast chemical exchanges are occurring. In summary, changes in linewidth of halide ion resonance are a powerful qualitative tool to demonstrate binding of the ion to a macromolecule. However, quantitative information is difficult to extract and usually some type of competition between the halide ion and other ligands will be investigated.

Halide ions may also be used as probes to investigate metal binding to proteins. The strong covalent bond formed between halide ions and metal ions such as Hg(II) create electric field gradients which significantly affect the relaxation of the halide nuclear spin state. An excellent example is the use of ^{35}Cl-NMR linewidth measurements to investigate mercury binding to —SH sites in proteins. Mercury bound to the available —SH sites may coordinate the chloride ion to form

$$-S-Hg-Cl$$

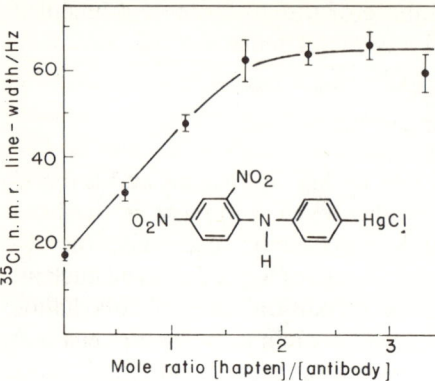

Fig. 7.15 ^{35}Cl n.m.r. titration of antidinitrophenyl antibody showing observed ^{35}Cl n.m.r. line-widths as a function of the DNP-mercurial: antibody mole ratio. (Reprinted with permission from R. P. Haughland, L. Stryer, T. R. Stengle, and J. D. Baldeschwieler, *Biochemistry*, **6**, 498 (1967). Copyright by the American Chemical Society.)

sites. The number of —SH groups available for coordination of the Hg(II) may be determined by plotting the ^{35}Cl-NMR linewidth versus the mole ratio of Hg(II) to protein. The linewidth will increase until all reactive sites are complexed by the mercury, then remain constant or increase only slightly with additional mercury concentration. These experiments were elucidated by Stengle and Baldeschwieler (177) who also suggested that the metal could be bound to a ligand rather than directly to the protein. The dominant mechanism for broadening of the halide line is the effect of the long molecular correlation time of the macromolecules. Therefore, it should make only little difference if the metal ion is bound directly to the macromolecule, or to a ligand which in turn is bound. This type of application is well illustrated by the titration of antidinitrophenyl antibody with a mercury-labeled hapten in $1M$ NaCl shown in Figure 7.15 (178). Addition of the labeled hapten causes a broadening of the ^{35}Cl-NMR resonance line until a hapten/antibody mole ratio of >2 is reached (178).

A knowledge of the relative efficiency and rate of different chelating agents toward the complete and selective removal of metals or "depoisoning" of proteins would be of great usefulness in the design and selection of chemotheropeutic agents for heavy metal poisoning. Sudmeier and Pesek (179) have made use of the decrease in the linewidth of the ^{35}Cl resonance to investigate the merits of several ligands toward the depoisoning of bovine serum albumin. They reported that EDTA removed cadmium but not mercury. 2,3-Dimercapto-1-propanol (BAL), D-cysteine, D,L-penicillamine, D-penicillamine, and 3-mercaptopropionic acid remove both cadmium and mercury. These studies appear to be very informative. However, quantitation of the data is difficult and a serious question may arise as to whether the metal is in fact removed from the protein. If the chelating ligand forms a mixed complex with the metal–protein complex, the chloride ion may not find sites available for coordination. The efficiency of the quadrupolar relaxation would decrease and the resonance line-width would decrease. This could be misinterpreted as a removal of the metal from the protein. This work is of great interest and potential, but will require a detailed investigation.

There are many other examples of the use of nuclei with quadrupole moments which could be cited. Sodium-23 and potassium-39 may serve as sources of important information on the mechanism of biological membrane systems. The advent of Fourier transform NMR spectrometers has encouraged the use of ^{14}N. This nucleus is present in virtually all biologically significant molecules. However, frequently its resonance line is too broad to observe.

7.4.1.3 Studies Using Paramagnetic Ions

Paramagnetic relaxation effects were recognized as methods of investigating structure and kinetic features in biochemical compounds about 1961. Much of the research since that time has been based on the increased relaxation rate of solvent water proton spins induced by the presence of paramagnetic ions. The effect of paramagnetic ions on the linewidth of ^1H-NMR lines is well known to anyone who has taken the NMR spectrum of any variety of samples. The presence of oxygen will broaden the resonance of protons in aromatic molecules. Traces of paramagnetic metal ions in aqueous solutions will broaden peaks. This line-broadening is a result of the interaction between an unpaired electron spin and the nuclear spin under observation by NMR. The electron has a magnetic moment about 10^3 times that of nuclear moments and generates strong local magnetic fields. It is the fluctuation of these fields that results in nuclear spin relaxation. The large field generated by the unpaired electrons create efficient relaxation processes of which there are two main mechanisms. The local field generated by the electron may couple through space with a nucleus by dipole–dipole interaction. The efficiency of this process depends in part on the distance between the electron and the nucleus, the magnetogyric ratio of the nucleus, the Larmor precessional frequencies of the electron and nucleus at the magnetic field strength of the instrument, and a correlation time of the interaction, τ_c. This correlation time may be defined as:

$$\frac{1}{\tau_c} = \frac{1}{\tau_s} + \frac{1}{\tau_m} + \frac{1}{\tau_R} \qquad (7.25)$$

where τ_s is the electron spin relaxation time, τ_m is the mean-lifetime of the nucleus in the paramagnetic site, and τ_R is the rotational correlation time of the paramagnetic site. The effectiveness of the dipole–dipole relaxation mechanism is then in part determined by the electron relaxation time, the chemical exchange rate of the nucleus or molecule bearing the nucleus, between the paramagnetic and some diamagnetic sites, and the rotational time of the molecule bearing the paramagnetic site.

A second mechanism for paramagnetic relaxation is the scalar coupling transmitted through chemical bonds. This process is the same as that which leads to spin–spin splitting in high resolution NMR and depends in part upon the nuclear–electron hyperfine coupling constant, the Lamor frequency of the electron, and a correlation time τ_e which is defined as

$$\frac{1}{\tau_e} = \frac{1}{\tau_s} + \frac{1}{\tau_m} \tag{7.26}$$

where τ_s and τ_m are as defined. We have written τ_s as a single electron relaxation time for simplicity. In fact in many cases, both the spin–lattice, $\tau_{1,s}$, and spin–spin, $\tau_{2,s}$, relaxation times must be considered (180). The discussion of the significance of τ_s, τ_R, and τ_m given here will be limited to qualitative statements. Dwek has given a more detailed discussion of the significance of these parameters (173).

To extract all of the potential information from studies using paramagnetic ions, all of the parameters must be considered. However, the nature of the observed results of NMR spectra in paramagnetic systems depend to a great extent upon the absolute and relative values of τ_s, τ_R, and τ_m. It is important to realize that this is an over-simplification but is useful in a qualitative discussion. The values of τ_c for the paramagnetic metal ions may be placed in two categories:

1. τ_c is mainly determined by the value of τ_R ($\tau_R < \tau_m < \tau_s$) and $\tau_c \simeq 10^{-10}$–10^{-11} sec.
2. τ_c is mainly determined by the value of τ_s ($\tau_s < \tau_R \leq \tau_m$) and $\tau_c \simeq 10^{-12}$–10^{-13} sec.

These classifications may also be stated as those paramagnetic ions which have long (1: Mn(II), Gd(III), Cu(II), Eu(II)) and short (2: Fe(II), Fe(III), Co(II), Ni(II), Lanthanide(III)ion) electron-spin relaxation times.

There is no discrete, sudden transition between these classes. However, as a general rule the class I metal ions result in very short relaxation times of nuclei under their influence and are therefore used as probes to create linebroadening of the resonance of such nuclei. However, these ions also cause large chemical shift changes in these nuclei, but with such an increase in linewidth that the resonances are rarely detectable. The shorter τ_c value of the class II ions makes them less efficient as relaxation stimulants, and therefore, even the lines shifted by the paramagnetic ion is sufficiently narrow to be

APPLICATIONS TO FUNDAMENTAL INVESTIGATIONS 373

observed. This class of metal ions is used for preparing shift reagents. The value of τ_m is limited by diffusion rates to $> 10^{-10}$ sec.

In summary, the two major classes of observations made in the presence of paramagnetic metal ions is broadening of the NMR spectral lines of those nuclei which come under the influence of the local field generated by unpaired electrons with long electron-spin relaxation times, and large chemical shift changes for those influenced by ions having unpaired electrons with shorter electron-spin relaxation times. The latter class has been discussed as shift reagents in Section 4.1.1.9.

Among the earliest application of paramagnetic relaxation probes to biochemical systems were simple experiments, performed by Li and associates (181). Copper(II) was added to solutions of amino acids and peptides. Under conditions of fast exchange, the nuclei nearest the site on the molecule which coordinates the copper ion shows the most broadening. If the spectrum of the molecule has been previously interpreted, this simple experiment demonstrates the site which coordinates best with the copper ion.

The most significant use of paramagnetic relaxation in biochemical studies has been applications to macromolecules. As with the quadrupolar nuclei, the use of paramagnetic ions provides a probe for indirect study of the chemistry of proteins and enzymes. Metalloenzymes are of particular interest in these studies as paramagnetic metal ions may be substituted for naturally occurring metal ions in the enzyme. The experiments consist of measuring the changes in the relaxation time of the bulk water in solutions of the enzyme containing the paramagnetic metal ion. The water provides a strong signal and if rapid chemical exchange between water molecules in the inner hydration shell of the bound metal ion occurs, then the rapid relaxation of the water protons stimulated by the unpaired electron is observed in the bulk water. The relaxation time of the water proton spins may be studied as a function of concentration, temperature, pH, frequency, and other parameters. Because the relaxation time may be dominated by one of several factors it is imperative that temperature and frequency variations be performed. Only in this way can one be certain that the results are properly interpreted. These experiments have been discussed in detail by Dwek (173). We shall give a qualitative example of a study. The reader should consult an in-depth treatment of the mathematics before designing experiments of this type.

Pyruvate kinase is a metalloenzyme containing Mg(II), which catalyzes the reaction between APP and phosphenol pyruvate to form ATP and pyruvate. Manganese(II) may be substituted for the Mg(II) with little or no loss in enzyme activity. This substitution can be done by dialysis. Navon (182) has investigated this enzyme using the paramagnetic relaxation effect on solvent water ^{1}H-NMR resonance line. Figure 7.16 shows a plot of $T_{1,P}$ and $T_{2,P}$ as a

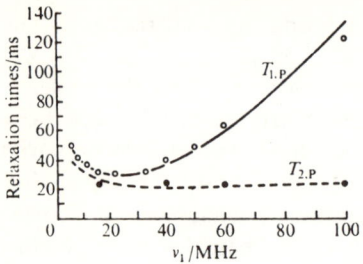

Fig. 7.16 Frequency dependence of the net relaxation times $T_{1,p}$ (○) and $T_{2,p}$ (●), of 6.16×10^{-5} M Mn-pyruvate kinase at 25°C. The solution contained 0.1 M KCl, 1.5×10^{-4} M excess Mn(II), 0.05 M Tris-HCl buffer pH 7.5. The continuous lines are computed. (G. Navon, *Chem. Phys. Lett.*, 7, 390 (1970), with permission.)

function of frequency. These values are the measured differences between T_1 and T_2 for identical solutions with and without the enzyme and are interpreted as the contribution of the bound paramagnetic ion to the respective relaxation mechanisms for the water protons. Figure 7.16 clearly shows that the values obtained depend on the frequency. Qualitatively, the differences between $T_{1,p}$ and $T_{2,p}$ may arise from the case that $T_{1,p}$ is dominated by the term $\omega_s \tau_R$ where ω_s is the frequency-dependent Larmor frequency of the electron, whereas $T_{2,p}$ may be dominated by τ_m.

A determination of whether slow or fast exchange obtains is required. This is done from a temperature study. Usually measurement of $T_{2,p}$ at the highest frequency will suffice. Figure 7.16 shows that $T_{2,p}$ at 100 MHz is the smallest (and differs most from $T_{1,p}$). The chemical exchange rate $1/\tau_m$ is independent of frequency and should make the largest relative contribution to the relaxation time at the higher frequency. A temperature study at 100 MHz showed that, in this example, $T_{2,p}$ increased with increasing temperature. This observation is indicative of fast exchange conditions.

Once the exchange condition is established, an analysis of the frequency dependence of the relaxation rates may be done. Inspection of the mathematical expressions for relaxation in paramagnetic systems (173) shows that when a minimum is observed in a plot of $T_{1,P}$ or $T_{2,P}$ versus frequency, τ_c for the respective relaxation term is frequency-dependent. A value of τ_c may be obtained by fitting theoretical equations to the data as in Figure 7.16. In this example $1/\tau_c$ was determined to be $(2.5 \pm 0.3) \times 10^8$ sec^{-1}. However, $1/\tau_c$ was defined earlier as

$$\frac{1}{\tau_c} = \frac{1}{\tau_R} + \frac{1}{\tau_m} + \frac{1}{\tau_s}$$

and one must settle for a combination of parameters, or deduce that one or more of the terms on the right are insignificant. In this example, the experimental results for $T_{2,P}$ may be fitted to the theoretical equations if it is assumed that $1/\tau_s \ll (1/\tau_R + 1/\tau_m)$. This implies that $\tau_e \simeq \tau_c$ and that both τ_e and τ_c are dominated by τ_m. The conclusion then is that the value of $1/\tau_c$

given is close to the chemical exchange rate of the water molecules coordinated to the Mn(II) bound to the enzyme.

This example shows only a general outline of the possible applications of paramagnetic relaxation probes in biochemical systems. These experiments may be expanded to measure the coordination number of the bound metal ion as well as the distance between the water molecule and the paramagnetic site. The same principles may be applied to ligands other than water. As the previous example indicates, NMR equipment must be available to permit T_1 and T_2 measurements to be made over a wide frequency and temperature range if certainty is to be assured in the interpretation of the results.

7.4.1.4 Applications of ^{13}C-NMR

The use of FT ^{13}C-NMR has been of great advantage to biochemists. The advantages of ^{13}C-NMR have already been pointed out earlier in this book. All of these advantages pertain to compounds of biological interest and some are of particular importance. The ^{13}C-NMR chemical shifts are much larger and occur over a wider range than ^{1}H-NMR chemical shifts. However, ^{13}C-NMR chemical shifts are determined primarily by the paramagnetic shielding term. This term depends much more on the orbital geometry about the nucleus than on the neighboring environment. Therefore, ^{13}C-NMR chemical shifts usually do not show large variations with changes in the environment. This result offers advantages and disadvantages to the biochemist. First, the ^{13}C-NMR resonances of amino acid constituents in peptides and proteins can be predicted from the ^{13}C-NMR chemical shifts of the amino acid. Lauterbur has demonstrated the potential usefulness of this approach (183). Figure 7.17 shows a ^{13}C-NMR spectrum taken on hen's egg white lysozyme. The bottom half of the figure is a simulated spectrum generated from ^{13}C-NMR chemical shifts obtained from the individual amino acids in the enzyme. Region *a* in the simulated spectrum contains all the carbonyl carbon resonances. Peak *b* results from the guanidinium carbons of the 11 arginyl residues and seems well matched with a similar peak in the experimental spectrum. The peaks *c*, *d*, and *e* represent, respectively, the number 4 carbons of three tyrosyl residues, the number 1 carbons of the three phenylalanyl residues, and the number 1 carbons of the three tyrosyl residues. Region *f* contains the remaining aromatic carbon resonances, region *g* contains all the α-carbon resonances except the glycyl residues. Regions *h* and *j* contain most of the aliphatic side-chain resonances. Although the comparison between the simulated and experimental spectra is not perfect, this early example shows the potential for this approach in the elucidation of protein structure by ^{13}C-NMR.

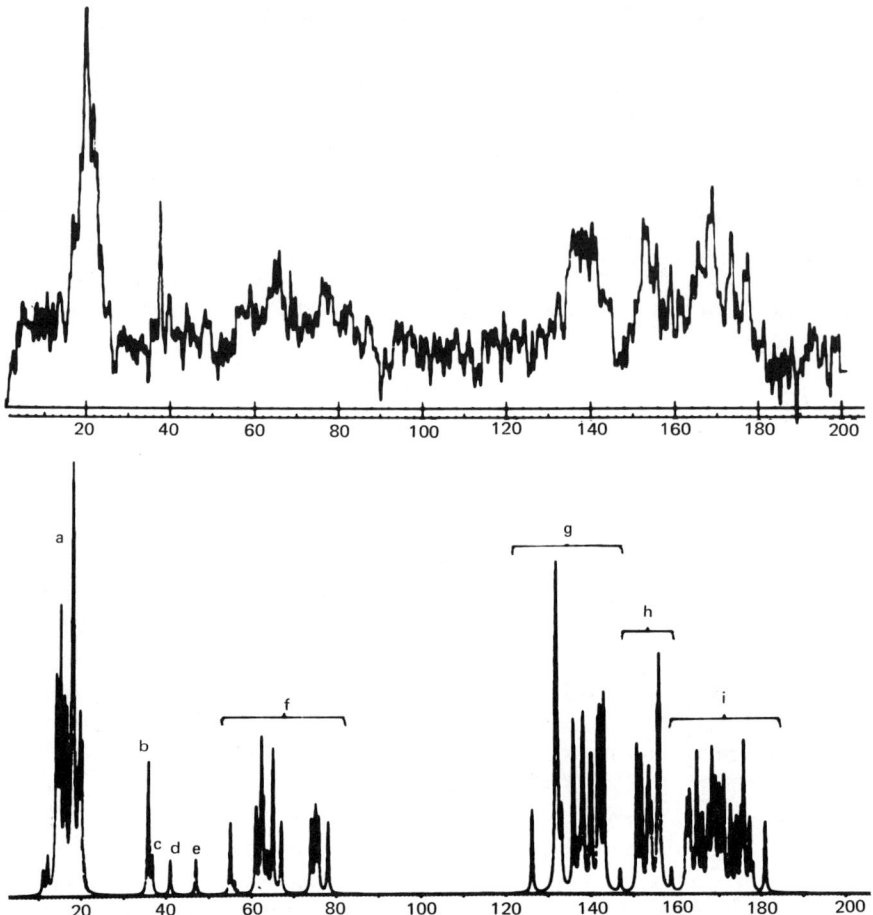

Fig. 7.17 Comparison of experimental (above) and simulated (below) natural abundance ^{13}C NMR spectra of hen's egg white lysozyme. The shift scales are in ppm to high field of external CS_2. (P. C. Lauterbur, *Applied Spectroscopy*, **24**, 450 (1970), with permission.)

The disadvantage of the nature of ^{13}C screening is that the potential of ^{13}C-NMR for studies of molecular interactions using chemical shift measurements may be limited. Chemical shift changes on binding, protonation, or complexation with diamagnetic metals may be too small to measure. There may be exceptions when the carbon atom is very close to the binding site.

The sensitivity enhancement of FT ^{13}C-NMR along with the relative ease of identification of peaks from model compounds provides an alternative to chemical and mass spectrometric methods of elucidating biosynthesis.

For example, Brown et al. (184) investigated the origin of the C-1 methyl group of the corrin ring A in vitamin B_{12}. The vitamin biosynthesis using enriched 5-$[^{13}C]$-δ-aminolaevulate did not show a ^{13}C-NMR resonance from the C-1 methyl group. This demonstrated that the C-1 methyl group does not originate from the previously suspected C-5 of δ-aminolaevulate. This ^{13}C-NMR method is faster and more convenient than chemical methods. It is also nondestructive which is not the case if mass spectrometry is used.

7.4.1.5 Biomolecular Conformation

Both ^{1}H-NMR and ^{13}C-NMR may be used effectively and jointly as tools for the elucidation of conformations in biomolecules. Chemical shift and the angular dependence of spin-spin coupling constants in ^{1}H-NMR are of value. Urry (185,186) has reported the use of these techniques in the elucidation of secondary structure in biomolecules. He used as one model the antibiotic gramicidin S, chosen for its rigid structure (185). We will discuss the results of the studies on this compound.

To determine secondary structure in peptide and protein conformation, one of the first steps is to determine the amide proton sites available for intramolecular hydrogen bonding. One basis for experimental design to locate these sites is to assume that an intramolecularly hydrogen bonded amide proton is screened or blocked from intermolecular interactions such as solvation. If this is true, it is expected that such protons will show the following: reduced chemical exchange rates with labile protons (or ^{2}H) in the solvent; a different temperature dependence of chemical shift than those protons that interact more freely with the solvent; and a different degree of chemical shift change when various solvents are used than those protons more exposed to the solvent. Therefore, exchange rate, temperature, and solvent studies are potentially useful in locating intramolecularly hydrogen bonded sites using ^{1}H-NMR.

Deuterium exchange studies were so effective for gramicidin S that two classes of amide proton exchange can be established by comparison of a spectrum taken in methanol with one taken in CH$_3$OD. The resonances resulting from protons labeled c and d in Figure 7.18, which are intramolecularly hydrogen bonded, are observed in CH$_3$OD. Those resonances for the remaining —NH protons observed in methanol are not seen in CH$_3$OD. If nothing else is learned, this simple experiment demonstrates that two different amide protons have slow exchange rates that probably result from intramolecular hydrogen bonding.

Hydrogen bonding of amide protons affect their chemical shift values. Changes in the extent of intramolecular hydrogen bonding with temperature

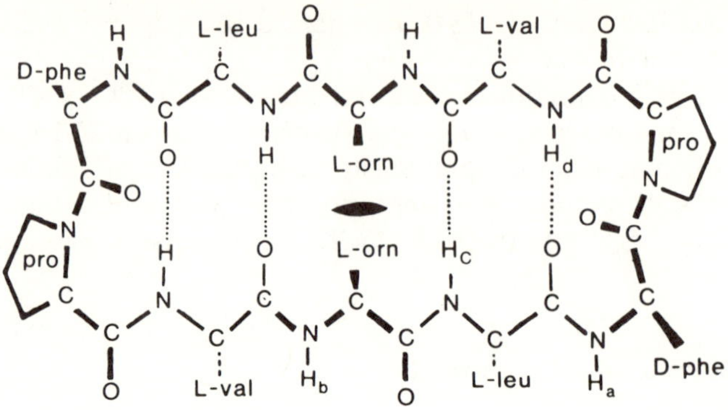

Fig. 7.18 Gramicidin S secondary structure. Molecule has twofold symmetry with two intramolecular hydrogen bonds per symmetric unit. This is cross β-type conformation. The 10 atom hydrogen-bonded ring at each end formed by proton H_d, is a β-turn. (D. W. Urry, *Res. Develop.*, **24**, 20 (1973), by permission.) Copyright © 1973, 1974 by Technical Publishing Co.

result in chemical shift changes. Figure 7.19 shows a plot of the chemical shift values of the four amide resonances of gramicidin S in methanol versus temperature (187); N-methyl acetamide is included for comparison. The protons *a* and *b* which were shown to undergo rapid chemical exchange show a higher temperature dependence and are similar to the N-methyl acetamide. In comparison, the amide protons *c* and *d* showed a markedly smaller temperature effect on their chemical shift. Again, a simple ^{1}H-NMR experiment indicates those protons (*c,d*), which are likely involved in intramolecular hydrogen bonding.

Solvent dependence of chemical shift is a third method of using ^{1}H-NMR to indicate the number of protons involved in intramolecular hydrogen bonding in biomolecules. Figure 7.20 shows a plot of the chemical shift of the four amide protons in gramicidin S versus the volume percent 2,2,2-trifluoroethanol (TFE) added to methanol. L-Ala-L-Ala-diketopiperazine is included in the plot for comparison. The plot shows that the two protons suspected of intramolecular hydrogen bonding (*c,d*) show slight down-field shifts, whereas the *a* and *b* protons show dramatic up-field shifts with increasing TFE.

Thus these three simple ^{1}H-NMR experiments can greatly assist in the determination of those sites in a peptide which exclude solvent interaction, likely because of intermolecular hydrogen bonding. It is important to note that the structure of gramicidin S is relatively rigid and all three methods gave clear-cut results. In practice, all three experiments should be performed, and the results should agree with other conformational dependent parameters

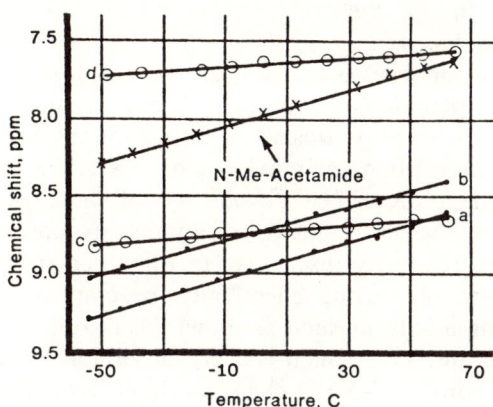

Fig. 7.19 Temperature dependence of chemical shift of the amide protons of gramicidin S in methanol. The two amide protons *c* and *d* indicated as hydrogen bonded in Fig. 7.18 and that are retained in CD_3OD have the lowest slope, i.e. the lowest temperature coefficient. The two amide protons *a* and *b* that are not hydrogen bonded and that exchange immediately in CD_3OD exhibit steeper slope, equivalent to that of N-methyl acetamide in CH_3OH. This provides second means of delineating amide protons. (M. Ohnishi and D. W. Urry, *Biochem. Biophys. Res. Commun.*, **36**, 194 (1969), with permission.)

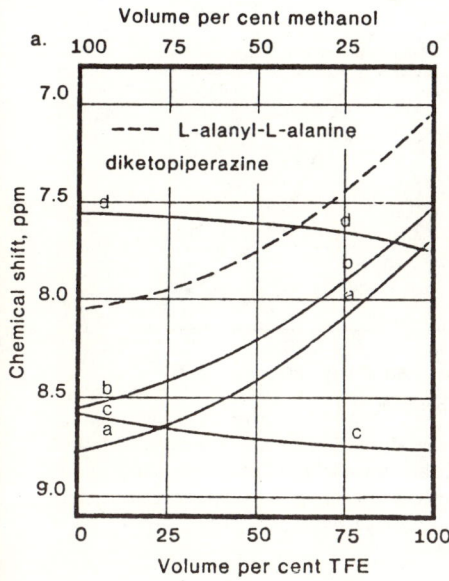

Fig. 7.20 Methanol-trifluoroethanol solvent mixture dependence of chemical shift of amide protons of gramicidin S. The two amide protons *c* (leucyl peptide) and *d* (valyl peptide), which are hydrogen bonded in Fig. 7.18 and exhibit lower temperature coefficient in Fig. 7.19, shift slightly downfield on addition of trifluoroethanol. Amide protons *a* (phenylalanyl peptide) and *b* (ornithyl peptide), which are not hydrogen bonded in Fig. 7.18, and have steeper slopes in Fig. 7.19, exhibit dramatic upfield shifts of greater than 1 ppm on addition of TFE. Protons *a* and *b* shift upfield in manner similar to solvent exposed amide protons of L-Ala—L-Ala dikeptopeperazine. This provides third means of delineating peptide protons in a way that can be correlated with secondary structure. (Reprinted with permission from T. P. Pitner and D. W. Urry, *J. Am. Chem. Soc.*, **94**, 1399 (1972). Copyright by the American Chemical Society.)

such as vicinal coupling constants and chemical shift values before any conclusions are drawn.

The ^{13}C-NMR resonances of carbonyl carbon atoms can also be used to assist in studies of peptide conformation. Again gramicidin S is used as a model. This cyclodicopeptide has two-fold symmetry and the ^{13}C-NMR shows five different peptide carbonyl resonances (187). According to the ^{1}H-NMR experiments described above, two of these five carbonyls are intramolecularly hydrogen bonded to amide protons. The ^{13}C-NMR chemical shifts are relatively insensitive to minor environmental changes. Therefore, a reasonably strong effect must be employed. One choice is solvent shifts using a solvent such as dimethylsulfoxide (DMSO) which is a proton acceptor and TFE which is a proton donor. Figure 7.21 shows a plot of the five carbonyl ^{13}C-NMR chemical shifts versus the volume percent TFE. The resonance labeled *a* was used as the chemical shift reference; positive values indicate a down-field shift relative to the *a* resonances. Although these data are less definitive than the ^{1}H-NMR solvent shift data,

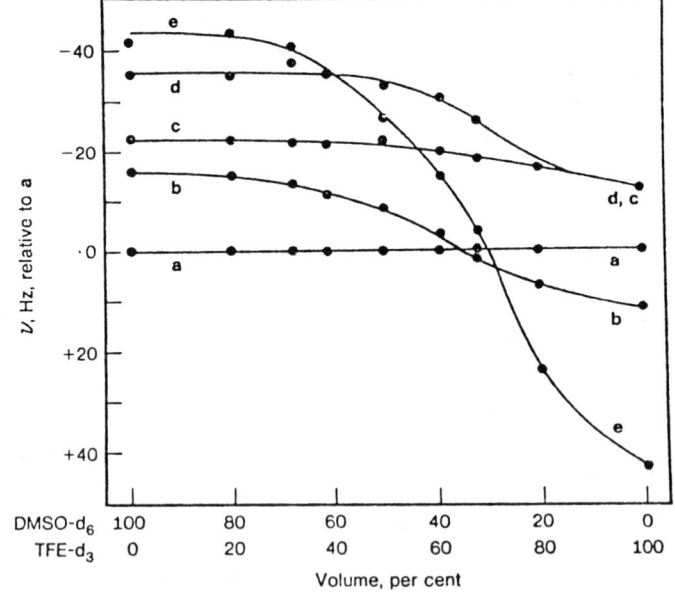

Fig. 7.21 The carbon-13 magnetic resonance chemical shift of gramicidin S peptide carbonyls in hertz relative to resonance *a*. Positive values indicate downfield. All resonances are seen to shift downfield relative to resonance *a* with resonance *c* exhibiting the smallest relative shift. (D. W. Urry, *Res. Develop.*, **25**, 18 (1974), with permission.) Copyright © 1973, 1974 by Technical Publishing Co.

APPLICATIONS TO FUNDAMENTAL INVESTIGATIONS 381

similar experiments in DMSO-methanol and TFE-methanol mixtures show that two of the five resonances (a, c) consistently show less solvent shift than the other three. At the time of this writing, the ^{13}C-NMR spectrum of gramicidin S has not been unambiguously assigned.

7.4.2 APPLICATIONS TO STUDIES OF STRUCTURE, EQUILIBRIA, AND KINETICS OF METAL COMPLEXES

Among the earliest applications of ^{1}H-NMR to the study of structure, equilibria and kinetics of compounds of direct analytical interest were those made in the 1960s using aminocarboxylate compounds. These types of compounds were well established as chelating reagents used primarily as titrants for volumetric determinations of metallic ions. The most widely used example is ethylenediaminetetracetic acid (EDTA). A discussion of the investigations of these compounds performed using ^{1}H-NMR is useful because several ^{1}H-NMR techniques were applied. The structure of the compounds are sufficiently simple that the principles and practices of these techniques are readily understood. An additional advantage of a discussion of these compounds is that their structural similarity to compounds of biological interest allows one to imagine extension of the experimental techniques to more complicated molecular systems. An aminocarboxylate compound may contain one or more amino and carboxylate sites capable of simultaneous coordination to a central metal metal ion to form a chelate. Because these functional groups participate in acid–base equilibria, the chelation equilibrium is pH dependent. It is understandable that the fundamental property of the acid–base equilibrium was among the first investigations to be made when NMR studies of chelating agents was undertaken.

7.4.2.1 Protonation of Free Ligands

Using the model that ethylenediaminetetraacetic acid (EDTA) represents a multifunctional α-amino acid, Schwarzenbach and Ackermann proposed a double zwitterion structure 7.XV (188). This proposition was based on

$$\begin{array}{c}
\text{O} \\
\parallel \\
^-\text{O}-\text{C}-\text{CH}_2 \\
\diagdown\overset{+}{\text{H}} \\
\text{N}-\text{CH}_2-\text{CH}_2-\text{N} \\
\diagup \\
\text{HO}-\text{C}-\text{CH}_2 \\
\parallel \\
\text{O}
\end{array}
\qquad
\begin{array}{c}
\text{O} \\
\parallel \\
\text{CH}_2-\text{C}-\text{O}^- \\
\overset{+}{\text{H}}\diagup \\
\diagdown \\
\text{CH}_2-\text{C}-\text{OH} \\
\parallel \\
\text{O}
\end{array}$$

7.XV

macroscopic ionization constants, and went unconfirmed for many years as a reasonable interpretation without serious challenge. Chapman et al. explicitly pointed out that a method for investigating protonation schemes must be capable of either giving a signal directly from the labile proton, or detecting its influence on a site near the functional group (189). It was also pointed out that NMR had the required capability. There are only two serious disadvantages encountered when using NMR for such studies. First, in the absence of special methods of signal enhancement, the technique requires a concentration of approximately 0.2 M for most ligands of interest. This is well above a desirable level for equilibrium studies. Second, the strong ^{1}H-NMR absorption by water is an occasional problem. This may be remedied in several ways, the most obvious of which is to utilize D_2O as the solvent. In spite of these two disadvantages, NMR can often provide information not obtainable by any other reasonable means.

The basis of the application of NMR to the study of microscopic protonation of multifunctional molecules lies in the influence the loss or gain of a labile proton has upon the chemical shift of adjacent nonlabile protons. The chemical shift of these nonlabile protons is determined by many factors including diamagnetic shielding by the electrons. If an adjacent site is protonated, the electron density about the nonlabile protons is reduced and the chemical shift will change to a lower magnetic field. The extent of the change in chemical shift is a function of the nature of the functional site, the distance from this site, and the degree of protonation of the site on a time average basis. The first two of these dependencies is easily accepted, at least qualitatively. The last, however, deserves some comment. If the labile protons did not undergo rapid exchange between all the molecules of the ligand, two spectra would be observed as protonation was carried out. One of these would represent those molecules protonated and would increase in intensity during titration. The second would represent those molecules not protonated and would decrease to zero intensity at the equivalence point. The degree of protonation would be measured from the relative intensity of the two spectra. However, in aqueous solution acidic protons undergo rapid exchange between solvent and ligand sites. The result is that only one NMR spectrum is observed which represents an average of the two described above. The position of the chemical shift between the two extremes may be interpreted as a measure of the fraction of time the site is protonated (f value). Identification of the proton(s) responsible for a peak whose shift is observed upon changing pH permits both the identification of the site being protonated and the determination of the f value for that site as a function of pH or the number of equivalents of acid added (n).

Chapman et al. were the first to apply these principles to elucidate protonation schemes by aminocarboxylates. This work will not be discussed in great detail because the conclusions were qualitative and were in essence confirmed by later quantitative studies. However, the concepts may be further illustrated by an example from this work. In order to assign the observed NMR peaks, the chemical shift changes for particular structural arrangements were obtained from model compounds and compared with more complicated compounds. For example, dimethylglycine was used as a model. On addition of the first proton a chemical shift change ($\Delta\delta$) of 0.68 ppm was observed for the CH_3 protons and 0.80 ppm for the CH_2 protons. When the second proton is added, $\Delta\delta_{CH_3}$ is 0.11 ppm, whereas $\Delta\delta_{CH_2}$ is 0.39 ppm. One would expect that protonation at the nitrogen would generate a large shift in both the CH_3 and CH_2 protons, while protonation at the carboxylate site would lead to a smaller shift in the more remote CH_3 protons than the adjacent CH_2 protons. Therefore, a set of values obtained from such models may be used to at least qualitatively elucidate protonation of multifunctional molecules. The conclusion of Chapman et al. was that amino-carboxylates protonate first at the nitrogen site. A further contribution was the suggestion that monoprotonated amino-polycarboxylates such as EDTA exist in a structure "stereochemically equivalent to a trifurcated hydrogen bond" as in 7.XVI. Further evidence for this structure obtained by Sudmeier and Senzel will be discussed later (190).

7.XVI

Following the general techniques of Chapman et al., a quantitative study of the protonation schemes of a series of polyamines and polyaminocarboxylates was performed by Sudmeier and Reilley (191). The chemical shift change of a given set of nonlabile protons upon protonation of an adjacent site is expressed as

$$\Delta\delta_i^c = \sum_{i=1}^{N} C_{ij} = f_i \qquad (7.28)$$

where $\Delta\delta_i^c$ is the total "protonation shift" (calculated) of the ith observed resonance, C_{ij} is the protonation shift constant for the ith resonance corresponding to total protonation of the jth basic site, and f_i is the fraction of time during which the jth site is protonated. There are two fundamental assumptions implied in equation 7.28. First the shift contribution of a particular basic site is linearly related to the fraction of time the site is protonated. Second, the contributions of protonating different sites are additive. The first assumption is reasonably well documented by the early work of Grünwald, Loewenstein, and Meiboom (192). The weakness in the method centers about the second assumption. In order to compare calculated values of $\Delta\delta_i^c$ with observed values ($\Delta\delta_i$), it is necessary to obtain the values of C_{ij}. If the protonation of one site perturbs the value of C_{ij} for the protonation of a second site, then these "protonation shift constants" will change as the titration proceeds. The effect of electronic and conformational changes can not be easily incorporated into functions for C_{ij}, therefore, one must

TABLE 7.11. Methylenic Substituent Shielding Constants (normal —CH_2— at 1.25 ppm vs. DSS)[a]

—CH_2—CH_2—$CH_2NH_3^+$	0.00
—CH_2—CH_2—NH_2	0.05
—CH_2—CH_2—NH_3^+	0.10
—CH_2—NH_2	0.15
—CH_2—NH_3^+	0.40(0.45)[a]
—NH_2	1.35
—NH_3^+	1.80(1.85)[a]
—CH_3	0.05
—CH_2—NRH	0.15(0.20)[a]
—CH_2—NRH_2^+	0.40(0.50)[a]
—NRH	1.30
—NRH_2^+	1.85(1.90)[a]
—CH_2—NR_2	0.15
—CH_2—NR_2H^+	0.50
—NR_2	1.15
—NR_2H^+	1.90
—NR_3^+	2.00
—CH_2—OH	0.20
—OH	2.25
—CH_2—OR	0.25
—OR	2.15
—CH_2—CO_2^-	0.30
—CH_2—COOH	0.35
—CO_2^-	0.90
—COOH	1.10

[a] Modified for application to polyamine compounds (J. L. Sudmeier and C. N. Reilley, *Anal. Chem.*, **36**, 1698 (1964), with permission).

resort to the use of model compounds containing substituents most commonly found in chelating agents. An example set of data for model compounds is shown in Table 7.11 (191). The normal or isolated methylenic group was arbitrarily set at 1.25 ppm versus sodium 3-(trimethylsilyl)-propane sulfonate (DSS) in this table. The values in parentheses are modifications to account for chain stiffening due to coulombic repulsion when polyamines are protonated. To calculate a protonation shift constant for a given methylenic group, one takes the difference between the corresponding two groups in the table. For example, the chemical shift change upon protonation of a primary amine beta to a methylene group whose resonance is observed is the difference between the chemical shift values given for the fifth and fourth entry in Table 7.11 or 0.25 ppm.

In addition to equation 7.29 one may normalize the total time spent on all sites by protons as

$$\sum_{j=1}^{N} f_i = n \tag{7.29}$$

where n is the number of equivalents of protons added. Equation 2 permits the elimination of one variable f_i. In many cases in which degenerate sites are present (such as EDTA with four equivalent carboxylate sites and two equivalent amino sites) the number of equations is greater than the number of unknowns and techniques to solve over-determined equations must be employed. Sudmeier and Reilley chose to minimize the squares of the difference between the calculated chemical shift and the observed chemical shift for each observed ith resonance as expressed in equation 7.30.

$$\sum_{i=1}^{N} (\Delta\delta_i^c - \Delta\delta_i)^2 = \sum_{i=1}^{N} d_i^2 = \text{minimum} \tag{7.30}$$

Using the sum of the squares of the ith deviations, standard methods were employed to estimate standard deviations in the results.

A plot of the observed chemical shift versus pH for the various methylenic protons in ethylether-diaminetetraacetic acid (EEDTA) is shown in Figure 7.22 (191). This example illustrates a case in which the addition of two equivalents of acid (note values of pK_3 and pK_4) results in a smooth change in the chemical shift to lower field. Notice that the a and b protons have approximately equal change in the chemical shift values indicating protonation at the nitrogen site, whereas the more remote c protons have a much smaller change in chemical shift. Figure 7.23 shows a similar plot for diethylenetriaminepentaacetic acid (DTPA) (191). The plot illustrates an example in which the central amino site (2) is protonated with the first equivalent of acid, however, presumably coulombic repulsion forces a redistribution upon the addition of the second equivalent of acid so that the

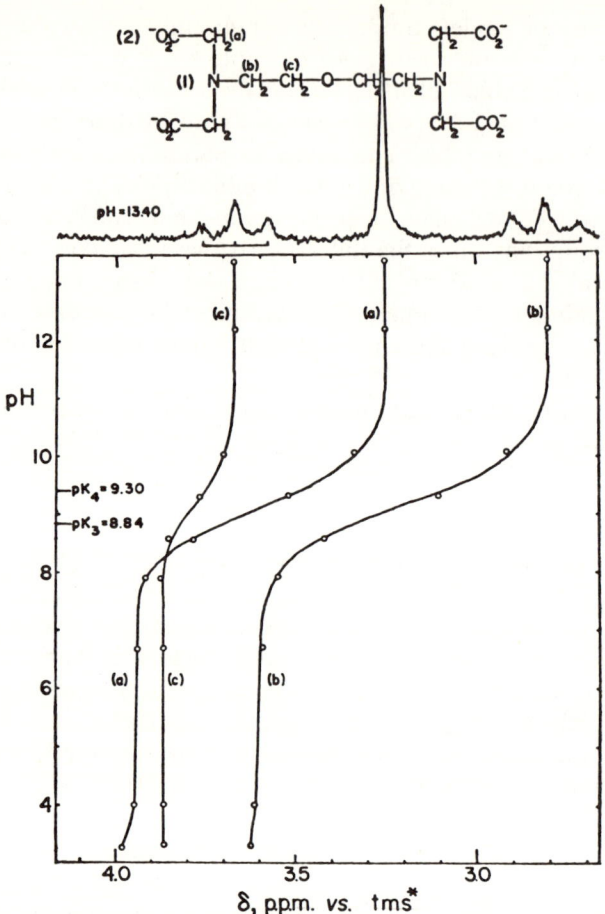

Fig. 7.22 Chemical shift of EEDTA at various pH values. (Reprinted with permission from J. L. Sudmeier and C. N. Reilley, *Anal. Chem.*, **36**, 1698 (1964). Copyright by the American Chemical Society.)

terminal amino sites (1) are the ones of major protonation. This interestingly vacillating chemical shift data for DTPA was reported almost simultaneously by Sudmeier and Reilley (191) and by Kula and Sawyer (193).

However, Sudmeier and Reilley applied equations 7.28 and 7.29 to calculate chemical shifts changes ($\Delta \delta_i^c$) and fit curves such as those shown in Figures 7.22 and 7.23 by means of equation 7.30. Table 7.12 shows the results for a series of polyamines and aminocarboxylates. The results shown in

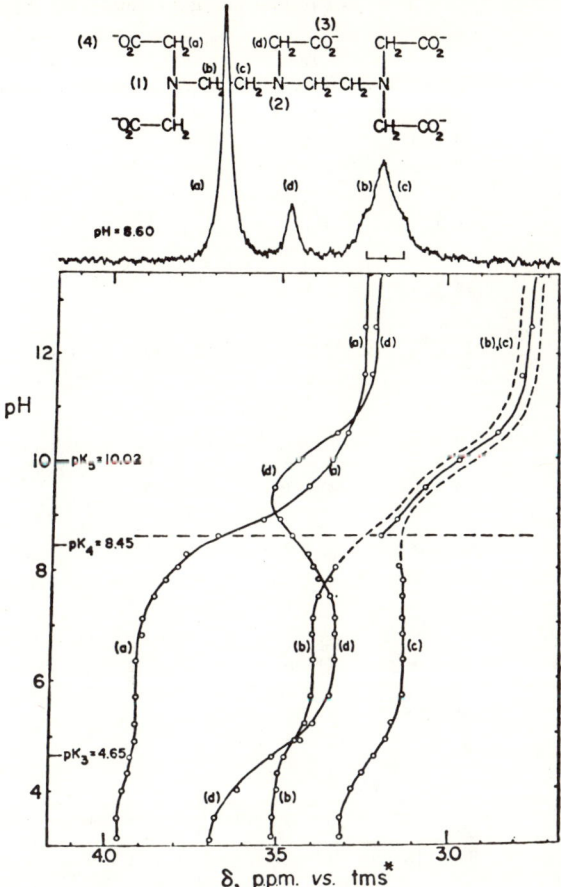

Fig. 7.23 Chemical shift of DTPA at various pH values. (Reprinted with permission from J. L. Sudmeier and C. N. Reilley, *Anal. Chem.*, **36**, 1698 (1964). Copyright by the American Chemical Society.)

Table 7.12 confirm with some extension the early work of Chapman. The rather large uncertainties are unfortunate. Most of the lack of precision can easily be attributed to the assumptions made on the constancy of the protonation shift parameter (C_{ij}). In addition, the values of these parameters are not large (0.05–0.5 ppm) and lack of precision is encountered in the measurement. However, these results leave little doubt that the general scheme for the protonation of aminocarboxylates is to protonate the amino groups first and then the carboxylate groups.

TABLE 7.12. Percent Protonation at Various Basic Sites[a]

Compound	f_1	f_2	f_3	f_4	σ
EDTA					
$n = 1$	53 ± 5	2 ± 3			±6.5
$n = 2$	96 ± 2	2 ± 1			±3.0
EEDTA					
$n = 2$	96 ± 11	2 ± 5			±11.3
EGTA					
$n = 2$	94 ± 6	3 ± 3			±6.4
HEDTA					
$n = 1$	42 ± 7	52 ± 6	3 ± 18	0 ± 21	±5.8
$n = 2$	87 ± 6	97 ± 5	0 ± 15	16 ± 18	±4.8
DTPA					
$n = 1$	26 ± 1	41 ± 1	7 ± 2	—	±1.0
$n = 2$	85 ± 5	15 ± 6	0 ± 10	4 ± 8	±3.7
$n = 3$	80 ± 11	64 ± 13	76 ± 23	—	±13.0
Dien					
$n = 1$	41 ± 4	18 ± 8			±3.9
$n = 2$	92 ± 12	16 ± 25			±11.9
$n = 3$	98 ± 6	104 ± 12			±5.9
Trien					
$n = 1$	36 ± 6	14 ± 6			±5.2
$n = 2$	76 ± 10	24 ± 10			±9.2
$n = 3$	99 ± 7	51 ± 7			±6.4
$n = 4$	100 ± 1	100 ± 1			±1.0
Tetren					
$n = 1$	29 ± 1	13 ± 1	16 ± 1		±0.6
$n = 2$	69 ± 11	11 ± 11	40 ± 15		±8.1
$n = 3$	99 ± 18	9 ± 19	84 ± 25		±13.7
$n = 4$	104 ± 6	67 ± 6	58 ± 8		±4.5
$n = 5$	99 ± 5	101 ± 5	100 ± 7		±3.8

(J. L. Sudmeier and C. N. Reilley, *Anal. Chem.*, **36**, 1698 (1964), with permission.)

Because of the small differences in the chemical shifts of the methylenic protons in polyamines, these results are even less certain than those for aminocarboxylates. This is unfortunate because of the need for answers to the aged question of the basicity of amines. An area of study which may answer some of the remaining questions of protonation schemes is the application of ^{13}C-NMR to the problem. The obvious disadvantage of ^{13}C-NMR is the need for signal enhancement. In the past isotopic enrichment was required. However, the use of FT techniques can provide spectra of ^{13}C nuclei at natural abundance with acceptable S/N ratios in approximately 5 min instrument time. The use of natural abundance ^{13}C is a major advantage because of the very low $^{13}C-{}^{13}C$ spin–spin coupling. $^1H-{}^{13}C$ coupling is removed by noise decoupling about the ^1H resonance frequency. Reilley and associates have investigated the ^{13}C-NMR protonation shifts of aqueous amines (194). Additivity rules were developed to calculate chemical shift and protonation shifts of α, β, γ, and δ ^{13}C-atoms.

The question of the nature of the proton in mono- and diprotonated amino-carboxylates have been discussed by Schwarzenbach (188) and Chapman (189). Perhaps the strongest evidence that monoprotated amino-carboxylates have structures equivalent to polyfurcated hydrogen bonds has been obtained by NMR studies. An early study by Sudmeier and Reilley suggested that the tetraanion of *trans*-1,2-cyclohexanediaminetetraacetic acid (CyDTA) prefered a conformation in which the iminodiacetate groups are diequatorial (191). Fujiwara and Reilley performed a detailed study of CyDTA and found that the AB quartet of the methylene protons in the acetate group is asymmetrical in the pH range 8–11; this is the range of monoprotonated CyDTA (195). The AB pattern results from the nonequivalence of the methylene protons in their configurationally different positions (191). After a careful series of investigations of the effects of pH, rf power, temperature, and cation on the spectra, it was concluded that in the stated pH range in which monoprotonated CyDTA is predominate a

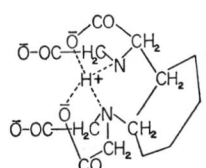

Structure 7.XVII (Reprinted with permission from Y. Fujiwara and C. N. Reilley, *Anal. Chem.*, **40**, 890 (1968). Copyright by the American Chemical Society.)

configuration represented by 7.XVII was preferred. This configuration is similar to one required for chelation which substantiates the suggestion by several researchers that the ligand is preoriented in a conformation suitable for chelation. The asymmetric AB pattern was explained as an interchange between the two types of methylene groups as a result of nitrogen inversion. One pair of acetate methylene protons apparently have accidental coincidence of chemical shift. The second pair exhibit a usual AB pattern. In the pH range in which monoprotonated CyDTA is predominate, an $AB \rightarrow CD$-type exchange occurs through nitrogen inversion at a sufficiently slow rate to cause line broadening. There was no evidence to indicate direct bimolecular proton exchange between the monoprotonated and tetraanion species.

To this point only the protonation of the amine and carboxylate sites has been considered. Recently, the protonation of the carboxyl oxygen in aminocarboxylates has been investigated by Sudmeier, Schwarz, and Senzel (29). An exceptionally strong acid such as $FSO_3H—SbF_5—SO_2$ is required. At $-80°C$ the $\diagdown C=\overset{+}{O}H$ protons are observed by NMR. This technique provides a rapid identification of the number, and in some cases, types of carbonyl groups in the molecule. For example, in hexadentate octrahedral complexes such as Co(III) EDTA there are a pair of in-plane and a pair of out-of-plane carbonyl groups. The NMR spectrum of the $\diagdown C=\overset{+}{O}H$ protons exhibits two peaks of equal intensity at 12.09 and 12.56 ppm versus TMS. This potentially useful tool for structure elucidation must be used with care. Co(III) EDTA has a C_2-symmetry axis whereas Co(III) PDTA (propylenediametetraccetic acid) has no C_2 axis and one should expect four carbonyl proton peaks. However, the fact that only two peaks are seen may be rationalized by the remoteness of the asymmetric sites (29).

7.4.2.2 Protonation of Metal Complexes

One of the early papers dealing with protonation studies of EDTA included several metal chelates (196). Although this paper did not present a quantitative treatment of the protonation data, there were several interesting points made which by in large confirmed earlier results or opinions concerning metal-EDTA chelate behavior as a function of pH. A plot of the chemical shift of the methylene and acetate protons versus pH showed two breaks when the EDTA salts of lithium, sodium and potassium were used. There were some variations in chemical shift in the presence of the different metal ions, but these were small compared with the pH effect. The fact that two steps were observed indicated very weak chelation of these ions. However, in the case of the alkaline earths, only one break is observed in the pH curve indicating much stronger chelation of the metal ion. Of significant interest is the result that in the presence of excess EDTA, two distinct sets of lines are observed over the pH range studied in the case of magnesium. Calcium under these conditions shows two sets of lines between pH 6 to 9.5 and one set above pH 9.5 which is at first broad and sharpens above pH 11.5. Strontium and barium show only single lines at all pH values studied with broadening of the spectra between pH 5 and 10. Other observations were that lead and zinc chelates gave two sets of lines over wide pH range and mercury-EDTA exhibited two sets between pH 4 and 9. The lower pH limits were usually forced by precipitation of the chelate. The distinct sets of lines were attributed to slow ligand exchange of the general type

$$MY + Y^* \longrightarrow MY^* + Y$$

where charge is ignored and the asterisks label the ligand molecules. Furthermore, the results of pH and chemical shift data permitted the authors to obtain a qualitative order of stability of the chelates. The order of stability was found to be Ca > Sr > Ba > Li > Na > · K ≃ Cs which is in agreement with the stability constants. The exact position of Mg in this series could not be determined because of precipitation below pH 5.5. The results confirmed that Pb, Al, Zn, and Hg all form more stable complexes than the alkali and alkaline earth metals.

Kula performed a more quantitative investigation of the formation of Zn(OH)-EDTA complexes (197). By observing the effect of pH upon the chemical shift of the Zn-EDTA complex much in the same way as the free ligands were studied, the equilibrium constant

$$K_1 = \frac{[Zn(OH)EDTA^{3-}]}{[ZnEDTA^{2-}][OH^-]} \tag{7.31}$$

was evaluated as $10^{2.0}$. A more detailed and complicated treatment of the line width of the complexed EDTA allowed the evaluation of the equilibrium constant

$$K_2 = \frac{[Zn(OH)_2EDTA^{5-}]}{[Zn(OH)EDTA^{3-}][OH^-]^2} \tag{7.32}$$

as $10^{-2.2}$. The concentration of the solutions required for NMR measurements and some of the small changes measured in the NMR data as well as inaccuracies in the pH measurement in alkaline solutions limit the accuracy of the data. However, calculation of the value

$$\beta_3 = \frac{[Zn(OH)_3^-]}{[Zn^{+2}][OH^-]^3} = 10^{16.3} \tag{7.33}$$

showed the NMR results to be compatible with the values of $10^{14.2}$–$10^{16.1}$ obtained by other means.

Whidby and Leyden have used NMR to evaluate constants for various protonated and hydroxy complexes of Cu(II), Ni(II), and Fe(III) (198). In these cases, advantage may be taken of the paramagnetic broadening of the solvent by rapid exchange of the proton and/or hydroxyl groups from the ligand. Figure 7.24 shows the effect of pH on the linewidth of the solvent. An inflection in the curve occurs at a pK for an acid–base equilibrium involving the complex. These pK values were confirmed by potentiometric titration. The advantage of this technique is that concentrations of approximately 0.05 M in the complex may be used. However, it is restricted to

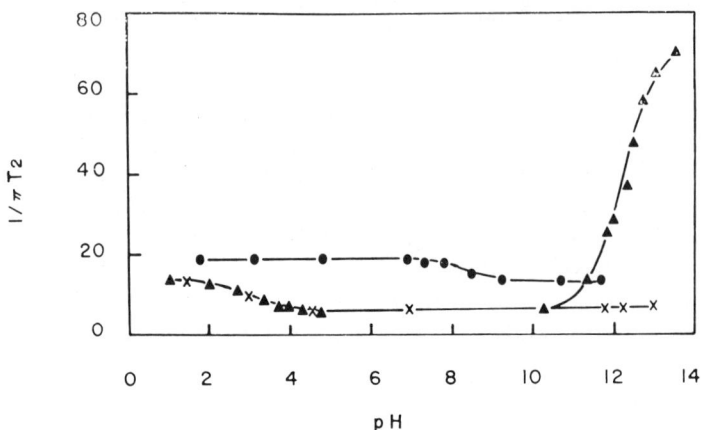

Fig. 7.24 $1/\pi T_2$ vs. pH for metal-EDTA solutions. (○) 0.013 M Fe(III)-EDTA; (△) 0.053 M Cu(II)-EDTA; (×) 0.070 M Ni(II)-EDTA. (J. F. Whidby and D. E. Leyden, *Anal. Chim. Acta*, **51**, 25 (1970), with permission.)

paramagnetic species. There is always some doubt as to the exact exchange process. For example, in hydroxy chelates, the OH⁻ or only the proton may exchange. Studies using O^{17} NMR would be required to determine the exchange process. The following equilibria were studied and the results are given in Table 7.13.

$$H_2MY \rightleftharpoons HMY + H^+ \qquad K_1 = \frac{[H^+][HMY]}{[H_2MY]} \qquad (7.34)$$

$$HMY \rightleftharpoons MY + H^+ \qquad K_2 = \frac{[H^+][MY]}{[HMY]} \qquad (7.35)$$

$$MY \rightleftharpoons M(OH)Y + H^+ \qquad K_3 = \frac{[H^+][M(OH)Y]}{[MY]} \qquad (7.36)$$

$$M(OH)Y \rightleftharpoons M(OH)_2Y + H^+ \qquad K_4 = \frac{[H^+][M(OH)_2Y]}{[M(OH)Y]} \qquad (7.37)$$

The effect of chemical shift change of a ligand upon coordination has also been applied to the detection of weak 2:1 chelates of calcium and strontium

TABLE 7.13. pK_a Values for Metal-EDTA Species

Metal ion	pK_1	pK_2	pK_3	pK_4
Cu(II)	2.5	3.3	12.2	—
Ni(II)	2.0	3.3	12.2	—
Fe(III)	—	3.0	8.0	12.0

with EDTA (199). If the equilibria represented by equations 7.38 and 7.39.

$$MY + M \rightleftharpoons M_2Y \quad (7.38)$$

$$M_2Y + M \rightleftharpoons M_3Y \quad (7.39)$$

are considered, the corresponding concentration equilibrium constants are

$$K_2 = \frac{[M_2Y]}{[M][MY]} \quad (7.40)$$

$$K_3 = \frac{[M_3Y]}{[M][M_2Y]} \quad (7.41)$$

The calculated chemical shift is then given by equation 7.42

$$\delta_{\text{calc}} = \left(\frac{[M_2Y]}{C}\right)\delta_{M_2Y} + \left(\frac{[M_3Y]}{C}\right)\delta_{M_3Y} \quad (7.42)$$

where C is the analytical concentration of MY, and $[M_2Y]$ and $[M_3Y]$ are the equilibrium concentrations of the respective species.

Each of these values is calculated with assumed values for K_2 and K_3 and the usual expressions for simultaneous equilibria. The values of δ_{M_2Y} and δ_{M_3Y} represent the chemical shift of the respective species in ppm measured from the resonance of MY. Estimates were made for δ_{M_2Y} and δ_{M_3Y} from limiting values for the observed chemical shift at high metal-to-ligand ratios. The function is relatively insensitive to δ_{M_2Y} and δ_{M_3Y} but quite sensitive to K_2 and K_3. This, of course, is advantageous inasmuch as the choice of the chemical shift parameters was not excessively critical. A true value of the chemical shift parameter cannot be obtained unless the species can be made to be predominate or the formation constants are previously known. The arguments for including the species M_3Y, which represents three metal ions per ligand, will be discussed later.

Figure 7.25 shows a plot of the relative chemical shift of the acetate protons of EDTA versus the mole percent of strontium in a strontium-Sr(EDTA) mixture. The circled points are the experimental data, whereas the solid line

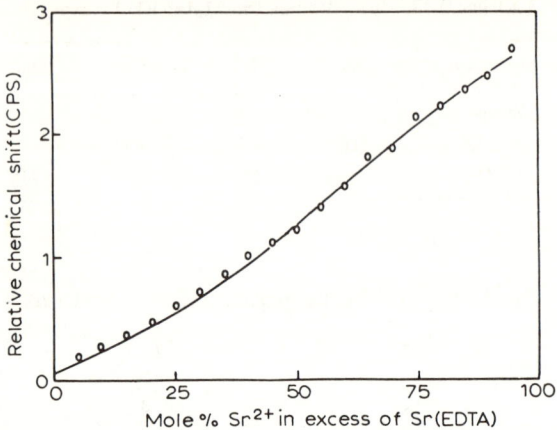

Fig. 7.25 Relative chemical shift of acetate protons vs. mol % strontium in a strontium-Sr(EDTA) mixture (pH 12.5). (D. E. Leyden and J. F. Whidby, *Anal. Chim. Acta*, **42**, 271 (1968), with permission.)

represents the calculated curve obtained from equation 7.42. The formation constants, K_2 and K_3 which gave the best comparison between the experimental and computed chemical shift values are given in Table 7.14.

The formation constants for the 2 : 1 metal-to-EDTA species are relatively small as expected. The similar numerical values for K_2 for Ca_2(EDTA) and Sr(EDTA) are also reasonable in view of the formation constants of the iminodiacetic acid chelates with these ions. In order to obtain the best correlation between the computed curves shown in Figure 7.25 the experimental data it was necessary to include a small value represented by K_3 which is larger than the estimated error of ±0.1 for all the values given in

TABLE 7.14. Stability Constants for Calcium and Strontium Complexes of EDTA

	log K_1	log K_2
Ca	10.57[a]	1.1 ± 0.1
Sr	8.63[a]	1.0 ± 0.1

[a] L. G. Sillen and A. E. Martel, *Stability Constants of Metal-Ion Complexes*, The Chemical Society, London, 1964, p. 634; D. E. Leyden and J. F. Whidby, *Anal. Chim. Acta.*, **42**, 271 (1968), with permission.

Table 7.14. However, there is no evidence at present to permit other than speculation about the structure of a species with a calcium-to-EDTA ratio of 3 : 1.

This work illustrates a further application of NMR to detect weak chelates in an equilibrium involving several species. However, as in this case, the chemical shift changes become smaller upon further coordination of the ligand and the precision suffers.

7.4.2.3 Protolysis Kinetics of Free Ligands

Just as the protonation equilibria play an important role in understanding the chemistry of aminocarboxylates and their behavior as ligands for chelate formation, so does the rate of the protolysis reactions. NMR is of moderate use in the study of these reactions. The reason for the limitation may be more easily understood if the methods of study are clear. There are two NMR techniques that may be applied with relative ease to the study of protolysis reaction rates. First, one may use a slow passage high-resolution method that involves the comparison of the shape of the experimental NMR lines with those computed from theory. For example, a simple molecule such as trimethylamine may be studied by the broadening and eventual coalescence of the methyl doublet arising from spin coupling with the NH proton in a highly acidic solution. As the acidity is lowered, the rate of proton exchange increases and the doublet is coalesced.

The second technique involves the use of pulsed NMR in which the relaxation time of the protons in the solvent is determined from the decay pattern of the NMR "echos" which follow a series of pulses of the radio frequency. This technique can usually measure faster rates than those measured by slow passage high resolution methods. However, both require acidic solutions in the case of amine sites so that the rate of proton exchange may be lowered to a measurable value. If the compound to be studied has more than one functional site, the lower pKa values may require extreme acidities to accomplish this slowing of the exchange rate.

The protolysis kinetics of the amino site of IDA (iminodiacetic acid) and MIDA (methyliminodiacetic acid) have been investigated by Leyden and Whidby utilizing the slow passage method (200). The acidity required to observe the spin coupling between the labile —NH proton and the —CH_2-protons of the acetate group in IDA and MIDA or the methyl protons in MIDA was achieved with approximately 80% sulfuric acid solutions. Protonation of the carboxyl groups to form C=OH^+ species was not present in these solutions. However, because of the medium, many difficulties were encountered. For example, the NMR lines were viscosity broadened which required elevated temperatures to minimize the effect. The concentrations of the solvent species were difficult to obtain and a reasonable representation of the acidity was in question. In spite of these difficulties, two significant exchange processes were found. The first is represented by a transfer of a proton to the solvent and values of 2.0 sec^{-1} and 2.1 sec^{-1} were obtained for IDA and MIDA, respectively. The second reaction may be illustrated by reaction 7.43 and rate constants of 4.9×10^4 M^{-1} sec^{-1} and 1.4×10^5

$$HSO_4^- + \overset{+}{NH}\underset{(CH_2COOH)_2}{\overset{R}{\diagup}} \longrightarrow H_2SO_4 + N\underset{(CH_2COOH)_2}{\overset{R}{\diagup}} \quad (7.43)$$

M^{-1} sec^{-1} were obtained for IDA and MIDA, respectively. All data are at 60°C. These studies also yielded qualitative results on the effect of acidity on the inversion of the nitrogen atom in MIDA. The inversion rate was determined from the AB pattern of the acetate protons. An attempt was made to investigate the protolysis of EDTA, CyDTA and other similar ligands. However, the compounds were not soluble in the acid concentrations required.

Because the protolysis reaction is so inherently involved in the chelation chemistry of aminocarboxylates, it would be of great interest to know more about these processes. However, it is likely that methods other than NMR must be used to obtain these data. Obviously one would like the data obtained in a medium similar to the ones in which chelation studies are made.

7.4.2.4 Protolysis Kinetics of Metal Complexes

Proton exchange reactions of aminocarboxylates are limited in their consideration to the labile protons of the amino and carboxyl sites. The acetate and ethylenic protons are inert, or at least exchange sufficiently slowly to be considered as inert. In certain chelates, however, this is not the case. Williams and Busch showed that hydrogen atoms on a carbon alpha to a nitrogen coordinated to Co(III) undergo deuteration in alkaline D$_2$O (201). Thus the protolysis reactions of amino-carboxylate chelates may be divided into these which involve —CH protons and those which involve protons of the amino and carboxyl groups.

Terrill and Reilley studied the base catalyzed isotopic exchange of the α-hydrogen in a series of EDTA chelates with bivalent metal ions (202). Only the acetate protons were observed to exchange. The rate of exchange was determined by following peak areas during isotopic substitution. The isotopic exchange of the alkaline earths were observed to follow the order

Mg > Ca > Sr > Ba. This order is rationalized on the basis of charge-to-radius ratio of the metal ion assuming ionic bonding predominates. On the other hand, the series $Cu^{+2} > Ni^{+2} > Co^{+2} > Zn^{+2}$ is rationalized on the basis of covalent bonding and ligand field considerations. It is interesting that the alkaline earth series is in qualitative agreement with the order of chemical shifts given by Kula et al. (196). Terrill and Reilley also pointed out that good correlation is obtained between the rate of exchange and the gross heat of ligation which suggests that the nature of the metal chelate in respect to coordinated water molecules is largely the same. That is, the presence of coordinated water molecules would alter the inductive effect of the metal ion on the exchange process. Only lead-EDTA could not be rationalized by these considerations.

An interesting observation was that uncomplexed EDTA undergoes isotopic exchange of the acetate protons which is zero order in base and with a rate faster than barium-EDTA in 1.0 M OD^-. The results suggest an intramolecular process in which the proton abstracting sites of the unassociated ligand are the amino and/or carboxylate groups on the same molecule. The more remote nitrogen was found to be more effective in removing the proton than the α-nitrogen. Mono-protonated EDTA exchange was immeasurably slow. The less basic carboxylate sites are apparently ineffective for proton abstraction, although studies of esters or alkylated amino sites were not performed. It should be mentioned that this study may be important in considering reactions other than hydrogen exchange at the acetate sites and illustrates an excellent method of preparation of acetate-deuterated ligands for other studies such as ligand exchange.

Terrill and Reilley extended their study to Co(III)–CyDTA (203). The NMR spectrum of this complex indicates there are four types of acetate protons. It was found that the acid-catalyzed deuteration was strikingly stereospecific in that certain hydrogen atoms on acetate groups out of the plane of the nitrogen metal bonds are deuterated more rapidly than their geminal neighbors. The base-catalyzed exchange was complicated by the formation of hydroxyl-chelates and reduction to paramagnetic Co(II). Sudmeier and Occupati found *trans*-Co(III) EDDA (en)$^+$ to be more useful where EDDA represents ethylenediaminediacetic acid and en represents ethylenediamine (204). Using this compound, which is stable in base, base-catalyzed exchange was found to be more rapid than acid-catalyzed exchange and approximately ten times more stereospecific. A series of compounds gave similar results and confirmed the findings of Terrill and Reilley. The major contribution of Sudmeier and Occupati was to eliminate a possible ambiguity in the spectral assignment of Terrill and Reilley. However, considerable data is available on the rate and activation parameters of α-hydrogen exchange in this paper.

A series of papers by Erickson and co-workers have considered the rates of N—H proton exchange and nitrogen inversion for some amino acid complexes of platinum(II) (205-208). The studies were done using both the slow passage method and by following slow spectral changes in D_2O solutions of the complexes. These exchange processes were found to be catalyzed by D_2O and OD^-, and in the pH range 4-7 intramolecular catalysis by the acetate group was found. The rates of exchange are in all cases much slower than that typical of the free ligand. A discussion of the correlation between the degree of nitrogen substitution and the macroscopic pKa of analogous free ligand sites with respect to the rate data is given. However, no report of an attempt to compare the rate law found by Erickson and associates for the complexes with those of free amino acids was given. The nitrogen inversion studies by Erickson and associates (207) and others (209) on platinum and cobalt complexes have shown that the rate of proton exchange catalyzed by hydroxide ion exceeds the rate of nitrogen inversion by $10-10^4$ fold. Erickson explains this by invoking a mechanism for inversion as follows

$$\text{N—H} + \text{OH}^- \underset{k_2}{\overset{k_1}{\rightleftarrows}} \text{N:}^- + \text{H}_2\text{O} \qquad (7.44)$$

$$\text{N:}^- \xrightarrow{k_3} {}^-\text{:N} \qquad (7.45)$$

where N:^- represents the deprotonated nitrogen and k_3 is a first-order rate constant for inversion. This mechanism is substantially the same as that suggested by Saunders and Yamada for the inversion of amines (210). Erickson and associates assumed $Ka \simeq 0.01\ K_w$ for the N—H discussion in the complexes, for few have been measured. Assuming that the value of the hydroxide catalyzed proton exchange rate constant for a series of complexes may be related to a simple Brönsted relation

$$k_{\text{OH}^-} = G(K_a)^\beta \qquad (7.46)$$

where G and β are constants, the authors estimated values of k_3 to be 10^5-10^6. These values compare remarkable well with those of nitrogen atoms contained in five and six membered rings (211). However, the results obtained by Erickson should be modified to include the mechanism for nitrogen inversion in aqueous acid proposed by Morgan and Leyden for acyclic amines (212). This mechanism considers the possibility that even though the amine is deprotonated, the strong hydrogen bond between the nitrogen atom and solvent water molecules inhibits inversion until that hydrogen bond is broken. This mechanism provides a qualitative explanation of faster proton exchange than inversion and yields a quantitative description of the rate of nitrogen inversion as a function of pH. The mechanism has been found by Pitner and Martin to hold for the inversion of nitrogen atoms in platinum complexes (213).

7.4.2.5 Structure and Bond Lability of Metal Complexes

The protonation studies have established reasonably well the structure of free aminopolycarboxylates in solution. The next step was to ascertain the gross structure of the various complexes. The extensive X-ray crystallographic studies by Hoard (214) and others had contributed a great deal to the elucidation of the structures of aminopolycarboxylate chelates in the solid state. However, there is always a reluctance to extrapolate these data to the solution phase. The solution chemist is interested in the dynamics of structural change as well as an average structure. The answers to many questions such as the effect of solvent, extraneous reagents and, of course, the central metal ion upon something as simple as an EDTA titration may lie in the lifetime of a metal ligand bond.

Day and Reilley (215) applied NMR to the investigation of individual metal ligand bond lifetimes. Before this it had long been known that in the solid state a given ligand such as EDTA formed different numbers of metal-ligand bonds depending on protonation of the ligand (216). Mixed aquo- and hydroxy-complexes were also known. However, no direct results had been obtained to determine the lifetime of these metal-ligand bonds.

Consider the predicted effect bond lifetime will have on the NMR spectrum of an aminopolycarboxylate chelate. If all metal-ligand bonds have short lifetimes, the spectrum of the nonlabile protons in the molecule will be relatively simple because of internal averaging. This is similar to the fast proton exchange in a partially protonated ligand. As in protonation, the chemical shift of the chelate will be different than the free ligand. An average spectrum would be observed as a result of rapid ligand exchange. On the other hand, if one or more metal-ligand bonds is sufficiently long so that it may be considered permanent by NMR detection, then chemical shifts which are identical in the free ligand may not average in the chelate and a more complex spectrum may result. As an example, two protons on a carbon atom adjacent to an asymmetric nitrogen are expected to be nonequivalent if the nitrogen atom does not undergo inversion. A structure such as

$$\begin{array}{c} \text{H} \quad R_1 \\ | \quad\ | \\ \text{R}-\text{C}-\text{N}-M- \\ | \quad\ | \\ \text{H} \quad R_2 \end{array}$$

where R_1, R_2 and RCH—$_2$ are all different and the nitrogen–metal bond has a long lifetime meets these requirements. The two protons on the carbon adjacent to the nitrogen are expected to yield the *AB*-type splitting pattern. The mere presence of an *AB* pattern confirms qualitatively a long nitrogen–metal bond lifetime. A more detailed examination of these *AB* patterns may yield quantitative rate data as well as information about the conformation of the structure and will be discussed later.

A further tool is the presence of abundant isotopes of the central metal ion which have nuclear spins greater than one half. Such isotopes may result in metal–proton spin coupling which is useful.

Day and Reilley proposed four situations to consider (215). The first of these is the case in which both the metal–oxygen and metal–nitrogen bond lifetimes are short. In such a case of labile bonds, there sould be two sharp peaks corresponding to the ethylenic and acetate protons. If there is a significant abundance of an isotope of the central metal ion with spin $\frac{1}{2}$ or greater, splitting of the ethylenic and acetate protons may occur unless there is complete ligand exchange. In the event of rapid bond breaking and reforming, the coupling would persist as long as the ligand bonds are reformed with the original metal ion. An example is ^{207}Pb which has a spin $\frac{1}{2}$ and a natural abundance of 21%. The NMR spectrum of Pb-EDTA is shown in Figure 7.26a. In some cases such as the alkali metals only an average spectrum of free and complexed ligand is observed indicating rapid ligand exchange. The alkaline earth metals on the other hand exhibit ligand exchange rates which are pH dependent (196).

The second situation is one in which the lifetime of the metal oxygen bonds is short while that of the metal–nitrogen bond is long. This case allows for rapid averaging of the acetate protons. However, the long lifetime of the metal–nitrogen bond inhibits inversion of the nitrogen atom and results in a nonequivalence of the two protons of the acetate groups. An *AB* pattern is thereby expected. Again, the presence of a metal isotope with a spin greater than one half would have persistent coupling with the protons on the ligand. An example of this case is cadmium-EDTA whose spectrum is given in Figure 7.26b which shows the acetate *AB* pattern and an ethylenic singlet. The spectrum is complicated by the presences of 25% isotopes of spin $\frac{1}{2}$. Figure 7.26c shows the *AB* pattern due to cadmium with $I = 0$.

The third possibility is one in which the lifetime of the metal–oxygen bond is long, whereas that of the metal–nitrogen bond is short. Consideration of the structure of metal-EDTA complexes suggests that this is an unlikely case. Although it is likely that certain metal–nitrogen bond lifetimes are inherently shorter than metal-oxygen bonds to the same metal ion, it is unlikely that the metal–nitrogen bond in an EDTA chelate could be broken without prior rupture of the metal–oxygen bonds. There have been no unambiguous confirmations of this type of situation. However, Kula has suggested this to be the case with some Mo(VI) chelates which will be discussed later (217).

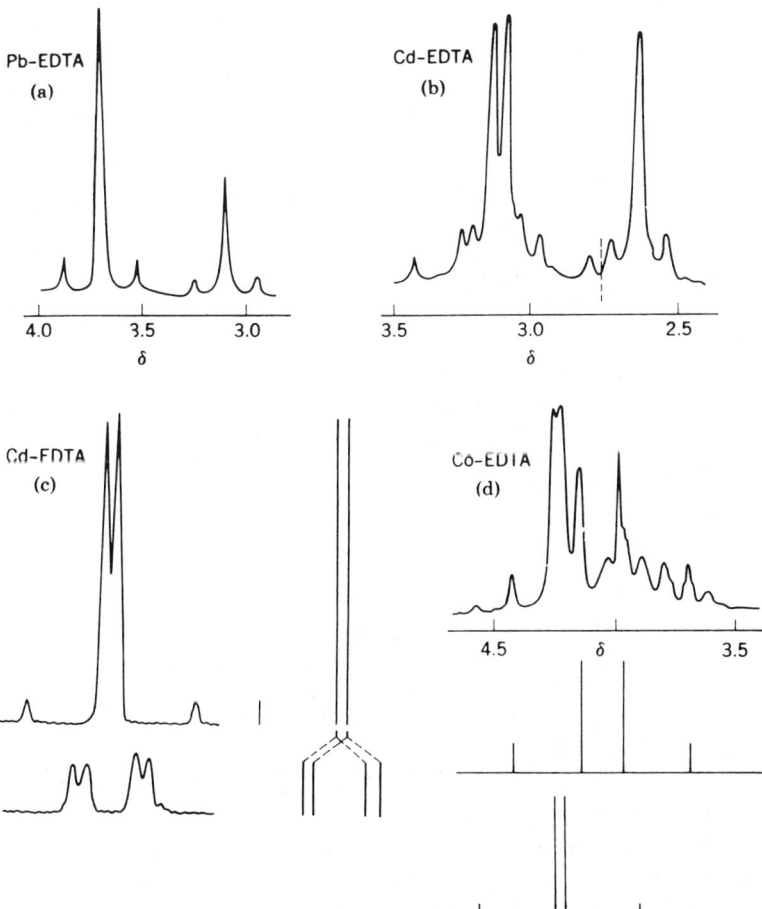

Fig. 7.26 PMR spectra at 60 mc sec^{-1} of 0.5 to 1.0 M aqueous solutions of metal-EDTA complexes; reference: sodium 3-(trimethylsilyl)-1-propanesulfonate. (a) Pb-EDTA; acetate and ethylenic proton resonances at ca. 3.7 and 3.1 δ, respectively, with side bands from ^{207}Pb shown. (b) Cd-EDTA; PMR for acetate protons to left of broken line and for ethylenic protons to right. (c) Cd-EDTA; acetate portion of PMR spectrum; on top of the AB pattern due to Cd with $I = 0$; on bottom ABX pattern due to Cd with $I = \frac{1}{2}$. (d) Co(III)-EDTA; two AB patterns for the acetate groups shown below spectrum; remaining peaks due to ethylenic protons. From *Spectroscopy and Structure of Metal Chelate Compounds*. (K. Nakamoto and P. J. McCarthy, Eds., Wiley, New York, 1968, with permission.)

The final combination is the case in which both the metal–oxygen and metal–nitrogen bond lifetimes are long. In this case, the most complex spectra are expected because the acetate groups are not equivalent. Furthermore the two protons on each acetate group are not equivalent because of the adjacent asymmetry about the nitrogen atom. The complex is substitution inert and in the case of an octahedral complex where all ligand atoms are bound, dl forms are present. For a given optical form, two of the acetate groups are "in plane" with the nitrogen-to-metal bonds and are identical, and the other two acetate groups are "out of plane" and are identical. An example of this case is cobalt(III)-EDTA whose spectrum is shown in Figure 7.26d. Two AB patterns are observed for the acetate protons. The ethylenic protons may not be equivalent because of the rigidity of the complex.

Chan, Kula, and Sawyer investigated the NMR spectra of molybdenum(VI) chelates of EDTA, NTA (nitrilotriacetic acid) and MIDA (methyliminodiacetic acid) (218). The results show that the metal to ligand ratios of the chelates are 1 : 1 for NTA and MIDA and 2 : 1 for EDTA. The acetate protons in the MIDA and EDTA complexes give rise to AB patterns. The authors interpret this as an indication that the carboxylate groups are "coordinated rigidly to the metal ion in these chelates." It is possible that rapid breaking and reforming of metal–oxygen bonds occurs without nitrogen inversion. Protonation of MIDA in 50% H_2SO_4 gives rise to an AB pattern for the acetate protons because of slow nitrogen inversion (200). Kula was able to demonstrate that J_{AX} and J_{BX} were different in the ABX pattern obtained by N—H proton coupling with the acetate protons in Mo(VI)–IDA (217). He calculates H_X—N—C—$H_{A \text{ or } B}$ angles to be 12° between AX and 110° between BX and interprets this as conclusive evidence that the bond lifetime of the Mo—O bonds were long (>0.2 sec). His interpretation was that this long Mo—O bond lifetime was required to preserve the distinct J_{AX} and B_{BX} values. He further concluded that nothing could be said with respect to the Mo—N bond lifetime because, assuming long Mo—O bond lifetimes, the spectrum would be the same regardless of the lability of the Mo—N bond: It is possible that this interpretation is correct. However, Kula's reasoning is based on the assumption that no, or very slow, rotation about the H_2C—N bond is required to observe the difference in J_{AX} and J_{BX}. If the Mo—N bond lifetime is long, on the other hand, rapid rupture and rotation about the H_2C—N bond could occur and still preserve the difference in J_{AX} and J_{BX} as long as there was a preferred rotamer on the time average. The preferred rotamer would likely be the one suitable for Mo—O bonding. The interpetation of the molybdenum(VI) chelate data should be reconsidered. A molybdenum(V)–EDTA chelate has been studied and combined X-ray and NMR suggest a bridged structure (219).

Rhodium(III)-EDTA is one of the more interesting chelates with EDTA to be investigated. Smith and Sawyer report that the NMR spectrum of the protonated complex Rh(HEDTA)(OH$_2$) indicates the presence of four different types of acetate protons in a chelate in which the ligand is pentadentate (220). Two acetate AB patterns are observed and two acetate singlets that result from accidental degeneracy of the two protons in each group. The ethylenic portion of the ligand gives rise to an $ABCD$ pattern. The pH dependence of the spectrum is attributed to the equilibria:

$$\begin{array}{ccc} \text{Rh(HEDTA)(OH}_2\text{)} & \rightleftharpoons & \text{Rh(EDTA)(OH}_2\text{)}^- \\ \updownarrow \text{pK} \sim 2 & \diagup & \updownarrow \\ \text{Rh(EDTA)}^- & \underset{\text{pK} \sim 9-10}{\rightleftharpoons} & \text{Rh(EDTA)(OH)}^{2-} \end{array} \quad (7.47)$$

By analogy with the spectra of Co(III)–EDTA, the sexadentate chelate is thought to be predominate at pH 5. In all cases, the metal–oxygen and metal–nitrogen bond lifetimes are long. Rh(EDTA)$^-$ is sensitive to ultraviolet radiation and optically active solutions of this chelate are rendered inactive after 2 hr exposure. There are changes observed in the NMR spectrum during this process which have been investigated recently.

Sudmeier and associates used Rh(III) PDTA to investigate the photolysis effect (221). The complex $(-)_{5461}$ Rh(III) D-(−)-PDTA lost optical activity upon exposure to ultraviolet light, but regained its optical activity after storage in the dark for several days. This cycle may be repeated many times. The NMR spectrum shows a new methyl doublet at the expense of the original methyl resonances following irradiation. The ratio of the areas of these peaks coincides with the loss of optical rotation. The CD and absorption spectra of the photolysis product showed no evidence of a change in the denticity of the ligand or coordination number of the rhodium ion. The authors proposed that the photolysis product contains a 1:3 ratio of diastereomers. The proposal is substantiated on the observation that the corresponding enantiomers of Rh(EDTA) would be equal and no dark recovery of optical activity would be expected as observed after three days. These interpretations are contrary to the early proposal of Dwyer and Garvan of a photo-equation reaction (222).

The type of ^{1}H-NMR spectrum observed for the chelates of EDTA have been shown to be very useful in evaluating individual bond lifetimes. Qualitatively, the more complicated the spectrum, the more inert the complex. For example, in an EDTA chelate, the presence of an AB pattern for the acetate protons implies a long metal–nitrogen bond lifetime. However, when more complicated ligands are used, the free ligand may give rise to complicated spectra. Day and Reilley investigated the effects of ligand

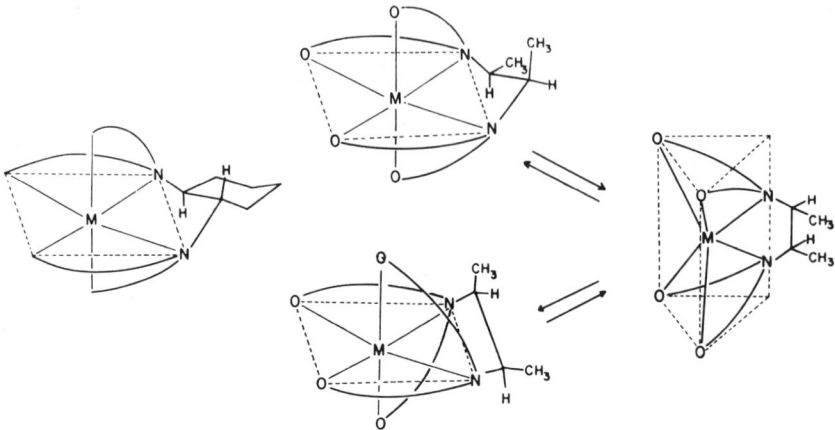

Fig. 7.27 Structures of octahedral metal complex of CyDTA and two octahedral configurations of meso-BDTA. Intermediate structure in interconversion of two configurations also shown. (Reprinted with permission from R. J. Day and C. N. Reilley, *Anal. Chem.*, **37**, 1326 (1965). Copyright by the American Chemical Society.)

structure on the bond lifetimes of chelates by using CyDTA (*trans*-1,2-cyclohexylenedinitrilo)-tetraacetic acid and BDTA (*meso*-(2,3-butylenedinitrilo)-tetraacetic acid in addition to EDTA (223). In molecules such as CyDTA and BDTA the presence of asymmetric carbon atoms within the ligands give rise to *AB* patterns for the acetate groups adjacent to these sites (191). Thus *AB* patterns may be present even in the free ligand. Caution must be used however, for the two acetate protons may be accidentally degenerate and give rise to a single peak. The spectrum of the chelate may be further complicated by slow interconversion between configurations. Figure 7.27 shows the structure of octahedral complexes of CyDTA and two configurations of *meso*-BDTA. The interconversion of all the BDTA chelates studied was found to be rapid as evidenced by a single sharp methyl resonance. The spectra of some of the complexes studied are shown in Figure 7.28 and 7.29. The total interpretation of these data is beyond the scope of this discussion. In brief, metal ions such as Mg, Ca, Zn, and Pb which show short metal–nitrogen bond lifetimes in EDTA chelates exhibit longer metal–nitrogen bond lifetimes with meso-BDTA. The metal–oxygen bond lifetimes remain short. Strontium and barium form chelates with both metal–nitrogen and metal–oxygen lifetimes short. Co(III) forms an inert chelate. For meso-BDTA, if both types of bonds are labile in the chelate, only one type of methyl group will be observed. If the metal–oxygen is labile but the metal nitrogen bond has a long lifetime, two acetate groups will be observed but

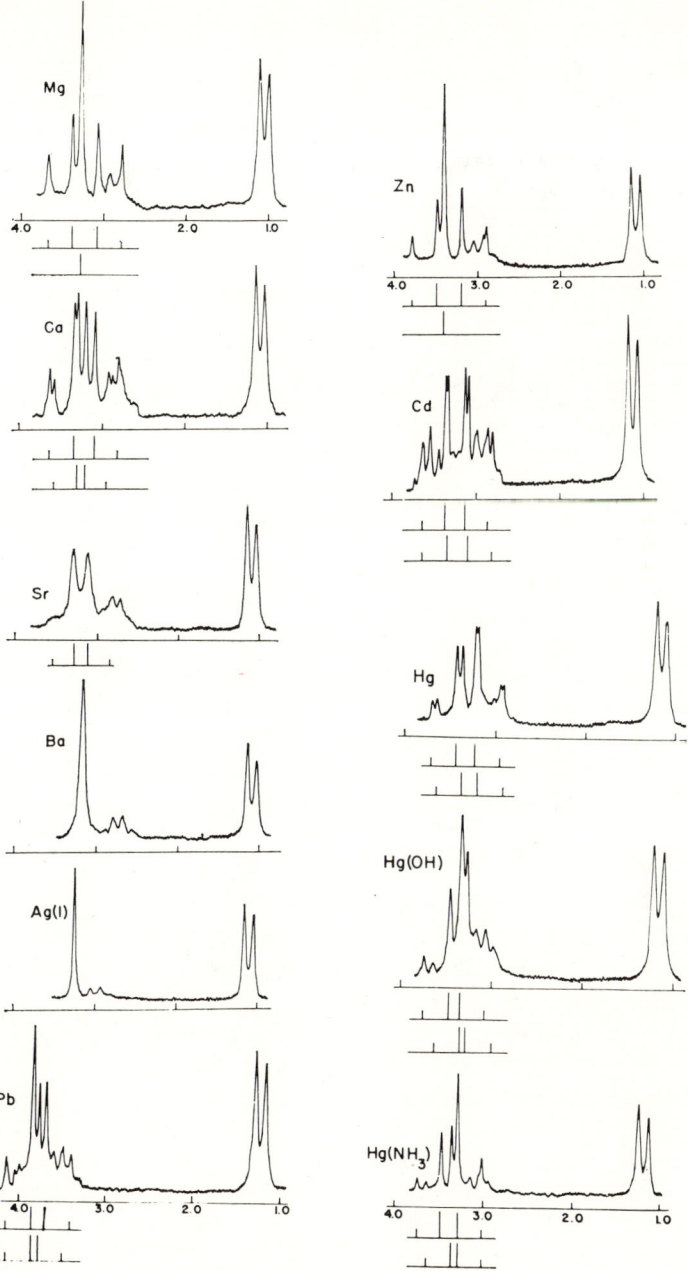

Fig. 7.28 Proton NMR spectra of some metal-BDIA complexes. Assignments for acetate *AB* patterns indicated below each spectrum. Chemical shifts. DSS. (Reprinted with permission from R. J. Day and C. N. Reilley, *Anal. Chem.*, **37**, 1326 (1965). Copyright by the American Chemical Society.)

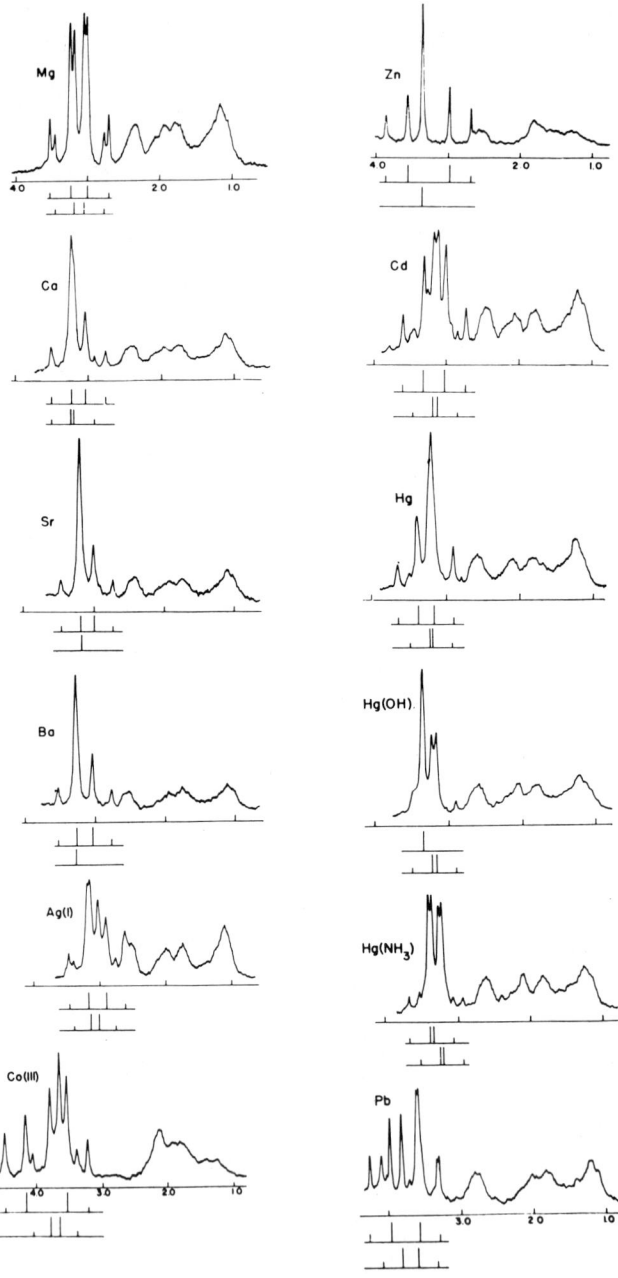

Fig. 7.29 Proton NMR spectra of some metal-CyDTA complexes. (Reprinted with permission from R. J. Day and C. N. Reilley, *Anal. Chem.*, **37**, 1326 (1965). Copyright by the American Chemical Society.)

only one type of methyl group will be observed. If both type of bonds are nonlabile, there will be four types of acetates and two types of methyl groups.

In the case of CyDTA, the conformation of the cyclohexane ring places additional restrictions upon the chelate which permits only one configuration. The symmetry of the octahedral chelate leads to the result that as long as the metal–nitrogen bond has a long lifetime, there will be two types of acetate protons (those in-plane with the nitrogen–metal bonds and those out-of-plane). In this case, the bond lifetime for the metal–oxygen bonds can only be evaluated by subtle changes in the value of J_{AB} resulting from variation in the strain upon the acetate groups. Kula and Sawyer suggested the addition of paramagnetic metal ions to the solution (224). A labile acetate group will spend some time coordinated to the paramagnetic ion which will broaden the $-CH_2-$ resonance of that group.

7.4.2.6 Exchange Reactions Involving Metal Complexes

The chemical dynamics of ligand exchange kinetics of aminocarboxylate chelates was the subject of extensive investigations utilizing techniques other than NMR (225). By these techniques, only unsymmetrical exchange reactions could be studied. That is, a ligand on a metal complex was exchanged for a chemically different though structurally similar ligand. For example, one ligand may be optically active (namely, PDTA) whereas the other (EDTA) optically inactive. The exchange process could then be studied by means of optical activity of the ligand molecules (226). However, care must be taken in relating these results to symmetrical exchange reactions. If the ligand exchange reaction is sufficiently slow, isotopic tagging may be employed to investigate symmetrical reactions by separation and isotopic analysis of the reaction mixture. However, most reactions involving ligand exchange of alkali, alkaline earth, and transition metal complexes of aminocarboxylates are much too fast for study by these means.

Ligand exchange reactions of multidentate ligands may be divided conveniently into two categories: (a) nucleophilic substitution reactions (S_N) in which excess ligand is added to a metal complex and the ensuing reaction represented as

$$MY_A + Y_B \rightleftharpoons [Y_B M Y_A] \rightleftharpoons MY_B + A \qquad (7.48)$$

where $Y_B M Y_A$ represents an intermediate in which both ligands are coordinated to the metal ion, and (b) electrophilic substitution reactions (S_E) represented as

$$M_A Y + M_B \rightleftharpoons [M_A Y M_B] \rightleftharpoons M_B Y + M_A \qquad (7.49)$$

in which an intermediate forms in which both metal ions coordinate to the ligand. A generalization of the S_E mechanism would include the dissociation of a protonated complex.

Unlike other techniques, NMR readily permits the study of symmetrical exchange reactions because the free and complexed ligands, and even parts of these ligands, are labeled by their nuclear spins. This permits the study of both intermolecular and intramolecular (such as individual bond lifetime) exchange reactions. An additional advantage is that because the measurement is based upon the shape of the NMR spectrum resulting from exchange, the studies are done using solutions which are at chemical equilibrium. Of course, slow exchange studies may be done by isotopically labeling the ligand with deuterium and observing changes in spectral areas with exchange. Both of these methods are easily applied using conventional NMR spectrometers. The limitations are essentially based on whether the exchange rate can be adjusted to fall into the range of rates which may be determined by NMR. This range, sometimes called the NMR kinetic "window," is determined by the separation of resonance frequencies of the protons in their different environments. This separation may manifest itself as a chemical shift or as a spin–spin splitting pattern. For example, the collapse of spin–spin multiplets ($J \sim 0.5$–20 Hz) would permit studies of rates on the order of 10^1–30^3 sec^{-1}, the exchange between diamagnetic environments in which chemical shift differences of 0.5–100 Hz are observed would permit rates of 10^1–10^4 sec^{-1} to be determined, and exchange between a diamagnetic environment and a paramagnetic one which gives rise to chemical shift changes of 10^2–10^4 Hz would permit rates of the order of 10^4–10^7 sec^{-1} to be investigated. Frequently the control of pH, reactant concentration, temperature, or other parameters can be used to bring the rate of the reaction which is to be studied into the appropriate range. The major disadvantage of NMR for kinetic and equilibrium studies is the high concentration of solute (e.g., 0.1 M) needed to obtain spectra with good S/N ratios. This problem may be eliminated by the use of signal averaging techniques such as FT-NMR. Occasionally, the NMR lines are broadened by the viscosity of the solutions. This is a result of a decrease in the transverse relaxation time and may require empirical correction, or in some cases may be removed by operating at an elevated temperature which lowers the viscosity.

Among the first studies of aminocarboxylate ligand exchange reaction rates done by NMR were those of Pearson and associates (227,228) who investigated amino acid complexes of paramagnetic ions. The exchange rates show the order Mn > Fe > Co > Ni ≪ Cu with the copper complexes showing a second-order (S_{N2}) mechanism and iron, cobalt, and nickel (and probably manganese) showing mixed first- and second-order rate laws. The octahedral complexes likely undergo a solvent-assisted dissociation (S_{N1}) mechanism.

Although Kula had given some qualitative values of exchange rates involving hydroxy complex of Zn(II) EDTA (197), the first detailed study of

APPLICATIONS TO FUNDAMENTAL INVESTIGATIONS 409

ligand exchange of an EDTA chelate using NMR was done by Sudmeier and Reilley (229). Cadmium-EDTA was chosen because it is diamagnetic and earlier work suggested its reaction rates fall in the NMR kinetic "window" (223). The metal–nitrogen bond lifetime was known to be long because AB patterns are observed for the acetate protons. Rupture of one of these metal–nitrogen bonds permits nitrogen inversion. On reforming the bond following inversion the A and B protons exchange their magnetic environments. If the rate of this process is rapid, a broadening of the AB pattern will result and the rate of the individual metal–nitrogen bond breaking may be determined (assuming the nitrogen inversion step is rapid). Intermolecular exchange rates are obtained from the line shape of excess (free) EDTA in the solution which becomes broadened by exchange between two diamagnetic environments: free and complexed.

The following exchange reactions may be considered:

$$MY^{-2} \underset{k_{-1}}{\overset{k_1}{\rightleftharpoons}} M^{+2} + Y^{-4} \tag{7.50}$$

$$MY^{-2} + \overset{*}{Y}^{-4} \underset{k_{-2}=k_2}{\overset{k_2}{\rightleftharpoons}} M\overset{*}{Y}^{-2} + Y^{-4} \tag{7.51}$$

$$MY^{-2} + HY^{-3} \underset{k_{-3}=k_3}{\overset{k_3}{\rightleftharpoons}} M\overset{*}{Y}^{-2} + HY^{-3} \tag{7.52}$$

$$MY^{-2} + H_2\overset{*}{Y}^{-2} \underset{k_{-4}=k_4}{\overset{k_4}{\rightleftharpoons}} M\overset{*}{Y}^{-2} + H_2Y^{-2} \tag{7.53}$$

$$MHY^{-} \underset{k_{-5}}{\overset{k_5}{\rightleftharpoons}} M^{+2} + HY^{-3} \tag{7.54}$$

$$MHY^{-} + H\overset{*}{Y}^{-3} \underset{k_{-6}=k_6}{\overset{k_6}{\rightleftharpoons}} MH\overset{*}{Y}^{-} + HY^{-3} \tag{7.55}$$

$$MHY^{-} + H_2\overset{*}{Y}^{-2} \underset{k_{-7}=k_7}{\overset{k_7}{\rightleftharpoons}} MH\overset{*}{Y}^{-} + H_2Y^{-2} \tag{7.56}$$

Table 7.15 shows a summary of the rate constants obtained for these reactions. A detailed consideration of the charge, size, and number of coordination sites on the attacking ligand in the S_N-type reactions suggested a mechanism for exchange not unlike that given by Pearson. A representation of this mechanism is shown in Figure 7.30. Figure 7.31 shows a proposed mechanism for the S_E-type reaction and invokes the 2:1 metal-complex intermediate. These compounds are known to be stable for molybdenum (218)

TABLE 7.15. Experimental Rate Constants for Ligand Exchange

	CdEDTA (28°C)(Ref. 229)		CaEDTA	SrEDTA	CaHEEDTA
			(32°C, $\mu \sim$ 1.6 M)(Refs. 230, 231)		
k_1 sec^{-1}	—	—	<1	3×10^1	3×10^1
k_{-1} M^{-1} sec^{-1}	—	—	$<2.5 \times 10^9$	8.1×10^8	1.3×10^8
k_2 M^{-1} sec^{-1}	1.3×10^2	($\mu = 2.5$)	1.2×10^2	1.1×10^3	1.1×10^3
k_3 M^{-1} sec^{-1}	<3	($\mu = 2.5$)	5	1.6×10^2	2.3×10^2
k_4 M^{-1} sec^{-1}	1.4	($\mu = 1.6$)	2×10^5	1×10^5	—
k_5 sec^{-1}	1.2	($\mu = 1.6$)	5×10^4	2×10^4	—
k_{-5} M^{-1} sec^{-1}	1×10^9	($\mu = 1.6$)	1×10^7	5×10^5	—

and evidence for a similar structure with calcium and strontium has been presented earlier (199).

Almost simultaneous with the publication of the Cd(II)-EDTA exchange studies, Kula and associates published the results of a similar study of Ca(II)-EDTA exchange (230) and later the results of Sr-EDTA and Ca-HEEDTA (N'-(2-hydroxyethyl)-ethylenediamine-N,N,N'-triacetic acid (231). The latter ligand, while structurally related to EDTA, is capable of only pentadendate coordination. The results of these studies are also given in Table 7.15. A less detailed but similar mechanism to that proposed by Sudmeier and Reilley were put forth by Kula and associates. The reader may compare the rate constants for these complexes as his interests direct. The most obvious trend shows the alkaline earth complexes to have much larger nucleophilic exchange rates, probably due to the inherent M—N

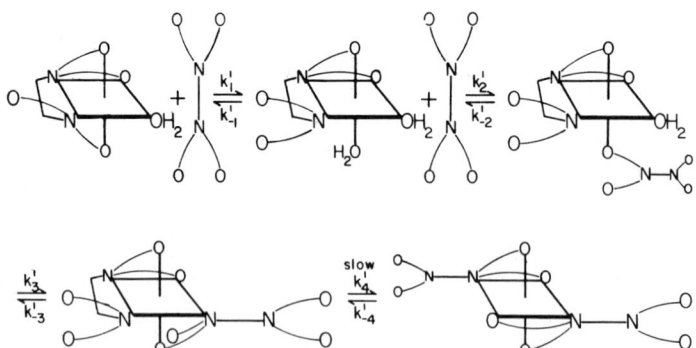

Fig. 7.30 Proposed mechanism of S_N reactions of Cd-EDTA and EDTA. (Reprinted with permission from J. L. Sudmeier, and C. N. Reilley *Inorg. Chem.*, **5**, 1047 (1966). Copyright by the American Chemical Society.)

APPLICATIONS TO FUNDAMENTAL INVESTIGATIONS   411

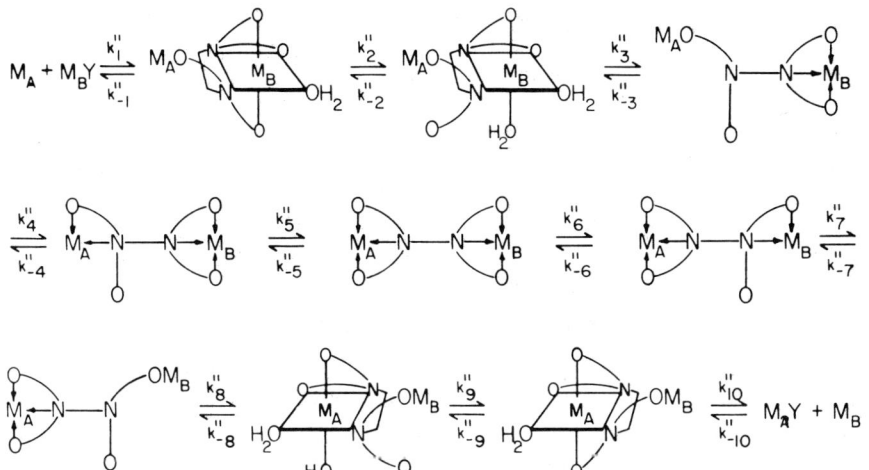

Fig. 7.31 Proposed mechanism of S_E reactions of Cd-EDTA. (Reprinted with permission from J. L. Sudmeier and C. N. Reilley, *Inorg. Chem.*, **5**, 1047 (1966). Copyright by the American Chemical Society.)

bond lability and the subsequent availability of coordination sites on metal for attacking ligands.

7.4.2.7 Conformational Analysis of Diamagnetic Chelates

The conformations of coordinated ligands have been a subject of interest for several years. The use of NMR had made it possible in some cases to determine the conformations of ligands in metal complexes by application of the Karplus relationship between vicinal proton coupling constants and the dihedral angle (232) and the large contact shifts observed with certain paramagnetic metal complexes. As an example of the use of the Karplus relationship, Sudmeier and Senzel performed a study of free and complexed PDTA (dl-propylenediaminetetraacetic acid) (190,233). The introduction of a methyl group into the ethylenic chain of EDTA to form PDTA also introduces asymmetry, which in turn leads to observable spin coupling between the ethylenic-CH_2— and —CH protons. Additional coupling to the —CH_3 protons produces a complex $ABCX_3$ spectrum, which of course contains more information related to conformation than the single peak observed for the ethylenic proton resonance in the case of EDTA. The point of this is to consider possible configurations of the ligand, and to evaluate their relative populations. Usually, three classical rotamers such as I, II, and III are proposed, as shown in Figure 7.32.

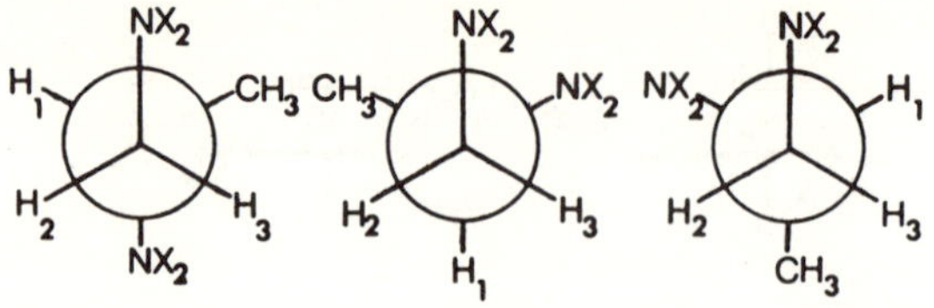

Fig. 7.32 Representations of the classical staggered rotamers of PDTA. (Reprinted with permission from J. L. Sudmeier and A. J. Senzel, *Anal. Chem.*, **40**, 1693 (1968). Copyright by the American Chemical Society.)

The determination of the relative populations of these rotamers is based upon the relationship between the vicinal coupling constants J_{12} and J_{13} and their corresponding dihedral angles. There is one serious limitation to this procedure. Although the experimental values of J_{12} and J_{13} may be readily and accurately determined, the parameters needed to calculate rotamer populations include the coupling constants for the *anti* (J_a) and *gauche* (J_g) configurations of the coupled protons. Obviously these cannot be obtained from the compound under study. Frequently, values obtained from rigid ring structures of similar functional groups are used. However, the coupling constants are subject to considerable variation with substituents. Certain empirical expressions have been proposed and used with some success to allow corrections on J_a and J_g for the electronegativity of substituents (234). The success is limited to the observation that qualitatively no shocking results have been obtained for relative rotamer populations. However, the significance of the numerical values frequently reported is an open question. In spite of these restrictions, Sudmeier and Senzel showed convincingly that monoprotonated PDTA had a conformation structurally equivalent to a "proton chelate" as represented by Structure II, which does not imply a polyfurcated hydrogen bond, but an "averaged" equivalent structure. In addition, these authors demonstrated the presence of weak 1:1 chelates of K^+, Rb^+, and $(CH_3)_4N^+$ in solution in which the metal ion (cation) is coordinated to both functional groups. Formation constants of 1.3, 0.14, and 0.12 respectively were estimated for these chelates. Cesium was used as the cation of the "free" PDTA, for it of all alkali metals showed the least tendency to orient the PDTA in a conformation equivalent to chelation. This technique is obviously valuable, at least qualitatively, in determining rotamer preference in aminocarboxylate ligands and detection of weak chelates.

The Karplus relationship may also be used to determine ligand conformation in substitution inert compounds such as Cobalt(III)-trisdiamines. For years the resolution of the NMR spectra of these compounds was of such low quality as to render accurate spectral analysis impossible. This lack of resolution has been attributed to intermediate inversion rates of the puckered five-membered chelate rings, to the presence of residual paramagnetism and to spin coupling of protons to long lived nuclear spin states of cobalt-59 ($I = \frac{1}{2}$, 100% abundant). By nuclear spin decoupling of the ^{59}Co, Sudmeier and Blackmer showed that the latter explanation is correct (235). Figure 7.33 shows the —CH$_2$—CH$_2$— portion of the spectrum of D-[Co(en)$_3$]Cl$_3$ at 100 MHz. Figure 7.33A shows the broad peak centered at 2.84 ppm versus DSS has a half-width of ~23 Hz. Figure 7.33B shows the spectrum after nitrogen deuteration in which the half width is reduced to ~18 Hz because of removal of H—N—C—H spin coupling. Figure 7.33C shows the spectrum of the N-deuterated complex after the addition of

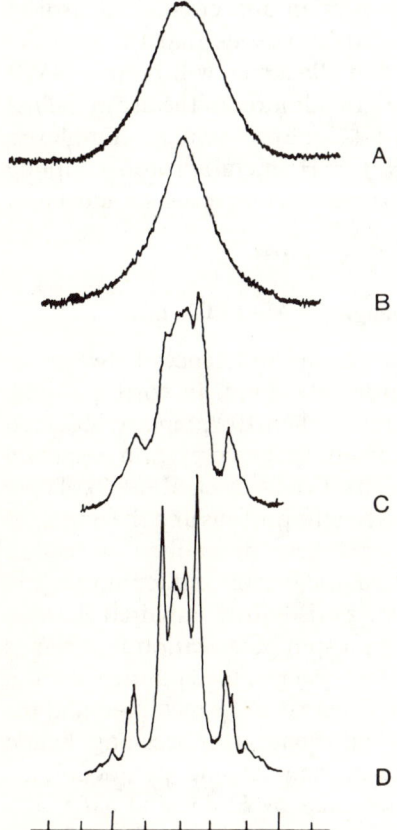

Fig. 7.33 Pmr spectra of D-[Co(en)$_3$]Cl$_3$ at 100 MHz: (A) in H$_2$O. (B) in D$_2$O. (C) in D$_2$O solution of 0.5 M K$_3$PO$_4$. (D) in D$_2$O solution potassium phosphate with Co-59 decoupling. (Reprinted with permission from J. L. Sudmeier and G. L. Blackmer, *J. Am. Chem. Soc.*, **92**, 5238 (1970). Copyright by the American Chemical Society.)

0.5 M K_3PO_4. The increased resolution is attributed to an increase of the rotational correlation time, τ_c, caused by phosphate complexing (which in turn reduces T_1 for cobalt-59 through increased efficiency in quadrupole relaxation). Figure 7.33D shows the effect of irradiation about the cobalt-59 resonance of the N-deuterated complex. The $AA'BB'$ pattern indicates the presence of only two types of protons and suggests rapid interconversion of the δ and λ puckered form of each diamine ring. A further conclusion is that the Co—N—C—H coupling constant is greater for the equatorial protons than for axial protons.

The application of the Karplus relationship to conformational studies, long known in organic compounds, shows considerable promise in metal chelate ligand conformation studies. Related research is needed to determine the effects of typical functional group substituents upon the coupling constants as well as the effect of protonation upon these coupling constants in the absence of conformational changes. Because so much research utilizing other methods such as optical rotatory dispersion and circular dichroism has been done on Co(III) complexes, it is advantageous that the spin decoupling methods reported by Sudmeier and Blackmer will permit NMR studies of these compounds to be performed. In addition to the use of vicinal proton–proton coupling to elucidate ligand conformation in complexes, recent data obtained for vicinal (M—N—C—H) metal–proton coupling constants for cobalt (235) and platinum (205,208) complexes should be of increasing value.

7.4.2.8 Conformational Analysis of Paramagnetic Metal Chelates

In diamagnetic complexes, proton chemical shift differences between the free and coordinated ligands are of the order of 1 ppm. In corresponding paramagnetic complexes, chemical shifts greater than 100 ppm are observed for protons close to the metal ion (236). Errors in the interpretation of spectra reported in Reference 236 were corrected by Erickson et al. (237). These large shifts result from spin interactions between the protons and the unpaired electron. The mechanism for these shifts have been mentioned in Section 4.1.1.9. Two mechanisms are to be considered, the Fermi contact interaction and the pseudocontact interaction. The former arises from unpaired electron spin density at the observed proton through a spin polarization in σ bonds or by delocalization in π bonds. In σ-bonded systems, the spin density at a given proton is sensitive to the dihedral angle between the metal ion and the proton (M—N—C—H), thus providing information concerning ligand conformation (238). The power of the Fermi contact and pseudocontact interactions has been convincingly demonstrated by Holm and associates on four-coordinate complexes with Ni(II) (239), and by Eaton, Phillips and associates on N-substituted Ni(II) aminotroponeimineates (240). When the Fermi contact term is dominate such as in many Ni(II) complexes, the interpretation of the data is reasonably straight-forward and the temperature dependence of the chemical shifts is useful in obtaining relative energies of conformations. The chemical shift of the ith group of protons, δ_i, (in ppm) is given by

$$\delta_i = -A_i\left(\frac{\gamma_e}{\gamma_H}\right)\frac{g\beta(S+1)}{6kT} \quad (7.57)$$

where A_i is the corresponding nuclear spin–electron spin hyperfine splitting constant (in gauss) and S is the total spin of the paramagnetic complex. Other symbols have their usual meaning. It is readily seen from this equation that the observed chemical shift should be inversely proportional to temperture. The only deviations from this with Ni(II) complexes have been shown to involve a configuration equilibrium between tetrahedral and square planar forms, the latter usually being diamagnetic (241). X-ray studies with ethylenediamine complexes indicate these coordination complexes could exist in two conformations designated as k and k' by Corey and Bailar (242). In the case of unsymmetrical N-substituted derivatives, the two conformations may not be equally populated, but rapid interconversion between these configurations may be expected (Figure 7.34). The temperature dependence of the contact shift is expected to follow equation 7.57 if there are no changes in configuration other than rapid interconversion between equally populated states of k and k'; this is the case with the Ni(II) complex of ethylenediamine. However, the temperature dependence of the contact shift of unsymmetrical N-alkyl derivatives is greater than that predicted, and may be explained as a shift in the $k \rightleftharpoons k'$ equilibrium with temperature (243). Evaluation of these data permits the calculation of relative free energies of the conformations, as well as the upper limits of the rate of nitrogen inversion which is slow ($< 10^4$ sec^{-1}) because of long lived Ni—N bonds.

The pseudocontact interaction arises from a non-zero dipolar interaction of electron spin and orbital contributions with the proton nuclear spin. Unlike Fermi contact interaction, this process is transmitted through space. Complexes in which pseudocontact interactions are significant present a more complicated situation because of a complex spin delocalization

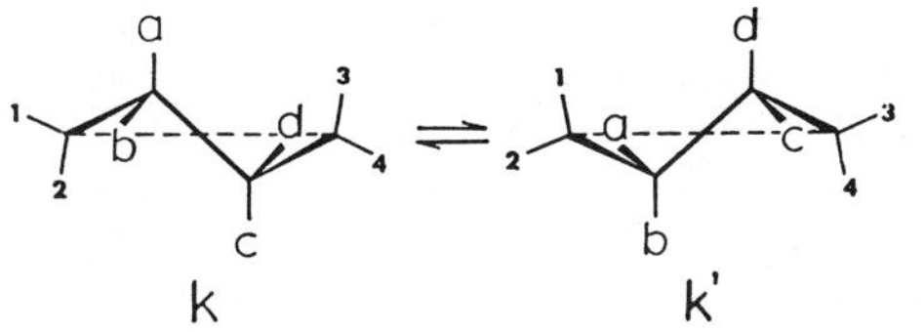

Fig. 7.34 k and k' forms of coordinated ethylenediamine rings.

mechanism or a significant temperature dependence of the population of the Kramer's doublets (244). Careful selection of complexes of the alkyldiamine variety may permit useful studies.

Reilley and associates have performed a detailed study of the configuration, conformation and interconversion dynamics of the EDTA, PDTA and CyDTA chelates of Ni(II) (237). The authors concerned themselves with the questions of both optical and coordination isomers with concern over the ligand-centered and metal-centered isomers. With ligands such as EDTA, only two equally probable isomers are possible (Δ or Λ) based on the absolute configuration of the ligand. However, in the case of ligands with asymmetric centers such as PDTA additional optical isomers are possible. The choice of nickel as the metal ion for use in the investigation is made because of its favorable paramagnetic properties. The dependence of the paramagnetic shift of the proton resonances on ring conformation and stereospecific deuteration of the acetate protons are of great aid in the interpretation of the spectra. For example, peaks which disappear on deuteration may be assigned to the acetate protons. From the differential rates of deuteration, in-plane and out-of-plane protons as well as axial and equitorial protons were not observed in any of the three complexes studied. No completely satisfactory explanation for this was given. A comparison of the observed chemical shift of the acetate and ethylenic protons with those expected for 5- and 6-coordinate species suggested that the EDTA and PDTA complexes were essentially 6-coordinate, but the CyDTA chelate was intermediate between 5 and 6 coordinate. Nitrogen inversion in the CyDTA complex was not observed after one week. However, nitrogen inversion was observed in the PDTA complex and both nitrogen inversion and Δ–Λ interconversion was observed for the EDTA chelate.

REFERENCES

1. J. W. Emsley, J. Feeney, and L. H. Sutcliffe, Eds., *Progress in Nuclear Magnetic Resonance Spectroscopy*, Pergamon, Oxford, Started in 1966.
2. J. S. Waugh, Ed., *Advances in Magnetic Resonance*, Academic, New York, Started in 1965.
3. R. K. Harris, Senior Reporter, *Nuclear Magnetic Resonance, A Specialist Report*, The Chemical Society, London, Started in 1971.
4. E. F. Mooney, Ed., *Annual Reports in NMR Spectroscopy*, Academic, New York, Started in 1968.
5. P. R. Masek, I. Sutherland, and B. R. Eggins, Eds., *Nuclear Magnetic Resonance Spectrometry Abstracts*, Science Technical Agency, London, Started in 1971.
6. T. G. Alexander and S. A. Koch, *J. Assoc. Offic. Agr. Chem.*, **48**, 618 (1965).
7. P. Nuhn, *Pharmazie*, **25**, 577 (1970).

8. R. T. Parfitt, *Pharm. J.*, 320 (1969).
9. N. C. Franklin, *Pharma. Int.*, 28 (1970).
10. J. J. Fischer, *Methods Pharmacol.*, **1**, 431 (1971).
11. D. P. Hollis, *Methods Pharmacol.*, **2**, 191 (1972).
12. N. F. Chamberlain, *Proc. Am. Petr. Inst. III*, **44**, 361 (1964).
13. N. F. Chamberlain, *Proc. 7th World Petrol. Congr. 1967*, Elsevier, New York.
14. A. Suzuki, *Nenryo Kyokai-shi*, **49**, 195 (1970).
15. L. H. Keith and A. L. Alford, *J. Assoc. Offic. Anal. Chem.*, **53**, 1018 (1970).
16. A. W. Garrison, L. H. Keith, and A. L. Alford, *Advan. Chem. Ser.*, **111**, 26 (1972).
17. R. A. Greff and P. W. Flanagan, *J. Am. Oil Chem. Soc.*, **40**, 118 (1963).
18. F. Kasler, *Quantitative Analysis by NMR Spectroscopy*, Academic, New York, 1973.
19. C. P. Poole, Jr., Ed., *Magnetic Resonance Review*, Started in 1972.
20. P. J. Paulsen and W. D. Cooke, *Anal. Chem.*, **36**, 721 (1964).
21. S. L. Manatt, *J. Amer. Chem. Soc.*, **88**, 1323 (1966).
22. J. S. Babiar, J. R. Barrante and G. C. Vickers, *Anal. Chem.*, **40**, 610 (1968).
23. O. L. Chapman and R. W. King, *J. Am. Chem. Soc.*, **86**, 1256 (1964).
24. J. G. Traynham and G. A. Knesel, *J. Am. Chem. Soc.*, **87**, 4220 (1965).
25. J. K. M. Sanders and D. H. Williams, *Chem. Commun.*, 422 (1970).
26. D. L. Rabenstein, *Anal. Chem.*, **43**, 1599 (1971).
27. M. W. Dietrich, J. S. Nash, and R. Kelbi, *Anal. Chem.*, **38**, 1479 (1966).
28. G. J. Karabatsos, F. M. Vane, R. A. Taller, and N. Hso, *J. Am. Chem. Soc.*, **86**, 3351 (1964).
29. J. L. Sudmeier, K. E. Schwartz, and A. J. Senzel, *Inorg. Chem.*, **8**, 2815 (1969).
30. J. L. Sudmeier and C. N. Reilley, *Anal. Chem.*, **38**, 1698 (1964).
31. J. L. Sudmeier and C. N. Reilley, *Anal. Chem.*, **36**, 1707 (1964).
32. D. E. Leyden, *Critical Rev. Anal. Chem.*, **2**, 383 (1971).
33. F. Kasler, *Quantitative Analysis by NMR Spectroscopy*, Academic, New York, 1973, p. 127.
34. J. V. Burakevich and J. O'Neill, Jr., *Anal. Chim. Acta*, **54**, 528 (1971).
35. R. O. Kan, *J. Amer. Chem. Soc.*, **86**, 5180 (1964).
36. G. R. Leader, *Anal. Chem.*, **42**, 16 (1970).
37. J. Feeney, A. Ledwith, and L. H. Sutcliffe, *J. Chem. Soc.*, 2021 (1962).
38. G. A. Ward and R. D. Main, *Anal. Chem.*, **41**, 538 (1969).
39. F. C. Stehling and K. W. Bartz, *Anal. Chem.*, **38**, 1467 (1966).
40. L. F. Johnson and J. N. Shoolery, *Anal. Chem.*, **34**, 1136 (1962).
41. J. N. Shoolery and L. H. Smithson, *J. Am. Oil Chem. Soc.*, **47**, 153 (1970).
42. W. Mueller and P. E. Butler, *J. Amer. Chem. Soc.*, **90**, 2075 (1968).
43. V. E. Diner and J. W. Lown, *Chem. Commun.*, 333 (1970).

44. K. Schaumberg, *Lipids*, **5**, 505 (1969).
45. M. M. Kreevoy, H. B. Charman, and D. R. Vinard, *J. Am. Chem. Soc.*, **83**, 1978 (1961).
46. E. Grünwald and E. K. Ralph, *Accounts Chem. Res.*, **4**, 107 (1971).
47. M. C. Caughman, A Review of N–H Proton Exchange Reactions of Nitrogen Compounds, M.S. Thesis, University of Georgia, Athens, Ga., 1970.
48. M. Saunders and F. Yamada, *J. Am. Chem. Soc.*, **85**, 1882 (1963).
49. W. R. Morgan and D. E. Leyden, *J. Am. Chem. Soc.*, **92**, 4527 (1970).
50. C. H. Bushweller and J. W. O'Neil, *J. Am. Chem. Soc.*, **92**, 2159 (1970).
51. D. E. Leyden and R. E. Channell, *J. Phys. Chem.*, **77**, 1562 (1973).
52. R. A. Bovey and G. V. D. Tiers, *J. Am. Chem. Soc.*, **81**, 2870 (1959).
53. T. R. Brunner, C. L. Wilkins, R. C. Williams, and P. J. McCombie, *Anal. Chem.*, **47**, 662 (1975).
54. B. R. Kowalski and C. A. Reilley, *J. Phys. Chem.*, **75**, 1402 (1971).
55. O. Yamamoto and M. Yanagisawa, *Anal. Chem.*, **47**, 697 (1975).
56. W. P. Slichter, *Fortsh. Hochpolym. Fersch.*, **1**, 35 (1958).
57. I. Ya. Slonim and A. N. Lyubimov, *The NMR of Polymers*, Translated from Russian by C. N. Turton and T. I. Turton, Plenum, New York, 1970.
58. F. A. Bovey, *High Resolution NMR of Macromolecules*, Academic, New York, 1972.
59. K. Matsuzaki, *Sen-i To Kogyo*, **3**, 620 (1970).
60. F. A. Bovey, *Prog. Polym. Sci.*, **3**, 1 (1971).
61. F. A. Bovey, *Nucl. Magn. Resonance*, **4**, 1 (1971).
62. H. Cheradame, *Bull. Soc. Chim. Fr.*, 2023 (1971).
63. C. E. Wilkes, *Macromolecules*, **4**, 443 (1971).
64. P. M. Borodin and F. I. Skripov, *Zavod. Lab.*, **29**, 164 (1963).
65. A. D. Ketley, Ed., *The Stereochemistry of Macromolecules*, 3 vols., Dekker, New York, 1967.
66. J. R. Dombroski, A. Sarko, and C. Schuerch, *Macromolecules*, **4**, 93 (1971).
67. J. R. Dombroski and C. Schuerch, *Macromolecules*, **4**, 449 (1971).
68. K. Matsuzaki, M. Okada, and K. Hosonuma, *J. Polymer Sci., Part A-1, Polymer Chem.*, **10**, 1179 (1972).
69. T. Suzaki, Y. Takeyami, J. Furukawa, and R. Hirai, *J. Polymer Sci., Part B, Polymer Letters*, **9**, 931 (1971).
70. B. Schneider, H. Pincova, and D. Doskocilova, *Macromolecules*, **5**, 120 (1972).
71. D. Doskocilova and B. Schneider, *Macromolecules*, **5**, 125 (1972).
72. F. P. Chlanda and L. G. Donaruma, *J. Appl. Polym. Sci.*, **15**, 1195 (1971).
73. K. Hatada, Y. Terawaki, H. Okuda, T. Niinomi, and H. Yuki, *Kobunshi Kagaku*, **28**, 293 (1971).
74. E. Klesper, W. Gronski, A. Johnsen, *Nucl. Magn. Resonance*, **4**, 47 (1971).

REFERENCES

75. R. H. Barber and R. A. Pittman, *High Polym.*, *Part 4*, **5**, 181 (1971).
76. A. Yamada and M. Yanagita, *J. Polym. Sci.*, *Part B*, **9**, 103 (1971).
77. P. Kuzay and W. Kimmer, *Plastic Kaut.*, **18**, 743 (1971).
78. V. C. Mochel and B. L. Johnson, *Int. Syn. Rubber*, *Symp.*, **3**, 74 (1969).
79. V. C. Mochel and W. E. Claxton, *J. Polym. Sci.*, *Part A-1*, **9**, 345 (1971).
80. F. J. Weigert, *J. Org. Magn. Resonance*, **3**, 373 (1971).
81. R. Bacskai, L. P. Lindeman, and J. Q. Adams, *J. Polym. Sci.*, *Part A-1*, **9**, 99! (1971).
82. V. J. McBrierty, D. W. McCall, D. C. Douglass, and D. R. Falcone, *Macromolecules*, **4**, 586 (1971).
83. R. E. Naylor and S. W. Lasoshi, Jr., *J. Polym. Sci.*, **44**, 1 (1960).
84. D. Doddrell, V. Glusko, and A. Allerhand, *J. Chem. Phys.*, **56**, 3683 (1972).
85. C. J. Carman, A. R. Tarply, Jr., and J. H. Goldstein, *Macromolecules*, **4**, 445 (1971).
86. Y. Inove, A. Nishioka, and R. Chujo, *Makromol. Chem.*, **152**, 15 (1972).
87. I. R. Peat and W. F. Reynolds, *Tetrahedron Lett.*, 1359 (1972).
88. Y. Inove, A. Nishioka, and R. Chujo, *Polymer J.*, **2**, 535 (1971).
89. Y. Inove and A. Nishioka, *Polymer J.*, **3**, 246 (1972).
90. A. M. Grotens, J. Smid, and E. de Boer, *Tetrahedron Lett.*, 4863 (1971).
91. F. F.-L. Ho, *J. Polym. Sci.*, *Part B*, *Polym. Lett.*, **9**, 491 (1971).
92. A. R. Katritzky and A. Smith, *Tetrahedron Lett.*, **21**, 1765 (1971).
93. D. F. G. Pusey, *Paint Manuf.*, **41**, 38 (1971).
94. L. C. Afremow, *J. Paint Tech.*, **40**, 503 (1968).
95. M. Gruenfield, *J. Paint Tech.*, **42**, 237 (1970).
96. J. P. Wineburg and D. Sivern, *J. Am. Oil Chem. Soc.*, **49**, 267 (1972).
97. J. P. Wineburg and D. Sivern, *J. Am. Oil Chem. Soc.*, **50**, 142 (1973).
98. *Perkin–Elmer*, *NMR Quarterly*, No. 11, 1974, Perkin–Elmer Ltd., Beaconsfield, England.
99. D. F. Percival and M. P. Stevens, *Anal. Chem.*, **39**, 1574 (1964).
100. M. L. Yeagle, *J. Paint Tech.*, **42**, 473 (1970).
101. C. P. Haney, F. A. Johnson, and M. G. Baldwin, *J. Polymer Sci.*, *Part A-1*, **4**, 1791 (1966).
102. R. A. Pittman and V. W. Tripp, *Appl. Spec.*, **25**, 235 (1971).
103. R. A. Pittman and V. W. Tripp, *J. Poly. Sci.*, **8**, 969 (1970).
104. P. H. Wiggall, A. D. Ince, and E. Walker, *J. Food Technol.*, **5**, 353 (1970).
105. S. Shanbhag, M. P. Steinberg, and A. Nelson, *J. Am. Oil Chem. Soc.*, **48**, 11 (1971).
106. E.-G. Samuelsson and J. Vikelsoe, *Milchwessenschaft*, **26**, 621 (1971); *Anal. Abstr.*, **23**, 846 (1972).
107. R. V. Harris, *Rep. Prog. Appl. Chem.*, **55**, 470 (1970).

108. T. F. Conway and F. R. Earl, *J. Am. Oil Chem. Soc.*, **40**, 265 (1963).
109. D. E. Alexander, L. Silvela, F. I. Collins and R. C. Rodgers, *J. Am. Oil Chem. Soc.*, **44**, 555 (1967).
110. F. I. Collins, D. E. Alexander, R. C. Rodgers, and L. Silvela, *J. Am. Oil Chem. Soc.*, **49**, 153 (1972).
111. A. J. Haighton, K. Van Putte, and L. F. Vermaas, *J. Am. Oil Chem. Soc.*, **49**, 153 (1972).
112. J. Schaefer and E. O. Stejskal, *J. Am. Oil Chem. Soc.*, **51**, 210 (1974).
113. J. Schaefer and E. O. Stejskal, *J. Am. Oil Chem. Soc.*, **51**, 562 (1974).
114. F. E. Barton, II, D. S. Himmelsbach, and D. Burdick, *J. Magn. Res.*, **18**, 167 (1975).
115. D. B. Walters and R. J. Horvat, *Anal. Chim. Acta*, **65**, 198 (1973).
116. R. B. Williams, *Spectrochim. Acta*, **14**, 24 (1959).
117. T. F. Yen and J. G. Erdman, *Preprints, Am. Chem. Soc. Div. Petrol. Chem.*, **7**, 99 (1962).
118. R. S. Winniford and M. Bershou, *Preprints, Amer. Chem. Soc. Div. Fuel Chem.*, 21 (September, 1962).
119. K. I. Zimina, A. A. Polyakova, N. I. Lulova, A. G. Siryuk, and S. A. Leont'era, *8th World Petrol. Congr., Proc. Moscow*, **6**, 211 (1971).
120. U. Lille, T. Pekh, H. Kundel, and L. Bitter, *Eestis NSV Tead. Akad. Toim., Keem. Geol.*, **22**, 86 (1973).
121. U. Lille, T. Pekh, T. Porre, and L. Bitter, *Eestis NSV Tead. Akad. Toim., Keem. Geol.*, **22**, 17 (1973).
122. A. W. Decora, J. P. Flaherty, F. R. McDonald, and G. L. Cook, Preprints, *Amer. Chem. Soc., Div. Fuel. Chem.*, **15**, 38 (1971).
123. E. W. Cook, *Fuel*, **53**, 16 (1973).
124. C. Trogolo, *Riv. Combust.*, **25**, 502 (1971).
125. J. W. Robinson and D. Truitt, *Spectrosc. Lett.*, **2**, 203 (1969).
126. J. V. Mengerhauser, *U.S. Clearinghouse Fed. Sci. Tech. Inform.*, AD-711892 (1970).
127. E. F. G. Herington and I. J. Lawrensen, *Ing. Quim*, (Mexico City), **15**, 36 (1970).
128. I. J. Lawrensen, *Chem. Ind. London*, 172 (1971).
129. B. Caser, *Riv. Combust.*, **25**, 404 (1971).
130. L. P. Lindeman and J. Q. Adams, *Anal. Chem.*, **43**, 1245 (1971).
131. D. M. Grant and E. G. Paul, *J. Am. Chem. Soc.*, **86**, 2984 (1964).
132. M. Novotny, M. L. Lee, and K. D. Bartle, *J. Chromatog. Sci.*, **12**, 606 (1974).
133. R. Haque and F. J. Biros, Eds., *Mass Spectrometry and NMR Spectroscopy in Pesticide Chemistry*, Plenum, New York, 1974.
134. L. H. Keith and A. L. Alford, *J. Assoc. of Offic. Anal. Chem.*, **53**, 1018 (1970).
135. R. Haque and D. R. Buhler, in *Annual Reviews of NMR Spectroscopy*, E. F. Mooney, Ed., Vol. 4, Academic, New York, 1971.

136. A. W. Garrison, L. H. Keith, and A. L. Alford, in *Fate of Organic Pesticides in the Aquatic Environment*, American Chemical Society Advances in Chemistry Series, No. 111, 1972.
137. L. H. Keith, A. W. Garrison, and A. L. Alford, *J. Assoc. of Offic. Anal. Chem.*, **51**, 1063 (1968).
138. L. H. Keith, A. L. Alford, and A. W. Garrison, *J. Assoc. of Offic. Anal. Chem.*, **52**, 1074 (1969).
139. L. H. Keith and A. L. Alford, *J. Assoc. of Offic. Anal. Chem.*, **53**, 157 (1970).
140. *Catalog of Pesticide NMR Spectra*, Water Pollution Control Research Series, 16020 EWC, Environmental Protection Agency, Washington, 1971.
141. N. Muller and J. Goldenson, *J. Am. Chem. Soc.*, **78**, 5182 (1956).
142. H. Babad and T. N. Taylor, *Anal. Chim. Acta*, **40**, 387 (1968).
143. J. I. Teasley, *Environ. Sci. Technol.*, **1**, 411 (1967).
144. E. R. White, W. W. Kilgore, and G. Mallett, *J. Agr. Food Chem.*, **17**, 585 (1969).
145. L. H. Keith, *Tetrah. Lett.*, 3 (1971).
146. H. A. Resing, in *Mass Spectrometry and NMR Spectroscopy in Pesticide Chemistry*, R. Haque and F. J. Biros, Eds., Plenum, New York, 1974, p. 273.
147. A. Pines, M. G. Gibby, and J. S. Waugh, *J. Chem. Phys.*, **59**, 569 (1973).
148. D. P. Hollis, *Methods Pharmacol.*, **2**, 191 (1972).
149. J. J. Fisher, *Methods Pharmacol.*, **1**, 431 (1971).
150. N. C. Franklin, *Pharm. Intl.*, **5**, 28 (1970).
151. D. P. Hollis, *Anal. Chem.*, **35**, 1682 (1963).
152. F. Kasler, *Quantitative Analysis by NMR Spectroscopy*, Academic, New York, 1973, Chap. 10.
153. H. W. Avdonich, P. Hanburg, and B. A. Lodge, *J. Pharm. Sci.*, **59**, 1164 (1970).
154. H. W. Avdonich, M. Bowron, and B. A. Lodge, *J. Pharm. Sci.*, **59**, 1821 (1970).
155. G. Rücker, *Z. Anal. Chem.*, **229**, 340 (1967).
156. G. Rücker and P. N. Natarajan, *Arch. Pharm.*, **300**, 276 (1967).
157. T. Rüll, *Bull. Soc. Chim. (France)*, 1897 (1963).
158. J. W. Turczan, T. C. Kram, *J. Pharm. Sci.*, **56**, 1643 (1967).
159. G. Slomp, R. H. Baker, and F. A. MacKellar, *Anal. Chem.*, **36**, 375 (1964).
160. R. J. Warren, D. B. Staiger, J. E. Zarembo, and A. Post, *Microchem. J.*, **20**, 242 (1975).
161. P. N. Giraldi, V. Mariotti, and I. de Carneri, *J. Med. Chem.*, **11**, 66 (1968).
162. P. Roffey and J. Verge, *J. Hetero. Chem.*, **6**, 497 (1969).
163. K. Katagiri, K. Tori, Y. Kimura, T. Yoshida, T. Nagasaki, and H. Minato, *J. Med. Chem.*, **10**, 1149 (1967).
164. R. J. Ovellette, *J. Am. Chem. Soc.*, **86**, 3089 (1964).
165. R. J. Ovellette, *J. Am. Chem. Soc.*, **86**, 4378 (1964).
166. R. J. Ovellette, G. E. Booth, and K. Liptak, *J. Am. Chem. Soc.*, **87**, 3436 (1965).

167. R. J. Ovellette, K. Liptak, and G. E. Booth, *J. Org. Chem.*, **31**, 546 (1966).
168. C. M. Huggins, G. C. Pimentel, and J. N. Shoolery, *J. Phys. Chem.*, **60**, 1311 (1956).
169. J. C. Jochims, G. Taigel, and A. Seeliger, *Tetrah. Lett.*, 1901 (1967).
170. J. J. Fischer and O. Jardetzky, *J. Am. Chem. Soc.*, **87**, 3237 (1965).
171. T. C. Farrar, JEOL Analytical Instruments Inc., personal communication, 1975.
172. T. L. James, *NMR in Biochemistry*, Academic, New York, 1975.
173. R. A. Dwek, *Nuclear Magnetic Resonance (N.M.R.) in Biochemistry*, Clarendon, Oxford, England, 1973.
174. M. A. Raftery, F. W. Dahlquist, S. I. Chan, and S. M. Parsons, *J. Biol. Chem.*, **243**, 4175 (1968).
175. T. Rand-Meir, F. W. Dahlquist, and M. A. Raftery, *Biochemistry*, **8**, 4206 (1969).
176. M. Zeppezauer, B. Lindman, S. Forsen, and J. Lindqvist, *Biochim. Biophys. Res. Commun.*, **37**, 137 (1969).
177. T. R. Stengle and J. D. Baldeschwieler, *J. Am. Chem. Soc.*, **89**, 3045 (1967).
178. R. P. Haughland, L. Stryer, T. R. Stengle, and J. D. Baldeschwieler, *Biochemistry*, **6**, 498 (1967).
179. J. L. Sudmeier and J. J. Pesek, *Anal. Biochem.*, **41**, 39 (1971).
180. J. Reuben, G. H. Reed, and M. Cohn, *J. Chem. Phys.*, **52**, 161 (1970).
181. N. C. Li, R. L. Scruggs, and E. D. Becker, *J. Am. Chem. Soc.*, **84**, 4650 (1962).
182. G. Navon, *Chem. Phys. Lett.*, **7**, 390 (1970).
183. P. C. Lauterbur, *Appl. Spec.*, **24**, 450 (1970).
184. C. E. Brown, J. J. Katz, and D. Shemin, *Proc. Natn. Acad. Sci. (U.S.)*, **69**, 2585 (1972).
185. D. W. Urry, *Res. Develop.*, **24**, 20 (1973).
186. D. W. Urry, *Res. Develop.*, **25**, 18 (1974).
187. M. Ohnishi and D. W. Urry, *Biochem. Biophys. Res. Commun.*, **36**, 194 (1969).
188. G. Schwarzenbach and H. Ackermann, *Helv. Chim. Acta.*, **30**, 1798 (1947).
189. D. Chapman, D. R. Lloyd, and R. H. Prince, *J. Chem. Soc.*, 3645 (1963).
190. J. L. Sudmeier and A. J. Senzel, *J. Am. Chem. Soc.*, **90**, 6860 (1968).
191. J. L. Sudmeier and C. N. Reilley, *Anal. Chem.*, **36**, 1689 (1964); **36**, 1707 (1964).
192. E. Grünwald, A. Loewenstein, and S. Meiboom, *J. Chem. Phys.*, **27**, 641 (1957).
193. R. J. Kula and D. T. Sawyer, *Inorg. Chem.*, **3**, 458 (1964).
194. J. E. Sarneski, H. L. Surprenant, F. K. Molen, and C. N. Reilley, *Anal. Chem.*, **47**, 2116 (1975).
195. Y. Fujiwara and C. N. Reilley, *Anal. Chem.*, **40**, 890 (1968).
196. R. J. Kula, D. T. Sawyer, S. I. Chan, and C. M. Finley, *J. Am. Chem. Soc.*, **85**, 2930 (1963).
197. R. J. Kula, *Anal. Chem.*, **37**, 989 (1965).
198. J. F. Whidby and D. E. Leyden, *Anal. Chim. Acta*, **51**, 25 (1970).

REFERENCES

199. D. E. Leyden and J. F. Whidby, *Anal. Chim. Acta*, **42**, 271 (1968).
200. D. E. Leyden and J. F. Whidby, *J. Phys. Chem.*, **73**, 3076 (1969).
201. D. H. Williams and D. H. Busch, *J. Am. Chem. Soc.*, **87**, 4644 (1965).
202. J. B. Terrill and C. N. Reilley, *Anal. Chem.*, **38**, 1876 (1966).
203. J. B. Terrill and C. N. Reilley, *Inorg. Chem.*, **5**, 1988 (1966).
204. J. L. Sudmeier and G. Occupati, *Inorg. Chem.*, **7**, 2524 (1968).
205. L. E. Erickson, J. W. McDonald, J. K. Howie, and R. P. Clow, *J. Am. Chem. Soc.*, **90**, 6371 (1968).
206. L. E. Erickson, A. J. Dappen, and J. C. Uhlenhopp, *J. Am. Chem. Soc.*, **91**, 2510 (1969).
207. L. E. Erickson, H. L. Fritz, R. J. May, and D. A. Wright, *J. Am. Chem. Soc.*, **91**, 2513 (1969).
208. L. E. Erickson, *J. Am. Chem. Soc.*, **91**, 6284 (1969).
209. D. A. Buckingham, L. G. Marzilli, and A. M. Sargeson, *J. Am. Chem. Soc.*, **89**, 3428 (1967) and references cited therein.
210. M. Saunders and F. Yamada, *J. Am. Chem. Soc.*, **85**, 1882 (1963).
211. L. B. Holzman, Ph.D. Dissertation, Yale University, 1968.
212. W. R. Morgan and D. E. Leyden, *J. Am. Chem. Soc.*, **92**, 4527 (1970).
213. T. P. Pitner and R. B. Martin, *J. Am. Chem. Soc.*, **93**, 4400 (1971).
214. J. L. Hoard, G. S. Smith, and M. Lind, *Advances in the Chemistry of Coordination Compounds*, New York, MacMillan, 1961.
215. R. J. Day and C. N. Reilley, *Anal. Chem.*, **36**, 1073 (1964).
216. F. P. Dwyer and D. P. Mellor, Eds., *Chelating Agents and Metal Chelates*, New York, Academic, 1964, Chap. 7.
217. R. J. Kula, *Anal. Chem.*, **39**, 1171 (1967).
218. S. I. Chan, R. J. Kula, and D. T. Sawyer, *J. Am. Chem. Soc.*, **86**, 377 (1964).
219. L. V. Haynes and D. T. Sawyer, *Inorg. Chem.*, **6**, 2146 (1967).
220. B. B. Smith and D. T. Sawyer, *Inorg. Chem.*, **7**, 2020 (1968).
221. G. L. Backmer, J. L. Sudmeier, R. Thibedeau, and R. M. Wing, Submitted for publication.
222. F. P. Dwyer and F. L. Garvan, *J. Am. Chem. Soc.*, **83**, 2610 (1961).
223. R. J. Day and C. N. Reilley, *Anal. Chem.*, **37**, 1326 (1965).
224. R. J. Kula and D. T. Sawyer, University of California, Riverside, Personal Communication to C. N. Reilley, 1965.
225. D. W. Margenum, *Record Chem. Progr.*, **24**, 237 (1963).
226. B. Bosnich, F. P. Dwyer, and A. M. Sargeson, *Nature*, **186**, 966 (1960).
227. R. G. Pearson, J. Palmer, M. M. Anderson, and A. L. Allred, *Z. Elektrochem.*, **64**, 110 (1960).
228. R. G. Pearson and R. D. Lanier, *J. Am. Chem. Soc.*, **86**, 765 (1964).
229. J. L. Sudmeier and C. N. Reilley, *Inorg. Chem.*, **5**, 1047 (1966).

230. R. J. Kula and G. H. Reed, *Anal. Chem.*, **38**, 697 (1966).
231. R. J. Kula and D. L. Rabenstein, *J. Am. Chem. Soc.*, **89**, 552 (1967).
232. M. Karplus, *J. Chem. Phys.*, **30**, 11 (1959).
233. J. L. Sudmeier and A. J. Senzel, *Anal. Chem.*, **40**, 1693 (1968).
234. K. G. R. Pachler, *Spectrochim. Acta*, **20**, 581 (1964).
235. J. L. Sudmeier and G. L. Blackmer, *J. Am. Chem. Soc.*, **92**, 5238 (1970).
236. R. S. Milner and L. Pratt, *Discussion Faraday Soc.*, **34**, 88 (1962).
237. L. E. Erickson, D. C. Young, F. F.-L. Ho, S. R. Watkins, J. B. Terrill, and C. N. Reilley, *Inorg. Chem.*, **10**, 441 (1971).
238. L. Pratt and B. B. Smith, *Trans. Faraday Soc.*, **65**, 915 (1969).
239. R. E. Ernst, M. J. O'Conner, and R. H. Holm, *J. Am. Chem. Soc.*, **90**, 5735 (1968).
240. D. R. Eaton, A. D. Josey, W. D. Phillips, and R. E. Benson, *J. Chem. Phys.*, **37**, 347 (1962).
241. L. Sacconi, in *Transition Metal Chemistry*, R. L. Carlin, Ed., Vol. 4, Dekker, New York, 1966, p. 199.
242. E. J. Corey and J. C. Bailar, Jr., *J. Am. Chem. Soc.*, **81**, 2620 (1959).
243. F. F.-L. Ho and C. N. Reilley, *Anal. Chem.*, **41**, 1835 (1969).
244. G. N. LaMar and G. R. Van Hecke, *J. Am. Chem. Soc.*, **92**, 3021 (1970).

APPENDIX

An attempt has been made in this book to provide references associated with each chapter that would permit the interested reader to pursue each topic discussed in the book to a greater depth than could be presented. This appendix provides additional references with brief annotations. The reports of applications of NMR number in the tens of thousands each year. This list is obviously small compared with that quantity. However, the intent is to supply a starting point for literature in the various fields. No attempt has been made to list the books and chapters which are devoted to the fundamental principles of NMR as there are such a large number of these, many of which were referenced in the text. The reader is encouraged to consult the many annual reviews for current reference material.

A. BOOKS

The books listed below are recent publications with titles explaining the content.

F. Kasler, *Quantitative Analysis by NMR Spectroscopy*, Academic, New York, 1973.

W. W. Simons and M. Zanger, The Sadtler Guide to the NMR Spectra of Polymers, Sadtler Research Laboratories, Philadelphia, Pa., 1973.

F. A. Bovey, *High Resolution NMR of Macromolecules*, Academic, New York, 1971.

I. Ya. Slonim and A. N. Lyubimov, *The NMR of Polymers*, Plenum, New York, 1970.

A. F. Casey, *PMR Spectroscopy in Medicinal and Biological Chemistry*, Academic, New York, 1971.

R. A. Dwek, *(N.M.R.) in Biochemistry*, Oxford University Press, 1973.

T. L. James, *NMR in Biochemistry*, Academic, New York, 1975.

J. T. Clerc, E. Pretsch, and S. Sternhell, *Carbon-13 Nuclear Resonance Spectroscopy*, Akad. Verlag, Frankfurt, Germany, 1973.

E. Breitmaier and W. Voelter, *Carbon-13 NMR Spectroscopy*, Verlag Chemie, Weinheim, Germany, 1974.

T. Axenrod and G. A. Webb, Ed., *Nuclear Magnetic Resonance Spectroscopy of Nuclei Other Than Protons*, Wiley, New York, 1974.

R. Hague and F. J. Biros, Ed., *Mass Spectrometry and NMR Spectroscopy in Pesticide Chemistry* (*Environmental Science Research Vol. 4*), Plenum, New York, 1974.

R. G. Lawler, *Chemically Induced Dynamic Nuclear Polarization* (*Prog. NMR Spectry, Vol. 9, Part 3*), Pergamon, Oxford, 1973.

C. Richard and P. Granger, *Chemically Induced Dynamic Nuclear and Electron Polarization, CIDNP and CIDEP*, Springer, New York, 1974

T. C. Farrar and E. D. Becker, *Pulse and Fourier Transform NMR*, Academic, New York, 1971.

J. H. Noggle and R. E. Shirmer, *The Nuclear Overhauser Effect*, Academic, New York, 1971.

L. M. Jackman and F. A. Cotton, Ed., *Dynamic Nuclear Magnetic Resonance Spectroscopy*, Academic, New York, 1975.

B. REVIEWS AND CHAPTERS

The following review articles or chapters in monographs or series pertain to various aspects of NMR and will serve as excellent sources for literature on those topics listed.

Analytical Applications (general)
    M. L. Martin, *Chim. Anal.* (*Paris*), **53**, 114 (1971).
    G. A. Ward, *Amer. Lab.*, **12**, 18, 21 24 (1970).
Microquantities of Organic Material
    G. M. Ayling, *Appl. Spectrosc. Rev.*, **8**, 147 (1974).
Hydrogen Bonding
    J. C. Davis, K. K. Deb, in *Advances in Magnetic Resonance*, Vol. 4, J. S. Waugh, Ed., Academic, New York, 1970, p. 201.
    S. V. Vinogradov and R. H. Linnell, *Hydrogen Bonding*, Van Nostrand Reinhold, New York, 1971, Chap. 4.
    A. I. Brodskii, V. D. Polhodenko, and V. S. Kuts, *Usp. Khim.*, **39**, 753 (1970).
    C. N. R. Rao and A. S. N. Murthy, *Develop. Appl. Spectrosc.*, **7B**, 54 (1968).
    M. D. Joesten and L. J. Schaad, *Hydrogen Bonding*, Dekker, New York, 1974.
    R. D. Green, *Hydrogen Bonding by C-H Groups*, Halsted, New York, 1974.
Conformational Analysis
    Ann. Rept. NMR Spectrosc., **3** (1970).
Organic Charge-transfer Complexes
    J. P. Heeschen, *Anal. Chem.*, **42**, 418R–451R (1970).

Studies of Metal Chelates
  D. E. Leyden, *Critical Rev. Anal. Chem.*, **2**, 383 (1971).
Organometallic Exchange Reactions
  N. S. Ham and T. Mole, in *Progress in Nuclear Magnetic Resonance Spectroscopy*, Vol. 4, J. W. Emsley, J. Feeney and L. H. Sutcliffe, Eds., Pergamon, London, 1969.
Paramagnetic Complexes
  G. A. Webb, *Ann. Repts. NMR Spectrosc.*, **3**, 211 (1970).
  H. J. Keller and K. E. Schwarzhans, *Angew. Chem., Int. Ed. Engl.*, **9**, 196 (1970).
  K. E. Schwarzhans, *Angew. Chem., Int. Ed. Engl.*, **9**, 946 (1970).
  B. Stalinski, *Metody Badaw. Chem. Koordynacyjnej*, 181–208 (1967).
  J. W. Dawson and L. M. Venanzi, *Trans. N.Y. Acad. Sci.*, **32**, 304 (1970).
Organometallic Compounds
  L. J. Todd, *J. Organometal. Chem.*, **77**, 1 (1974).
  B. E. Mann, *Adv. Organometal. Chem.*, **12**, 135 (1974).
  A. Maercker, *Chem.-Ztg.*, **97**, 361 (1973).
  L. A. Gedorov, *Usp. Khim.*, **42**, 1481 (1973).
  V. S. Petrosyan and O. A. Reutov, *J. Organometal. Chem.*, **76**, 123 (1974).
  N. M. Sergeyev, *Progr. NMR Spectrosc.* **9**, (Pt. 2) (1973).
  V. S. Petrosyan and O. A. Reutov, *Pure Appl. Chem.*, **37**, 147 (1974).
  B. E. Mann, *Spectrosc. Prop. Inorg. Organometal. Compounds*, **6**, 1 (1973).
Organolithium Compounds
  L. D. McKeever, in *Ions and Ion Pairs in Organic Reaction*, M. Sware, Ed., Vol. 1 Wiley-Interscience, New York, 1971.
Sulfur Compounds
  S. R. Heller, *Intra-Sci. Chem. Rep.*, **2**, 321 (1968).
Carbohydrates
  T. D. Inch, *Ann. Rev. NMR Spectrosc.*, Vol. 2, Academic, New York, 1969 p. 35.
  T. D. Inch, *Annu. Rep. NMR Spectrosc.*, **5A**, 305 (1972).
  B. Coxon, *Adv. Carbohyd. Chem.*, **27**, 7 (1972).
  L. D. Hall, *Adv. Carbohyd. Chem. Biochem.*, **29**, 11 (1974).
Steroids
  J. E. Page, *Ann. Rept. NMR Spectrosc.*, Vol. 3, Academic, New York, 1970, p. 149.
Proteins
  O. Jardetzky, *Proc. Int. Conf. Stable Isotop. Chem. Biol., Med., 1st. 1973*, P. D. Klein, Ed., NTIS, Springfield, Va., 1973, pp. 99–102.

K. Wuethrich, *Experientia*, **30**, 577 (1974).

W. D. Phillips, in *Methods Enzymology*, C. H. W. Hirs, Ed., Vol. 27, Academic, New York, 1973, pp. 825–836.

J. C. Metcalfe, N. J. M. Birdsall, and A. G. Lee, *Companion Biochem.*, A. T. Bull, Ed., Longman Group, London, 1974, pp. 139–162.

Drug-protein Interactions

B. D. Sykes and W. E. Hull, *Ann. N.Y. Acad. Sci.*, **226**, 60 (1973).

Peptides

I. C. P. Smith, R. Deslauriers, H. Saito, R. Walter, C. Garrigou-Lagrange, H. McGregor, and D. Sarantakis, *Ann. N.Y. Acad. Sci.*, **222**, 597 (1973).

V. F. Bystrov, S. L. Portnova, et al., *Pure Appl. Chem.*, **36**, 19 (1973).

E. M. Bradbury, P. D. Cary, C. Crane-Robinson, and P. G. Hartman, *Pure Appl. Chem.*, **36**, 53 (1973).

J. P. Meraldi, H. Moeschler, R. Schwyzer, A. Tun-Kyi, and K. Wuethrich, *J. Phys. (Paris) Colloq.*, **41** (1973).

K. Nagayama, *Seibutsu Butsuri*, **14**, 66 (1974).

Nucleic Acids

K. Masatune, *Tampakushitsu Kagusan Koso, Bessatsu*, 238 (1974).

T. Wilczok and B. Lubas, *Stud. Biophys.*, **38**, 85 (1973).

L. B. Townsend, *Syn. Proced. Nuclei Acid Chem.*, W. W. Zorbach, Ed., Vol. 2, Wiley, New York, 1973, pp. 267–398.

L. S. Kan and P. O. P. Ts'o, *Bot. Bull. Acad. Sinica*, **14**, 180 (1973).

R. G. Shulman, D. R. Kearns, B. R. Reid, and Y. P. Wong, *Dyn. Aspects Conform. Changes Biol. Macromol., Proc. Ann. Meet. Soc. Chim. Phys. 23rd, 1972*, C. Sadron, Ed., Reidel, Dordrecht, Netherlands, 1973, pp. 165–169.

Metal–enzymes

P. J. Quilley and G. A. Webb, *Coord. Chem. Rev.*, **12**, 407 (1974).

Biopolymers

J. J. M. Rowe, J. Hinton, and K. L. Rowe, *Chem. Rev.*, **70**, 1 (1970).

A. Wada, Ed., *NMR No Seitai Kobunshi Eno Oyo*, Kyoritsu Shuppansha, Tokyo, Japan, 1973.

Polymers

I. Ya. Slonim, *Usp. Khim. Fiz. Polim.*, 386 (1970).

M. E. A. Cudby and H. A. Willis, *Annu. Rep. NMR Spectrosc.*, **4**, 363 (1971).

I. D. Robb and G. J. T. Tiddy, *Nucl. Magn. Reson.*, **3**, 279 (1974).

V. J. McBriety, *Polymer*, **15**, 503 (1974).

Polymer Configuration and Conformation

F. A. Bovey, *Polym. Prepr. Am. Chem. Soc., Div. Polym. Chem.*, **10**, 9 (1969).

Molecular Relaxation in Polymers
   D. W. McCall, *Accounts Chem. Res.*, **4**, 223 (1971).
Application of NMR to Kinetics
   D. N. Hague, *Compr. Chem. Kinet.*, **1**, 112 (1969).
   F. Garnier, *Mises Jour Sci.*, **3**, 395 (1968).
NMR of Dynamic Processes
   A. A. Bothner-By, B. R. Appleman and B. P. Dailey, *Critical Evaluation of Chemical and Physical Structural Information, Proc., Conf. 1973*, D. R. Lide, Ed., NAS, Wash., D.C.
   K. M. L'vov and A. G. Sukhourudrenko, *Biofizika*, **19**, 576 (1974).
   W. D. Ollis, J. F. Stoddart, and I. O. Sutherland, *Tetrahedron*, **30**, 1903 (1974).
   F. A. Cotton and B. A. Frenz, *Tetrahedron*, **30**, 1587 (1974).
   N. M. Sergeev, *Usp. Khim.*, **42**, 769 (1973).
   K. Pihlaja, *Kem.-Kemi*, **1**, 492 (1974).
   F. A. L. Anet, *Determination Org. Struct. Phys. Methods*, **3**, 343 (1971).
   M. J. T. Robinson, *Tetrahedron*, **30**, 1971 (1974).
   E. L. Eliel, *Chem.-Ztg.*, **97**, 582 (1973).
   F. G. Riddell, *Aliphatic, Alicyclic, Saturated Heterocycl. Chem.*, (Pt. 3), **1**, 69 (1973).
   D. M. Grant, *Pure Appl. Chem.*, **37**, 61 (1974).
   S. Mager, *Rev. Chim. (Bucharest)*, **24**, 861 (1974).
   I. O. Sutherland, *Annu. Rep. NMR Spectrosc.*, **4**, 71 (1971).
Anisotropic Rotation of Molecules in Liquids
   W. T. Hunfress, Jr., in *Advances in Magnetic Resonance*, J. S. Waugh, Ed., Vol. 4, Academic, New York, 1970, p. 2.
NMR Line Shapes
   R. A. Hoffman, in *Advances in Magnetic Resonance*, J. S. Waugh, Ed., Vol. 4, Academic, New York, 1970, p. 88.
Computer Usage in NMR
   C. W. Haigh, *Annu. Rep. NMR Spectrosc.*, **4**, 311 (1971).
   L. Newman, *Comput. Chem. Instrum.*, **3**, 3 (1973).
   T. Clerc and F. Erni, *Fortschr. Chem. Forsch.*, **39**, 91 (1973).
NMR Pattern Recognition
   B. R. Kowalski, *Comput. Chem. Biochem. Res.*, **2**, 1 (1974).
Fourier, Hadamard and Pulse Techniques
   M. Imanari, *Kagaku No Ryoiki*, **27**, 287 (1973).
   D. Ziessow and B. Bluemich, *Ber. Bunsenges. Phys. Chem.*, **78**, 1168 (1974)
   R. R. Ernst, *Pulsed Nucl. Magn. Res. Spin Dyn. Solids, Proc. Spec. Colloq. Ampere 1st 1973*, J. W. Hennel, Ed., Inst. Nucl. Phys., Krakow, Poland, 1973, pp. 40–52.

D. Shaw, *Nucl. Magn. Reson.*, **3**, 249 (1974).

Y. Utsumi, *New Methods Environ. Chem. Toxicol., Collect. Pap. Res. Conf. New Methodol. Ecol. Chem. 1973*, F. Coulston, Ed., Int. Acad. Print Co., Ltd., Tokyo, Japan, 1973, pp. 51–55.

R. R. Ernst, W. P. Aue, E. Bartholdi, A. Hoehener, and S. Schaeublin, *Pure Appl. Chem.*, **37**, 47 (1974).

R. Radeglia, *Z. Chem.*, **14**, 82 (1974).

J. A. Glasel, *Fed. Proc., Fed. Am. Soc. Exp. Biol.*, **33**, 1973 (1974).

E. D. Becker, *Ann. N.Y. Acad. Sci.*, **222**, 724 (1973).

D. G. Gilles, *Annu. Rep. NMR Spectrosc.*, **5A**, 557 (1972).

I. P. Biryukov, A. F. Babkin, and A. Ya. Deich, *Latv. PSR Zinat. Akad. Vestis, Fiz, Teh. Zinat. Ser.*, 78 (1973).

N. Boden, in *Determination of Organic Structures by Physical Methods*, F. C. Nachod and J. J. Zuckerman, Eds., Vol. 4, Academic, New York, 1971, p. 51.

A. G. Marshal and M. B. Comisarow, *Anal. Chem.*, **47**, 491A, 494A, 496A, 498A, 500A, 502A, 504A (1975).

CIDNP

R. G. Lawler and H. R. Ward, *Determination Org. Struct. Phys. Methods*, **5**, 99 (1973).

G. L. Closs, in *Advances in Magnetic Resonance*, J. S. Waugh, Ed., Vol. 7, Academic, New York, 1974, p. 157.

Water in Hydrate Crystals

L. W. Reeves, in *Progress in Nuclear Magnetic Resonance*, J. W. Emsley, J. Feeney and L. H. Sutcliffe, Eds., Vol. 4, Pergamon, London, 1969.

Liquid Crystals

L. C. Snyder and S. Meiboom, *Mol. Cryst. Liquid Cryst.*, **7**, 181 (1969).

P. Diehl and C. L. Khetrapal, *Nucl. Magn. Resonance*, 1 (1969).

S. Meiboom and L. C. Snyder, *Accounts Chem. Res.*, **4**, 81 (1971).

R. Briere, *Commiss. Energ. At. [Fr.] Serv. Doc., Ser. 'Bibliogr.,'* 1969, *CEA-BIB-167*, 27pp.

P. Diehl and W. Niederberger, *Nucl. Magn. Reson.*, **3**, 368 (1974).

A. Loesche, *Comments Solid State Phys.*, **5**, 119 (1973).

L. W. Reeves, *Cienc. Cult. (Sao Paulo)*, **25**, 516 (1973).

J. Jonas, *Magn. Reson. Rev.*, **2**, 203 (1973).

A. Loesche, *Postepy Fiz.*, **24**, 407 (1973).

A. Loesche, *Magn. Resonance Relat. Phenomena, Proc. Congr. AMPERE, 17th, 1972* (Publ. 1973), V. Hovi, Ed., North-Holland, Amsterdam, pp. 101–117.

Glasses

H. Dutz and W. Poch, *Fachausschussber. Dtsch. Glastech. Ges.*, **70**, 219 (1974).

Solids
- W. Derbyshire, *Nucl. Magn. Reson.*, **2**, 323 (1973).
- T. Tsuda, *Nippon Butsuri Gakkaishi*, **27**, 394 (1972).
- R. Orbach, M. Peter, and D. Shaltiel, *Arch. Sci.*, **27**, 141 (1974).
- T. Fujito, *Nippon Butsuri Gakkaishi*, **29**, 935 (1974).
- W. E. E. Stone, *Silic. Ind.*, **39**, 255 (1974).
- H. A. Hertz, *Mol. Motions Liq.*, *Proc. Ann. Met. Soc. Chim. Phys.*, 24th 1972, J. Lascombe., Ed., Reidel, Dordrecht, Netherlands, 1974, pp. 337–357.
- H. Yasuoka, *Nippon Butsuri Gakkaishi*, **29**, 696 (1974).
- P. G. Frith and K. A. McLauchlan, *Nucl. Magn. Reson.*, **3**, 378 (1974).
- H. R. Ward, R. G. Lawler, and S. M. Rosenfeld, *Ann. N.Y. Acad. Sci.*, **222**, 740 (1973).

NMR Shift Reagents
- B. C. Mayo, *Chem. Soc. Rev.*, **2**, 49 (1973).
- A. Yamazaki, *Kagaku (Kyoto)*, **29**, 349, 435 (1974).
- Z. Ksandr and M. Hajek, *Chem. Listy*, **68**, 129 (1974).
- B. Danieli and G. Palmisano, R. Ugo, *Relaz. Corso Teor.-Prat. Risonanza Magn. Nucl.*, A. Frigerio, Ed., Tamburini, Milan, Italy, 1973, pp. 231–290
- M. Kainosho and K. Ajisaka, *Yuki Gosei Kagaku Kyokai Shi*, **31**, 126 (1973).
- C. C. Hinckley, in *Modern Methods of Steroid Analysis*, E. Heftmann, Ed., Academic Press, New York, 1973, pp. 265–279.
- J. Reuben, *Progr. NMR Spectrosc.*, **9**, Pt. 1 (1973).
- J. A. Glasel, *Progr. Inorg. Chem.*, **18**, 383 (1973).
- I. Ya. Slonim and A. Kh. Bulai, *Usp. Khim.*, **42**, 1976 (1973).

Correlation of NMR and ESCA Chemical Shifts
- B. J. Lindberg, *J. Electron Spectrosc. Relat. Phenom.*, **5**, 149 (1974).

Solvent Effects
- *Ann. Rev. NMR Spectrosc.*, **2** (1969).
- J. Homer, *Appl. Spectrosc. Rev.*, **9**, 1 (1975).
- M. I. Goreman, *Nucl. Magn. Reson.*, **2**, 355 (1973).
- B. Danieli, G. Palmisano, and R. Ugo, *Relaz. Corso Teor.-Prat. Risonanza Magn. Nucl.*, A. Frigerio, Ed., Tamburini, Milan, Italy, 1973, pp. 61–89.

Electrolyte Solutions
- C. Deverell, in *Progress in Nuclear Magnetic Resonance Spectroscopy*, Vol. 4, J. W. Emsley, J. Feeney and L. H. Sutcliffe, Eds., Pergamon, London, 1969.
- V. I. Chizhik, *Mol. Fiz. Biofiz. Vod. Sist.*, **1**, 108 (1973).

Solvation Number of Ions
- J. F. Hinton and E. S. Amis, *Chem. Rev.*, **71**, 627 (1971).

P. H. Kasai, *Solid State Chem. Phys.*, P. F. Weller, Ed., Vol. 1, Dekker, N.Y., 1973, pp. 357–410.

C. J. Ford and G. A. Styles, *Prop. Liquid Metals, Proc. Int. Conf. 2nd 1972*, S. Takeuchi, Ed., Taylor and Francis, London, 1973, pp. 189–196.

E. R. Andrew, *Pulsed NMR Spin Dyn. Solids, Prod. Spec. Colloq. Ampere, 1st 1973*, J. W. Hennel, Ed., Inst. Nucl. Phys., Krakow, Poland, 1973, pp. 3–15.

W. W. Warren, Jr., *Charge Transfer/Electron. Structure Alloys, Twin Symp. 1973*, L. H. Bennet and R. H. Willens, Ed., AIME, New York, 1974, pp. 223–270.

W. Derbyshire, *Nucl. Magn. Reson.*, **3**, 311 (1974).

J. L. Bjorkstam, in *Advances in Magnetic Resonance*, J. S. Waugh, Ed., Vol. 7, Academic, New York, 1974.

Pharmaceutical Problems

M. Plat, *Farmaco, Ed. Prat.*, **25**, 143 (1970).

D. M. Rackham, *Talanta*, **17**, 895 (1970).

A. S. V. Burgen and J. C. Metcalfe, *J. Pharm. Pharmacol.*, **22**, 153 (1970).

M. Cohn and J. Reuben, *Acct. Chem. Res.*, **4**, 214 (1971).

R. T. Parfitt, *Pharm. J.*, 203, 300, and 320, (1969).

R. T. Parfitt, *Instrum. News*, **20**, 8 (1970).

D. K. Banerjee, *J. Indian Chem. Soc.*, **47**, 199 (1970).

H. Kessler, *Mitt. Deut. Pharm. Ges.*, **39**, 177 (1969).

P. O. P. Ts'o, M. P. Schweizer and D. P. Hollis, *Ann. N.Y. Acad. Sci.*, **158**, 256 (1969).

J. Wozniak, *Farm. Pol.*, **24**, 907 (1968).

A. Haemers, *Pharm. Tijdschr. Belg.*, **46**, 21 (1969).

M. Vlassa, *Stud. Cercet. Chim.*, **18**, 1109 (1970).

G. Heinisch, *Oesterr. Apoth. Ztg.*, **28**, 275, 295, 361 (1974).

Perfumery Industry

O. Okuda and T. Imura, *Koryo*, **104**, 35 (1973).

Paints and Coatings Industry

O. Kinoshita, *Toso to Toryo*, **235**, 31 (1974).

Pesticides

L. H. Keith and A. L. Alford, *J. Ass. Offic. Anal. Chem.*, **53**, 1018 (1970).

R. Hague and D. R. Buhler, *Annu. Rep. NMR Spectrosc.*, **4**, 237 (1971).

H. A. Resing in *Mass Spectrom. NMR Spectrosc. Pesticide Chem., Proc. Symp. 1973*, R. Hague, Ed., Plenum, New York, 1974, pp. 273–277.

Petroleum Products

N. F. Chamberlain, *Proc. 7th World Petrol Congr. 1967*, Elsevier, New York, N.Y.

A. Suzuki, *Nenryo Kyokai-shi*, **49**, 175 (1970).

^{13}C-NMR

F. G. Riddell, *Aliphatic, Alicyclic, Saturated Heterocycl. Chem.*, **1**, (Part 3), 69 (1973).

H. Guenther, *Chem. Unsererer Zeit.*, **8**, 45, 84 (1974).

E. Wenkert, J. S. Bindra, C.-J. Chan, D. W. Cochran, and F. M. Schell, *Acc. Chem. Res.*, **7**, 46 (1974).

B. Casn, R. Ugo, *Relaz. Corso Teor.-Prat. Risonanza Magn. Nucl.*, A. Frigerio, Ed., Tamburini, Milan, Italy, 1973, pp. 3–15.

I. C. P. Smith, H. J. Jennings, and R. Deslauriers, *Acc. Chem. Res.*, **8**, 306 (1975).

A. G. McInnes and J. L. C. Wright, *Acc. Chem. Res.*, **8**, 313 (1975).

D. Shaw, R. Ugo, *Relaz. Corso Teor.-Prat. Risonanza Magn. Nucl.*, A. Frigerio, Ed., Tamburini, Milan, Italy, 1973, pp. 193–205.

W. Bremser, *Chem.-Ztg.*, **97**, 248 (1973).

D. Deininger and D. Michel, *Wiss. Z. Karl-Marx-Univ.*, Leipzig, *Math.-Naturiwiss Reihe*, **22**, 551 (1973).

L. J. Todd, *J. Organometal. Chem.*, **77**, 1 (1974).

J. R. Llinas, E. J., Vincent, and G. Peiffer, *Bull. Soc. Chim. Fr.*, 3209 (1973).

M. Tsuda, *Garumashia*, **9**, 756 (1973).

P. S. Pregosin and E. W. Randall, *Determination Org. Struct. Phys. Methods*, **4**, 263 (1971).

U. Sequin and A. I. Scott, *Science*, **186**, 101 (1974).

W. Bremser, *Chem.-Zig.*, **97**, 259 (1973).

B. E. Mann, *Adv. Organometal. Chem.*, **12**, 135 (1974).

^{19}F-NMR

T. N. Huckerby, *Annu. Rep. NMR Spectrosc.*, **5A**, 1 (1972).

K. Jones and E. F. Mooney, *Annu. Rep. NMR Spectrosc.*, **4**, 391 (1971).

R. Fields, *Annu. Rep. NMR Spectrosc.*, **5A**, 99 (1972).

M. G. Barlow, *Fluorocarbon Relat. Chem.*, **2**, 456 (1974).

Halogen-NMR

C. Hall, *Quart. Rev. Soc.* (London), **25**, 87 (1971).

^{14}N-NMR

J. D. Memory, in *Analytical Chemistry of Nitrogen and Its Compounds*, C. A. Streuli, Ed., Interscience, New York, 1970, (Part 1), p. 29.

^{77}Se-NMR

M. A. Lardon, *Organic Selenium Compounds: Their Chemistry and Biology*, D. L. Klayman, Ed., Wiley New York, 1973, pp. 933–939.

^{119}Sn-NMR

P. J. Smith, *Inorg. Chim. Acta Rev.*, **7**, 11 (1973).

C. REFERENCES ON SPECIFIC TOPICS

The following references are intended as an initial source for those who wish to pursue a specific topic. They are presented as examples of various applications of nuclear magnetic resonance in chemically related fields.

a. QUANTITATIVE METHODS

The effect of operating conditions on the accuracy of spectral measurements.	K. Hatada, Y. Terawaki, and H. Yuki, *Kogyo Kagaku Zasshi*, **71**, 1163, 1168 (1968).
Dual standard addition for quantitative NMR analysis.	A. F. Cockerill, R. C. Harden, G. L. O. Davies, and D. M. Rackham, *Org. Magn. Reson.*, **6**, 452 (1974).
Magnetic titrations by an NMR method.	D. F. Evans, *J. Chem. Soc., Dalton Trans.*, 2587 (1973).
A small general purpose computer for use with a high resolution NMR spectrometer for quantitative analytical chemistry.	J. N. Shoolery and L. H. Smithson, *J. Am. Chem. Soc.*, **47**, 153 (1970).
Application of NMR to semicontinuous reaction monitoring.	K. Parker and D. K. Das Gupta, *Proc. Inst. Elec. Engr.*, **116**, 1060 (1969).
The use of hexafluoroacetone as an NMR reagent.	G. R. Leader, *Anal. Chem.*, **42**, 16 (1970).
Determination of water in coal.	H. Reinhardt, *Bergbautechnik*, **20**, 29, 88 (1970).
Automatic moisture measurements of lignites.	H. Reinhardt, *Exp. Tech. Phys.*, **17**, 517 (1969).

b. DETERMINATION OF FUNCTIONAL GROUPS

Submicrodetermination of functional groups by time-averaged NMR	F. Kasler, *Mikrochim. Acta*, 702 (1970)
Microdetermination of methyl groups.	F. Kasler, *Mikrochim. Acta*, 1065 (1967).
Determination of α- and β-methyl naphthalenes.	S. Fujiwara and T. Wainai, *Anal. Chem.*, **33**, 1085 (1961).

Determation of methylation of D-glycopyranosides.
D. Gagnaire and L. Odier, *Carbohyd. Res.*, **11**, 33 (1969).

Determination of amines in trifluoroacetic acid.
W. R. Anderson, Jr. and R. M. Silverstein, *Anal. Chem.*, **37**, 1417 (1965).

Determination of N-methyl amines.
J. C. N. Ma and E. W. Warnhoff, *Canad. J. Chem.*, **43**, 1849 (1965).

S. A. Koch and T. D. Doyle, *Anal. Chem.*, **39**, 1273 (1967).

Methyl groups on a carbon bearing a hydroxyl group.
C. R. Narayanau and A. K. Kulkarni, *Curr. Sci.*, **42**, 367 (1973).

Determination of hydroxyl groups.
S. Siggia and J. G. Hannain, *Chemistry of the Hydroxyl Group*, S. Patai, Ed., Wiley, New York, 1971, Part 1, pp. 295–325.

Simultaneous quantitation of water and hydroxyl groups.
F. F. L. Ho and R. R. Kohler, *Anal. Chem.*, **46**, 1302 (1974).

Determination of alcohols using shift reagents.
D. L. Rabenstein, *Anal. Chem.*, **43**, 1599 (1971).

F. J Smentowski and R. D. Stipanovic, *J. Amer. Oil Chem. Soc.*, **49**, 48 (1972).

Determination of alcohols by acetylation.
A. Mathias, *Anal. Chim. Acta*, **31**, 598 (1964).

Determination of alcohols using dichloroacetic anhydride.
J. S. Babiec, Jr., J. R. Barrante and D. G. Vickers, *Anal. Chem.*, **40**, 610 (1968).

Classification of alcohols in dimethylsulfoxide.
O. L. Chapman and R. W. King, *J. Am. Chem. Soc.*, **86**, 1256 (1964).

J. G. Traynham and G. A. Knesel, *J. Amer. Chem. Soc.*, **87**, 4220 (1965).

R. B. Bass and M. J. Sewell, *Tetrahedron Lett.*, 1941 (1969).

Determination of thiols by derivatization.
P. E. Butler and W. H. Mueller, *Anal. Chem.*, **38**, 1407 (1966).

Determination of carboxylic acids.
N. van Meurs, *Z. Anal. Chem.*, **295**, 194 (1964).

Determination of Δ' alkene-sulphonates in the presence of other alkensulphonates.	T. Nagai, I. Tamai, S. Hashimoto, I. Yamane, and A. Mori, *Kogyo Kagaku Zasshi*, **74**, 32 (1971).
Determination of olefin impurities in alpha olefins.	P. W. Flanagan and H. F. Smith, *Anal. Chem.*, **37**, 1969 (1965).
Aromatics in dry acetic acid.	F. F. Caserio, Jr., *Anal. Chem.*, **38**, 1802 (1966).

c. DETERMINATION AND CHARACTERIZATION OF ORGANIC COMPOUNDS AND MIXTURES

NMR of enantiomers of α-amino acid methyl esters in optically active 2,2,2-trifluorophenyl-ethanol solvent.	W. H. Pirkle and S. D. Beare, *J. Am. Chem. Soc.*, **91**, 5150 (1969).
J (P–C–H) coupling constants for the identification of cyclic phosphines.	J. P. Albrand, D. Gagnaire, M. Picard, and J. B. Robert, *Tetrahedron Lett.*, 4593 (1970).
2,4-dichlorophenol in dichloro-phenol isomers.	N. Esumi, T. Suzuki, and S. Hayashi, *Org. Magn. Resonance*, **2**, 397 (1970).
Fatty acid derivatives.	K. Yamanaka, *Yukaguku*, **22**, 533 (1973).
Automated identification of mono-alkenes.	S. Ochias, Y. Hirota, Y. Kudo, and S. Sasaki, *Bunseki Kagaku*, **22**, 399 (1973).
Determination of 4-methylimi-dazole as the 1-acetyl derivative.	G. Fuchs and S. Sundell, *J. Agric. Food Chem.*, **23**, 120 (1975).
Isomers of cresol, toluic acid, and toluidine.	S. Yasuda and H. Kakiyama, *Bunseki Kagaku*, **23**, 615 (1974).
Hydroxy (acetoxy) alkanes.	N. S. Nikitina, A. D. Mysak, T. N. Veretenova, V. P. Tikhouov, and E. V. Lebedev, *Neftepererab, Neftekhim.* (Kiev), 88 (1973).
Ketoximes	R. E. Rondeau and B. L. Fox, U.S. Patent 3,756,779 (1973); *Chem. Abstr.*, **79**, 1146311t, (1973).
Hydroperoxides in methyl linoleate and sunflower oil.	K. Sohde, T. Ogawa, and S. Matsushita, *Yukagaku*, **23**, 228 (1974).

Mixtures of arenemono- or disulfonic acids.	H. Cerfoutain, A. Toeberg-Telder, C. Kruk, and C. Ris, *Anal. Chem.*, **46**, 272 (1974).
Isomeric toluenesulfonic acids.	K. Mashimo and T. Wainai, *Bunseki Kagaku*, **21**, 1079 (1972).
	K. Mashimo and T. Wainai, *Anal. Chem.*, **45**, 2424 (1973).
1,2- and 1,3-diglyceride mixtures.	R. J. Warren and J. E. Zarembo, *J. Pharm. Sci.*, **59**, 840 (1970).
Adamantadine hydrochloride in soft capsules and syrups.	J. W. Turczau and T. Medwick, *J. Pharm. Sci.*, **63**, 425 (1974).
Mixtures of organic peroxides.	G. A. Ward and R. D. Mair, *Anal. Chem.*, **41**, 538 (1969).
Water in benzenesulfonic acid.	T. Wainai and K. Mashimo, *Bunseki Kagaku*, **19**, 1629 (1970).
Organosilicon compounds containing Si—F bonds.	A. P. Kreshkov V. F. Andronov, and V. A. Drazdov, *Zh. Anal. Khim.*, **25**, 2218 (1970).
Analysis of fatty acid methyl ester mixtures with tris (dipivaloylmethanato) europium (111).	D. B. Walters, *Anal. Chim. Acta*, **66**, 134 (1973).
Naphthalenesulfonic acids.	K. Mashimo and T. Wainai, *Bunseki Kagaku*, **23**, 750 (1974).
α-olefinsulfonic acids.	S. Hashimoto, H. Tokuwaka, and T. Nagai, *Nippon Kagaku Kaishi*, 2384 (1973).
Determination of phenols by fluorine-19 NMR of hexafluroacetone derivatives.	E. F. L. Ho, *Anal. Chem.*, **46**, 496 (1974).
Disulfuram determination.	E. B. Sheinin, W. R. Bensen, and M. M. Smith, Jr., *J. Assoc. Off. Anal. Chem.*, **56**, 124 (1973).
Mixtures of di- and trichlorophenols.	E. Naohumi, T. Suzuki and S. Hayashi, *Org. Magn. Res.*, **2**, 397 (1970).
Determination of meso and racemic 2,3-diaminobutane mixtures.	J. E. Sarneski and C. N. Reilley, *Anal. Chem.*, **48**, 1303 (1976).
Carbon-13 NMR as a technique for distinguishing between cis and trans dianionobis (ethylendiamine) cobalt (111) complexes.	D. A. House and J. W. Blunt, *Inorg. Nucl. Chem. Lett.*, **11**, 219 (1975).

Determination of Grignard solutions.	R. Jones, *J. Organometal. Chem.*, **18**, 15 (1969).
Applications to the determination of isotope distribution.	G. J. Martin, M. T. Quemeneur, and M. L. Martin, *Bull. Soc. Chim. Fr.*, 4082 (1970).
	H. Frischleder, H. Sprinz, and M. Wahren, *Isotopenpraxis*, **5**, 273 (1969).
	S. Fujiwara, Y. Yano, and K. Nagashima, *Chem. Instrum.*, **2**, 103 (1969).
pK_{BH^+} for aliphatic ketones.	D. G. Lee, *Can. J. Chem.*, **49**, 1919 (1970).

d. DETERMINATION OF DRUGS

Pharmaceutical analysis.	M. H. Bowen, F. K. O'Neill, and M. A. Pringuer, *Proc. Soc. Anal. Chem.*, **11**, 294 (1974).
Hallucinogenic drugs.	S. W. Bellman, J. W. Turczan, and T. C. Kram, *J. Forensic Sci.*, **15**, 261 (1970).
Mixtures of hypnotic agents.	G. Rücker and P. N. Natarajan, *Arch. Pharm.*, **300**, 276 (1967).
Barbituates.	G. Rücker, *Z. Anal. Chem.*, **229**, 340 (1967).
	G. Rücker, G. Bohn and A. F. Fell, *Arch. Toxikol.*, **27**, 168 (1971).
	H. W. Avdovich and G. A. Neville, *Can. J. Pharm. Sci.*, **4**, 51 (1969).
	H. Lackner and G. Doering, *Arch. Toxikol.*, **26**, 220 (1970).
Identifications of penicillins and cephalosporins.	W. L. Wilson, H. W. Avdovich, and D. W. Hughes, *J. Assoc. Off. Anal. Chem.*, **57**, 1300 (1974).
Bacteriostats in commercial products.	M. W. Dietrich and R. E. Keller, *J. Amer. Oil Chem. Soc.*, **44**, 491 (1967).
	E. Jungerman and E. C. Beck, *J. Amer. Oil Chem. Soc.*, **38**, 513 (1961).

Alkyl-*p*-hydroxybenzoate anti-fungal preparations.	F. Shihab, W. Sheffield, J. Sprowls and J. Nematolahi, *J. Pharm. Sci.*, **59**, 1182 (1970).
Synthetic corticosteroids.	H. W. Avdovich, P. Hanbury and B. A. Lodge, *J. Pharm. Sci.*, **59**, 1164 (1970).
Methylxanthine mixtures.	K. Rehse, *Deut. Apoth.*, **107**, 1530 (1967).
Analgesic mixtures.	K. Rehse, *Arch. Pharm.*, **303**, 617 (1970).
Aminophylline.	J. W. Turczan, B. A. Goldwitz and J. J. Nelson, *Talanta*, **19**, 1549 (1972).
Sulphonamides.	K. Rehse, *Z. Anal. Chem.*, **246**, 22 (1969).
Cryptopine in thebaine	T. Rüll, *Bull. Soc. Chim. France*, 1897 (1963).
Dimethylsulfoxide in pharmaceutical preparations.	T. C. Kram and J. W. Turczan, *J. Pharm. Sci.*, **57**, 651 (1968).
An oral estrogen.	H. W. Avdovich, M. Bowron and B. A. Lodge, *J. Pharm. Sci.*, **59**, 1821 (1970).
Cyclandetate.	C. van der Vlies, G. A. Bakker and R. F. Rekker, *Parm. Weekblad.*, **101**, 93 (1966).
Maolate.	G. Slomp, R. H. Baker, F. A. MacKellar, *Anal. Chem.*, **36**, 375 (1964).
Meprobamate.	J. W. Turczan and T. C. Kram, *J. Pharm. Sci.*, **56**, 1643 (1967).

e. DETERMINATION OF PESTICIDES AND INSECTICIDES

Pesticides.	L. H. Keith and A. L. Alford, *J. Ass. Offic. Anal. Chem.*, **53**, 1018 (1970).
DDT-type pesticides.	L. H. Keith, A. L. Alford and A. W. Garrison, *J. Ass. Offic. Anal. Chem.*, **52**, 1074 (1969).
Carbamate pesticides.	L. H. Keith, A. L. Alford, *J. Ass. Offic. Anal. Chem.* **53**, 157 (1970).

DDT mixtures.   F. J. Biros, *J. Ass. Offic. Anal. Chem.*, **53**, 733 (1970).

Systox insecticides.   H. Babad and T. N. Taylor, *Anal. Chim. Acta*, **40**, 387 (1968).

f. DETERMINATION OF POLYMERS

Analysis of microstructure of polymers.   H. Inoue, *Kayaku To Kogyo (Osada)*, **47**, 196 (1973).

The end group analysis of some polyether polyols and polyester polyols by NMR spectroscopy.   Y. Chokki, M. Nakabayashi, K. Kodama, and M. Sumi, *Nippon Kagaku Kaishi*, 1662 (1974).

Cellulose acetates.   V. W. Goodlett, J. T. Dougherty and H. W. Patton, *Polym. Sci. A-1*, **9**, 155 (1971).

Copolyamides.   H. R. Kricheldorf, E. Leppert, and G. Schilling, *Makromol. Chem.*, **176**, 81 (1975).

Poly-γ-benzyl-*L*-glutamate.   E. Brosio, M. Delfini, A. Depaolis, M. Paci, and F. Conti, *Biopolymers*, **13**, 745 (1974).

1,4-*cis*-polybutadiene.   R. Lenk, *Chem. Phys. Lett.*, **28**, 398 (1974).

K. E. Elgert, G. Quack, and B. Stuetze, *Makromol. Chem.*, **176**, 759 (1975).

Poly-N_5-(3-hydroxypropyl)-*L*-glutamine   R. DiBlasi and A. S. Verdini, *Biopolymers*, **13**, 765 (1974).

Ethylene-vinyl acetate copolymers.   F. Keller, E. Kowasch, and J. Sobotka, *Z. Phys. Chem. (Leipzig)*, **255**, 1166 (1974).

Melamine derivatives.   D. G. Anderson, D. A. Netzel and D. J. Tessari, *J. Appl. Polym. Sci.*, **14**, 3021 (1970).

Methyl acrylate.   S. Y. Kulkarni and U. S. Pansare, *J. Appl. Chem. Biotechnol.*, **23**, 479 (1973).

Polypentenamers.   H. Y. Chen, *J. Polym. Sci., Polym. Lett. Ed.*, **12**, 85 (1974).

Molecular weight of polyethylene glycol.   T. F. Page, Jr., and W. E. Bresler, *Anal. Chem.*, **36**, 1981 (1964).

	Liu Kang-Jen, *Makromol. Chem.*, **116**, 146 (1968).
	J. E. Tanner and Liu Kang-Jen, *Makromol. Chem.*, **142**, 309 (1971).
Use of shift reagents for molecular weight determination of polypropylene glycol.	F. F. L. Ho, *Polymer Lett.*, **9**, 491 (1971).
Pure styrene *n*-mers.	S. Fujishige and N. Ohguri, *Makromol. Chem.*, **176**, 233 (1975).
Poly(methyl acrylates) and stereoregular poly(allylic alcohols).	H. Girard and P. Ponjol, *C.R. Hebd. Seances Acad. Sci., Ser. C*, **279**, 553 (1974).
Styrene-methylmethacrylate copolymers.	I. Ando, A. Nishioka, and T. Asakura, *Makromol. Chem.*, **176**, 411 (1975).
Poly(styrene sulfone).	M. Lino, K. Katagiri, and M. Matsuda, *Macromolecules*, **7**, 439 (1974).
Polyurethane elastomers.	Y. Chokki, *Nippon Kagaku Kaishi*, **1** (1975).
Poly(vinylchloride).	I. Ando, A Nishioka, and T. Asakura, *Makromol. Chem.*, **176**, 411 (1975).
THF polymers.	K. Matyjaszewski and S. Penczek, *J. Polym. Sci., Polym. Chem. Ed.*, **12**, 1905 (1974).
Pulsed and wide line methods of polymer molecular weight determination.	R. Liepins, B. Crist and H. G. Olf, *J. Polym. Sci. A-1*, **8**, 2049 (1970).
	B. Crist, *ACS Polym. Preprints*, **12**, No. 2, 794 (1971).
Rate of polymerization.	C. P. Haney, F. A. Johnson and M. G. Baldwin, *J. Polym. Sci. A-1*, **4**, 1791 (1966).

g. VARIOUS EXAMPLES

Estimation of potential oil yields from oil shales by pulsed NMR studies.	F. P. Miknis, A. W. Decora, and G. L. Cook, *U.S. Bur. Mines, Rep. Invest.*, R1-7984 (1974).
Gasolines.	B. K. Cernicki, *Nafta (Zagreb)*, **19**, 219 (1968).

Impurities in crude TNT.	D. G. Gehring, *Anal. Chem.*, **42**, 898 (1970).
Applications in coatings.	L. C. Afremow, *J. Paint Technol.*, **40**, 503 (1968).
Toxic solvents in paints, varnish removers and sewage solvents.	M. Gruenfeld, *J. Paint Technol.*, **42**, 237 (1970).
Organic coatings on activated carbons.	W. B. Moniz and C. F. Poranski, Jr., Naval Research Laboratory Report 6566, NRL, Washington, D.C. (1967).
Detergents.	H. Koenig, *Fresenius' Z. Anal. Chem.*, **251**, 225 (1971).
	M. M. Crutchfield, R. R. Irani and J. T. Yoder, *J. Am. Oil Chem. Soc.*, **41**, 129 (1964).
Nonionic surfactants.	C. McDonald and B. Robinson, *J. Pharm. Pharmac. Canad.*, **22**, 727 (1970).
Broad-line NMR measurement of water accessibility in cotton and wood pulp celluloses.	T. F. Child and D. W. Jones, *Cell. Chem. Technol.*, **7**, 525 (1973).
The amount of solid phase in cottonseed oil hydrogenation products.	I. P. Nazarova, A. I. Glushenkova, A. L. Markman, V. P. Tatarskii, and O. Ya. Vadzhipov, *Khim. Prir. Svedin*, **10**, 246 (1974).
Flavor components.	A. Kobayashi, *Yukagaku*, **22**, 591 (1973).
Certifiable food colors.	D. M. Marmino, *J. Assoc. Off. Anal. Chem.*, **57**, 495 (1974).
Hydroxypropyl groups in modified starch.	H. Stahl and P. R. McNaught, *Cereal Chem.*, **47**, 345 (1970).
	F. F. L. Ho, R. R. Kohler and G. A. Ward, *Anal. Chem.*, **44**, 178 (1972).
	C. J. Clemett, *Anal. Chem.*, **45**, 186 (1973).
Molecular weights of plasticizers.	Ya. G. Urman, T. S. Khramova, V. G. Gorbunova, R. S. Barshtein and I. Y. Slonim, *Vysokomol. Soedin*, **A12**, No. 1, 160 (1970).

Applications of NMR spectroscopy described for assigning hydrogen bonds.
P. H. Von Dreele and I. A. Stanhouse, *J. Am. Chem. Soc.*, **96**, 7546 (1974).

Irradiation dosimetry based on magnetic resonance spectra.
J. R. Manambelona, *Dosim. Agr., Ind., Biol., Med., Proc. Symp., 1972*, **28**, 293 (1973).

Use of trifluoroacetic acid in magnetic susceptibility measurements.
A. J. I. Ward, A. Morris, and M. A. Healy, *J. Magn. Reson.*, **16**, 357 (1974).

COMPOUND INDEX

Acepleiadiene, 220
Acepleiadylene, 220
Acetaldehyde, 60, 61, 163
Acetamide, 309
Acetanilide, 219
Acetic acid, 41, 65, 163, 304
Acetone, 24, 65, 241
Acetonitrile, 65, 163, 230
Acetophenone, 172, 219, 225
Acetophenone oxime, 309
Acetylacetone, 266, 277, 280, 281
Acetylene, 24, 224, 226, 228, 312
n-Acetylglucosamine, 361
Acrylamide, 309
Acrylonitrile-methyl methacrylate copolymer, 324
Adenisine triphosphate, 373
L-Ala-L-Ala diketopiperazine, 378
Allene, 168, 184
Allyl alcohol, 306
p-Aminobenzoic acid, 308
δ-Amino laeuvlate, 377
p-Aminophenol, 308
3-Amino-l-propanol, 306, 308
2-Amino pyridine, 308
3-Amino pyridine, 308
4-Amino pyridine, 308
Ammonia, 309
t-Amyl alcohol, 307
Aniline, 172, 219, 308
Anisole, 172, 219
Anthracene, 174
Asperlin, 248
Asprin, 268, 269, 270, 355
7-Azaindole, 79, 86
Azulene, 175

Barbiturates, 355
Benzaldehyde, 172, 219, 225, 357
Benzamide, 172, 309
Benzene, 24, 31, 41, 65, 145, 186, 199, 219, 220, 226, 313
Benzimidazole, 174, 222
Benzofuran, 174
Benzoic acid, 172, 225

Benzonitrile, 172, 219
Benzophenone, 172
Benzothiophene, 174
Benzotrifluoride, 189
Benzoyl chloride, 172
Benzyl alcohol, 302, 306
Benzylamine, 308
Benzyl mercaptan, 307
Bicyclobutane, 227
Bicyclo [4.1.0] heptane, 168
Bicyclo [2.1.1] hexane, 178, 185
Bidrin, 350
Biphenyl, 172
Bovine serum albumin, 359, 360, 370
p-Bromo aniline, 136
Bromo benzene, 100, 172, 219
l-Bromo-2-chloroethane, 100
Bromo chloroethylene, 100
l-Bromo-3-chloropropane, 100
Butanal, 163
Butane, 203
1,4-Butanediol, 306
1-Butanethiol, 307
2-Butanethiol, 307
1-Butanol, 306
2-Butanol, 306
Butanoic acid, 163
2-Butanone, 225, 326
Butanonitrile, 163
cis-2-Butene, 216
$trans$-2-Butene, 216
1,4-Butenediol, 306
t-Butyl alcohol, 306
t-Butyl benzene, 172
4-t-Butylcyclohexyl alcohol, 302
$meso$-(2,3-Butylenedinitrilo)-tetraacetic acid, 404, 405
t-Butyl mercaptan, 307
Di-t-Butyl nitroxide, 241
1-Butyne, 223
Butyrolactone, 170, 178

Caffeine, 269, 270, 355
Carbon disulfide, 65, 154, 199
Carbon tetrachloride, 65

COMPOUND INDEX

Carbon tetraiodide, 209
Cellulose, 324, 328
Chloroacetamide, 309
o-Chloroaniline, 308
p-Chloroaniline, 308
Chlorobenzene, 172, 219
Chloroform, 65, 144, 145, 199, 357
4-Chloro-3-methylphenol, 307
p-Chloronitrobenzene, 97, 100
o-Chlorophenol, 307
p-Chlorophenol, 307
Chlorophenosin-1-carbamate, 356
Chlorophenosin-2-carbamate, 356
l-chloro-2-phenylethane, 136
m-Chlorotoluene, 4
o-Chlorotoluene, 4
p-Chlorotoluene, 4
Ciodrin, 350
Citric acid, 307
Coconut oil, 311
Co-Ral, 352
Corticosteroids, 355
o-Cresol, 307
p-Cresol, 307
Crude oil, 333
Cryptopine, 355
Cyclobutane, 168, 178, 180, 205, 227
Cyclobutanone, 169, 178
Cyclobutene, 168, 180, 217
Cyclodecane, 205
cis-Cyclododecene, 217
Cycloheptane, 168
1,3,5-Cycloheptatriene, 33
1,3-Cyclohexadiene, 184
Cyclohexane, 29, 65, 168, 178, 180, 199, 205, 227
trans-1,2-Cyclohexanediaminetetraacetic acid, 389, 404, 406
1,4-Cyclohexanediol, 306
Cyclohexanol, 214, 306
Cyclohexanone, 169, 178, 210
Cyclohexene, 168, 180, 217
Cyclohexylcarboxaldehyde, 225
Cyclohexylcarboxylic acid, 225
Cyclohexyl methyl ketone, 225
Cyclononane, 205
1,5-Cyclooctadiene, 245
Cyclooctane, 205, 245
Cyclooctatetraene, 33
Cyclooctatetraenyl dianion, 175

cis-Cyclooctene, 217
trans-Cyclooctene, 217
Cyclopentadiene, 33
Cyclopentadienyl anion, 175
Cyclopentane, 168, 178, 180, 205
Cyclopentanol, 306
Cyclopentanone, 169, 178
Cyclopentene, 168, 180, 217
Cyclopentenone, 184
Cyclopropane, 24, 33, 145, 168, 178, 180, 205, 227, 313
Cyclopropanone, 168
Cyclopropene, 168, 180
Cygon, 350, 351
Cysteine, 314, 370

p,p'-DDT[1,1,1-trichloro-2,2-bis(p-chlorophenyl)ethane], 352
DDVP, 350
DEF, 351
Delnav, 351, 352
Diacetone alcohol, 307
15,16-Dialkyldihydropyrenes, 33
Diazinon, 352
Diazoxon, 351
Dibrom, 350
o-Di-t-butylbenzene, 187
Di-n-butylmercury, 229
Dicapthon, 350
3,6-Dicarbomethoxy-S-tetrazine, 357
Dichloroacetic anhydride, 301
o-Dichlorobenzene, 100, 317
1,3-Dichloro-2-propanol, 306
2,6-Dichlorotoluene, 120, 121
Diethanolamine, 306, 308
Diethylamine, 308
Diethylenetriamine pentaacetic acid, 385, 387
Diethylhydroxylamine, 309
3,3-Diethylpentane, 343
2,3-Dihydropyran, 357
2,3-Dimercapto-l-propanol, 370
Dimethoxon, 350, 351
1,1-Diemthylallene, 228
Diemthylamine, 308
Dimethylaminoethanol, 306
N,N-Dimethylaniline, 172, 219
2,2-Dimethylbutane, 202, 203, 336
2,3-Dimethylbutane, 202, 203, 336
Dimethyl cadmium, 230

COMPOUND INDEX

cis-1,2-Dimethylcyclohexane, 206, 245
cis-1,4-Dimethylcyclohexane, 206, 245
0,0-Dimethyl-5-(1,2-dicarbethoxyethyl)-
   phosphorodithionate, 346
2,4-Dimethyl-3-ethylpentane, 343
Dimethylglycine, 383
Dimethylglyoxime, 309
2,2-Dimethylheptane, 341
2,3-Dimethylheptane, 340
2,4-Dimethylheptane, 341
2,5-Dimethylheptane, 341
2,6-Dimethylheptane, 341
3,3-Dimethylheptane, 341
3,4-Dimethylheptane, 341
3,5-Dimethylheptane, 203, 341
4,4-Dimethylheptane, 341
2,2-Dimethylhexane, 339
2,3-Dimethylhexane, 338
2,4-Dimethylhexane, 338
2,5-Dimethylhexane, 338
3,3-Dimethylhexane, 339
3,4-Dimethylhexane, 203, 338
1,1-Dimethylhydrazine, 309
Dimethylmercury, 230
2,2-Dimethylpentane, 337
2,3-Dimethylpentane, 203, 337
2,4-Dimethylpentane, 337
3,3-Dimethylpentane, 337
2,2-Dimethylpropane, 336
Dimethylselinium, 230
2,2-Dimethyl-2-silapentane-5-sulphonic
   acid, 66, 385
Dimethyl sulfoxide, 65
Dimethyl tellurium, 230
N,N-Dimethyl trichloroacetamide, 245, 292
o-Dinitrobenzene, 187
2,4-Dinitrophenylhydrazine, 309
1,3-Dioxalane, 169
1,3-Dioxane, 169
1,4-Dioxane, 65, 169, 199, 241
Diphenylacetylene, 172
1,4-Diphenylazetidinone, 125, 151, 153
Diphenyl ether, 172, 237
Diphenyl mercury, 229
meso-2,4-Diphenyl pentane, 321, 322
tris-Dipivaloymethanotes, 158, 302
Di-N-propylamine, 308
Di-syston, 351, 352
1,3-Dithiane, 169
4,8,13-Duvatriene-1,3-diol, 241, 242

EDTA-metal chelates, 390–416
EPN, 352
Estrogens, 355
Ethane, 24, 29, 163, 203, 226, 228
Ethane thiol mercaptoethane, 307
Ethanolamine, 306, 308
Ethion, 351, 352
β-Ethoxyethylamine, 308
Ethyl acetate, 148
Ethyl alcohol, 163, 306
Ethyl amine, 163
Ethylbenzene, 70, 163, 235, 274
Ethyl bromide, 99, 163, 227
Ethyl chloride, 163, 227
Ethyl cyanoacetate, 247
Ethylene, 24, 31, 145, 165, 180, 216, 218,
   226, 228, 313
Ethylenediamine, 308, 415
Ethylenediaminetetraacetic acid, 260, 381,
   383
Ethylene glycol, 209, 306
Ethylene glycol monomethyl ether, 306
Ethylene imine, 169, 178
Ethylene oxide, 168, 178
Ethylene sulfide, 169, 178
Ethylether diaminetetraacetic acid, 385, 386
Ethyl fluoride, 163
3-Ethylheptane, 343
3-Ethylhexane, 340
Ethyl iodide, 163
Ethyl mercaptan, 163
3-Ethyl-2-methylpentane, 340
3-Ethyl-3-methylpentane, 340
Ethyl parathion, 352
3-Ethylpentane, 337
3-Ethyl-3-pentanol, 227
Ethyl thioglycolate, 307
Europium nitrate, 158

Fluorine, 24
Fluorobenzene, 172, 188, 189, 219, 229
Fluorocyclohexane, 189
1-Fluoropropene, 189
Fluorosulfonic acid, 390
Formaldehyde, 24, 30, 178
Formic acid, 304
Furan, 174, 186, 221
Furanonycin, 358

Glycerol, 306

Glycine, 308, 314
Glycolic acid, 306
Gramicidin S, 377–379
Guthion, 350
Guthion oxygen analog, 350, 351

Hen white lysozyme, 364, 375, 376
tris-1,1,1,2,2,3,3,-Heptafluoro-7,7-octane-
   dionates, 158
Heptane, 203, 337
1-Heptyne, 223
2-Heptyne, 223
3-Heptyne, 223
Hexafluoroacetone, 304, 313
Hexafluoro-2-propanol, 306
Hexamethyl phosphoramide, 302
Hexanal, 225
Hexane, 202, 203, 336
Hexanoic acid, 225
Hexanol, 83, 157, 239, 302
n-Hexyl fluoride, 229
1-Hexyne, 223
2-Hexyne, 223
3-Hexyne, 223
Hydrazine, 309
Hydrogen peroxide, 309
Hydrogen sulfide, 309
Hydroquinone, 307
N'-(2-Hydroxyethyl)-ethylenediamine-
   N,N,N'-triacetic acid, 410
Hydroxylamine, 309
3-Hydroxypropanonitrile, 131

Imidan, 350, 351
Imidazole, 174, 186, 221, 222
Iminodiacetic acid, 385
Indole, 174, 221
Iodobenzene, 172, 219
Isoleucine, 358
Isophthalic acid, 327
Isopropyl alcohol, 5, 163, 306
Isopropylamine, 163, 308
Isopropylbenzene, 163
Isopropylbromide, 163
Isopropyl chloride, 163
Isopropyliodide, 163
Isopropylmercaptan, 163
Isoquinoline, 174, 222

Linoleic acid, 330, 331

Linolenic acid, 330, 331
Linseed oil, 311
Lithium deuterioxide, 301, 313
Lysozyme, 364, 369

Malaoxon, 350, 351
Malathion, 347, 348, 350, 351
Maleic anhydride, 327
Malononitrile, 178
Maolate, 356
Mebromate, 355
2-Mercaptoethanol, 306, 307
2-Mercaptopropanol, 306
3-Mercaptopropanol, 307
3-Mercaptopropionic acid, 370
Methane, 24, 29, 30, 145, 163, 178, 203
Methane sulfonamide, 309
Methionine, 349
3-Methoxypropylamine, 308
N-Methyl acetamide, 378
Methyl acetate, 304
Methyl N-acetyl-β-D-glucosamine, 364, 365
Methyl alcohol, 45, 65, 163, 306, 323, 377
Methyl amine, 163, 278, 279, 308
Methyl o-aminobenzoate, 308
2-Methylaminoethanol, 306, 308
N-Methylaniline, 308
Methyl benzoate, 172, 219, 225
α-Methylbenzyl amine, 308
Methyl bromide, 163, 226
2-Methylbutane, 336
Methyl chloride, 163
trans-Methyl cinnamate, 116
Methyl cyanoacetate, 247
Methylcyclohexane, 206, 245
Methylcyclohexanone, 247
Methyl cyclohexylcarboxylate, 225
Methyl eladiate, 331, 332
Methylene chloride, 65, 258, 334
Methylene cyclobutane, 168
Methylene cyclohexane, 169
Methylene cyclopentane, 168
Methylenecyclopropane, 168
Methylene fluoride, 99, 100
Methyl ethyl ketoxime, 309
Methyl fluoride, 163.
2-Methylheptane, 338
3-Methylheptane, 338
4-Methylheptane, 338

COMPOUND INDEX 449

2-Methylhexane, 337
3-Methylhexane, 337
Methyl hexanoate, 225
Methylhydrazine, 309
Methyl 12-hydroxysterate, 326
Methyliminodiacetic acid, 385, 402
Methyl iodide, 41, 163
Methyl isocyanide, 230
Methyllithium, 230
Methyl mercaptan, 163
2-Methyl-2-nitro-1,3-propanediol, 306
2-Methyl-2-nitropropanol, 306
2-Methylnorbornan-2-01, 234
2-Methyloctane, 340
3-Methyloctane, 340
4-Methyloctane, 340
Methyl oleate, 326
Methyl parathion, 350
2-Methylpentane, 202, 203, 336
3-Methylpentane, 202, 203, 336
Methyl petroselinate, 326
2-Methylpropanal, 163, 306
2-Methylpropanethiol, 307
Methyl propanoate, 225
2-Methylpropanoic acid, 163
2-Methylpropanonitrile, 163
Methyl propenoate, 225
Methylpyrene, 344
Methyl ricinoleate, 326
Methyl thiourea, 309
Methyl triphenylphosphonium iodide, 229
Methyl trithion, 350, 351
Methyl urea, 309
Methyl vinyl ketone, 225
Morpholine, 308

Naphthalene, 173, 174, 186
Neguvon, 350
Neopentane, 24
Nitrilotriacetic acid, 402
Nitro benzene, 172, 219
Nitroethane, 163
Nitroimidazole, 357
Nitromethane, 65, 163, 258, 334
o-Nitrophenol, 187
p-Nitrophenyl-4-0-(2-deoxy-2-acetamido-
    β-D-glucopyranosyl)-β-D-glucopyrano-
    side, 367
1-Nitropropane, 163
2-Nitropropane, 163

p-Nitrotoluene, 36
Nonane, 203, 340
Norbornadiene, 33, 168
Norbornane, 168, 178, 180, 184, 227
Norbornene, 168, 178, 184

Octane, 203, 338
2-Octanone, 225
1-Octene, 216
cis-2-Octene, 216
trans-2-Octene, 216
cis-3-Octene, 216
trans-3-Octene, 216
cis-4-Octene, 216
trans-4-Octene, 216
1-Octyne, 223
2-Octyne, 223
3-Octyne, 223
4-Octyne, 223
Oleic acid, 330
Olive oil, 311, 312

Pencillamine, 370
Penicillin G, 359, 360
Penicillin V, 361
Pentane, 203, 336
1-Pentanol, 306
3-Pentanol, 227
1-Pentyne, 223
Perhydrotriphenylene, 243, 244
Phenacetin, 268, 269, 270, 355
Phenanthrene, 173, 175, 186
Phencapton, 351, 352
Phenol, 172, 219, 307
Phenyl acetate, 219
Phenylacetylene, 172, 313
Phenyl benzoate, 172
2-Phenylethanol, 306
Phenyl lithium, 172
Phenyl magnesium bromide, 172
Phenyl trimethyltin, 229
Phenyl urea, 309
Phosdrin, 350
Phosphamidon, 350
Phosphenol pyrurate, 373
Phthalazine, 174
Phygon, 353
Piperazine, 308
Piperidine, 169, 308
Polyacrylonitrile, 325

Poly(isopropyl acrylate), 323
Poly(methyl acrylate), 323
Poly(methyl methacrylate), 324, 325, 327
Polypropylene, 325
Poly(propylene-butadiene), 323
Polystyrene, 321
Polytetrafluoroethylene, 325
Polyvinylchloride, 317, 325
Polyvinylfluoride, 325
Potassium bromide, 369
Pristane, 335
Prodigiosin, 248
Propanal, 163, 225
Propane, 29, 145, 163, 203
1,2-Propanediol, 306
1,3-Propanediol, 306
1-Propanethiol, 307
2-Propanethiol, 307
Propanoic acid, 163, 225
Propanonitrile, 163
Propenal, 225
Propenoic acid, 225
Propiolactone, 170
n-Propyl alcohol, 5, 163, 227, 306
n-Propylamine, 163, 308
n-Propylbenzene, 163
n-Propyl bromide, 163
n-Propyl chloride, 163
Propylenediaminetetraacetic acid, 390, 411
411
Propylene oxide, 100
n-Propyl mercaptan, 163
Propyne, 228
Pyrazine, 174, 186, 221
Pyrazole, 174, 186, 221, 222
Pyridazine, 174, 186, 221
Pyridine, 65, 174, 186, 189, 220, 221, 301, 313
Pyrimidine, 174, 186, 220, 221
Pyrrocaline, 222
Pyrrole, 174, 186, 221, 222
Pyrrole nitranion, 189
Pyrrolidine, 169
2-Pyrrolidinone, 170
Pyruvate kinase, 373
Purine, 221

Quadricyclane, 227
Quinoline, 174, 221, 222

Quinoxaline, 174

Resorcinol, 307
Ronnel, 350
Rubidium bromide, 369
Ruelene, 350

Safflower seed oil, 310, 311, 312
Selenophene, 174, 221, 222
Sodium chloride, 370
Soybean oil, 311, 321
Styrene, 172
Styrene-butadiene copolymer, 324
Sulfanilamide, 308, 309
Sulfur dioxide, 65
Sunflower seed oil, 311
meta-Sysfox-R, 350

Tetra-n-butyl phosphonium bromide, 229
Tetra-n-butyltin, 229
1,2,4,5 - Tetrachloroperfluoropentane, 321
0,0,0,0-Tetraethyldithiopyrophosphate, 349
Tetrahydrofuran, 65, 168
Tetrahydrofurfuryl alcohol, 306
Tetrahydropyran, 169
Tetrahydrothiophene, 169
2,2,3,3-Tetramethylbutane, 339
2,2,4,4-Tetramethyl-1,3-cyclobutanediol, 306
Tetramethylene solfone, 170
Tetramethyl lead, 230
2,2,3,3-Tetramethylpentane, 343
2,2,3,4-Tetramethylpentane, 342
2,3,3,4-Tetramethylpentane, 342
2,2,4,4-Tetramethylpentane, 343
Tetramethylsilane, 66, 199, 226, 230
Tetramethyltin, 230
Tetraphenylborate, 229
Tetraphenyl lead, 229
Thebaine, 355
Thiacyclohexane, 170
Thiazole, 186
Thimet, 351, 352
Thiophene, 174, 186, 221, 222
Thiophenol, 172, 307
Tiguvon, 350
Toluene, 24, 41, 65, 163, 172, 219, 228, 326
p-Toluene sulfonamide, 309
Tri-n-butylphosphine, 229
Tri-n-butylphosphite, 229

COMPOUND INDEX

Tri-*n*-butylthiophosphate, 353
Tri-*n*-butylthiophosphite, 353
Trichlorofluoromethane, 301
Trifluoroacetic acid, 65, 313, 314
Trifluoroacetic anhydride, 301
2,2,2-Trifluoroethanol, 306, 378
Trimethyl anilinium iodide, 172
2,2,3-Trimethylbutane, 337
Trimethyleneamine, 169
Trimethyleneoxide, 169
Trimethylene sulfide, 170
2,2,4-Trimethylhexane, 342
2,2,5-Trimethylhexane, 342
2,3,3-Trimethylhexane, 342
2,3,5-Trimethylhexane, 341
2,2,3-Trimethylpentane, 339
2,2,4-Trimethylpentane, 339
2,3,3-Trimethylpentane, 339
2,3,4-Trimethylpentane, 339
Trimethyl phenylphosphonate, 172
1,3,5-Trioxacyclohexane, 169

Triphenyl lead chloride, 172
Triphenyl phosphine, 229
Triphenyl phosphite, 229
Triphenylsilyl chloride, 172
5-Trithiane, 170
Trithion, 351, 352
Tropylium cation, 175
Tung oil, 311

Valerolactone, 170
Vinyl acetate, 233
Vinyl fluoride, 189
Vinyl isobutyl ether, 327

Warfarin, 243
Water, 41, 65, 285, 301, 327
 in margarine, 276
Whale oil, 311

Zinophos, 352
Zytron, 350

SUBJECT INDEX

Absorption signal, U-mode, 21
  V-mode, 21
Analysis of CMR Spectra, aids in, 230
  spin-lattice relaxation, 238
  undecoupled spectra, 236
Analysis of NMR spectra, aids in, 142
  heteronuclear double resonance, 155
  homonuclear spin decoupling, 147
  indor, 152
  isotopic substitution, 143
  Lanthanide shift reagents, 156
  spin tickling, 149
  variations in magnetic field, 142
complex, 104
first-order, 101
four nuclei ($AA'BB'$), repeated spacings, 133
  transition frequencies, 132
  types of spectra, 135
  (A_2B_2), 137
  ($AA'XX'$), 128
    deceptively simple spectra, 131
intramolecular nuclear Overhauser effect, 154
peaks, number of, 101
  spacing between, 102
quantum mechanical formalism, 104
Summary of rules for, 111
Three nuclei, AB_2, 118
  AB_2 or AX_2, 117
  ABC, 127
  ABX, 122
    deceptively simple spectra, 127
    repeated spacings, 124
  AMX, 120
Two equivalent nuclei (A_2), 113
Two nuclei (AB), 114
Two-spin system, 108
Angular momentum, 10
Area, measurements of, 17, 69, 260
Assignments of CMR spectra, see CMR spectra, assignments

Biochemistry, 362
  applications of ^{13}C-NMR, 375

binding studies, 363
binding studies using quadropolar nuclei, 367
biomolecular conformation, 377
  studies using paramagentic ions, 371
Biosynthetic studies using CMR, 247
Bloch equations, 21
Bloch formalism, 20
Boltzmann equation, 15
Bond lability of metal complexes, 399

Carbon-13, chemical shifts, 200
  alkanes, 202
    parameters for calculating, 204
  alkenes, 215
    cyclic, 217
    substituted, 218
  alkyl derivatives, substituent effects on, 208
  alkynes, 222
    parameters for calculating, 223
  benzenes, substituted, 218
  cycloalkanes, 205
  of functional groups, 224
  heterocycles, aromatic, 220
    saturated, 214
  olefinic carbons, for calculation of, 217
  substituent effects on, 201
NMR, 194
  calibration of spectra, 199
  observation of spectra, 197
Chemical kinetics, 279
  experimental procedures, 293
  line-shape analysis, complete, 289
  reactions, fast, 281
    slow, 279
  transient methods, 291
  two-site exchange, 283
Chemical shifts, 22
  additivity scheme for, 165
  additivity scheme substituted benzene, 172
  aromatic compounds, 172
  average energy approx., 27
  diamagnetic term, 35

15,16-Dialkyldihydropyrenes, 32
frequency calibration, 68
measurements, 66
methylene protons, 165
monosubstituted benzenes, 172
Olefinic protons, 166
paramagnetic term, 25
protons, in aliphatic compounds, 158
  bonded to heteroatoms, 173
quantitation by, 256
reference compounds, 66
see also Carbon-13, chemical shifts
CMR spectra, aids in analysis of, 230
  spin-lattice relaxation, 238
  undecoupled spectra, 238
  assignments of, 230
    off-resonance and selective decoupling, 232
    spectral comparison, 234
    specific labeling, 236
    use of chemical shift, 231
  biosynthetic studies using, 247
  conformational analysis using, 244
  mechanistic studies using, 243
  quantitative analysis by, 240
  structural determinations using, 241
Coatings, 326
Conformational analysis using CMR, 244
Coupling constants, 176
  absolute sign, 36
  allylic coupling, 183
  in aromatic and heteroaromatic, 185
  carbon-coupling with other nuclei, 229
  carbon-hydrogen coupling, 226
    in benzene, 228
    long range, 228
  coupling along W-path, 185
  disubstituted benzenes, 187
    an additivity scheme for, 187
  geminal, 177
    substituent effect on, 177
  homoallylic long-range, 185
  long-range, 183
  practical considerations, 38
  relative sign, 36
  vicinal, 179
    dihedral angle dependence of, 179
    in cyclic compounds, 180
  with other nuclei, 189

Digital data smoothing, 275

Electric quadrupole moment, 10
Electronic integrator, 260
  accuracy, 260, 270
  amplitude, 261
  precision, 260, 270
Energy levels, 14, 17
  Boltzmann distribution, 15
Equilibrium studies, 277
Equivalence, 99
  chemical shift equivalent, 97
  magnetically equivalent, 97

Filter, 82
  digital smoothing, 275
  smoothing, 82
Foods, 327
  high-resolution ^{1}H and ^{13}C methods, 330
  pulsed methods, 329
  wide line methods, 327
Fourier transform NMR, 73
  data acquisition, 80
  phase corrections, 85
  pulsed FT-NMR, 75
Functional group analysis, 300
  amines, 313
  carbonyl groups, 302
  carboxylic acids and related compounds, 303
  ethers, epoxides, peroxides, 305
  hydroxyl groups, 301
  olefins, 305

Industry related applications, 316
  synthetic polymers, 317
integrated intensity (area), 17, 260
Integration, 17, 260
  accuracy, 260, 270
  precision, 260, 270

Lanthanide shift reagents, 156
Larmor frequency, 12
Line shape, analysis of kinetics, 289
  effect of exchange processes on, 46
  factors influencing, 59
  Lorentzian, 14

Magnetic moment, 10, 11, 12, 21

SUBJECT INDEX

Magnetic susceptibility, 28
Magnetization, 20, 77
Magnetogyric ratio, 10, 12
Mechanistic studies using CMR, 243
Metal chelates, conformational analysis of, 411
Metal complexes, 381
   bond lability of, 399
   protolysis kinetics of, 396

NMR spectra analysis, see Analysis of NMR spectra
NMR spectrometers, 50
   commercial, 89
   field/frequency stabilization, 57
   integrator, 56
   probe, 53
   radiofrequency detection, 55
   sweep units, 57
   transmitter, 52

Peak, area, effect of ^{13}C, 269
   instrumental effects on, 260
   quantitation by, 260
   height, 258
   quantitation by, 258
   width, factors influencing, 59
   quantitation by, 259
Pesticides, 344
   DDT-type, 352
   in environment, 353
   organophosphorus, 346
Petroleum, 331
   CMR chemical shift parameters, 335
   crude oil, 333
   hydrocarbons, 333
Pharmaceuticals, 354
   molecular interactions, 359
   quantitative, 355
   structure, 357
Processional frequency, 12
Process control monitors, 276
Protolysis kinetics, 395
   of free ligands, 395
   of metal complexes, 396
Protonation, of free ligands, 381
   of metal complexes, 390
Proton counting, 265
Pulsed NMR, Fourier transform, 73
   quantitation by, 275

Quadrature FT-NMR, 87
Quantitative analysis by CMR, 240

Rapid-scan or correlation FT-NMR, 88
Reference compound, 23
   DSS, 66
   TMS for ^{1}H-NMR, 23, 66
References for nuclei other than ^{1}H and ^{13}C, 249
Relaxation mechanism, 39
   chemical shift anisotropy, 41
   dipole-dipole, 40
   quadrupole, 44
   scalar coupling, 42
   spin rotation, 43
   T_1, 16
   T_2, 16
   with paramagnetic species, 44
Ring currents, 25
   diamagnetic, 31
   in nonaromatic rings, 33
   paramagnetic, 32

Sample preparation, 64
   for quantitative measurements, 263
Saturation, 17
Screening constant, 23
Sensitivity, enhancement, 73
   factors influencing, 70
Shielding, carbon-carbon and carbon-hydrogen bonds, 29
   carbon-carbon double and triple bonds, 29
   carbonyl group, 30
   miscellaneous groups, 31
   other atoms in molecule, 27
Signal averaging, quantitation with, 271
Signal to noise ratio, 70
Solvent effects, 33
   bulk susceptibility of medium, van der Waals interactions, anisotropy of susceptibilities of surrounding molecules, reaction field of medium, specific solute-solute interactions, 33
Solvents, 65
   shift with aromatic, 34
Spin decoupling, heteronuclear, 155
   homonuclear, 147
   INDOR, 152
   Spin-tickling, 149
Spin lattice relaxation, 16

Spinning, "Sidebands," 54
Spin quantum number, 10
Spin system, 99
  nomenclature, 101
Stochastic resonance, 87
Structural determinations using CMR, 241
Synthetic polymers, 317
  quantitative determinations, 317
  structure determinations, 318
  nuclei other than ^{1}H, 324
  shift reagents, 325

T_1, 16, 19
  effect on pulsed-FTNMR, 77
  spin-lattice relaxation time, 16
  table of, 14
  use in structure determination, 67a
T_2, 16, 19
  spin-spin relaxation time, 16

U mode, 85

V mode, 84

Volume magnetic susceptibility, 67